Phosphorus Recovery and Recycling

Hisao Ohtake • Satoshi Tsuneda
Editors

Phosphorus Recovery and Recycling

Springer

Editors
Hisao Ohtake
Phosphorus Atlas Research Institute
Waseda University
Shinjuku-ku, Tokyo, Japan

Satoshi Tsuneda
Phosphorus Atlas Research Institute
Waseda University
Shinjuku-ku, Tokyo, Japan

ISBN 978-981-10-8030-2 ISBN 978-981-10-8031-9 (eBook)
https://doi.org/10.1007/978-981-10-8031-9

Library of Congress Control Number: 2018942182

Printed on acid-free paper

This Springer imprint is published by the registered company Springer Nature Singapore Pte Ltd.
The registered company address is: 152 Beach Road, #21-01/04 Gateway East, Singapore 189721, Singapore

Foreword: Phosphorus Recycling — Mending a Broken Biogeochemical Cycle by Roland W. Scholz

Understanding the Dissipative Nature of Phosphorus

The book *Phosphorus Recovery and Recycling* is a milestone in the course of taking global action to sustainably mend an *anthropogenically broken natural-phosphorus cycle*. Phosphorus is a *bioessential* and *unsubstitutable element of life* and a necessity for food production. In most ecosystems of the world in its pre-agricultural age, weathered phosphorus atoms cycled many times on a small scale before erosion and runoff transported particulate and soluble phosphorus to the ocean or it became unavailable as a result of lithification or leaching processes. For atoms that end up on the seafloor, it takes 10–100 million years before they become re-exposed by tectonic uplift. This has been fundamentally changed by agricultural activity, particularly the arable land covering approximately 10% of Earth's terrestrial area (Smil 2000; Scholz et al. 2014). On agricultural land, a larger number of phosphorus atoms do not cycle a single time.

Today, 50% of worldwide food production is estimated to be based on mineral fertilizers (Galloway et al. 2008). There are no reliable estimates of global runoffs and erosion, but phosphorus flows have a dissipative nature. According to the literature, global estimations suggest that about one third of mineral fertilizers are lost by runoff and erosion (from weathered and mineral fertilizers) annually (Scholz et al. 2014). An assessment of hidden or virtual phosphorus flows is needed, e.g., for food production, and thus is an important factor for sustainable food management.

If we look at the supply side of the mineral phosphorus supply, there is no technology on the horizon or an alternative compartment (such as aquatic systems) to higher concentrated but finite phosphate rock deposits that could substitute for phosphate mining from ores. If we were to consider covering the current annual mineral phosphorus production from (average) seawater, this would require us to extract phosphorus from seawater at an average volume of the magnitude of 100 times the Amazonas effluent (Scholz and Wellmer 2013a). Alternatively, seabed mining could provide access to additional phosphate rock deposits. However, from an environmental perspective, this is an ambiguous and highly uncertain option.

Thus, the long-term supply security of phosphorus of mineral origin for humankind is questionable. For example, the 2017 USGS Mineral Commodity Summaries on *Phosphate Rock* (Jasinski 2017) states "**Recycling:** None."[1] Let me explain why this is a critical issue and why the present book is an overly timely, important, and – in my opinion – key contribution to sustainable resource management.

Phosphorus Recycling, a Key Feature of Sustainable Agriculture

As described by Franklin Hiram King (1848–1911) in his book *Farmers of Forty Centuries* (King 1911/2004), the Asian agricultural system has developed technologies for recycling almost everything in order to enable "permanent agriculture." This has included the use of night soil or mud from rivers where dissipated phosphorus from agriculture, sewage, etc., have accumulated. King notes: "Manure of all kinds, human and animal, is religiously saved and applied to the fields in a manner which secures an efficiency far above our own practice" (p. 9). Asian farmers were "scrupulous in their care to make everything that can possibly serve as fertilizer for the soil" (p. 23). This phrasing may also be applied to desperate farmers' raiding of Napoleonic battlefields such as Waterloo (1815) to access potential fertilizers from the bodies of the dead (Hillel 1991). The recycling of organic fertilizers has become a fundamental characteristic of the stable, viable Asian model of agriculture. Furthermore, historically, we find similar patterns in other parts of the world.

Reshaping the Derailed Phosphorus Cycle of the Industrial Age

Since the start of the Industrial Age, humankind has faced strong growth of national and, thus, global GDP is linked to – in many domains – an almost exponential increase in the consumption of energy carriers, construction aggregates, industrial minerals, and metal ores. Technological innovation has been a key driver of this increase. The foundation for the increase of mineral phosphorus use is based on Sir John Bennet Lawes' (1803–1873) patent for producing superphosphate by treating phosphate rock with sulfuric acid. We can see this as a cornerstone of increasing the world's carrying capacity of population size. Industrialization was accompanied by a rapid increase in urbanization. The small-scale, largely closed nutrient-recycling scheme on the scale of farms, villages, and small towns (including the night-soil industry) was broken. Today, most of the phosphorus in sewage, for instance, is not

[1] There are certainly some recycling streams in agriculture and other fields. However, no statistically relevant phosphorus recycling of phosphorus which can be used in idustrial processes takes place from sewage, manure, food waste, industrial slags, or other streams from a mineral economics perspective.

recycled. Globally, only a small share of households and industries are connected to sewerage systems. Estimates suggest that "it is likely that over 80% of wastewater is released" (Connor et al. 2017) without adequate centralized or decentralized treatment. In Croatia, a member state of the European Union, approximately 70% of wastewater is released without treatment (International Bank for Reconstruction and Development / The World Bank). This situation leads to severe environmental impacts, primarily for the world's *poor*, i.e., lower-income countries. But in addition, the situation in developed, high-income countries has become critical, for instance, as the direct recycling of phosphorus via centralized sewage plants has become uneconomic (due to the costs of long-distance transportation) and therefore has been discontinued. In other developed countries, direct recycling has been halted for environmental (heavy metals) and health (pathogens) reasons. Germany, however, has recently continued allowing the direct application of sewage from small-scale – mostly rural – wastewater sewage plants (Bundesregierung Deutschland 2017). Manure flows from animal factory production are another significant stream whose recycling potential is not fully used, and it is much larger than sewage flows (globally of factor 2–3).

From Recovery to Recycling

Phosphorus is a Janus-faced element and has been called the devil's element (Emsley 2000). Phosphate minerals are the source of a key nutrient, but in its elemental form, white phosphorus (P_4) is highly toxic and was banned from use in matches (then called Lucifers) by the Berne Convention in 1906. Phosphorus is an essential raw material of high value adding industrial products, including EV, electronics, pharmaceuticals, and high-functional plastics. Furthermore, although phosphorus is an excellent additive for effective detergents, wastewater from laundry, like phosphorus loads in human urine and feces, causes environmental damage and dead zones in aquatic systems. This situation calls for phosphorus recycling of wastewater in highly populated areas such as urban systems. In order to attain this ambitious goal on a global level, humankind has a long road ahead. Increasing the current low percentage of centralized and decentralized sewage treatment is a major goal and must precede recovery and recycling. Besides sewage, phosphorus inherent to slaughter and food waste, manure from various animal species and types of farming, industrial slag, and industrial wastes including phosphorus must be identified and, ultimately, recovered. The benefit of recovery is the reduction of environmental damage – and thereby costs – to lakes, rivers, coastal zones, etc. Estimates for the United States suggest that nutrient pollution costs about $2.2 billion annually in diminished property values (Dodds et al. 2008).

Recovery from within the supply chain as well as at the end of the pipe frame is feasible. However, as the world of tomorrow may see 80% of the world's population living in cities, proper technologies and policy instruments are needed to promote, foster, and establish economic phosphorus-recycling schemes today for the sewage and human biowaste.

Moving Toward a Circular Economy

Japan, countries of the European Union, and several other highly developed nations are on the cusp of advancing from recovery to recycling. The European Union demonstrated its strong commitment to phosphorus in 2014 when it put phosphate rock on a list of critical raw materials and referred to recycling as a potential option in terms of economic importance and supply risk. This significant step was taken although, at the present time, there is no physical scarcity of phosphate rock supply in the short term, i.e., in this century and most likely not in this millennium (Scholz and Wellmer 2013b). In a recent paper (Scholz and Wellmer 2018), we analyzed the reasons this route was taken and identified three. *First*, recycling would increase supply security. This relates to the 2007/2008 price peak of fertilizers and agrogoods, as well as the fact that a large share of the globally traded phosphorus is sourced from countries with political stability rated as low. As such, recycling can foster increasing diversity among phosphorus producers and thereby reduce supply risk. A *second* reason is linked to the principle of intergenerational justice, a basic regulative concept that was agreed upon at the 1992 United Nations Rio Convention. In order to reduce opportunity costs for future generations and provide long-term access to high-grade deposits, the broken phosphorus cycle has to be closed. *Third*, the key ideas of a *circular economy* (Pearce and Turner 1990; Ghisellini et al. 2016), or *cradle to cradle* (Braungart and McDonough 2002), which are less than 30 years old, increasingly acknowledge that recycling and closing loops in the supply–demand chain are at the competitive edge of economic development.

Toward Technology Development for Sustainable Resource Use

The book *Phosphorus Recovery and Recycling* focuses on the perspective of highly developed European and Asian countries (in particular, Japan). Many chapters strongly emphasize high-quality phosphorus recovery from sewage plants (in which phosphorus precipitation by chemical or biological means has been achieved) or slags from steelmaking. This is certainly one important line of sustainable technology development that may contribute to a responsible use of phosphorus through its recovery from different types of wastes. However, we should not forget that much phosphorus could be recovered from mining wastes. If we take a view of the entire supply-chain perspective (this view has been taken in Scholz et al. 2014), about 30–50% of the phosphorus in mines does not enter the value chain (Scholz and Wellmer 2015). Here, secondary mining, i.e., extracting the phosphorus from (lower-grade) mining waste by floatation, would be beneficial from a number of perspectives. We suggest that mineral flow analysis and mineral flow management could help identify the most efficient recycling potential in globalized, highly technologically developed countries and in the potentially digitally monitored world of the twenty-first century. If we take this perspective, we can argue that the recovery of phosphorus from sewage is not the most important issue from a mass-flow perspective. Sewage includes only about 5% of the anthropogenic flows (Scholz and Wellmer 2015), and only a small amount of this small percentage is accessible via wastewater treatment.

However, the recycling of phosphorus from sewage is undoubtedly at the cutting edge of technology development and may pave an important section of the road toward phosphorus recycling. There are a couple of thermal, chemical, and presumably biological technologies that are on the cusp of becoming economically competitive by substituting other – obviously unsustainable – pathways, such as disposing (phosphorus of) dried sewage in the incineration processes of cement production or coal-fired power plants. Finally, urban sewage and biowaste (e.g., food waste) can be economically recycled, as they are available in a highly geographically concentrated form. These phosphorus-recycling technologies should be supported by proper political or legislative actions. The 2017 amendment of the German sewage sludge ordinance serves as a good example. This amendment demands that – after a transition phase of 12–15 years – any larger German sewage treatment plant (i.e., 500 of 9300 existing plants treating two thirds of Germany's sewage) will be required to recover 50% of the phosphorus in sewage water (if P concentrations in sewage sludge are not below 20 g P per kg of dry matter) (Scholz and Wellmer 2018). This promotes what I like to call the *sustainability competition of technologies*, in this case, for phosphorus recycling. I am confident that the present book, like this German law, should be considered significant "catalysts" or steps forward on the journey toward a sustainable future of resource management.

Several main challenges of sustainable phosphorus recycling were identified at the end of the Global TraPs project (2010–2014), a transdisciplinary process that included key actors of all nodes of the supply chain. At that time, two of seven "hot spot" areas identified for gaining better insight into sustainable phosphorus management were related to recycling. One was the "economic extraction of heavy metals/radionuclides," particularly from phosphate fertilizers. The other was "phosphorus recycling technologies for integrated soil fertilization" (Scholz et al. 2014). The latter includes nutrient utilization from sewage ash and slags in the recycling stream as well as the melioration and processing of sewage, e.g., by contamination-free mineral fertilizers. The present book elucidates, deepens, and broadens these issues in a comprehensive and substantiated manner.

Roland W. Scholz

Department for Knowledge and Communication Management, Danube University Krems, Krems, Austria

Fraunhofer Institute for Interfacial Engineering and Biotechnology (IGB), Stuttgart, Germany

Department of Environmental System Sciences, Swiss Federal Institute of Technology (ETH) Zurich, Zurich, Switzerland

References

Braungart M, McDonough W (2002) Cradle to cradle. North Point Press, New York

Bundesregierung Deutschland (2017) Klärschlammverordnung vom 27.9. 2017, BGBl. I S. 3465. Bundesministeriums der Justiz und für Verbraucherschutz, Berlin

Connor R et al (2017) The United Nations World Water Development Report 2017. Wastewater: the untapped resource.The United Nations World Water Development Report

Dodds WK et al (2008) Eutrophication of US freshwaters: analysis of potential economic damages. ACS Publications

Emsley J (2000) The shocking history of phosphorus. The biography of the devil's element. MacMillan, London

Galloway JN et al (2008) Transformation of the nitrogen cycle: recent trends, questions, and potential solutions. Science 320(5878):889–892

Ghisellini P, Cialani C, Ulgiati S (2016) A review on circular economy: the expected transition to a balanced interplay of environmental and economic systems. J Clean Prod 114:11–32

Hillel D (1991) Out of Earth: civilization an the life of the soil. University of California, Berkeley

International Bank for Reconstruction and Development / The World Bank (2015) Wastewater treatment. [cited 2018 Jan. 25]; Available from: http://sos.danubis.org/eng/report/iv-access-to-services/c-wastewater-treatment/

Jasinski SM (2017) In: US Geological Survey (ed) Phosphate rock, in Mineral Commodity Summaries 2017. USGS, St. Louis, pp 124–127

King FH (1911/2004) Farmers of forty centuries: organic farming in China, Korea and Japan. Dover Publications, Mineola

Pearce DW, Turner RK (1990) Economics of natural resources and the environment. John Hopkins University Press, Baltimore

Scholz RW, Wellmer FW (2013a) Supplementary information to interactive comment on "Recent revisions of phosphate rock reserves and resources: reassuring or misleading? An in-depth literature review of global estimates of phosphate rock reserves and resources". In: Edixhoven JD et al. Earth System Dynamics Discussion, vol 4

Scholz RW, Wellmer F-W (2013b) Approaching a dynamic view on the availability of mineral resources: what we may learn from the case of phosphorus? Glob Environ Chang 23:11–27

Scholz RW, Wellmer F-W (2015) Losses and use efficiencies along the phosphorus cycle. Part 1: dilemmata and losses on the mines and other nodes of the supply chain. Resources Conservation & Recycling 105:216–234

Scholz RW, Wellmer F-W (2018) Although not physical scarce in the short-term, improvements in resource efficiency are needed. J Ind Ecol https://doi.org/10.111(jiec.1250)

Scholz RW, Roy AH, Hellums DT (2014) Sustainable phosphorus management: a global transdisciplinary challenge. In: Scholz RW et al (eds) Sustainable phosphorus management: a global transdisciplinary roadmap. Springer, Berlin, pp 1–129

Smil V (2000) Phosphorus in the environment: natural flows and human interferences. Annua Rev Energy Environ 25:53–88

Preface

Phosphorus (P) is an essential element for all living organisms. Since P cannot be replaced by any other element in biochemical processes, humans ultimately rely on P availability. Today, P is mostly obtained from mined phosphate (P_i) rock. However, global P_i rock reserves are concentrated in a limited number of countries such as Morocco, Iraq, China, and Algeria (USGS 2017). More than 90% of countries in the world have no significant reserve of P_i rock and are most entirely dependent on imported P. Nevertheless, P_i rock is a nonrenewable resource on a human time scale, and high-grade P_i rock is being depleted by ongoing mining worldwide.

The current use of P is inefficient and leaky, particularly in the agricultural sector, thereby causing eutrophication problems in natural bodies of water. Moreover, since P_i rock is generally contaminated with toxic heavy metals and radioactive elements, there is some concern that they may cause health and environmental issues through the long-term application of chemical fertilizers to farmland. In this sense, the sustainable management of P is becoming an increasing global concern on food and environmental securities. International and national platforms to address issues surrounding the sustainable management of P have been founded in Europe, North America, and Japan.

P recovery and recycling can play a critical role in closing the anthropogenic cycle of P, thereby contributing to the sustainable management of P. This book presents the latest research advances, innovations, and applications of P recovery and recycling which is becoming a rapidly evolving and challenging field of engineering. Full-scale applications are highlighted to illustrate the performance and effectiveness of the new technologies. This book provides advanced engineering knowledge, information, and new ideas in the field of P recovery and recycling practices. The intended audience for this book includes engineers, planners, consultants, municipal staff, and resource managers, among others.

This book is structured in seven main parts. Part I first reviews the enabling/disabling legal and social framework on P recovery and recycling. The individual six chapters of Part I deal with a different aspect of the framework, including P flow, circular economy, life cycle assessment, P reserves, production and applications, and success factors for full-scale implementation. In particular, Chap. 1 gives a comparative overview of the full-scale applications and their importance in relation to the potential in Europe and Japan which are front-runners in P recovery and recycling practices in the world. Both Europe and Japan have a multitude of reasons for proactively seeking strategies for sustainable P management: high P import dependency, soil nutrient imbalance, eutrophication, and agricultural sustainability. Chapter 1 also introduces the innovative P value chain concept that maximizes the synergies of P recycling across the biosphere and the technosphere. The technology and business innovation based on P recycling, called P innovation, is expected to make a great contribution not only to the sustainable agricultural production but also to the secured supply of high-purity P to the manufacturing industry.

The twenty three chapters of Parts II to V cover recent research advances and applications of P recovery and recycling from secondary resources such as incinerated sludge ash (Part II), sewage sludge and night soil (Part III), steelmaking slag (Part IV), and animal manure (Part V). Each chapter covers a different technology option for P recovery and recycling from secondary resources. In particular, Chaps. 8, 9, 17, 18, and 25 highlight the performance and effectiveness of the full-scale applications. Chapters 7 and 14 also introduce full-scale processes for manufacturing phosphoric acid from sewage sludge ash, while Chap. 16 sheds light of the full-scale production of fertilizers from secondary resources. Finally, Parts VI and VII cover emerging technologies for P recovery and recycling using solid adsorbents and biotechnology.

We hope that this book provides a cutting-edge resource of information in this rapidly evolving field for a variety of readers, including students, researchers, and professionals interested in and responsible for sustainable use of P. This book can also be considered, in a way, the follow-up of the previous book entitled *Sustainable Phosphorus Management*, published by Springer in 2014. We would like to encourage readers to go through this previous book for a better understanding of the sustainable management of P. We also hope that this publication will promote further progress in the research and development of P recovery and recycling. Finally, we would like to thank all the contributors for their willingness to share their knowledge, know-how, and experiences in this book. We must appreciate the time and efforts of all the contributors for committing their time to the publication of this book.

Tokyo, Japan
2018

Hisao Ohtake
Satoshi Tsuneda

Contents

Part I
Phosphorus Flow and Recycling

Chapter 1
Development of Phosphorus Recycling in Europe and Japan

Anders Nättorp, Christian Kabbe, Kazuyo Matsubae, and Hisao Ohtake

Abstract This chapter will explore whether the differences in the process development for phosphorus recovery and recycling and its implementation in Europe and Japan are linked to the waste flows and the regulative framework. The main waste flows and their qualities are summarized for the two geographical areas. Then a comparative overview of the full-scale applications and their importance in relation to the potential is presented. The drivers for phosphorus recycling and the expected further development in Europe and Japan are described.

Keywords P flows · Full-scale P recovery · P recycling · Regulatory framework · Value chain

1.1 Introduction

Phosphorus (P) has important biological functions as the building block of DNA, cell membranes, energy-storing ATP, and bones. It is necessary for growth and development of all living organisms and is present in most organic materials. More than 80% of mined phosphate rock is used for fertilizer, providing about half of the fertilizer P consumed globally. The remaining P comes from recycled organic waste,

---

A. Nättorp (✉)
Institute for Ecopreneurship, School of Life Sciences, University of Applied Sciences and Arts Northwestern Switzerland, Muttenz, Switzerland
e-mail: anders.naettorp@fhnw.ch

C. Kabbe
P-REX® Environment, Am Goldmannpark 43, Berlin 12587, Germany

K. Matsubae
Department of Environmental Studies for Advanced Society, Graduate School of Environmental Studies, Tohoku University, Sendai, Japan

H. Ohtake
Phosphorus Atlas Research Institute, Waseda University, Shinjuku-ku, Tokyo, Japan

H. Ohtake, S. Tsuneda (eds.), *Phosphorus Recovery and Recycling*,
https://doi.org/10.1007/978-981-10-8031-9_1

mainly manure (Scholz et al. 2014). The fossil phosphorus reserves are concentrated in Morocco and Western Sahara (>70%) (USGS 2018), and their share of reserves is increasing as other countries mine their deposits at a higher rate. Therefore, numerous initiatives are emerging which aim at increasing the fraction of recycled phosphorus. On a global scale, Japan and Europe are front-runners in phosphorus recovery technologies.

From a material flow perspective, phosphorus recycling requires to bridge the gap from several distinctive P-containing waste streams with characteristic impurity profiles to well-defined outputs responding to market demand. Numerous technical processes have been proposed which achieve one or several of the following objectives:

- Purification
- Improvement of plant availability
- Concentration

This book presents a selection of processes which were developed in Europe or Japan and which reached a certain level of technical maturity, mostly at least pilot scale. By processing, the P can in many cases be recycled more effectively and diminish the consumption of limited and geographically critical fossil P reserves. In addition, often costs and other resources can be saved. Through the extraction of P from the waste and the reduced input of fossil P, the risk of diffuse emissions into the environment diminishes, and especially the risk for eutrophication can be reduced.

This chapter will explore whether the differences in recovery process development and implementation in Europe and Japan is linked to the waste flows and the regulative framework. The main waste flows and their qualities are summarized for the two geographical areas. Then a comparative overview of the full-scale applications and their importance in relation to the potential is presented. The drivers for P recycling and the expected further development in Europe and Japan are also described.

1.2 P Flow and Hot Spots

1.2.1 Europe

The European P flows were studied by various authors (Ott and Rechberger 2012; van Dijk et al. 2016). In 2015, EU28 used 1480 kt of mineral P (76% fertilizer, 17% feed, 7% others (including phaseout of detergents)) (Hukari et al. 2015). Ninety-two percent of the mineral P inputs such as phosphoric acid, phosphate (P_i) rock, and fertilizers are imported (European Commission 2013a). White phosphorus, which is used as a starting point for specialty chemicals synthesis and high-purity phosphoric acid (see also Chap. 5), is 100% imported.

The main P-rich waste streams are manure, sewage sludge, slaughterhouse waste (P almost exclusively in meat and bone meal), and food waste. Food waste from household and retail represents a large amount of P (187 kt/a). However, the sources are diffuse and heterogeneous, and the concentration is also usually low (average 0.4% on dry matter (DM)). Therefore, there has been little development of processes for recovery of mineral P materials from food waste.

The composition and volumes of the other three main waste streams are described in Table 1.1. Manure is the largest stream (1810 kt P/a) and is today mostly recycled directly to fields. Sewage sludge (374 kt P/a) is also spread on fields to a large extent (39%). Meat and bone meal (312 kt P/a) was used in animal feed, but because of the

Table 1.1 The composition of three streams with the largest potential for technical phosphorus recovery in Europe

		Manure	Sewage sludge	Meat and bone meal
Amount (kt P/a)		1810	374	312
Recycled (%)		97	39	6[a]
DM (%)		5–30[c]	~25%	90–95[b]
Phosphorus (% on DM)		Cattle 0.5; poultry 1–2; pig 0.3–2.7[c]	3.7[d]	12–15[e]
Organic contaminants, pathogens		High use and excretion (30–90%)[f] of pharmaceuticals followed by land application	PFA, PFC, etc. have bioaccumulating and persistent properties. Pharmaceuticals much less critical[g]	BSE risk. Concentration of other contaminants in tissue is low
Concentration ash	DüMV (DE)	Average ashes and charcoal[h]	Germany, average[i]	Ash from sample of pig MBM, kat. 3[j]
P (%)		4	9 (municipal ash) 5 (mixed sludge ash)	18
As (ppm)	40	n.d.	17	n.d.
Cd (ppm)	1.5/50	2	3	0.3
Cr (ppm)	2 (Cr^{VI})	27	267	21
Cu (ppm)	900	274	916	47
Ni (ppm)	80	32	75	79
Pb (ppm)	150	11	151	<19
Zn (ppm)	5000		2500	373

[a]Many parts of the slaughtered animal are not eaten but used for pet food etc. These are not classified as waste
[b]Campoy et al. (2014)
[c]Schoumans et al. (2010)
[d]Milieu Ltd, WRc (2008) and van Dijk et al. (2016)
[e]Buckwell et al. (2016)
[f]Lianfeng and Wenke (2012)
[g]Clarke and Smith (2011)
[h]Foged et al. (2011)
[i]Krüger and Adam (2014)
[j]Coutand et al. (2008) and Hua et al. (2005)

mad cow (BSE) disease outbreak, this use was prohibited by the EU Animal By-Products (ABP) Regulation (1774/2002). Sewage sludge and manure pose less danger of pathogens if used correctly but contain other organic contaminants that might cause concern.

In EU28 most of the meat and bone meal (MBM), about a quarter of the sewage sludge and a small fraction of the manure are incinerated. The MBM ash has the highest P content (18%), the municipal sewage sludge ash about half as much and the manure ash about a quarter (3.9%). The MBM has also the advantage of low heavy metal content. The sewage sludge ash has much higher heavy metal content. Yet, sewage sludge ash on German average just fulfills the limits of Cd, Cu, Ni, and Pb in the German fertilizer legislation, which is one of the strictest in Europe.

Expressed in absolute figures, manure ash contains less heavy metals, but Cd, Cu, and Ni levels are relatively high in comparison to its limited P content. Manure is mostly recycled to fields, but there are large surpluses in some regions. Pig factories cause most of these problems. Poultry factories also produce large amounts of manure, but it has a much higher dry matter content and can thus be transported further. In contrast, cattle is usually raised more in balance with cropping agriculture and pastures and does not cause nutrient hot spots (Schoumans et al. 2010).

In Europe, waste streams present a large unused potential for P recycling. Unused manure, sewage sludge, and MBM together contain over 500 kt P/a. Several properties of these streams can be improved by processing:

- Removing contaminants is important for part of the sewage sludge (heavy metals), for meat and bone meal (prions), and to some extent for manure (heavy metals).
- Improving plant availability is important since sewage sludge (unless EBPR) and ashes excepting chicken litter ash have low plant availability which makes processing worthwhile.
- Finally, the dry matter and nutrient concentrations are low compared to mineral fertilizers, and the materials are biologically active and with varying concentrations with regard to both the source and the time period. Thus, processing is useful to facilitate storage and transport and reach a more uniform quality.

1.2.2 Japan

Figure 1.1 shows the estimated P flows in Japan in 2016. As the primary resource of P, P_i rock is a significant starting material for P-related products. Since Japan has no significant P_i rock reserve, all the ore is imported from abroad. The imported ore is mostly used in the chemical industry as a raw material, amounting to an estimated 41 ktP/a. The fertilizer industry produces commercial and special fertilizers by using raw materials and commodity chemicals produced by the chemical industry. This P throughput in the fertilizer industry is reported as 126.9 ktP/a.

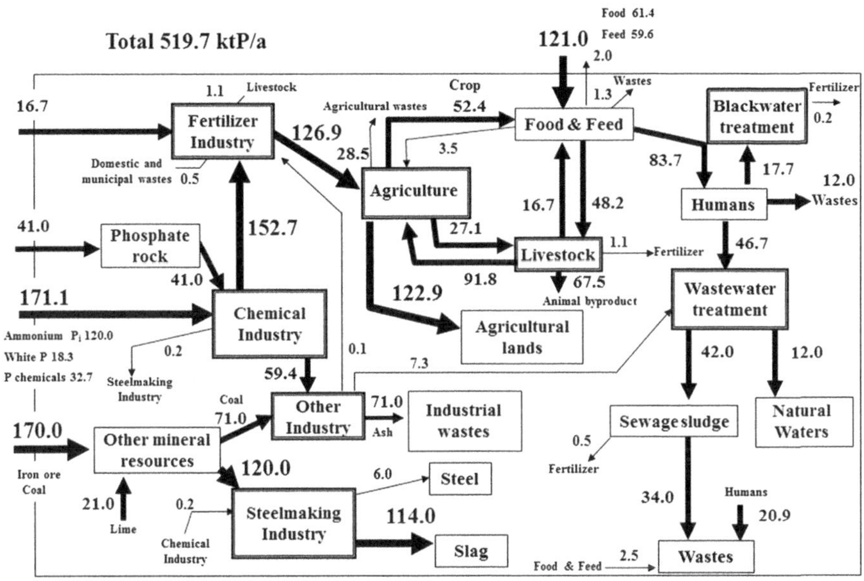

Fig. 1.1 Annual P flows in Japan (2016) (kP/a)

The main input flows of P in the agricultural sector originate from fertilizers used in farms and ranches. About 127 ktP/a was applied to farms and ranches in the form of fertilizer. This value is one of the largest input flows in the entire domestic P flow. Input flows to the food and feed sector also have large values. They flow in mainly from world imports and marine resources (121 ktP/a) and domestic crop production from farmlands (52 ktP/a). This P is mainly consumed by humankind and livestock (84 ktP/a and 48 ktP/a, respectively). The P flows associated with cattle, pigs, chickens, horses, sheep, goats, and other livestock are considered in this sector. Livestock grow by eating grass and feed on ranches, and the P in livestock manure ends up accumulating in the soil. The amount of P accumulated in the soil (123 ktP/a) is nearly equal to the input from fertilizer to farms and ranches.

The main path for P return from the economy to farms and ranches is in the form of compost and livestock manure, amounting to about 95 ktP/a. Almost all the P in livestock manure is likely utilized as fertilizing materials in the farm and ranch sector intentionally or not, only 1% is clearly utilized as a resource of the fertilizer production sector, and the rest is assumed to be treated in well-managed landfill site. Other relevant P flows come from the iron and steel industry in Japan. Iron ore, coal, and limestone are major mineral resources for the iron and steel industry, amounting to 120 ktP/a. Japan produces more than 100 million tons of steel materials annually, while generating a huge amount of steelmaking slag at the same time. Of the several types of steelmaking slag, the slags which are generated after dephosphorization in torpedo cars and basic oxygen furnaces (BOFs) have approximated 5.0 and 3.0 mass% of P_2O_5, respectively (Matsubae-Yokoyama et al. 2009).

The total P input to Japanese society is estimated to be 520 ktP/a. Of this, 227 ktP/a is supplied to the economic activities as a valuable resource, while the remainder is not recognized as a resource flow. Most of the P resources are used in fertilizers. Though approximately 80–100 ktP/a of the P cycle is involved in the farm/ranch sector, the remaining P diffuses into the environment. The P flow into the steel industry is estimated to be approximately 114 ktP/a, most of which is accumulated in steelmaking slag through the dephosphorization of molten iron. Although the main flows of P finally reach the soil, river, and sea, approximately 42 and 114 ktP/a end up in sewage sludge and steelmaking slag, respectively. Concerning the restricted supplies of P resource, it is important to consider the quantity and availability of P resources that currently remain untapped.

1.3 Full-Scale Plants

Examples of full-scale implementation of recovery processes are presented in this section. In the text below and in this book, they are grouped according to raw materials and process type (Table 1.2).

1.3.1 Europe

Previous summaries of European processes were made by Egle et al. (2016) and Kabbe et al. (2015). A few key characteristics of each group in Table 1.2 are given below together with a short description of an implementation example.

Table 1.2 Full-scale implementation of recovery processes grouped according to raw materials and process type. Reference to chapters describing such processes

Process type	Full-scale example Europe	Full-scale example Japan	Chapters
Ash for direct fertilizer use	×	×	–
Ash as input for fertilizer production	×		16
Ash leaching	×	×	8, 9, 10, 14
Wastewater and sludge precipitation	×	×	17, 18
Sludge leaching	×		19, 20, 21
Manure processing	×	×	25, 26, 27, 28
Recovery with absorbents	×		30, 31, 32, 33
Purification of spent phosphoric acid		×	–

1.3.1.1 Ash for Direct Fertilizer Use

Ashes from for example poultry litter (PLA) and MBM can be directly used to produce a fertilizer.[1]

Advantages are that incineration is a simple and relatively low-cost process that eliminates pathogens and organic contaminants and provides a more concentrated and storable output.

Limitations are that most ashes have limited plant availability and some have high heavy metal loads. MBM combustion ash produced by the Saria group in the UK is recycled in agriculture, having gained end of waste status, since 2014. About 12,000 t/year with the with average NPK content 0-21-4[2] (and 37% CaO) is currently sold under name "Kalfos" in the form of a coarse powder that is best spread using a lime spreading machine. The product is not soluble in water, but about 20% of the phosphate is soluble in neutral ammonium citrate. It has been demonstrated over several years to be effective at promoting grass growth in a range of farm and controlled laboratory conditions.

1.3.1.2 Ash as a Feedstock in Fertilizer Production

Ashes and other mineral P-rich materials can be combined with fossil feedstocks in fertilizer production.

Advantage is a simple process that uses existing infrastructure and distribution channels in the fertilizer industry.

Limitation is that such a process can increase plant availability through acidulation, but does not concentrate phosphorus or reduce contaminant load. The iron and aluminum coagulants used in wastewater treatment (unless EBPR is used) create processing problems with these ashes.

Cost is close to break-even, making the process economically viable in certain cases depending on ash phosphorus concentration, contaminants as well as waste disposal and transport costs. ICL, Amsterdam processed a few hundred tons of MBM ash in 2016.[3]

---

[1] Current production and marketing of poultry litter ash includes among others: Fibrophos UK, BHSL Ireland, BMC Moerdijk Netherlands, Sanders France. MBM ash: SARIA UK, Fibrophos UK, Wykes Engineering UK.

[2] 0% N, 21% P_2O_5, 4% K_2O.

[3] Anthony Zanelli, personal communication, August 2017

1.3.1.3 Ash Leaching

Sewage sludge ash, MBM ash, and other ashes can be leached to produce a purified, concentrated, bioavailable material.

Advantages are that the processes have several possible outputs (CaP, H_3PO_4, etc.) which can be tailored to different markets (fertilizer, feed, technical).

Limitation is the process complexity, the availability of ash, and the dilution of some ashes through co-incineration of phosphorus-rich and phosphorus-poor materials.

Cost of the process is close to breakeven, as indicated by the construction of full-scale production lines and cost estimations (Nättorp et al. 2017).

Ecophos is constructing a plant for production of 200 kt of dicalcium phosphate (DCP) for the feed market in Dunkerque, France. Production should start in autumn 2017, and in 2019 a production line from sewage ash should follow. Ecophos has a supply contract with the Dutch incineration operators SNB and HVC for 60 kt/a of sewage sludge ash, approximately 4 kt P/a. The Ecophos process uses sulfuric acid, which is reacted with calcium chloride to gypsum and hydrochloric acid, which in turn is used in the ash leaching. The leachate is purified, and phosphate is precipitated with calcium carbonate yielding also calcium chloride (see Chap. 14).

Hamburg water and REMONDIS are planning a plant with capacity to process 20 kt of sewage sludge ash/a using the REMONDIS TetraPhos® leaching process with intended start in 2019.[4]

1.3.1.4 Wastewater and Sludge Precipitation

Struvite or calcium phosphate is precipitated from an orthophosphate-rich process stream (concentration usually >100 g P/m^3).

Advantages of struvite precipitation are operational benefits in WWTP with EBPR and anaerobic digestion. Precipitation is also used to valorize P from industrial wastewater rich in P (e.g., potato processing, dairy).

Limitations are that the process is limited to the few WWTP which use EBPR (e.g., <10% of the wastewater in Germany) and to certain industrial streams. In municipal wastewater applications, only a small fraction of the phosphorus is recovered (5%–15%), but processes with higher yield are in the process of being implemented.

[4] Patrick Herr, REMONDIS Aqua, personal communication, August 2017

Cost of struvite precipitation processes is in general compensated by the aforementioned operational benefits. In industrial streams precipitation is a viable treatment method for high phosphorus concentrations.

Volume in Europe is approximately 1 kt P/a recovered as struvite at >20 WWTP and more than 40 full-scale installations worldwide. The first European high-yield precipitation process installed in Amersfoort, Netherlands[5], combines WASSTRIP, Lysotherm thermal sludge disintegration, and an Ostara PEARL precipitation. The plant should reach 30–40% recovery from the sludge according to a first estimation by the operator. It is dimensioned for 900 t/struvite per year (~100 t P/a). The PEARL process produces well-defined struvite prills which Ostara buys from the WWTP for several hundreds of euro per ton and sells as premium fertilizer Crystal Green.

1.3.1.5 Sludge Leaching

Sewage sludge is leached with acid and phosphorus precipitated as struvite or calcium phosphate, mainly depending on ammonia availability. A demonstration and a full-scale plant are in operation.

Advantages of sludge dissolution are firstly the higher recovery yield (~50%) compared to precipitation processes (5–15%), although these are catching up. Secondly, these processes, in contrast to precipitation processes, can be applied to sludges containing phosphorus precipitated with iron- or aluminum-based coagulants. Finally, P-depleted sludge allows co-incineration, which is not possible for the higher-yield ash-based routes.

Limitations are their high demand of mineral acid and large effort for purification. Combined with medium yield, this makes the process comparatively expensive (~10 EUR/kg P).

Seaborne Environmental Laboratory developed and from 2000 onwards piloted a process for nutrient recovery for a great variety of organic materials. The process simultaneously eliminates pollutants of the treated biomass (mainly heavy metals from municipal sewage sludge) and utilizes the energy content of the organic materials. The only full-scale plant of this type was built at the WWTP Gifhorn, Germany, in 2005 (50,000 p.e.). It has a capacity of approx. 1000 tons of dry solids per year (120 m^3 digested sludge per day). Full-scale investigations and laboratory scale experiments were conducted to overcome operational problems and to increase the economic efficiency of the plant. To increase robustness, the process was simplified to what became known as the "Gifhorn process." An efficient separation of iron allowed a safe operation of the ammonia stripping unit and a high phosphorus recovery yield (~50%) with moderate demands of auxiliary chemicals. Currently

[5] http://phosphorusplatform.eu/scope-in-print/news/1219-ostara-struvite-recovery-amersfoort

the ammonia stripping unit and heavy metal separation are not in operation due to economic reasons. As a consequence, the phosphorus yield rate is significantly lower, and the plant availability of the resulting product is reduced (Niewersch et al. 2014).

1.3.1.6 Manure Processing

When regional manure concentration is high, which is in general due to pig factories, the transport cost for distribution to fields where the manure is allowed to be spread becomes expensive. Approximately 8% of European manure is processed corresponding to 139 kt P/a. Main processes are solid-liquid separation and anaerobic digestion as well as some drying but very little incineration (Foged et al. 2011).

Advantages of the manure processing are simple low-cost processes to concentrate nutrients and facilitate transport.

Limitations of these processes are that they do not remove contaminants.

Cost for separation of pig slurry, drying, and transport of solid fraction is 24–31 EUR/m^3 slurry (14–18 EUR/kg P), approximately half for the treatment of the liquid fraction (Schoumans et al. 2010).

Cooperl in Brittany, France, processes 400,000 t/a manure to organic fertilizer products adapted to specific crops. The products are storable and transportable and can thus be sold in all of France.[6] An example of mineral output is the 50 t P/a recovered as K-struvite[7] from calf urine, the liquid fraction of manure through addition of magnesium in a CSTR (Schuiling and Andrade 1999).

1.3.1.7 Recovery with Adsorbents

Solid adsorbents can remove phosphorus from liquid streams. The sorbents can either be regenerated, used as a fertilizer or as a soil improver, or be landfilled.

Advantages of non-regeneration adsorbents are that they can be made from waste or by-products. Small-scale applications with little operator know-how are possible.

Limitations of non-regeneration adsorbents are that the loaded material has low nutrient content making it a less efficient fertilizer.

One example of adsorption is filters with Polonite mineral-based reactive filter media marketed by Ecofiltration Nordic AB for households, WWTP, and agricul-

---

[6] http://www.cooperl.com/en/environmental-solutions

[7] $KMg(PO_4)\bullet 6(H_2O)$

tural runoff. A household filter has a capacity of 3–5 kg of phosphorus for 500 kg of adsorber at a cost of approximately 500 EUR (100–150 EUR/kg P), and 5000–6000 units are currently used, predominantly in Sweden. The adsorber itself is harmless, and approximately 90% of spent filters are recycled as soil improver. The phosphorus mineral phases quantified by X-ray diffraction show about 60% easily mobilizable phosphorus, which is comparable to manure.[8]

1.3.1.8 Summary

To sum up, over 10 kt P/a is recycled in inorganic form. The largest volumes are by far the direct use of ash (>10,000 t/a of PLA, also MBM ash and others). About 1 kt P/a is recycled as struvite. Two plants for SSA leaching of a total of approximately 5 kt P/a are in planning and construction. This is still low compared to 500 kt P/a from sewage sludge, MBM, and manure that are wasted.

1.3.2 Japan

Japan has long experience in commercial operation of full-scale plants to recover P from secondary resources such as sewage sludge, sewage sludge ash (SSA), animal manure, and industrial wastewater (Fig. 1.2; Ohtake and Okano 2015). The direct application of sewage sludge to agricultural land is not acceptable in Japan, because of environmental and health securities as well as a shortage of land. Exceptionally, sewage sludge from small WWTP is used as a fertilizing material in some rural areas after being subjected to anaerobic digestion and composting. Hence, in order to recycle P from sewage to farmland, it is essentially necessary to recover it from digested sludge, dewatering liquor, or SSA in WWTP. Meanwhile, manufacturing companies have begun to consider P recycling as an economically beneficial option in waste management, because it leads to saving sludge disposal costs.

1.3.2.1 Ash Leaching for Fertilizer Production

Approximately 75% of sewage sludge is mono-incinerated to reduce volume and for energy recovery. Mono-incinerated SSA has a high P content, typically ranging between 4 and 8 wt% (10–20 wt% as P_2O_5). Chemical phosphate (P_i) leaching is a technology option for recycling P in SSA. Full-scale plants for recovering P_i from SSA have been implemented at Gifu and Tottori WWTP (see Fig. 1.2; Chaps. 8 and 9). The plants started operation in 2010 and 2013, respectively. Both plants adopted alkaline (NaOH) leaching with subsequent P_i precipitation with Ca^{2+}. Alkaline leaching was chosen because acid leaching inevitably dissolves heavy metals,

[8] Ecofiltration information sheet "Avsättning änvänd Polonite" Version 2017.

Fig. 1.2 Locations of full-scale plants operating for P recovery and recycling in Japan

thereby contaminating recovered products. Recovered P products (i.e., calcium hydroxyapatite (HAP)) are dewatered, dried, and granulated into pellets to generate P_i fertilizer. Approximately 35–40% of P_i (mainly from $AlPO_4$) is recovered from SSA. The P_i fertilizer, whose efficacy has been demonstrated in farmland, is sold to farmers through fertilizer companies. Solid residues after P_i recovery are subjected to weak-acid cleaning to remove heavy metals and then used as soil amendment for street trees.

The Gifu plant supplies approximately 300 t/a (c. 80 t P/a) of P_i fertilizer to local farmers. The P recovery cost accounts for approximately 3% of the total operating cost of Gifu WWTP. The Tottori plant recovers approximately 150 t/a (c. 40 t P/a) of HAP from 500 t/a of SSA with 25–30 wt% P_2O_5 (Fig. 1.3). Stable channels for the distribution and sale of recovered HAP have been established within the local area. However, mainly due to the expense of chemicals, the cost of P recovery is high compared to the disposal of SSA as industrial waste.

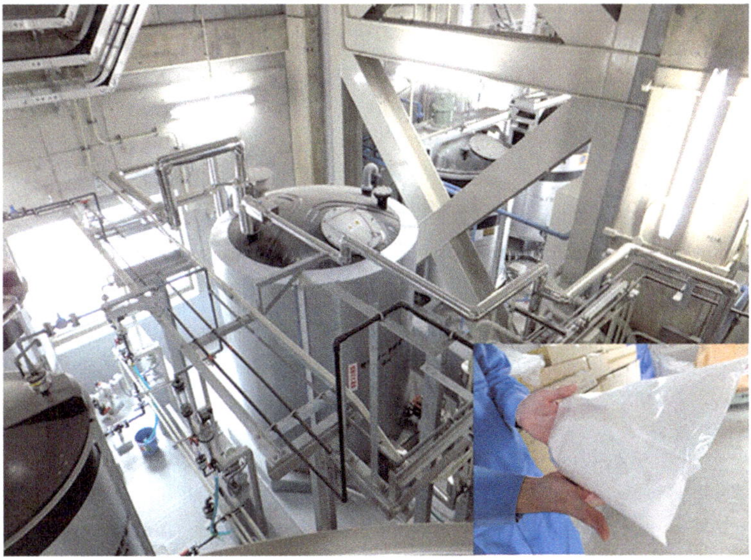

Fig. 1.3 Full-scale plant for recovering P from SSA at Tottori WWTP (*inset*, recovered P product, HAP) (Photograph courtesy of M. Hashimoto)

1.3.2.2 Ash Leaching for Phosphoric Acid Production

Nippon Phosphoric Acid Co. (NPA) produces approximately 90 kt P_2O_5/a of merchant-grade phosphoric acid (PA) (see Chap. 7). NPA adopts a wet acid process in which sulfuric acid is used to extract PA from P_i rock. NPA started to use SSA, which was transported from a local WWTP, as part of raw materials to generate merchant-grade PA and by-product gypsum ($CaSO_4$) in 2012. NPA accepts SSA from the WWTP equipped with a Bio-P process to remove P_i without using Fe- or Al-based flocculants. The transportation cost is covered by the WWTP. In the PA-manufacturing plant, SSA is blended with roller-milled P_i rock. The mixture is then dissolved in concentrated sulfuric acid to generate phosphogypsum slurry which is continuously filtered to separate PA and gypsum. The by-product gypsum is used as a raw material for the manufacture of plasterboards and cement.

Currently, NPA uses a total of 1000 t/a of SSA for the manufacture of PA. In full operation, the manufacturing plant has the capacity to accept 5000 t/a of SSA. The blend ratio of SSA with roller-milled P_i rock is currently less than 2.5% (<97.5% P_i rock). This is needed to guarantee the quality of both PA and by-product gypsum. Importantly, it was not necessary for NPA to modify its existing facilities to accept SSA as a raw material. NPA could save raw material expenses in proportion to the amount of SSA accepted from the WWTP.

1.3.2.3 Precipitation from Sewage

Fukuoka and Matsue WWTPs have implemented struvite recovery from rejected water from a membrane-type solid-liquid separator for digested sludge. Struvite particles are formed by the addition of $Mg(OH)_2$ in a fluidized-bed tower reactor (Fig. 1.4). The Fukuoka and Matsue plants started operation in 1997 and 1998, respectively. Both of the plants have been implemented mainly to prevent struvite from forming a damaging build-up in pipe lines downstream of anaerobic sludge digesters (Ueno and Fujii 2001).

The Fukuoka plant recovers a total of 140 t/a (c. 14 t P/a) of struvite which is sold to fertilizer companies. However, the pioneering plant has suffered from the unwanted wall growth of struvite in the fluidized-bed tower reactor from the very beginning. Owing to this obstacle, only one-third of the rejected water is currently processed for struvite recovery. Accordingly, only 2% of influent P is recovered as struvite at the Fukuoka WWTP. On the other hand, the Matsue plant recovers approximately 150 t/a (c. 15 t P/a) of struvite, accounting for about 20% of P flowing into the WWTP. The recovered struvite is sold as a slow-release P_i fertilizer. The sales of struvite cannot cover the expenses of chemicals required for struvite recovery at both Fukuoka and Matsue WWTP.

Another full-scale plant was installed at Kobe WWTP (see Chap. 17) to recover struvite directly from digested sludge in 2012. The Kobe WWTP has three anaerobic sludge digesters. Digested sludge contains P_i at concentrations as high as 600 mgP/L. $Mg(OH)_2$ is supplied to the reactor as a source of magnesium. Digested

Fig. 1.4 Full-scale plant for recovering struvite from rejected water from a membrane-type solid-liquid separator for digested sludge at Matsue WWTP

sludge in the reactor is mechanically mixed to enhance the contact between solid and liquid. The mechanical mixing also causes decarbonization and collision of struvite crystals, thereby controlling their sizes in the reactor. The recovered struvite is washed with water to obtain high-purity products. This plant recovers approximately 360 kg/day of struvite from 239 m^3/day of digested sludge with P recovery efficiencies of 30–40%. This amounts to about 130 t/a (c. 13 t P/a) of struvite, which is equivalent to one-tenth of the annual consumption of P_i fertilizer in the Kobe area.

1.3.2.4 Precipitation from Blackwater

The sludge treatment center at Senboku city (see Fig. 1.2) has a capacity of treating approximately 60 m^3/day of blackwater (i.e., night soil and johkasou sludge in Chap. 18). The plant started operation in 2009 to treat blackwater collected from households in a local area where no sewage service was available. It is equipped for nitrogen (N) and P removal in addition to mainstream sludge treatment processes. N is removed from undiluted sludge liquor using a jet aeration system. P_i is precipitated with $CaCl_2$ from the liquid rejected by a membrane-type solid-liquid separator after N removal. Approximately 72% of P_i can be removed from blackwater, thereby reducing the effluent P_i concentration well below the regional P_i discharge standard.

The recovered product (HAP) contains approximately 30 wt% citrate-soluble P_2O_5 on a dry weight basis. It is shipped out in wet form containing about 30–40% water. The production rate of HAP is approximately 50 kg/day (about 10–15 t/a). Approximately 20% of P is recovered as HAP, whereas the rest ends up in dewatered sludge. Currently, the P recovery cost is less than one-tenth of the total cost of plant operation. However, it is still high compared to simple P_i removal without recycling. Up to now, ten local municipalities have embarked on HAP recovery from blackwater (see Chap. 18).

1.3.2.5 Precipitation from Industrial Wastewater

Japan Synthetic Alcohol Co., Ltd. (JSA) is a principal supplier of chemically synthesized ethanol for industrial purposes. This company manufactures 95 v/v% ethanol by a direct ethylene hydration process, in which phosphoric acid (PA) is used as a chemical catalyst. Fresh ethylene gas is fed into a tower reactor, where PA is impregnated in silica beads, and converted to ethanol as it passes through the tower reactor. Wastewater from the tower reactor contains high levels of P_i, which leaches from the packed silica beads. JSA needs to remove P_i from the wastewater to meet P discharge regulations in the Tokyo Bay area (less than 18.2 kgP/day). This company implemented P recovery from the wastewater in 1998. Approximately 75% of P_i is recovered from the wastewater as HAP, whose pellets (30 wt% P_2O_5 and 10% water) are sold to a fertilizer company at a low price. JSA can save sludge disposal costs by turning the waste to a resource, while the fertilizer company benefits from the low-price fertilizing material.

Kyowa Hakko Bio Co., Ltd. (KHB) is a leading fermentation company in Japan. This company manufactures nucleic acids, amino acids and other products for foods, seasonings and pharmaceutical agents. The wastewater from the fermentation processes contains P_i as high as 1000 mg P/L. KHB began to recover P_i from the wastewater to meet regulation standards for total P emissions to the Seto Inland Sea (see Fig. 1.2) in 2006. About 90% of P_i is removed from the wastewater by precipitation with $Ca(OH)_2$. The recovered product is HAP (typically 29 wt% P_2O_5 and 59 wt% CaO; Fig. 1.5). After drying, HAP is sold to a fertilizer company. At present, approximately 60–90 t/a of P is recovered from fermentation wastewater in KHB.

J-Oil Mills, Inc. (JOM) is a principal supplier of cooking oil in Japan. Edible oil is purified from crude oil extracted from soybeans, grapes, peanuts, and walnuts. Crude vegetable oil contains various phosphatides such as phosphatidylcholine, phosphatidylinositol, phosphatidylethanolamine, and phosphatidic acid. They need to be removed from crude oil, because they are responsible for refining losses due to emulsion formation (van Nieuwenhuyzen and Tomas 2008). The elimination of phosphatides from crude vegetable oil (called degumming) is a key step in vegetable oil refining. To remove nonhydratable phosphatides, high-grade PA is added as an 85% solution to crude oil at a concentration of 0.05–0.15 v/v%. Hence, wastewater from the acid degumming process contains high levels of spent PA. JOM recovers P_i from the wastewater by chemical precipitation with $Ca(OH)_2$. Since the wastewater contains essentially no harmful substances, recovered product (HAP) is suitable for the manufacture of fertilizer. JOM sells the recovered HAP as a by-product P_i fertilizer (approximately 260 t/a).

(A) (B)

Fig. 1.5 P recovery from fermentation wastewater in KHB. (**a**) P_i precipitation tank; (**b**) recovered P product (HAP) (Photograph courtesy of H. Hidaka, KHB)

1.3.2.6 Manure Processing

Miyazaki Biomass Recycle Co. (MBR) implemented a power plant to generate electricity by incinerating broiler chicken manure in 2005 (Fig. 1.6). Broiler chicken manure with a water content of typically 43 wt% can produce about 1900 kcal/kg when incinerated. The incineration of chicken manure can contribute to solving regional problems such as illegal dumping, offensive odor emission, and groundwater nitrate contamination. The MBR power plant has an incineration capacity of 132,000 t/a of broiler chicken manure. It generates electricity of 74,943 MWh, of which about 80% is sold to a local power company. About 500 t/a of broiler chicken manure (more than 50% water) is transported by road from 400 poultry houses in the Miyazaki area (see Fig. 1.2) to the MBR plant. The collected manure is incinerated in a self-sustaining combustion furnace at 7.43 MPa and 468 °C. After incineration, approximately 13,000 t/a of ash (about 10% of the original weight) is transported by road to a fertilizer company. It is used as a fertilizing material (20 wt% P_2O_5 and 15 wt% K on average) for compound fertilizers.

1.3.2.7 Purification of Spent Phosphoric Acid

In the manufacture of liquid crystal glass substrates, phosphoric acid (PA) is commonly used as an aluminum etching agent. Wastewater from wet etching processes contains P_i at concentrations as high as 5000 mgP$_i$/L. Kurita Water Industries Ltd.

Fig. 1.6 Power plant for generating electricity by incinerating broiler chicken manure. (Photograph courtesy of K. Yano, MBR)

(KWI) has developed a system for recovering P_i from wastewater in liquid crystal manufacturing processes. After removing suspended solids (SS) by filtration, Al^{3+} is removed from a wet etching wastewater using an H-type cation exchange column. Then, P_i is concentrated to 4–7% using a two-step reverse osmosis membrane system at 0.7–2.0 and 4 MPa, respectively. P_i can be further condensed to 50% or higher by reduced-pressure distillation. This process has been installed in three liquid crystal manufacturing plants (two in Japan and one in Taiwan).Typically, about 2.7 m^3 of 50% P_i solution can be generated from 800 m^3/day of wastewater containing 3000 mg P_i/L. More than 1000 t P/a can be recovered, leading to a reduction of 120 t/day of waste product (60% water).

Spent acid etchant (or mixed waste acid) in semiconductor-manufacturing plants typically consists of PA (60–80%), acetic acid (3–5%), nitric acid (3–5%), water (10–30%), and various metallic impurities (Kim 2006). Sanwa Yuka Kogyo Co. (SYK) has developed a commercial process for purifying PA from mixed waste acid using the solvent extraction method (Shin et al. 2009). In this process, mixed waste acid is countercurrently contacted with organic solvent (tri-2-ethylhexyl-phosphate (TOP)) in a multistage mixer-settler, thereby extracting acetic and nitric acids from the aqueous solution. TOP is selected as an organic solvent, because of its high tolerance to hydrolysis. Linear saturated hydrocarbon (C_6–C_{13}) is used as the diluent for TOP. The formation of NO_3^--TOP complex enables the extraction of nitric acid by TOP from the aqueous phase. Nitric acid and the protonated form of acetic acid ($pKa = 4.7$) are extracted by TOP, while PA remains in the aqueous phase at the operating pH. After extraction, organic solvent is washed with water to strip acetic and nitric acids and returned to the extraction process. The commercial process generates approximately 2000 t/a of PA (60%) which is recycled for various industrial applications.

1.4 Perspectives

1.4.1 Europe

What are the drivers influencing technical P recycling in Europe today and how are they developing? Switzerland and Germany have introduced legislation requiring recycling of P. The Swiss legislation (Schweizerischer Bundesrat 2015) requires recovery from sewage sludge (6 kt P/a) and meat and bone meal (the 1.5 kt P/a that is currently lost) from 2026. The new German sewage sludge ordinance (German federal government 2017) enforces P recovery for all German WWTP not being able to dispose their sludge on arable land. Disposal on arable land is prohibited for all WWTP with more than 50,000 p.e. capacity, producing about two-thirds of the German sludge representing 40 kt P/a. after a transition period. They are thus required to implement P recovery before 2029 (large WWTP >100,000 p.e.) and 2032 (medium-size WWTP >50,000 p.e.). For sludge with more than 2% P on dry matter, a P recovery or mono-incineration of the sludge is required. The recovery from the sludge must have a yield of at least 50% or reduce the concentration below 2%. Recovery from mono-incineration ash must have a yield of at least 80%. The ash may also be separately stored in a landfill for future recovery. Today, 55% of the Swiss sludge and 24% of the German sludge are mono-incinerated. Other countries such as Denmark and Sweden have set more general targets for recycling of P in sludge and other wastes, which include use of sludge in agriculture and thus are not expected to act as drivers for technical recycling of P. Phosphorus rock was added to the EU list of critical raw materials in 2014. In 2017 white phosphorus (P_4) was added. These acts by themselves have no legal implications but may serve as a rationale for related policies.

The circular economy package launched by the European Commission in 2015 targets the areas of product design, production processes, consumption, secondary raw materials, and innovation investment. It is an important signal that Europe sees circular economy as a key to prosperity and a potential solution to many resource-related problems such as resource scarcity, resource criticality, and emissions. One prominent part of the circular economy is the revision of the fertilizer regulation, which is currently under way and will bring important drivers for technical phosphorus recycling and recycling of organic phosphorus materials:

- Product function categories (PFC) and component material categories (CMC) are defined. If these are fulfilled, the outputs have product status and thus are no longer considered waste. This removes an important legal hurdle and facilitates marketing.
- Several new nutrient-rich CMC are introduced: digestates, compost and food industry by-products, and animal by-products.
- Additional CMC are in the process of being defined so they can be added: crystallisates like struvite, biochar, and ash.

The Urban Waste Water Treatment Directive (91/271/EEC) (European Commission 2013b) lays out guidelines for treatment of wastewater. Its implementation report reveals a high level of sewerage connection (average 94%) and of treatment more stringent than secondary treatment (77%) in EU27. This category includes not only phosphorus elimination but also nitrogen elimination and reduction of bacteriological pollution. Thus, the level of phosphorus elimination is lower than 77%. The increase in secondary treatment between 2005 and 2009 is <5%, which is hard to distinguish from the methodological uncertainties. Thus, it is not certain if the amount of P ending up in sludge increases. The EU Nitrates Directive (1991), which limits the amount of nitrogen from manure which may be applied, is a constant pressure on the regions with a large amount of livestock, mainly concerning pig factories.

To sum up, the European regulative framework for recovery from the high-potential stream sewage sludge, meat and bone meal, and manure is positive and improving. P is recognized to be of critical importance, and technical recovery has recently been made obligatory in two countries.

1.4.2 *Japan*

Although more than ten full-scale P recovery plants are operating in Japan, P recycling is no more than a secondary objective to prevent a damaging buildup of struvite on plant equipment (see Chap. 17) or for saving sludge disposal costs in WWTP (Chap. 9). No regulation requires P recovery and recycling for the wastewater treatment sector. This allows the wastewater treatment sector to consider P recovery as an extra service. On the other hand, fertilizer companies cannot accept recovered P products unless they bring some economic benefits to their business. The low market price of recovered P products makes P recycling business unattractive, thereby demotivating P recovery and recycling in the wastewater treatment sector. P recycling should hold great potential but thus far also significant challenges to establish stable channels for the distribution and sale of recovered P products. While technologies exist, lack of incentives impedes their implementation. To make P recycling business more attractive, it is likely imperative to develop a new value chain that can extract the maximum vale from secondary P resources while taking sustainability and safety into account.

1.4.2.1 High-Purity P Supply

Although the sustainable management of P is becoming a global concern on food and environmental securities, it is not well recognized that high-purity P becomes increasingly difficult to obtain in the global market. Unlike P_i fertilizer, high-purity P is not a low-cost commodity. In particular, elemental P (P_4, also known as white phosphorus) is a strategically important starting material for the synthesis of highly valuable P compounds. They are crucial for the manufacture of high-value added industrial products, including automobiles, electronics, fuel batteries, processed foods, medicines, and fire-retardant plastics (Fertilizer International 2015).

However, since quality matters more than quantity for technical applications, quantitative P flow studies have often failed to find the true importance in the global economy. For example, some researchers have argued that non-fertilizer P applications are in part substitutable and have even advocated seeking to restrict technical use of P (Reijnders 2014). However, it is not easy to find a substitute for P in technical applications in light of its high versatility and safety in handling and use. Needless to say, manufacturing companies are reluctant to substitute P by other elements, because it unavoidably causes serious impacts on cost, quality, and applicable technology in their manufacturing processes. The substitution may also fail to provide the same performance as it is and satisfy consumer demands to products. In reality, the restriction of technical P use is much more difficult than one might imagine.

A more serious but unrecognized challenge is that true risks relevant to P_4 supply have not been properly understood by the manufacturing sector. Currently, P_4 is available only from China, the USA, Kazakhstan, and Vietnam in the global market.

After the world oil shock of the 1970s, Japan ceased the production of P_4 because of energy shortages and became entirely dependent on the import from the USA and China (see Fig. 1.7). However, the USA stopped the export of P_4 by 2002 to secure domestic supply. The P_4 import from China also dramatically declined after the P price shock of 2008 due to the export restriction imposed by China. In addition, Thermophos International B.V., which was the only one P_4 manufacturer in Europe, went bankrupt in 2012. To secure the supply P_4 for the manufacturing industry, Japan has expanded P_4 import from Vietnam since 2007. However, as high-grade P_i rock becomes increasingly scarce in Vietnam, concerns are being raised over the future secured supply of P_4. Obviously, this import reliance makes Japan's industry vulnerable to disruptions in the supply of P_4 as it happened in 2008.

The production of P_4 is an energy-intensive process which is strongly dependent on locally sourced electricity, P_i rock, and cheap labor forces. To generate P_4, P_i rock needs to be carbothermally reduced in an electric furnace at temperatures higher than 1300 °C. This requires approximately 14 MWh of electricity for each ton of P_4 produced. Furthermore, environmental pollution such as the dispersion of radioactive dust from P_i rock is another problem related to the production of P_4 from P_i rock. While P_4 becomes increasingly difficult to obtain in the global market, Japan's industry continues to consume approximately 20,000 tons of P_4 annually. Nevertheless, since high-purity P is used only in small quantity per each industrial product, the true value of P_4 has not been well recognized by the manufacturing sector. Except the phosphate industry, manufacturing companies are barely motivated by concerns about the looming risk of P_4 availability.

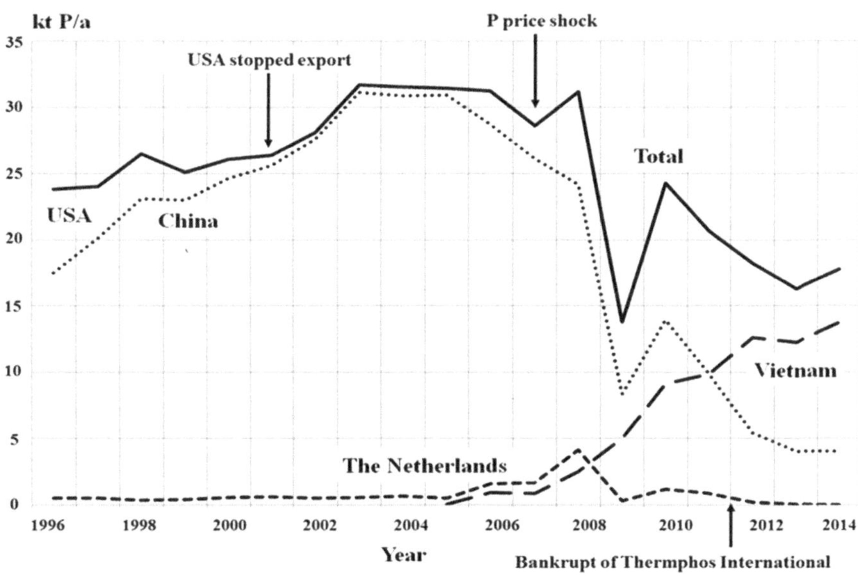

Fig. 1.7 Year-to-year change of P_4 import to Japan

1.4.2.2 Innovative P Value Chain

Since the legal, social, and economic structures are not fully developed for the sustainable use of P in Japan, it is critical to consider how to call on industry to participate in P recycling business. The future risk for secured P supply alone is not enough to get industry involved in P recycling business. To make P recycling business more attractive, it is necessary to redefine existing P value chains in order to boost technology and business innovation based on P recycling (Fig. 1.8). As mentioned above, P has a wide variety of technical applications. From industrial perspectives, P can be viewed as a technical nutrient which plays a critical role in the manufacturing industry. Accordingly, a new value chain needs to be designed for realizing P recycling not only in the biosphere but also in the technosphere (Fig. 1.8). The innovative P value chain should never restrict the technical use of P but rather maximize synergies of P recycling across biosphere and technosphere. The innovative P value chain needs to boost technology innovation on:

- Highly efficient P recovery from secondary P resources
- Wet acid process to generate crude PA from recovered P
- Carbothermal reduction of crude PA to P_4 with minimum electricity consumption and low environmental burden
- New processes for the manufacture of high-value added specialty P compounds

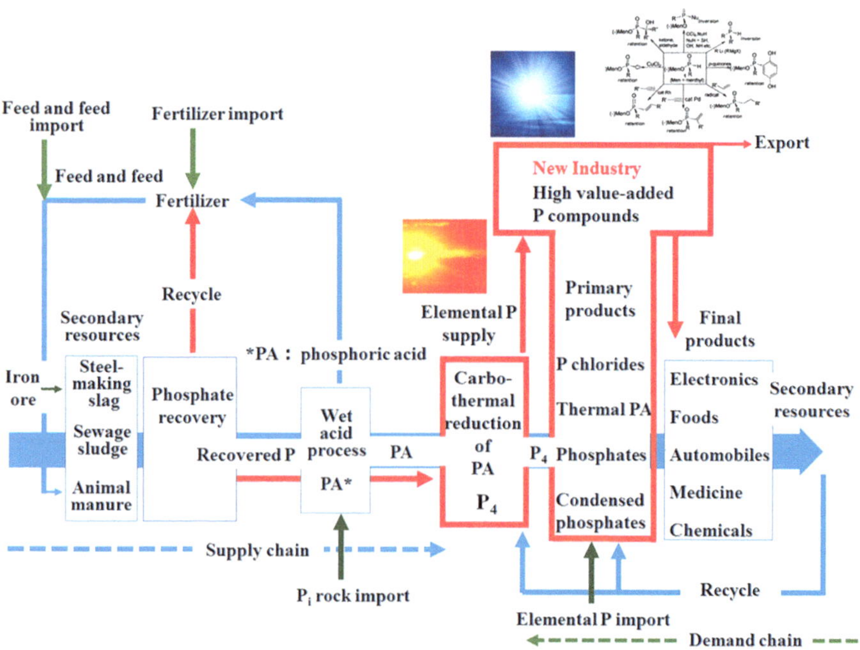

Fig. 1.8 Innovative P value chain in Japan

The highest priority should be given to the development of innovative carbothermal reduction of low-grade phosphoric acid (PA), which enables P_4 production with minimum electricity consumption and low environmental burden. The carbothermal reduction of P_i rock in conventional arc processes is prohibitively difficult in Japan because of energy shortage and environmental issues. Without this disruptive technology development, it would be difficult to make the innovative P value chain a reality. Of course, low-grade PA must be generated using P recovered from domestic secondary resources. Regarding secondary P resources, special attention should be given to steelmaking slag. In the steelmaking process, P is present in raw materials such as iron ore, coal, and limestone at concentrations as low as 0.1–0.2 wt% P_2O_5 (Matsubae-Yokoyama et al. 2009). Since P has detrimental effects on the mechanical properties of steel, it must be removed into hot metal pretreatment slag (dephosphorization slag). Dephosphorization slag contains approximately 2–10 wt% P_2O_5. In Japan alone, P emitted into dephosphorization slag amounts to 100 kt P/a, which is about two times more than that of sewage sludge. Hence, dephosphorization slag should be a quantitatively important secondary P resource, but it is currently landfilled or used as construction materials.

1.5 Conclusions

The main three waste streams with large P potential, namely, manure, sewage sludge, and meat and bone meal, have different characteristics regarding contaminants plant availability and nutrient concentration. Together they represent a large unused potential, in the order of 500 kt P/a in Europe, about a third of the mineral P consumption. In Japan similarly these three streams are estimated to approximately 115 kt P/a, steelmaking slag another 115 kt/a, each about a fifth of the total input (520 kt P/a). The different types of P recovery processes implemented in full scale in Europe and Japan have been summarized. These are diverse and can treat the three organic waste streams, whereas processes for steelmaking slag are still in development.

In Europe, the installed capacity is less than 10% of the available potential because the development has so far been limited to profitable niches. Switzerland and Germany have passed legislation in 2016 and 2017, respectively, that makes recovery obligatory, and the fertilizer regulation is expected to open new possibilities for recycling of especially ashes. These regulatory developments together with the fact that ash processes are close to break-even is expected to lead to increased recycling fractions in the future. In Japan, on the other hand, the secured supply of high-purity P is essential for the high-tech industry. The government funded a new research project to develop an innovative carbothermal reduction of low-grade phosphoric acid to P_4 with minimum electricity consumption and low environmental burden in 2017.

Technology and business innovation based on P recycling has the potential to make a great contribution not only to the sustainable agricultural production but also to the secured supply of high-purity P for high-tech industries. There should be no significant trade-off between fertilizer use and technical application, because the later needs only a small portion of recovered P. Rather, redefining the supply chain of high-purity P could provide economic incentives to P recovery from secondary resources and help promote P recycling.

The European Sustainable Phosphorus Platform (ESPP) was founded in 2013 to:

- Define and promote a long-term vision for phosphorus sustainability in Europe
- Identify and facilitate collaborative actions, policies, business cases, and value chains
- Address short-term obstacles, propose solutions, identify, and communicate opportunities

ESPP (www.phosphorusplatform.eu) has in the last years gained a solid reputation as an information source brokering ideas between different parties. This is facilitated by the fact that ESPP has numerous important members in all three areas shaping sustainable P use in Europe, namely, industry, science, and policy.

Meanwhile, to promote P recycling in Japan, the Phosphorus Recycling Promotion Council of Japan (PRPCJ) was founded in 2008 with the support from relevant ministries of Japan. Currently, PRPCJ has more than 150 members, including 13 large stakeholder associations and more than 70 private companies. PRPCJ members share their knowledge and information to develop a comprehensive understanding of P recycling, taking into account the social dimension, resource efficiency, and environmental protection. PRPCJ is opening a window of opportunity for getting policy support for P recovery and recycling.

References

Buckwell A, Nadeu E, Six L, Van Keer K, Williams A (2016) Nutrient Recovery and Reuse (NRR) in European agriculture

Campoy M, Gómez-barea A, Ollero P, Nilsson S (2014) Gasification of wastes in a pilot fluidized bed gasifier. 121:63–69

Clarke BO, Smith SR (2011) Review of " emerging " organic contaminants in biosolids and assessment of international research priorities for the agricultural use of biosolids. Environ Int 37(1):226–247

Coutand M, Cyr M, Deydier E, Guilet R, Clastres P (2008) Characteristics of industrial and laboratory meat and bone meal ashes and their potential applications. J Hazard Mater 150:522–532

Egle L, Rechberger H, Krampe J, Zessner M (2016) Phosphorus recovery from municipal wastewater: an integrated comparative technological, environmental and economic assessment of P recovery technologies. Sci Total Environ 571:522–542

European Commission (2013a) Consultative communication on the sustainable use of phosphorus. http://ec.europa.eu/environment/consultations/phosphorus_en.htm

European Commission (2013b) Seventh report on the implementation of the urban waste water treatment directive (91/271/EEC)

Fertilizer International (2015) Elemental phosphorus market review. 468:52–57

Foged HL et al (2011) Inventory of manure processing

Hua K, Liu L, Bureau DP (2005) Determination of phosphorus fractions in animal protein ingredients. Agric Food Chem 53(5):1571–1574

Hukari S, Nättorp A, Kabbe C (eds) (2015) Phosphorus recycling-now! Building on full-scale practical experiences to tap the potential in European municipal wastewater. P-REX Sustainable sewage sludge management fostering phosphorus recovery and energy efficiency. European Commission no. 308645, P-REX. (2017, February 13). Main P-REX deliverables. Zenodo. https://doi.org/10.5281/zenodo.242550

Kabbe C, Remy C, Kraus F (2015) Review of promising methods for phosphorus recovery and recycling from wastewater. In 763. International Fertiliser Society

Kim K-J (2006) Purification of phosphoric acid from waste acid etchant using layer melt crystallization. Chem Eng Technol 29:271–276

Krüger O, Adam C (2014) Monitoring von Klärschlammmonoverbrennungsaschen hinsichtlich ihrer Zusammensetzung zur Ermittlung ihrer Rohstoffrückgewinnungspotentiale und zur Erstellung von Referenzmaterial für die Überwachungsanalytik (Monitoring of sewage sludge ashes from mono-inc. https://www.umweltbundesamt.de/sites/default/files/medien/378/publikationen/texte_49_2015_monitoring_von_klaerschlammverbrennungsaschen.pdf

Lianfeng D, Wenke L (2012) Occurrence, fate, and ecotoxicity of antibiotics in agro-ecosystems. A review. Agron Sustain Dev 32:309–327

Matsubae-Yokoyama K, Kubo H, Nakajima K, Nagasaka T (2009) A material flow analysis of phosphorus in Japan. J Ind Ecol 13:687–705

Milieu Ltd, WRc, and R (2008) Environmental, economic and social impacts of the use of sewage sludge on land Draft Summary Report 2 Baseline Scenario, Analysis of Risk and Opportunities

Nättorp A, Remmen K, Remy C (2017) Cost assessment of different routes for phosphorus recovery from wastewater using data from pilot and production plants. Water Sci Technol 76:413–424

Niewersch C, Ewert W, Hermanussen O, Kabbe C, Mêlé C, Paillard H, …, Wagenbach A (2014) Description of sludge related processes. P-REX Sustainable sewage sludge management fostering phosphorus recovery and energy efficiency. European Commission no. 308645. P-REX. (2017, February 13). Main P-REX deliverables. Zenodo. https://doi.org/10.5281/zenodo.242550

Ohtake H, Okano K (2015) Development and implementation of technologies for recycling phosphorus in secondary resources in Japan. Glob Environ Res 19:49–65

Ott C, Rechberger H (2012) The European phosphorus balance. Resour Conserv Recycl 60:159–172

Reijnders L (2014) Phosphorus resources, their depletion and conservation, a review. Resour Conserv Recycl 93:32–49

Scholz RW, Roy AH, Brand FS, Hellums DT, Ulrich AE (2014) Sustainable phosphorus management

Schoumans OF, Rulkens WH, Oenema O, Ehlert P. a I (2010) Phosphorus recovery from animal manure. Alterra Report. 2158

Schuiling RD, Andrade A (1999) Recovery of struvite from calf manure. Environ Technol 20(7):765–768

Schweizerischer Bundesrat (2015) Verordnung über die Vermeidung und die Entsorgung von Abfällen (Abfallverordnung, VVEA), 1–46. Retrieved from https://www.admin.ch/opc/de/classified-compilation/20141858/

Shin C-H, Kim J-Y, Kim J-Y, Kim H-S, Lee H-S, Mohapatra D, Ahn J-G, Bae W (2009) Recovery of nitric acid from waste etching solution using solvent extraction. J Hazard Mat 163:729–734

Ueno Y, Fujii M (2001) Three-year experience of operating and selling recovered struvite from full-scale plant. Environ Technol 22:1373–1381

USGS. (2018). Phosphate rock. USGS mineral commodities summaries. Retrieved from http://minerals.usgs.gov/minerals/pubs/mcs/

van Dijk KC, Lesschen JP, Oenema O (2016) Phosphorus flows and balances of the European Union member states. Sci Total Environ 542:1078–1093

van Nieuwenhuyzen W, Tomas MC (2008) Update on vegetable lecithin and phospholipid technologies. Eur J Lipid Sci Technol 110:472–486

Chapter 2
Phosphorus Flows in Asia

Kazuyo Matsubae and Elizabeth Webeck

Abstract In this chapter, phosphorus resource flow was analyzed to determine the extent of the dependency of Asian countries on other countries to meet their phosphorus requirements. Additionally, a virtual phosphorus flow analysis was conducted to determine the future phosphorus demand in various Asian countries. The future phosphorus scenarios in Asia are also discussed based on the findings of these analyses. The findings on the basis of virtual phosphorus analysis indicated that from the life cycle perspective, Asian growing economy became to need more and more virtual phosphorus and many Asian countries rely strongly on phosphorus imports via the international trade supply chain. Thus, they realize that high economic and political risks are associated with the secured supply of phosphorus. Another important finding is that besides population growth, urbanization and economic growth in Asia are likely to be strong driving forces for higher phosphorus demand in the future. Considering the untapped P in waste streams, it was clarified that the steelmaking slag should be one of a good target to introduce P recycling technology for two big Asian steel producing countries, China and India. Approximately 7% and 5% of the virtual P-requirement for plant-based food and feed production was lost in China and India, respectively.

Keywords Asia · Food diet · Virtual phosphorus · Herfindahl–Hirschman index · Steelmaking slag

---

K. Matsubae (✉)
Department of Environmental Studies for Advanced Society, Graduate School of Environmental Studies, Tohoku University, Sendai, Japan
e-mail: matsubae@m.tohoku.ac.jp

E. Webeck
Department of Metallurgy, Graduate School of Engineering, Tohoku University, Sendai, Japan

H. Ohtake, S. Tsuneda (eds.), *Phosphorus Recovery and Recycling*,
https://doi.org/10.1007/978-981-10-8031-9_2

2.1 Introduction

Phosphorus (P) security is a particularly pressing issue because there is no alternative to phosphorus in agriculture. Cell growth does not occur in the absence of phosphorus. The three largest producers of phosphorus ore, which account for two-thirds of all the phosphate rock mined, are Morocco, China, and the United States. It is estimated that Moroccan reserves account for more than two-third of all phosphate rock reserves in the world. While China has considerable phosphate ore reserves, the majority of Asian countries have no phosphate reserves of their own and are completely dependent on imports. As a result, countries around the world rely on phosphorus imports to meet their respective demands. In the case of Japan, its phosphorus needs are mostly met by China, Morocco, and the United States.

Although each country has its own unique phosphorus story, the need to feed the burgeoning populations in the booming economies of India and China makes them worthy of special attention. India has limited phosphate reserves, phosphorus poor soil, and a heavy dependence on phosphorus imports. Eighty-five percent of India's phosphorus demand is met with imports. China requires even more phosphorus than India. Although their populations are relatively the same size, the effort to satisfy what seems to be an insatiable appetite for meat among the Chinese population threatens to bring about the day when China no longer exports phosphate ore or phosphate product. The possibility that China might no longer have the capacity to export phosphate products poses a serious problem for all of Asia.

In order to improve P efficiency at the regional/country level, regional information/data are essential. The stocks and flows of phosphorous have been well summarized from global perspectives (Bouwman et al. 2009; Elser and Bennett 2011; Liu et al. 2007; Mihelcic et al. 2011). Both national and region-wide case studies that have charted P flows have been conducted in Japan (Matsubae et al. 2011; Matsubae-Yokoyama et al. 2009), China (Yuan et al. 2011; Qiao et al. 2011), Korea (Jeong et al. 2009), and other regions of Asia (Cooper and Carliell-Marquet 2013). Although these studies are all relatively recent, there is still a clear need for further studies to be done, particularly in the countries and regions with burgeoning populations and rapidly changing demographics.

In this chapter, phosphorus resource flow was analyzed to determine the extent of the dependency of Asian countries on other countries to meet their phosphorus requirements. Additionally, a virtual phosphorus flow analysis was conducted to determine the future phosphorus demand in various Asian countries. The future phosphorus scenarios in Asia are also discussed based on the findings of these analyses.

2.2 Phosphorus Demand Analysis and Primary Phosphorus Resources in Asia

In this section, P flows were evaluated from an international trade perspective using trade statistics. In order to determine the extent of the dependency of one country on another to meet its phosphorus demand, we employed the Herfindahl–Hirschman index (HHI). The HHI is an economic concept widely applied in competition law and in antitrust and market investigations. The HHI is based on the sum of the squares of the market shares; the sum ranges from 0 to 1.0 and correlates to the market concentration, which ranges from many very small-scale shareholders to a single monopolistic market.

Several other considerations can also influence markets and demand. For example, in the case of phosphate ore, the contamination of naturally occurring radioactive materials (NORM) is an important consideration when selecting the origin of resources. However, it should be noted that the HHI does not reflect the quality of traded commodities.

2.2.1 *Phosphate Ore*

According to the estimated results, more than two-thirds of all the phosphate ore mined worldwide originate from the three largest producers. China supplied 37% of the phosphorus traded, and the United States and Morocco both supply 15%. According to a recent USGS report, more than one-third of the phosphate rock reserves worldwide are in Morocco. The two largest suppliers of phosphorus to the Asian region are China and Jordan.

Figure 2.1 shows the quantities of phosphate ore imported by countries including Asia's five largest importers: India, Indonesia, Japan, South Korea, and Malaysia. According to the trade data in 2016, India imported 7500 kt (kilo tons) of phosphate ore which was almost doubled in 2000. The second largest importer of phosphate ore was Indonesia, which imported 1693 kt in 2016.

Figure 2.2 shows the HHI of phosphate ore imported by Asia's five biggest importers. The bar shows the amount, while the dots and lines show the HHI. As shown in Fig. 2.2, higher HHIs in 2016 were found in Indonesia (HHI = 0.48) and Malaysia (HHI = 0.32), but the values in others were less than 0.3, India (0.26), South Korea (0.27), and Japan (0.20). With no phosphate reserves of their own, Japan and South Korea also imported all the phosphate ore required to meet their domestic needs from overseas. The ore imported to Indonesia originated from three countries, including Jordan, Morocco, and Egypt, whereas Japan imported ore from seven countries, including China, Jordan, and South Africa in 2016.

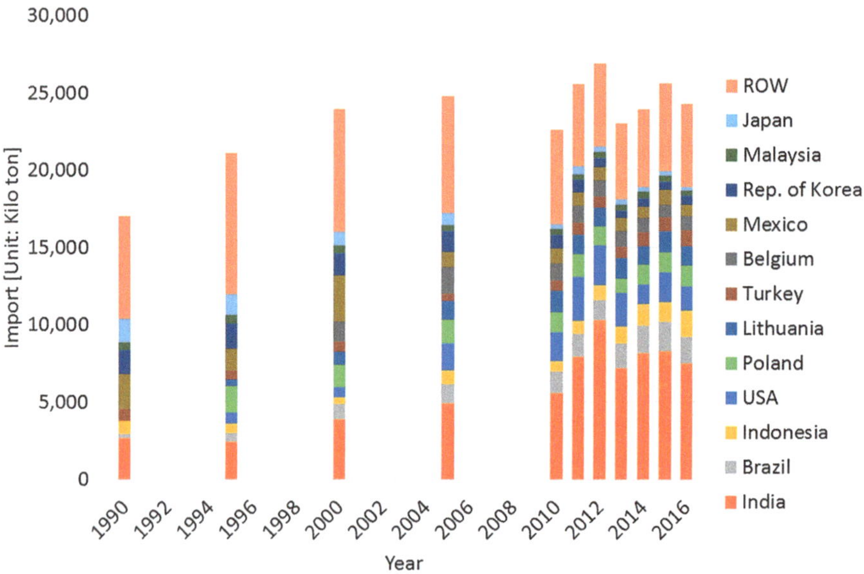

Fig. 2.1 Imports of phosphate rock by various countries. *ROW* rest of the world

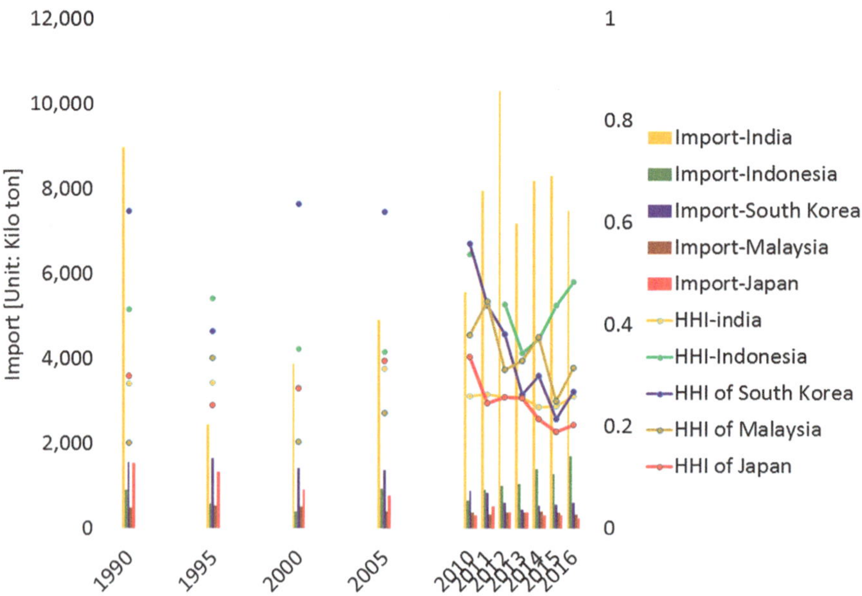

Fig. 2.2 HHI related to P rock imports by Asian countries

The NORM contamination of Chinese phosphate ore is relatively low. For this reason alone, the chemical and fertilizer industries prefer to use Chinese ore or use it in mixtures to dilute the high NORM content of other phosphate ores, such as Moroccan ore. This suggests that even in the case of countries with a low HHI, the dependency on other countries for natural resources cannot easily be changed, since each country imports according to the abilities and limits of the refinery technologies, infrastructure, and the environmental regulations in place in these countries for the treatment of waste.

2.2.2 *Phosphoric Acid and Fertilizer*

Approximately two-thirds of the phosphorus-containing chemicals produced by China were exported to Asian countries. South Korea, Japan, and Taiwan were its biggest importers in the Asian market. Almost every country in Asia imported phosphorus products from China. India imported its supplies from a wide range of sources with Morocco and South Africa both accounting for more than a quarter of its imports. India also imported phosphorus-containing chemicals from the United States, Australia, Tunisia, Senegal, and Jordan and smaller quantities from Panama, Lebanon, and Russia. While most countries imported their phosphorus products from a variety of sources, as indicated by the relatively low dependency value, many countries in Asia depended on China for a considerable portion of their phosphorus in chemicals.

Approximately 40% of all the fertilizer traded around the world is imported by countries in the Asian region, and roughly half of all the phosphorus imported in Asia in 2016 (c. 25%) was imported by Indonesia. Indonesia, Brazil, and Malaysia were the world's three biggest fertilizer importers in the past decade. Just five countries controlled 75% of the international trade market in phosphate fertilizers in 2016 with China commanding more than quarter of the global market. The second largest exporter was Morocco, while Israel ranked as the world's third largest exporter of phosphate fertilizer in 2016, exporting a total of 5469 kt of phosphate fertilizer, approximately more than half of which was destined for the Asian market. While China was the largest exporter of phosphate fertilizer in Asia, its imports exceeded its exports. Indeed, China was the world's largest importer of phosphate fertilizers in 2005, importing more than 550 kt of phosphate fertilizers that year, approximately 200 kt P more than it exported. Over 60% of China's imported phosphate fertilizers were sourced from the United States with the remaining 40% obtained from a number of other countries. In 2016, China became the biggest net exporter of P fertilizer in the world.

The amount of phosphate fertilizers imported by various countries and their dependencies is shown in Fig. 2.3. The first and third largest importers of phosphate fertilizers were Indonesia and Malaysia, accounting for more than 30% of the global market in fertilizers. While they imported huge quantity of phosphate fertilizer, Indonesia and Malaysia did not depend strongly on one country to meet their rela-

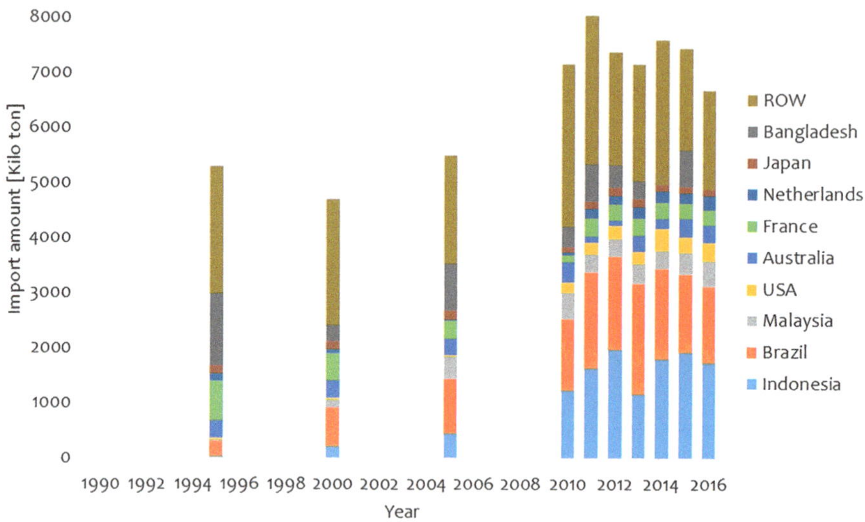

Fig. 2.3 Fertilizer imported by various countries

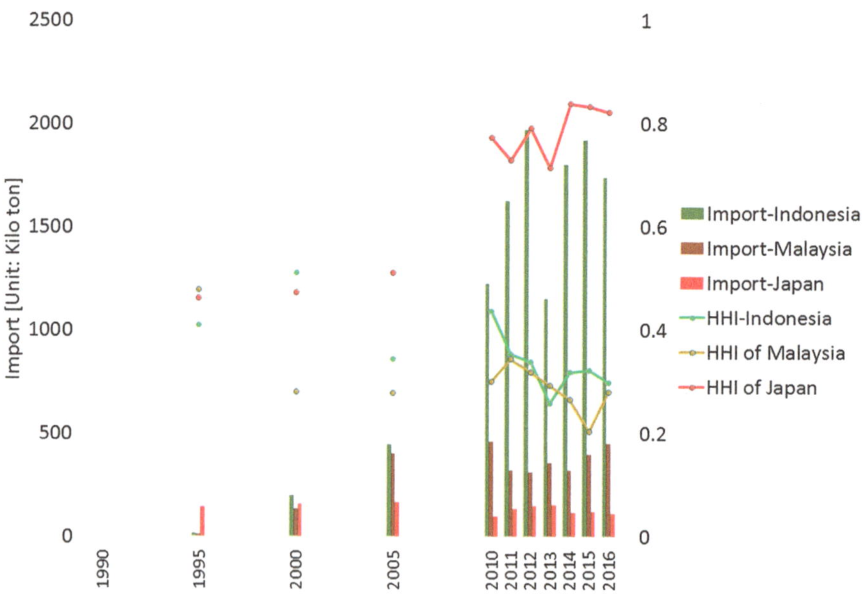

Fig. 2.4 HHI referring to the import of phosphate fertilizer by major Asian countries

tive phosphate fertilizer needs, as shown in Fig. 2.4. However Japan, the third biggest importer in the Asian region, demonstrated a strong reliance on China to meet its local demand for phosphate fertilizers.

2.2.3 Elemental Phosphorus

Approximately 67% of the elemental phosphorus (also known as white phosphorus) provided by China and Vietnam for the international market in 2016 was consumed in Japan and India. Approximately three-fourths of all the elemental phosphorus in the international trade market was consumed in five countries, Germany (20%), India (20%), Poland (14%), Japan (13%), and Brazil (10%), as shown in Fig. 2.5. The automobile industry uses high-purity derivatives of elemental phosphorus to pretreat the surface of steel before the vehicles are painted, and they are also used to galvanize machine parts and as an etching agent in the semiconductor industry. Japan imported 32 kt of phosphorus as elemental phosphorus from China in 2005, which accounted for 30% of Chinese exports, and 16 kt P from Vietnam in 2015. As shown in Fig. 2.6, the high HHI for elemental phosphorus in Japan (0.95 in 2005 and 0.75 in 2016) shows that Japanese industry is almost totally reliant on China in 2005 (Webeck et al. 2014a) and currently in Vietnam for its high-purity phosphorus.

In 2015, the largest consumer of elemental phosphorus in Asia was India. India accounted for approximately 35% of all the elemental phosphorus exported from Vietnam in 2015. India imported approximately 20% of elemental phosphorus exported by China in 2005. The manufacturing sector in India has a strong demand for high-purity phosphorus as a galvanizing agent. It can be expected that the Indian demand for elemental phosphorus will grow considerably over time as its industry develops. The demand from the growing Indian automobile industry is expected to

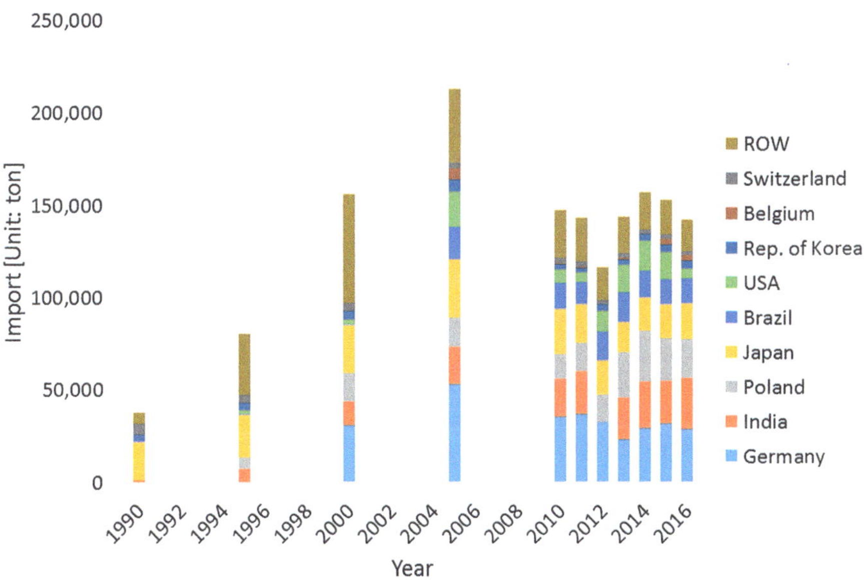

Fig. 2.5 Import of elemental P by major countries

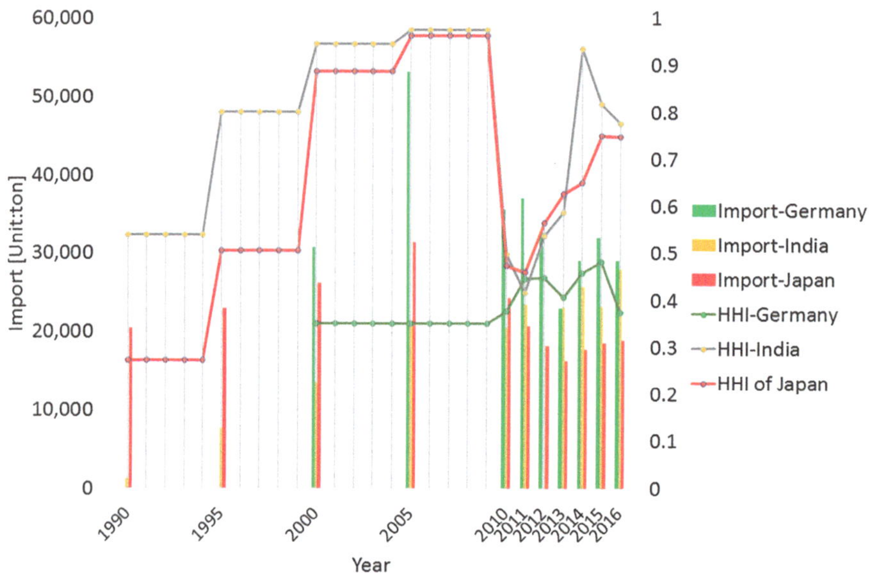

Fig. 2.6 HHI referring to the import of elemental P by Asian countries

require a particularly high amount of elemental phosphorus over time. Even though the demand for elemental phosphorus dropped significantly by 2012 in Japan and South Korea, the Indian demand has been continuously increasing, as shown in Fig. 2.6. China and Vietnam were the key providers of elemental phosphorus for the international market over this period.

2.3 Virtual Phosphorus

Virtual phosphorus (VP) is used as an indicator to consider the direct and indirect phosphorus requirements for our food consumption (Webeck et al. 2014b; Matsubae et al. 2011). This index reflects the direct and indirect demands of phosphorus transformed into agricultural products. As in the case of virtual water (Hoekstra and Chapagain 2007; Hoekstra and Mekonnen 2012), our consumption of agricultural products implies that we indirectly demand phosphorus as a nutrient; in other words, the food chain creates an indirect phosphorus demand itself.

Further, in the vegetation process, fertilizer is used for plant growth. However, not all the phosphorus in fertilizer is transformed into harvested products. Loss is caused by adsorption in soil, waste in residual portions of agricultural products, and the leftover portions of agricultural products. Material flow analyses focusing on the phosphorus contained in products tend to neglect such phosphorus flow. To estimate the VP, this hidden phosphorus flow was considered.

The hidden or virtual phosphorus is the total phosphorus requirement or the amount contained in agricultural products, including the loss to soil, non-edible parts, and feedstuff for livestock. Therefore, VP is defined as the sum of the hidden phosphorus input and the phosphorus contained in products.

2.3.1 Virtual Phosphorus Flows for China, India, and Japan

The projections for the per capita food consumption in China, India, and Japan shown in Fig. 2.7a are based on a linear projection of per capita food consumption trends over the last two decades (Webeck et al. 2014b). Taking into account their rapidly expanding economies and rapid urbanization, it is also assumed that per

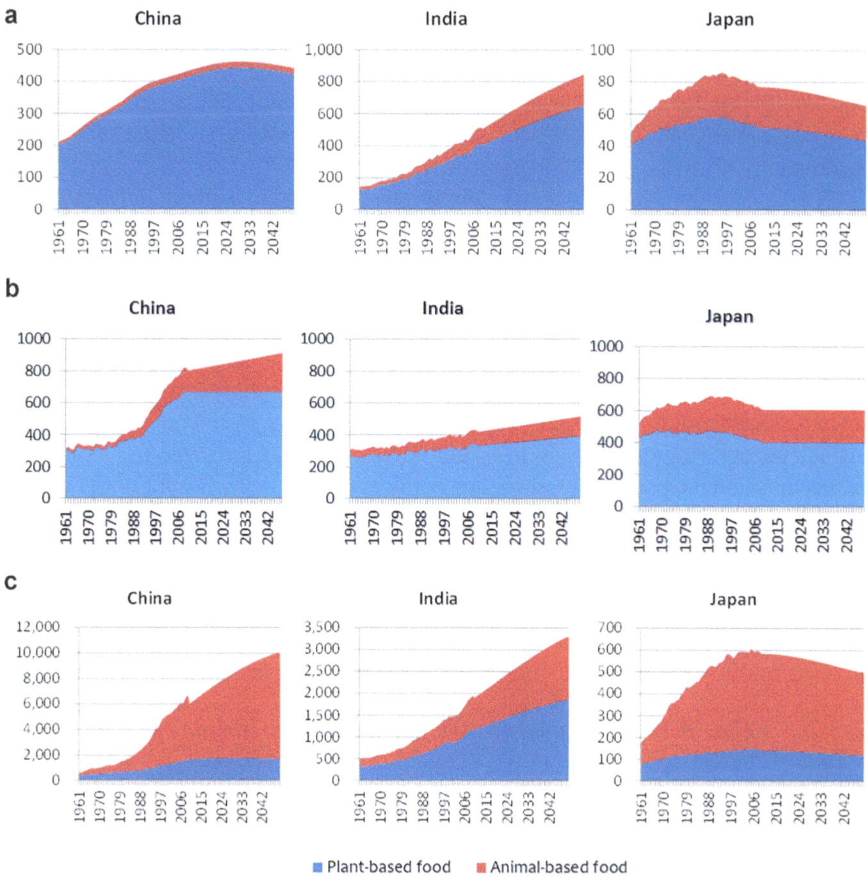

Fig. 2.7 Projection of the amount of phosphorus in food supply per capita (Unit g-P) (**a**), the amount of phosphorus total food supply (Unit: kt P) (**b**), and the amount of virtual phosphorus in total food supply (Unit: kt P) (**c**) (Webeck et al. 2014b)

capita consumption will continue to grow for China and India. On the other hand, the per capita consumption pattern in Japan is assumed to remain largely unchanged in the future.

The projected total food supply requirements for these countries, as shown in Fig. 2.7b, are based on the projected per capita consumption data and the United Nations population growth forecasts for these respective countries (UNDESA 2017). It can be seen that the forecast peak and then gradual decline in China's population will affect the amount of food required. Additionally, by 2050, India's unabated population growth will result in a huge demand for food, equivalent to almost double that required in China by 2050. Japan, on the other hand, has a rapidly aging population that is expected to peak and then decline from 2025, resulting also in a significant decline in the demand for food. The projected virtual phosphorus requirement of China, India, and Japan until 2050 is shown in Fig. 2.7c. While India is expected to require twice the amount of food than required by China in 2050, the per capita consumption pattern of China is expected to become triple that of India. The preference for a low-meat diet in India will translate into a significantly lower virtual phosphorus demand. With the assumed unchanging per capita consumption pattern of Japan, mainly due to the stagnant economy and shrinking population, the virtual phosphorus demand in Japan is expected to taper off slowly, declining by about 20% by 2050 compared to that in 2009.

These projections indicate that the phosphorus demand to feed the future populations of China, India, and Japan will vary enormously and that the discrepancies between these countries will grow over time. It is of vital importance that governments consider ways to ensure that phosphorus is available for producing the food required to feed their future populations. It is also important that the changing food preferences of their populations be taken into account. Unless there is a significant slowdown in the growing demand for meat, the virtual phosphorus demand will increase dramatically with increasing population. In order to slow down the future virtual phosphorus demand, it has been suggested that fertilizer use could be reduced, in effect potentially slow down the future virtual phosphorus demand. Additionally, using phosphate fertilizers more efficiently in the agricultural production phase, combined with a phosphorus recycling system, would contribute to reducing the need for obtaining natural resource origin phosphorus to meet the requirements for the future food supply.

2.3.2 VP and Phosphorus in the Steelmaking Industry

In addition to the primary resources of phosphorus, a significant amount of phosphorus flows into our society as hidden phosphorus in other forms. One example is the phosphorus that enters steelmaking nations as an element accompanying the iron ore import. Many of these sources of phosphorus are not regarded as phosphorus resources but rather, they are regarded as impurities in more useful materials.

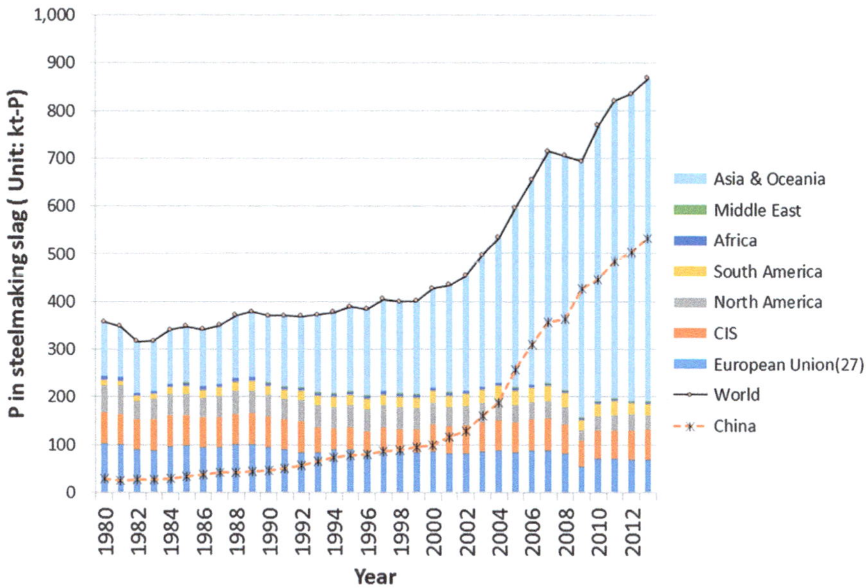

Fig. 2.8 Estimated amount of hidden phosphorus in steelmaking slag (Matsubae et al. 2016)

Most of these resources flow out from our economic sphere with little or no effort to recover the phosphorus that enters the waste stream.

The estimated amount of phosphorus contained in various steelmaking slags since 1980 is indicated in Fig. 2.8. The regional classification as indicated in this figure follows the UN statistics on phosphorus in this sector. The 1649 million tons of crude steel produced in 2013 were more than twice that produced in 1980. Behind the scenes, the production of this huge amount of steel has been accompanied by a significant amount of hidden phosphorus. This phosphorus has been accumulated into steelmaking slag and has been lost or dissipated in our subeconomic field, where it is economically difficult to mine. According to the USGS, the global production of phosphate ore was 135,000 kt and 228,000 kt in 1980 and 2013 in the world, respectively (USGS 2014). If we suppose that phosphorus rock has a P_2O_5 content of 30%, the hidden phosphorus flow from the steelmaking process was equivalent to 2% of the global phosphorus supply in 1980 and 4% in 2013. It is worth noting that the hidden phosphorus flow in China was more than 500 kt P, which was more than ten times the amount of phosphate rock exported by China in 2013.

The variation in the P-content from ore to ore is significant. Australian iron ore has the highest P-content, an average of 0.057%, with the P-content varying from 0.040% to 0.084% depending on the mining site. The average P-content of South American ore (c. 0.037%) is significantly lower than that of Australian ore, while Venezuelan ore is considerably high (c. 0.068%), and Chilean ore has a low P-content (c. 0.035%). The P-content of Brazilian ore ranges between 0.027% and 0.062%, averaging 0.038%. Indian ore has a low P-content of 0.037%, and that of South African ore is in the range of 0.055%. Clearly, the potential for P-recovery at any steelmaking facility will depend on the origin of the ore used in the blast oxygen furnace (BOF).

It should be noted that the P-content of iron ore from any mining site, regardless of whether powder ore or lump ore, tends to increase as the high-grade ore becomes exhausted over time. For example, although Indian ore is currently estimated to have a P-content of 0.037%, this value will unavoidably increase in the near future. With the huge demand for steel in India's burgeoning economy, high-grade ore is gradually becoming depleted, and ores with a higher P-content will be mined for use in the local steel industry.

Approximately 100 kt of P ended up in slag in 2015 was essentially equivalent to all the P required for food production in Japan. While Japan's demand for food and feed is expected to decrease due to the falling population, the P output from the steel industry is expected to remain stable or rather increase. Hence, the recovery and recycling of P from steelmaking slag have the potential to supply sufficient P for the food production in Japan in years to come. This would mitigate the risk that accompanies reliance on the imported P.

Similarly, approximately 100 kt of P was lost in the steel industry of India in 2015. The amount of potentially recoverable P was essentially equivalent to 5% of the virtual P-requirement for plant-based food and feed production in India. This is by no means insignificant in a country where the consumption of meat is comparatively low, and the food production relies exclusively on the import of P. Since the virtual P-requirement for plant-based food and feed products is steadily increasing in India, the steelmaking industry should have a high potential to significantly reduce its reliance on P import and thus improve food security of the country. Meanwhile, China lost approximately 800 kt on P into slag in 2015. This was roughly 7% of its projected P-requirement for the production of plant-based food and feedstock that year. The steelmaking industry in China also has a good potential to supply P to meet the domestic demand.

2.4 Challenges for Phosphorus Governance in Asia

2.4.1 Population Growth

As mentioned above, Asia's population accounts for 60% of the current world population and is estimated to grow in the coming decades. It is of vital importance that governments in Asian countries consider ways to ensure that phosphorus supply is secured for the production of food to feed their future populations. Nutrient emissions from human activities should be managed in appropriate ways. Wastewater treatment and utilization of sewage sludge are keys for phosphorus recycling and the mitigation of future phosphorus resource use. However, the cost of waste treatment is still higher than their acceptable levels for many Asian developing countries. Technology transfer and international collaborative innovations are necessary to promote nutrient recovery and recycling in Asia.

2.4.2 Economic Growth

Economic growth has enriched dietary life as individual preferences of food shift to foods of higher quality. Economic growth has also led to an increase in demand of processed or pre-prepared foods. The demand of processed food consumption requires more food additives, which include phosphoric acid and other artificial substances to prolong the food life and to soften food texture. This type of phosphoric acid is mainly made from elemental phosphorus, and China is the principal supplier in Asia. Taking into account the current phosphorus demand and supply market of high-grade phosphoric acid as well as the rapid economic growth, Asian countries will unavoidably increase their dependency on China for phosphorus.

Additionally, economic growth is expected to increase the phosphorus loss in the agricultural sector. As was shown in Fig. 2.5, the fertilizer input in India and China increased despite the doubling in the fertilizer price from 2005 to 2012. Obviously, economic growth made it possible for farmers to purchase more fertilizer and this accelerated fertilizer consumption in Asia. This is usually the case in areas with the low phosphorus accumulation in soil. To make matters worse, the farmers tend to apply fertilizer to farmland as far as their budget allows.

2.4.3 Urbanization

The increase of population generally drives the extent of urbanization. According to UN statistics, the urban areas in India increased from 18% in 1961 to 31% in 2010. By 2025, more than 37% of India is expected to be urban. The urbanization in China has progressed at a remarkable rate. While about 16% of the population was urban in 1960, the urban area increased to 49.5% in 2010 and is projected to be 65.4% by 2025 (UNDESA 2017). In China and India, their rapidly expanding economy and urbanization could cause the positive and negative effects for phosphorus resource management.

The positive effects of urbanization include the development of infrastructure to treat wastewater. This can gradually be achieved as urbanization progresses, making it possible to recover phosphorus from human waste streams and to prevent water pollution. Needless to say, phosphorus recovery requires new investment for the construction of infrastructure in the wastewater treatment sector.

On the other hand, one of the negative effects of urbanization is the inevitable change in the diet from the traditional to the "global" one. This change of diet does not mean that traditional meals are always better than global one. However, Asian traditional meals generally contain less meat compare to "westernized" meals. Therefore, unless there is a significant slowdown in the growing demand for meat, the virtual phosphorus demand will increase dramatically with increasing population.

An additional negative effect of urbanization may be the increase of food transport distance between supplier and consumer, thereby breaking the local nutrient cycle. This may also contribute to the increase of carbon emissions (Satterthwaite et al. 2010). The reduction of fertilizer use is an option to slow down the future virtual phosphorus demand. The more efficient use of phosphate fertilizers would contribute to reducing the demand for natural resource origin phosphorus to meet the requirements for the future food supply.

2.4.4 Challenges for P-Governance in Asia

Firstly, the reduction of fertilizer use and appropriate management of nutrients should be promoted in the agricultural sector. Excess nutrients not only cause harmful effects in ecosystems but also degrade the quality of drinking water and pose additional risks to human health. Many Asian countries have been facing water scarcity issues. Additionally, land resource management is now becoming increasingly critical as the urbanization progresses, adding pressure to the agricultural sector on the productivity per unit of arable land. To meet this demand, developing countries are bound to require higher fertilizer inputs. Furthermore, in the case of Japan, the increase in the number of farmers practicing part-time agriculture makes it difficult to maintain a healthy nutrient cycle. For example, farmers tend to value their time over practicing sustainable agriculture and because they can afford to purchase fertilizers and pesticides, they do not think deeply about how they use them. These circumstances have promoted the fertilizer-intensive agricultural production in Asia today. Thus, achieving appropriate and sustainable nutrient management poses a sizable challenge for agricultural production in Asia.

Phosphorus recovery and recycling are much-needed options to conserve phosphorus resources. However, they require the use of advanced technologies, social acceptance, and infrastructure investment. Such discussions typically touch upon additional costs for resource recovery, which is not acceptable for most stakeholders. Thus, it is vital to dissolve these barriers via effective knowledge sharing and information dissemination among key stakeholders. Such agreements are hard to achieve even in small communities. Moreover, achieving a sound phosphorus cycle requires international collaboration as well as efficient and effective technology transfer among developed and developing countries. Additionally, options to improve the phosphorus resource efficiency would also help decrease phosphorus dependency on external sources, thereby lowering the geopolitical and environmental risks inherent in phosphorus trade.

2.5 Conclusion

Phosphorus flows in Asia were investigated using statistical/trade data for 2005. Then, the dependency on external phosphorus resources for Asian countries was evaluated using the Herfindahl–Hirschman index. This was followed by a discussion of future phosphorus requirements for food supply as well as virtual phosphorus flows for two populated countries in Asia, China, and India.

The findings indicate that Asian countries currently consume more than 60% of the global fertilizer production. Moreover, from the life cycle perspective, they require more virtual phosphorus from the international food trade. Many Asian countries rely strongly on phosphorus imports via the international trade supply chain. Thus, they realize that high economic and political risks are associated with the secured supply of phosphorus. Another important finding is that besides population growth, urbanization and economic growth in Asia are likely to be strong driving forces for higher phosphorus demand in the future.

The utilization of secondary phosphorus resources has the potential to improve external phosphorus dependency for many countries. In order to promote phosphorus recovery and recycling, however, some significant bottlenecks need to be overcome. For example, developing countries lack relevant environmental regulations and the necessary infrastructure to collect and treat phosphorus-rich waste streams such as animal manure, sewage sludge, and incineration ash. For developed countries, the market itself poses a problem. For instance, consumer preference is a barrier with regard to marketing recovered phosphorus. Additionally, there are considerable technology and knowledge gaps among countries in Asia. International technology transfer, including social skill upgradation, will help close these gaps. To achieve higher resource efficiency, improved public awareness of the importance of phosphorus is required in this hour of need.

In order to achieve sustainable phosphorus resource management on a global scale, it is necessary to recognize the importance of resource governance in Asia. There is a need to develop and tap into international collaborative networks for technology transfer pertaining to the recovery and recycling of phosphorus, the development of an Asia-wide market to trade secondary phosphorus resources, and the establishment and implementation of stringent regulations.

References

Bouwman A, Beusen AHW, Billen G (2009) Human alteration of the global nitrogen and phosphorus soil balances for the period 1970–2050. Glob Biogeochem Cycles 23:16

Cooper J, Carliell-Marquet C (2013) A substance flow analysis of phosphorus in the UK food production and consumption system. Resour Conserv Recycl 74:82–100. https://doi.org/10.1016/j.resconrec.2013.03.001

Elser J, Bennett E (2011) A broken biogeochemical cycle. Nature 478:29–31

Hoekstra AY, Chapagain AK (2007) Water footprints of nations: water use by people as a function of their consumption pattern. Water Resour Manag 21:35

Hoekstra AY, Mekonnen MM (2012) The water footprint of humanity. Proc Natl Acad Sci U S A 109:3232–3237

International Nitrogen Initiative [Online] (n.d.) Available: http://www.initrogen.org/. Accessed 2014/08/1

Jeong YS, Matsubae-Yokoyama K, Kubo H, Pak JJ, Nagasaka T (2009) Substance flow analysis of phosphorus and manganese correlated with South Korean steel industry. Resour Conserv Recycl 53:479–489

Liu Y, Chen JN, Mol APJ, Ayres RU (2007) Comparative analysis of phosphorus use within national and local economies in China. Resour Conserv Recycl 51:454–474

Matsubae K, Kajiyama J, Hiraki T, Nagasaka T (2011) Virtual phosphorus ore requirement of Japanese economy. Chemosphere 84:767–772

Matsubae K, Yamasue E, Inazumi T, Webeck E, Miki T, Nagasaka T (2016) Innovations in steelmaking technology and hidden phosphorus flows. Sci Total Environ 542:1162–1168

Matsubae-Yokoyama K, Kubo H, Nakajima K, Nagasaka T (2009) A material flow analysis of phosphorus in Japan: the iron and steel industry as a major phosphorus source. J Ind Ecol 13:687–705

Mihelcic JR, Fry LM, SHAW R (2011) Global potential of phosphorus recovery from human urine and feces. Chemosphere 84:832–839

Qiao M, Zheng YM, Zhu YG (2011) Material flow analysis of phosphorus through food consumption in two megacities in northern China. Chemosphere 84:773–778

Satterthwaite D, McGranahan G, Tacoli C (2010) Urbanization and its implications for food and farming. Philos Trans R Soc Lond Ser B Biol Sci 365:2809–2820

United Nations Department of Economic and Social Affairs (UNDESA) (2017) World economic and social survey 2017: reflecting on 70 years of development policy analysis

Webeck E, Matsubae K, Nakajima K, Nansai K, Nagasaka T (2014a) Analysis of phosphorus dependency in Asia. Sociotechnica 11:119–126

Webeck E, Matsubae K, Nagasaka T (2014b) Phosphorus requirements for the changing diets of China, India and Japan. Environ Econ Policy Stud

Yuan ZW, Shi JK, Wu HJ, Zhang L, Bi J (2011) Understanding the anthropogenic phosphorus pathway with substance flow analysis at the city level. J Environ Manag 92:2021–2028

Chapter 3
Circular Economy: Bridging the Gap Between Phosphorus Recovery and Recycling

Christian Kabbe

Abstract Circular economy for nutrients! How to transfer buzzwords into solid results? So far the potential to recover and recycle phosphorus remains untapped or is just inefficiently used as in the case of sewage sludge, manure and food waste. To provide alternatives to argued traditional nutrient recycling routes, various technical solutions have been developed in recent years. They allow recovery of phosphorus minerals suitable as raw material for industries like fertilizer production or even as ready-to-use renewable or next-generation fertilizer. This contribution focuses on mineral phosphorus-containing materials recovered from wastewater. It discusses legal aspects and market opportunities regarding their valorization in Europe. It has to be kept in mind that there are many other recovery/recycling options out there to allow sustainable nutrient management, especially when it comes to organic wastes and their recycling. A frequently updated inventory is provided on the European Sustainable Phosphorus Platform's website (http://www.phosphorusplatform.eu/). The current revision of the European fertilizer regulation within the European Commission's circular economy package provides a concrete example, what issues have to be coped with and what measures have to be taken to create a level playing field for both primary- and secondary-based materials destined for fertilizer use. Some EU member states have started to enforce phosphorus recovery from relevant wastes but are lagging behind in enabling efficient recycling, be it in mineral or organic form. Still, the so-called technical nutrient recovery is missing a demand-side driven market pull for recovered (secondary) nutrients and the biggest challenge will be bridging the gap between supply (recovery) and demand (recycling), especially when it comes to new types of materials or products, not already established on the market. Whereas in the past, the focus of nutrient recovery technologies was laid upon high recovery rates for single nutrients, now energy efficiency, synergies and cost become more and more important. What about value chains? We have to look for easy to implement rather integrative solutions instead of reinventing the wheel, creating fancy parallel (infra)structures.

C. Kabbe (✉)
P-REX® Environment, Am Goldmannpark 43, Berlin 12587, Germany
e-mail: CKabbe@p-rex.eu

H. Ohtake, S. Tsuneda (eds.), *Phosphorus Recovery and Recycling*,
https://doi.org/10.1007/978-981-10-8031-9_3

Keywords P recovery · Nutrient recycling · Value chain, regulatory aspects · End-of-waste

3.1 Introduction

Whether or not there will be a phosphorus (P) peak within decades, centuries or millennia (Cordell and White 2011; Scholz and Wellmer 2013), one thing is for sure – phosphorus is a limited and, in its function as a nutrient, an essential and irreplaceable resource (Asimov 1959). As Isaac Asimov stated in April 1959, in his essay 'Life's bottleneck', phosphorus limits the biomass potential on Earth. Essentially, all phosphorus in fertilizers and feed is originally mined from phosphorus-rich rocks, which are finite and distributed in just a few places on the planet. From the European perspective and in the light of having just one small mine in Finland, the geopolitics and economic vulnerability are issues to be taken seriously. Europe is highly dependent on phosphorus imports (de Ridder et al. 2012; van Dijk et al. 2016). Concerns about the reliability of global P data related to reserves, mining and processing capacities already led to several proposals to establish a global committee for independent monitoring (Wellmer and Scholz 2015; Acatech 2017).

Recovery and recycling can and have to play an important role in improving resource efficiency and sustainable nutrient management. Although both steps, recovery and recycling, are explicit stages in the European waste hierarchy, the legal framework as it is in place today still poses the impression to discriminate nutrients derived from secondary, renewable sources vs. primary sources.

On our way towards circular economy, traditional terms like waste, raw material and product deserve redefinition. Although nutrient recycling was well established and efficient without harm and low risk to human health and environment, industrialization, population growth and excessive mass production became a threat for those end points. In consequence and thanks to precautious thinking, waste criteria have been defined and implemented at the end of the last century. The implementation of regulatory barriers intended to reduce the distribution and circulation of hazardous substances from all kinds and fields of anthropogenic activity contributed to reduced water, soil and air pollution since the end of the last century. Also the prohibition of production and application of the so-called priority substances (chemicals considered threads to human health and environment) lead to improvements where implemented. Depending on the member state, either very fragmented or cross-environmental media approaches can be observed in the regulatory framework. The latter represent already a rudimental harmonization of requirements and restrictions for a certain set of environmental compartments (water, soil and air) – the integral approach.

Within the EU, there are two main levels of regulation: the European level and the member state level. Depending on the member state, there might also be provinces, federal states and even municipalities mandated to regulate certain issues. In general, one can state that the European level sets the general frame, and member states implement accordingly within certain flexibility, allowing stricter requirements, but not less strict requirements. Depending on type of EU level legislation, we distinguish between directives and regulations implementing the issues addressed in directives.

But the legal framework is only one side of the medal. To place a product onto a market also strongly, if not mainly, depends on non-legal criteria. It is always harder to find customers for new products, directly competing with a high-volume commodity or bulk market. Here new and better product properties, functionalities and performances that meet or even excel customer's expectations can play a key role. But also sufficient quantities and reliable supply, homogeneity, handling and other parameters are decisive. If quantities are rather limited, a niche market provides a good option for market implementation. If the product is tailored to enter an exclusive premium market, it is better to start there instead of trying to compete with a low- or medium-price market segment. It is much easier to lower sales prices, if the premium market uptake fails, than to increase prices once the product has been marketed for low or medium prices.

It makes a difference if the product is sold as a renewable, high-quality product or as waste-derived material with potential contamination and uncertain risks linked to the waste origin. I call these the opportunity-focused and the problem-focused marketing strategies. It is easy to conclude which strategy will be more successful. Some examples shall serve to provide more insights to the aspects briefly introduced.

3.2 The Legal Framework: Enabler or Disabler?

Before diving into details, it can be stated that any environmental protection measure would not have been implemented without law enforcement, especially when it's linked with additional cost. So, in general, it can be claimed that the legal framework, especially the environmental regulations, is a key innovation motor of our society.

Looking at Switzerland (2016) and Germany (2017), being the first European countries to really set and not just announce legal requirements for phosphorus recovery, an acceleration for P recovery technology development and implementation can be expected. Other member states already announced to follow the given examples of both. But, recovery alone is not yet recycling! Sure, there will be transition phases for implementation, but the clock is already ticking for the legislators to enable actual recycling.

The existing heterogeneity of legislation between the different 'domains' and between member states still poses barriers for EU-wide marketing of recovered nutrients. Even the recovery, when it comes to large-scale operations, as in the case of sewage sludge ashes, is hampered, since the border crossing transport of waste is very challenging from the operator's point of view. Here, initiatives or green deals like the North Sea Resources Roundabout (http://www.greendeals.nl/ north-sea-resources-roundabout/) can provide a good template on how these obstacles can be resolved on regional and multinational level.

Recycling and recovery are explicitly addressed in the five-stage waste hierarchy of the European Waste Framework Directive (2008/98/EC) setting the frame for all EU member states. The waste hierarchy is defined as follows, also setting the borderline between waste and non-waste status:

1. Prevention of waste	Non-waste
2. Preparing for reuse	Waste
3. Recycling	
4. Recovery	
5. Disposal	

Stage 1 represents the non-waste domain, whereas all other stages 2–5 represent the waste domain. Regaining product or end-of-waste status is the prerequisite for all materials to be allowed to be marketed in Europe as a product. The specific product or end-of-waste criteria for materials destined for a certain use are defined in the related regulations or ordinances. So, in the case of fertilizers, the European Fertiliser Regulation (2003/2003/EC), currently under revision, has to define the criteria for substances or materials to be used as fertilizers (PFC, product function category) or raw materials for fertilizer production (CMC, component material category). The European Commission follows the ambition to create a level playing field for both primary/virgin materials and secondary/renewable materials recovered from wastes. Although there are already various value chains established to enable the recycling of recovered nutrients, the implementation and interpretation of EU legislation can vary from member state to member state. Depending on the pragmatism of authorities, it can be observed that recovered materials allowed to be used as fertilizer or fertilizer raw materials in one member state can face a ban from this application route in another member state. Hopefully, this heterogeneity will be phased out after the revised EU Fertiliser Regulation entered into force. But this is not to be expected before 2019.

The national implementation of the Waste Framework Directive in Germany is the Circular Economy Act (Kreislaufwirtschaftsgesetz, KrWG), implemented in 2012. End-of-waste status in general is defined under Sect. 3.5:

1. The waste status of a material ends once having been processed/valorized and fulfilling the following basic requirements:

   (i) Being used for specific purposes.
   (ii) Has a market or demand.
   (iii) Meets all technical requirements for the intended application and complies with all legal requirements and standards for products (also implicating that it is registered under REACH).
   (iv) Its use does not lead to harm for human health and environment.

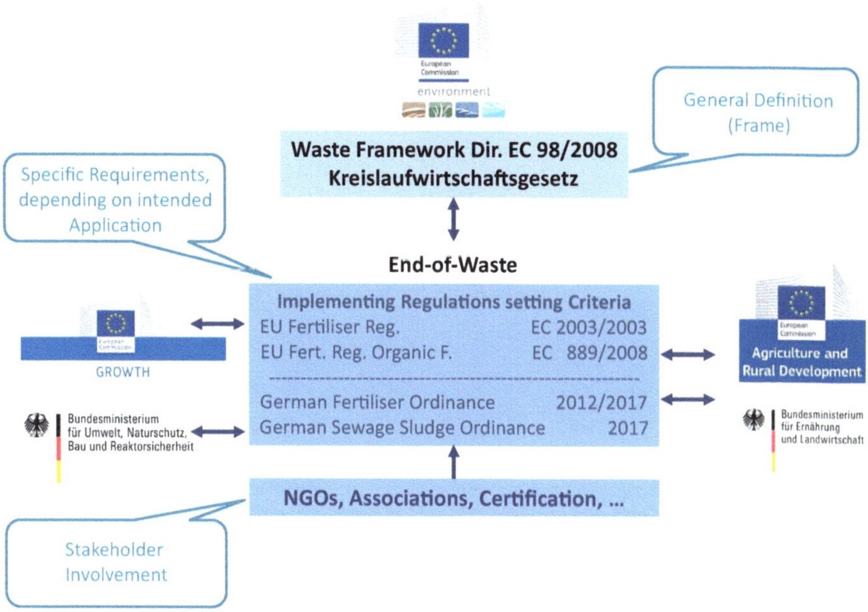

Fig. 3.1 Interaction of waste, fertilizer and sewage sludge regulation relevant for nutrient recycling in the EU and Germany

2. The federal government is mandated to define specific EoW criteria and implement specific ordinances after approval by the federal council including limit values for contaminants, etc. Both the sewage sludge ordinance and the fertilizer ordinance are such references.

Figure 3.1 intends to reflect the most relevant regulations for the valorization of recovered nutrients as mineral fertilizer. Table 3.1 provides an overview on contaminant limits set in different regulations or being currently under discussion for implementation. These are to be considered as part of the end-of-waste criteria. A big impact for recyclates, since most of them contain far less Cd compared to some fossil-based mineral phosphate derivates, can be expected from the implementation of cadmium limits under the EU Fertiliser Regulation. The European Commission's ambition is not only to introduce a Cd limitation but also to tighten the limit over the coming years. The starting point is intended to be set at 60 mg Cd per kg P_2O_5. It is still uncertain if and where the tighter limit will be set in the regulation after passing all political and juridical stages in the coming years, no matter if the EU still consists of 28 or 27 member states.

Most urgent is to set up the enabling framework allowing the fertilizer use for the following materials, representing the most prominent in the field of P recovery from wastewater/sewage sludge.

Struvite ($MgNH_4PO_4$ $6H_2O$) is the global champion in terms of quantity (several 10,000 tons) and number of installations (more than 40 operational) for mineral phosphates crystallized from the wastewater/sewage sludge stream. Similar to struvite, several Ca-based phosphates are recovered. Struvite as such is already mar-

Table 3.1 Legal requirements for sewage sludge and mineral recyclates

Parameters in mg/kg DM	EU municipal sludge framework dir. 1986	EU mineral fertilizer reg. draft 2016 PFC 1	STRUBIAS proposed PFC levels for inorg. macronut. fertilizer	Germany sludge ordinance		Germany Fertiliser Ord. 2012 last amended 2017
				1992	2017	
Heavy metals						
As – arsenic	–	60	60	–	–	40
Tl – thallium	–	–	–	–	–	1
Cd – cadmium	20–40	3	–	5/10	DüMV	1.5
Cd for $P_2O_5 > 5\%$		60–40 mg/kg P_2O_5	60-40-20			50 mg/kg P_2O_5
Cu – copper	1000–1750	–	–	800	DÜMV	900
Hg – mercury	16–25	2	–	8	DÜMV	1
Ni – nickel	300–400	120	120	200	DÜMV	80
Pb – lead	750–1200	150	150	900	DÜMV	150
Zn – zinc	2500–4000	–	–	2500	4000	–
Cr – chromium (Total)	–	–		900	DÜMV	–
Cr – chromium (VI)	–	2	2			2

Organic compounds						
Organic carbon			3%			
PCB 28, 52, 101, 138, 153, 180 each				0.2	0.1	–
AOX				500	400	–
Benzo[a]pyrene				–	1	–
PAH_{16}			6?			
PFC (PFOA + PFOS)			tbc	–	DÜMV	0.1
I-TE Dioxins and dl PCB (WHO-TEQ 2005)				0.1	DÜMV	0.03
Biuret ($C_2H_5N_3O_2$)		1200				
Perchlorate (ClO_4^-)		1				

DÜMV fertilizer regulation

keted as fertilizer or can be used to customize the nutrient content, i.e. organic materials used as fertilizer or soil improver. Ca phosphates are mainly used as a fertilizer component.

P-rich incineration ashes, like from animal meat and bone meal, sewage sludge and other P-rich wastes, provide mineral concentrates, low or even free in organic contaminants suitable for fertilizer products (meeting CMC requirements) or even as fertilizer ready to use (meeting PFC requirements). Besides contaminant levels, the P availability for plant uptake will make the difference in case of ashes. Of course, as trends already indicate, the fertilizer route might not be the only one for ash-based materials, especially when considering merchant grade phosphoric acid or even white phosphorus (P_4), allowing valorization also in higher value markets.

A global inventory of installations has been recently published by (Walker 2017; Kabbe and Kraus 2017). The latter will be frequently updated and put online under www.p-rex.eu.

3.3 No Recycling Without Value Chains

Technologies to recover phosphorus from P-rich wastes like sewage sludge, ashes, manures and biowastes are already there (Schoumans et al. 2015; Kabbe et al. 2015; Ohtake and Okano 2015). Some of them are well advanced and already considerable as state-of-the-art, and others still need optimization not only in technical but also in economic terms. The major challenge will be the bridging of the gap between recovery and actual recycling (Fig. 3.2).

Closing the loop will only work with value chains. Otherwise, stockpiles of recovered nutrients will grow without being valorized. It also means that there needs to be customers for the materials recovered, be it for direct use as a product (fertilizer) or as raw material for further treatment and processing. Every processing step is linked with additional efforts (labour, energy, chemicals), finally meaning costs. The challenge will be, once the value chain is legally allowed to find customers, seeing the positive value in these materials and being able and willing to pay an adequate price. If the process chain fails to generate a positive value, only law enforcement can foster rather artificially a market for these recovered materials.

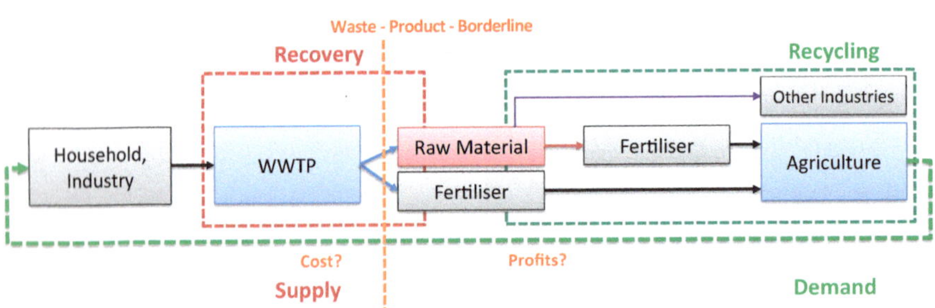

Fig. 3.2 Bridging the gap between phosphorus recovery and recycling – fertilizer value chain

The worst case would be that recovered materials end up as waste or that certain industries make profits by getting 'cheap' secondary raw materials in the end always heavily subsidized by the taxpayer. It's never the government; it's always us, who are paying taxes! Then of course a pressing question will arise: Is it more sustainable to invest in waste prevention instead of producing more and more waste that needs to be recycled? Still economic growth is kind of dogmatically linked with growing and exhaustive resource consumption. More and more stuff and waste are produced with less and less labour. Circular economy promises to make a difference, but so far, the so-called great transformation remains hidden behind the curtain, waiting for reasonable thinking to mature into reasonable action.

Technologies cannot be more than just one pillar to switch towards sustainable nutrient management and circular economy. Preventing waste is at least as important if not the most important approach. In the end, it is a societal challenge to define how and how fast we move forward and become circular.

3.4 Opportunities

Since not all value chains can be covered and discussed in this chapter and some of them are represented in other chapters, only the following may serve as a good example for the valorization of struvite.

Struvite has been recognized as an effective fertilizer providing the opportunity to reduce fertilizer consumption and nutrient losses to the environment. It is often called slow-release fertilizers. To reflect new functions or performance parameters overcoming the dogma of high water solubility, the term 'next-generation fertilizer' is becoming more and more popular. The following figure reflects how the Canadian company Ostara is bridging the gap between phosphorus recovery and recycling. A good technology and a good fertilizer product combined with a good marketing strategy are the key ingredients of Ostara success with already 14 recovery units up and running worldwide and recovered struvite sold as the premium fertilizer Crystal Green® (Fig. 3.3).

The combination of Ostara Nutrient Recovery Technologies' WASSTRIP and PEARL with LysoTherm in Amersfoort (NL, Waterboard Vallei en Veluwe) is one good example of the second-generation nutrient recovery with enhanced carbon management. Additional operational benefits besides prevented struvite scaling in the sludge train and reduced return load are improved biodegradation and increased biogas yield. The increased ortho-P concentration in the aqueous phase providing higher recovery rates and the reduced sludge solids to be disposed are strong arguments for implementation at WWTPs with enhanced biological phosphorus removal and anaerobic digestion. The operational cost can be reduced by several hundreds of thousands euros per year even without selling the struvite. Ostara's third installation in Europe just started operation in Madrid.

But, besides these operational benefits, Ostara's business case provides another selling point for potential customers. The offtake guarantee generates a source of

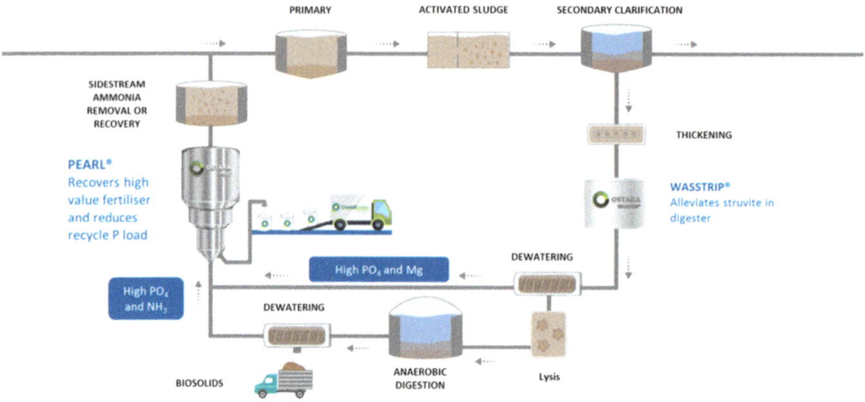

Fig. 3.3 Struvite recovery to recycling – the Ostara value chain (Source: Ostara)

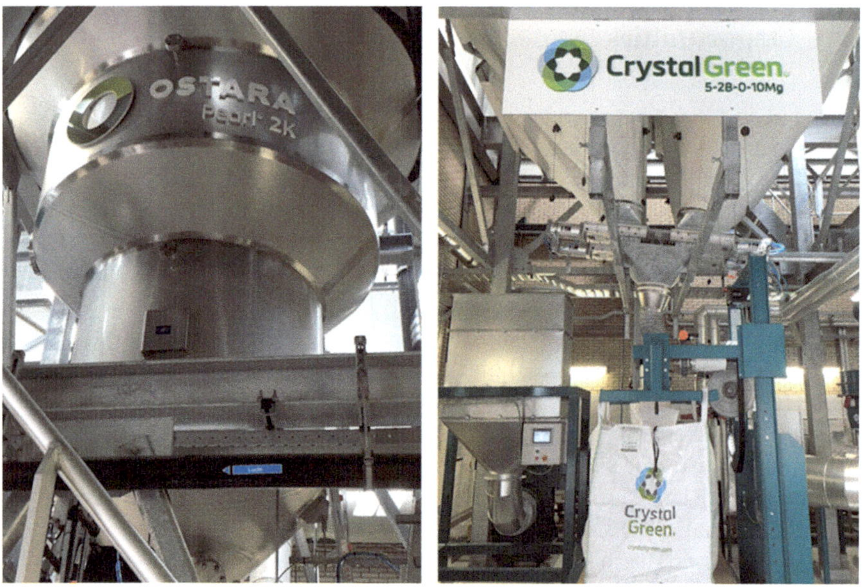

Fig. 3.4 PEARL struvite recovery reactor and Crystal Green bagging station in Amersfoort (source: Kabbe)

income, since long-term contracts make sure the struvite produced on-site the WWTP will be taken off by Ostara and marketed as premium fertilizer Crystal Green®. Selling the Crystal Green® as premium product enables Ostara to pay their customers a reasonably good offtake price that is so far several times higher compared to other struvite valorizors. Since WWTP operators are often not familiar with fertilizer marketing and related bureaucratic burden, this business model provides a kind of appealing all-inclusive package (Fig. 3.4).

So it is not the technology alone that counts. Rather, the whole value chain has to be considered when deciding on a viable solution for nutrient recovery and reuse. Advantageously, struvite crystallization from wastewater also functions as a purification step. Struvite recovered in this way can even be classed as suitable for organic farming, the expert technical group on organic production (EGTOP 2016) has recommended.

3.5 Challenges

- The legal framework is tailored for existing structures and is very slow at adapting to future challenges. In relation to resource efficiency and sustainability, we are still a long way from implementing what is being discussed. For example, the upgrading of recovered material from being treated as a waste to being considered a resource or a product is proving to be a challenge with regard to legal barriers in place. The redefinition of end-of-waste criteria is a tough process but is a prerequisite to enable value chains to bridge the gap between recovery (supply) and recycling (demand) and make a circular economy really happen. Therefore, the revision of the EU Fertiliser Regulation (2003/2003/EC) needs to be progressed to provide a level playing field for fertilizers, irrespective of whether they are produced from fossil or secondary sources (Hukari et al. 2016). Another issue that deserves to be considered is the application of appropriate products for use in organic farming, for instance, by adding recovered struvite and calcined phosphates to the list of approved fertilizers in 889/2008/EC, as it was recommended by the Expert Group for Technical Advice on Organic Production (EGTOP 2016). Projects like the EU project IMPROVE-P (2013–2016) or the German project nurec4org (2017–2018) have been shaped to provide further fact-based information to facilitate stakeholder acceptance and cooperation along the relevant value chains from recovery to recycling.
- In some countries, decision-makers focus only on the 'highest hanging fruits' in terms of recovery and potentially recycling rates, instead of starting with the 'lowest hanging fruits' and allowing technical evolution. How can a market for material including P from secondary sources develop, if the already viable options are ignored and non-feasible options are favoured? A market starts with a product that is available and of use to someone. The same applies to the technology itself. Market penetration and replication will only happen with full-scale demonstrations. Instead of broadening the range of technologies, the focus should now be on setting up full-scale demonstrations of the most promising options. This should be augmented by making the most out of the existing infrastructure.
- If sludge incineration becomes the favoured or even only routed, the reason for improving the quality of wastewater and sludge will decrease, which can bear the risk that the sludge and finally ash quality will become poorer and poorer calling for increasing efforts to be taken for resource recovery. So far, the whole recov-

ery discussion is focused on sludge only, being an end-of-pipe approach. Wouldn't it be more reasonable also to trigger innovation upstream the WWTP or sludge train? What is the point to recover precious resources after having them contaminated?
- But also downstream of WWTPs, there is still potential to improve P recovery and recycling, just by applying better or more reasonable (authors avoid the term 'smart' here by purpose) sludge disposal or incineration logistics. As the German sewage sludge ash monitoring revealed, a big share of the German mono-incinerators burn municipal sludge mixed with industrial sludge. Why not exclusively dedicate the existing mono-incineration capacities for municipal sludge only? Instead of calling for more mono-incinerators, the decision-makers should foster first 'making the best and most out of the existing infrastructure'. This would save the taxpayer a lot of money and would prevent avoidable expenditures for surplus capacities.

3.6 Summary and Outlook

There is no doubt that phosphorus is a limited essential resource. Efforts should be taken to increase the resource efficiency of phosphorus while we have a choice. In the light of existing and feasible technologies, attention should be focused on bringing these to the market, rather than increasing the range of technologies. Also, the existing infrastructure already provides the opportunity to recover and recycle substantial quantities of P, including from ash. Smarter sludge management will help to make the most out of the existing infrastructure without the need for huge investments. Of course, recovery alone will not work. Feasible value chains are needed to bridge the gap between recovery and recycling. The current legal framework and the low prices for raw materials have to be considered as market barriers. At current price levels for phosphate rock and other raw materials, only legal requirements are likely to boost a widespread implementation of phosphorus recovery and recycling especially from the wastewater stream. A 'level playing field' is needed for fertilizers so that it does not matter if they are made from primary or from secondary sources. The definition of end-of-waste criteria for recovered nutrients is a crucial element, and binding recycling targets (comparable to the CO_2 emission reductions goals) based on achievable goals should be developed. The uncertainty-based precautionary dogma needs to be replaced by risk assessment-based requirements. Otherwise, no recovery technology and recycling value chain will be implemented in Europe if it is not providing benefits to its operators under current conditions. When it comes to data and their availability and reliability, we have to transform uncertainties into certain and fact-based knowledge. The future will be shaped by the ones who dare (to take a risk), not by the ones who fear! In the end only technologies yielding homogenous products or raw materials, independent from input material quality and meeting both criteria, energy efficiency and resource efficiency in unity, will have a chance for widespread application under sustainability aspects in a circular economy.

There is not only a lot of know-how already waiting to be shared with huge potential to be creatively transformed into innovation. Recovery to recycling value chains are already implemented here and there just waiting to be replicated. Think forward, act circular!

References

Acatech, Leopoldina, Union der deutschen Akademien der Wissenschaften (2017) Rohstoffe für die Energiewende: Wege zu einer sicheren und nachhaltigen Versorgung. Berlin. February 2017, ISBN 978-3-8047-3664-1

Asimov I (1959) Life's bottleneck. In: Mills RP (ed) The magazine of fantasy and science fiction. Mercury Press, Oklahoma City

Cordell D, White S (2011) Peak phosphorus: clarifying the key issues of a vigorous debate about long-term phosphorus security. Sustainability 3:2027–2049

De Ridder M, De Jong S, Polchar J, Lingemann S (2012) Risks and opportunities in the global phosphate rock market: robust strategies in times of uncertainty. The Hague Centre for Strategic Studies (HCSS) report. The Hague Centre for Strategic Studies (HCSS), The Hague

EG TOP (2016) Final report on organic fertilizers and soil conditioners (II). Expert Group for Technical Advice on Organic Production, Brussels. 2 Feb 2016

Hukari S, Hermann L, Nättorp A (2016) From wastewater to fertilisers—technical overview and critical review of European legislation governing phosphorus recycling. Sci Total Environ 542B:1127–1135

Kabbe C, Kraus F (2017) P recovery: from evolution to revolution. Fertil Int 479:37–41

Kabbe C, Remy C, Kraus F (2015) Review of promising methods for phosphorus recovery and recycling from wastewater. Proc Int Fertil Soc 763:1–29

Ohtake H, Okano K (2015) Development and implementation of technologies for recycling phosphorus: in secondary resources in Japan. Global Environ Res 19:49–65

Scholz RW, Wellmer FW (2013) Approaching a dynamic view on the availability of mineral resources: what we may learn from the case of phosphorus? Glob Environ Chang 23:11–27

Schoumans OF, Bouraoui F, Kabbe C, Oenema O, van Dijk KC (2015) Phosphorus management in Europe in a changing world. AMBIO 44:180–192

van Dijk KC, Lesschen JP, Oenema O (2016) Phosphorus flows and balances of the European Union member states. Sci Total Environ 542B:1078–1093

Walker C (2017) Market map – beating the burn rate for resource and energy recovery from sludge. Glob Water Intell Mag 1:40–47

Wellmer FW, Scholz RW (2015) The right to know the geopotential of minerals for ensuring food supply security: the case of phosphorus. J Ind Ecol 19(1):3–6

Chapter 4
Life Cycle Assessment of Processes for P Recycling

Christian Remy and Fabian Kraus

Abstract Recent developments and innovations in the field of P recovery and recycling from municipal sewage sludge claim to provide a sustainable and more efficient alternative to the traditional sludge valorization in agriculture. The method of life cycle assessment (LCA) offers a detailed analysis of the potential environmental impacts associated with different technologies, but it needs to be based on sound definitions and validated input data, not only for the specific technologies but also for the methodological framework. Since the relevant ISO standards 14040/44 provide methodological guidance without specifically fixed definitions, the application of LCA leaves a lot of potential for interpretation of results. Within the European research project P-REX, a methodological framework was developed to assess various available technologies for P recovery from sewage sludge, sludge liquor or incineration ash. Decisive definitions are the setting of adequate system boundaries and functional unit, the selection of LCA indicators and their interpretation. The following chapter discusses important definitions of the LCA methodology and provides recommendations towards a consolidated approach for future LCA studies in the field of P recovery from sewage sludge.

Keywords Life cycle assessment · Phosphorus recovery · Sewage sludge · Environmental impact assessment

4.1 Introduction

A significant proportion of phosphorus (P) used for food production is contained in municipal wastewater and ends up in sewage sludge (Milieu 2010; van Dijk et al. 2016). Hence, the recycling of P from sewage sludge to agriculture has been realized for decades with the agricultural disposal of this sludge onto farmlands. However, recent years have seen a reduction of this recycling route in many EU

C. Remy (✉) · F. Kraus
Berlin Centre of Competence for Water, Berlin, Germany
e-mail: Christian.remy@kompetenz-wasser.de

H. Ohtake, S. Tsuneda (eds.), *Phosphorus Recovery and Recycling*,
https://doi.org/10.1007/978-981-10-8031-9_4

countries due to concerns about inorganic and organic pollutants found in the sludge and also due to indications of limited plant availability of P in sewage sludge depending on its Fe content. Moving towards thermal disposal of sewage sludge in incineration plants (either as mono-incineration in dedicated facilities for sludge disposal or as co-incineration in power plants, municipal waste incineration plants or cement kilns), the direct recycling of P from sewage sludge into agriculture is usually not feasible with the residual ash from incineration due to limited plant availability of P and relatively high contamination of the ash with, e.g. heavy metals.

To overcome this drawback of limited P recycling with sludge incineration and close the P management cycle again, different pathways and processes of P recovery from sewage sludge or incineration ash have been developed in recent years. These processes claim to provide a sustainable and more efficient alternative to traditional sludge valorization in agriculture. However, they also require additional resources (e.g. electricity, chemicals, infrastructure) and will thus have indirect impacts on the overall environmental profile of the secondary P fertilizer products. Consequently, a comprehensive assessment of environmental impacts and benefits of these recycling processes should be targeted to identify and characterize potential benefits and drawbacks of the different routes, reveal potentials for optimization and direct future research and development activities towards promising routes of P recovery and recycling.

4.1.1 Using Life Cycle Assessment for P Recovery Processes

For assessing technical processes and systems in their potential environmental impacts, the method of life cycle assessment (LCA) has been developed and applied widely within research and industry (Klöpffer and Grahl 2009) and also in the field of wastewater and sludge treatment as reviewed in literature (Corominas et al. 2013; Yoshida et al. 2013). Taking the life cycle perspective into account, this method quantifies potential environmental impacts of a defined process or system based on quantitative information on resource needs and emissions of all relevant processes which are affected by the system under study. This includes on-site effects of the process or system ("foreground system") and also indirect effects in the related production systems for upstream (e.g. electricity production) and downstream (e.g. waste disposal) processes ("background system"). Building on substance flow models of the foreground system and database information for the background system, all relevant inputs and outputs across the system boundary from or into the environment are quantified and summarized. This inventory information is then evaluated with a defined set of environmental indicators, relating to specific areas of environmental concern (e.g. anthropogenic climate change, acidification or ecotoxicity) and describing potential negative impacts of emissions and resource use.

The general framework and steps of an LCA are defined in the ISO standard 14040/44 (ISO 14040 2006; ISO 14044 2006) and include four distinctive working steps: (1) definition of goal and scope, (2) life cycle inventory, (3) life cycle impact assessment and (4) interpretation. The standard gives methodological guidance, but the LCA practitioner can freely choose adequate definitions to define the system under study depending on the respective goal and scope of the assessment, provided that they are well justified in the LCA report. Hence, LCA studies in the same field of technology often lack comparable definitions and may show divergent outcomes for the same processes, thus leaving a wide potential for contrasting interpretation of results, leading to low trust in their validity and credibility of the LCA results and method. This shortcoming may be overcome by a consolidated approach for LCA studies in a specific field of interest, so that future LCA can build upon a comparable and validated framework for environmental assessments and increase the credibility of their results.

4.1.2 Features of LCA

Per definition, the method of LCA can be used to compare different technical processes or systems in their potential environmental impacts, following a defined methodological approach. It quantifies the environmental profile of a specific technology in terms of emissions and resource use, revealing hotspots of environmental concern and optimization potentials towards reduction of emissions and resource consumption. However, the approach of LCA is characterized by specific features which have developed since its first application in the 1980s. Using LCA for a comparative environmental assessment of different processes or systems, these features have to be clearly identified and kept in mind while using the tool and interpreting its results.

LCA modelling of flows between the technical system and the environment is based on "steady-state" substance flow models of the technical system, typically describing a non-dynamic average operational state of the processes. In addition, flows of resources and emissions into the environment are aggregated over time and space, and these aggregates are used to characterize the potential environmental impact associated with a process or service. Hence, LCA does not provide any temporal or spatial resolution of emissions by definition. Consequently, effects of peak loads and temporal or spatial variations in emission profiles cannot be accounted for in LCA. Future developments in LCA data management and impact characterization may include more temporal and spatial aspects, but these are still under development.

Impact characterization in LCA is based on global or regional impact factors for emissions and resource use. For emissions, these factors take into account fate and effect of pollutants in multimedia transport and effect models, representing a generic environmental system. For resource use, impact factors typically relate available resources to consumption profiles to account for relative resource availability and potential scarcity. Hence, LCA indicator results are clearly identified as "potential" environmental impacts caused by these flows rather than actual predictions of local impacts on specific ecosystems. LCA is not useful for predicting if any specific environmental impact of a certain process or service is actually realized for a specific ecosystem (e.g. eutrophication in lake XY). This task has to include a localized environmental impact assessment, taking into account background concentrations, pollution loads and temporal and spatial aspects of emissions, which are not part of LCA.

Interpretation of LCA indicator results can be difficult to communicate to non-experts. Describing different potential environmental impacts on a global perspective, LCA will give information on multiple aspects of environmental concern based on quantitative input data. However, LCA will not tell the user automatically which process or system is to be preferred or rejected from an environmental perspective. As different categories of environmental impact are characterized in LCA (e.g. global warming, primary energy demand, eutrophication, ecotoxicity), there has to be a subjective weighting of the interested stakeholders between these environmental indicators to develop a conclusive decision. Although weighting schemes are proposed in selected LCA frameworks based on predefined weighting factors (e.g. using the scheme of Swiss eco-points Frischknecht et al. 2008), it is clearly recommended in the ISO standard to report the results of an LCA in non-weighted categories of environmental impact. Thus, each stakeholder can draw his/her own conclusions according to local, regional or other conditions and valuations.

Keeping in mind these inherent limitations of the LCA approach, it is clear that LCA cannot take the full responsibility for the decision from the user or stakeholder that is deciding between different processes or systems from an environmental point of view. However, LCA is based on a quantitative substance flow model of the relevant process or service, and it often serves as a focal point of process development and optimization, facilitating discussions on process improvements or optimization of the value chain, including relevant processes upstream and downstream of the core process or service. It can also draw attention to side effects or trade-offs in environmental impacts, for example, reducing a specific impact while shifting burden into another category of environmental concern.

Within the European research project P-REX (www.p-rex.eu), a methodological framework was developed to assess various processes and pathways for P recovery from sewage sludge, sludge liquor from dewatering or incineration ash (Remy and Jossa 2015). This chapter introduces this methodological framework defined in P-REX in terms of system boundaries, functional units, selection of LCA indicators and their interpretation. The following sections will discuss important definitions of the LCA methodology and provide recommendations towards a consolidated approach for future LCA studies in this field. It is structured into different parts according to the ISO framework for LCA to elaborate on important aspects of LCA goal and scope definitions and application within the context of P recovery in technical processes.

4.2 Definition of Goal and Scope

4.2.1 Goal of an LCA

The definition of a specific goal for an LCA study seems to be redundant at first; however it can provide valuable insights and help to formulate this goal in a precise way, so that the definitions can be set to fulfil this goal in an adequate manner. The goal will give information on the nature and purpose of the LCA study and the intended use of its outcome.

Some examples for proper goal definitions in the field of P recycling from municipal wastewater are listed below:

1. The goal of this LCA is to analyse and compare potential environmental impacts of different P recovery technologies from sewage sludge, sludge liquor or mono-incineration ash against conventional phosphorus fertilizer production.
2. The goal of this LCA is to analyse and compare potential environmental impacts of different P recovery technologies from sewage sludge, sludge liquor or mono-incineration ash and the recycling of obtained products against conventional P recycling via sludge application to arable land.
3. The goal of this LCA is to analyse and compare potential environmental impacts of different P recovery technologies from mono-incineration ash and the recycling of obtained products to arable land against the usage of conventional phosphorus fertilizers.

Although all listed examples describe in principle the use of LCA for phosphorus recovery, the exact goal definition has decisive impact on LCA scoping such as the choice of adequate system boundaries, functions and functional units, LCA inventory, results, interpretation and conclusion that can be obtained from the LCA:

1. The first goal definition allows comparability between different types of phosphorus recovery processes and fertilizer production; however the application of P products on arable land (e.g. environmental impacts of different heavy metal contamination or leaching of N and P into groundwater) is NOT within the scope of this LCA.
2. The second goal definition allows comparability between different types of phosphorus recycling routes, including P recovery processes and traditional direct application of sludge on farmland. This LCA outcome focusses, e.g. on the question whether it is worth to invest energy, etc. for phosphorus recovery technologies compared to traditional sludge recycling from an environmental point of view.
3. The third goal definition only considers technologies aiming for phosphorus recovery from ash; consequentially system boundaries will neither include sludge treatment nor ash production via mono-incineration. A comparison of ash-derived P products and conventional phosphorus fertilizers and the impacts

of recovered and conventional phosphorus products on arable land can be included. However, no conclusion can be obtained if it is environmentally preferable to invest in sludge mono-incineration or not, since sludge disposal is not within the system boundaries.

The following sections show conclusive options which have to be considered by the LCA practitioner referring to the goal definition 1 ("compare P recovery processes and mineral fertilizer production").

4.2.2 Scoping of LCA: System Boundaries, Functions and Functional Unit

LCA is based on two main principles: (1) account all relevant processes which are affected by the process under study ("life-cycle perspective") and (2) account for all relevant environmental impacts to reveal synergies and trade-offs between different categories of environmental impact. To reflect these principles, the adequate definition of system boundaries is essential for developing a reasonable framework for the LCA. These system boundaries include all processes that will be modelled with their inputs and outputs towards the environment. They should be defined to include all relevant processes upstream and downstream of the core process, thus delivering a comprehensive picture of the process and its effects. This relates both to the "foreground system" and to the "background system" (e.g. electricity production in power plants, chemical production, disposal of wastes). For keeping the LCA model manageable, certain processes and material flows can be cut off from the assessment, e.g. by defining reasonable cut-off criteria in mass or relevance. However, this "cut-off" has to be carefully justified to prevent important system parts or flows to be excluded from the assessment.

For scenario definitions of comparative LCA studies, the systems under study should target the same system functions to allow for a fair comparison between systems of equal purpose. In case of a sludge treatment line of a WWTP, this function could be defined as "treatment and disposal of raw sludge from a 1 Mio pe (million population equivalents) WWTP per year". To reach functional equivalency between systems, it is suitable to expand the systems with additional functions ("system expansion") to reach comparability with other systems (e.g. recovery of secondary P from sewage sludge). Secondary functions can be accounted by subtracting the avoided process for secondary functions (i.e. the production of mineral P fertilizer) and thus crediting the avoided impacts ("avoided burden approach").

Functional equivalency should also be reflected in the suitable selection of a functional unit (FU) to which all environmental impacts are related to. This FU has to be carefully defined, as it may affect the comparison between different processes

Table 4.1 Different perspectives of LCA results and suitable functional units (scenario A, reference system; scenarios B and C, systems with different P recovery processes)

Perspective	Description	Calculation	Example for functional unit
System	Total environmental impacts of the entire system	A, B, C	[Sludge treatment and disposal for WWTP serving 1 Mio population equivalents and year]$^{-1}$
System change	Total environmental impacts of P recovery process	B-A, C-A	[P recovery from WWTP serving 1 Mio population equivalents and year]$^{-1}$
Product	Total environmental impacts per mass of recovered P	(B-A)/mass of $P_{recovered}$	[Mass of $P_{recovered}$]$^{-1}$

or services. Following the system function and perspective of the study, different FU can be defined for P recovery from sewage sludge (Table 4.1):

- The "system perspective" describes the total impacts of the entire system of sludge disposal (e.g. sludge disposal for a specific WWTP serving a defined number of inhabitants), giving information on the environmental profile of sludge treatment and disposal with or without P recovery options (e.g. scenarios A, B, C), including potential credits for substitution of mineral P production, electricity and heat. This perspective is useful to picture the overall environmental impacts of sludge treatment and disposal with and without P recovery, thus revealing also the contribution of P recovery processes to the total impacts of sludge disposal.
- The "system change perspective" quantifies all changes in environmental impacts that can be associated with the introduction of a P recovery process in the defined system. For revealing these effects, a reference scenario (A) without dedicated P recovery should be compared against scenarios with P recovery (B, C). The difference between both scenarios (e.g. B-A. C-A) can then be fully attributed to the P recovery process, allocating all changes to the function of P recovery.
- The "product perspective" describes the environmental impacts that are associated relative to the amount of recovered P. It should be calculated from the difference of the reference scenario (A) and the scenario for P recovery (B), related to the total amount of recovered P (e.g. (B-A)/mass of $P_{recovered}$). The product perspective gives information about the environmental profile of the recovered P product, enabling the comparison of different secondary P products from different sites, pathways and processes.

All these perspectives can be useful for the interpretation of the LCA results. The LCA documentation should at least include suitable information to enable the transfer between these functional units, if not all results for these perspectives are reported explicitly. Thus, results from LCA studies with different functional units can be compared against each other.

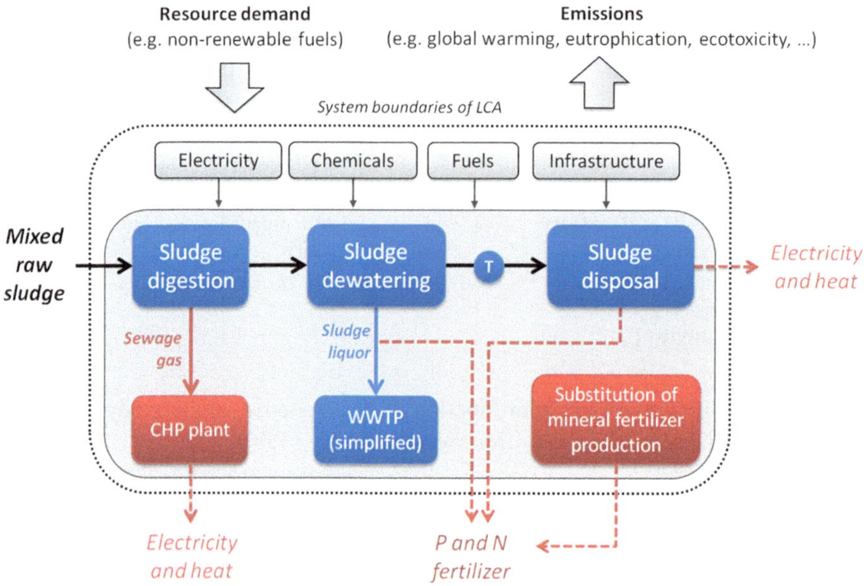

Fig. 4.1 Exemplary system boundaries of comparative LCA studies of P recovery from sewage sludge

Regarding the definition of system boundaries of an LCA for P recovery from sewage sludge or incineration ash, the following system parts should be considered as a minimum to reflect all side effects of P recovery options on sludge treatment and disposal (Fig. 4.1):

- Starting with raw mixed sludge coming from the WWTP ("reference flow")
- Sludge stabilization in digestion and valorization of products (e.g. sewage gas utilization in combined heat and power (CHP) plant)
- Sludge dewatering
- Treatment of sludge dewatering liquor which is usually recycled back to the influent of the WWTP ("return load")
- Sludge disposal (e.g. drying, transport, application in agriculture, incineration, disposal of incineration ash)
- Credits from P recovery (e.g. avoided production of mineral P fertilizer) and energy recovery (avoided production of electricity, heat) in sludge digestion and disposal
- All background processes relevant for the system function (i.e. production of electricity, chemicals, fuels and – if relevant – materials for infrastructure).

In particular, LCA of P recovery processes should include all processes downstream of the P recovery process which could be affected by the extraction of P. As P is often extracted from the liquid phase, the impact of the return of sludge liquor to the WWTP inlet ("return load") is affected by P recovery and thus has to be included in the LCA definitions. Removing P and other substances (e.g. N, COD) from the return load will lead to a lower effort of treatment in the main line of the WWTP and can also affect WWTP effluent quality, which is naturally a major environmental effect of the system. The effect of changing return loads may be estimated by a simplified WWTP model (i.e. modelling only electricity demand and effluent quality of the mainline) to limit model complexity of the LCA.

Similarly, the dewatering and disposal of sludge can be affected by upstream P recovery due to changes in sludge properties (e.g. dewatering efficiency and dry matter content), affecting also transport volumes, incineration characteristics and related potential for energy recovery. Finally, disposal of any residuals from sludge incineration (e.g. fly ashes) or subsequent P recovery processes should also be included, as quality and quantity of these residuals can be affected by upstream P recovery from incineration ashes.

4.3 Life Cycle Inventory: Collecting Data for LCA

Quality of the inventory data for the LCA model will be decisive for the validity of the LCA results. In general, LCA input data should fulfil several criteria to meet the requirements of the study goal and scope. Particular attention has to be paid to the representativeness of the collected data for the system under study, the completeness in data collection so no relevant flows are overlooked and the consistency of the data collection procedure between different components of the analysis. In addition, information on uncertainty and precision (e.g. variation) of the collected data can be useful for sensitivity analysis and stability check of the LCA results. For the inventory, it can be distinguished between foreground data (process-specific flows and emissions of the system under study) and background data (resource use and emissions for background processes such as electricity production, chemicals production, transport, etc.).

4.3.1 Data Requirements for an LCA of P Recovery Processes

A minimum set of LCA foreground data for P recovery processes should include information on product yield (related to input flow of the process and P content), product quality (inorganic and organic pollutants, P content, content of important metals such as Fe), and demand for auxiliaries such as electricity, heat, fuels and chemicals. In addition, side effects of P recovery on upstream or downstream processes should be reflected as precisely as possible, e.g. changes in return load or sludge dewatering properties.

To adequately show potential effects of P recovery and the change in sludge properties, it is mandatory to consider the main parameters for sludge properties within flow modelling, which are at least volume, dry matter, organic dry matter (or COD), nitrogen and phosphorus. Further the consideration of inorganic matrix elements as calcium, silicon, iron, aluminium or magnesium is recommended. The consideration of further trace elements as heavy metals or trace organics is not necessary in terms of process performance. However a simplified approach for different contaminations of products should be considered if the application to arable land is within the scope of LCA.

4.3.2 Data Quality and Validation

In LCA studies of technical processes which are still under development and optimization (such as many evolving P recovery processes), input data has to be typically collected from process suppliers or operators based on test installations in laboratory or pilot scale. Here, it has to be kept in mind that results of lab or pilot tests may not reflect the actual technical full-scale realization in terms of energy and chemicals demand, product yield and product quality. Hence, upscaling of lab or pilot data towards a projected full-scale installation has to be carefully justified in close cooperation with suppliers and operators to end up with reasonable data for the LCA. Typically, process performance (product yield, product quality) and chemical demand (e.g. acids, complexing agents) of P recovery processes can be transferred from pilot-scale to full-scale process using suitable relations, whereas energy demand (electricity demand, heat management) has to be estimated or extrapolated from detailed engineering concepts. Potential biases between different processes with different stages of development (e.g. comparing lab-scale tests of process A with pilot or even full-scale results of process B) have to be addressed in the LCA documentation and also in the interpretation of the results of a comparative analysis.

For a comparative LCA of different P recovery processes, data collection will most probably include process data from different sites and treatment plants, thus introducing another potential bias in the comparison. In fact, sewage sludge composition and other relevant boundary conditions may vary in a wider range for important parameters (e.g. P concentration), which can have a high impact on efforts for P recovery as well as product yield and quality. To overcome this bias for comparative studies, LCA input data should be transferred to a reference model of a defined WWTP to enable the comparison of different processes based on the same boundary conditions such as sludge or liquor quality. For the transfer, a suitable relation of relevant process data to input characteristics (e.g. molar dosing of Mg as precipitation agent in relation to PO_4-P content, P yield in % of dissolved P) should be targeted wherever possible.

For the background data, LCA databases should be used which reflect the current knowledge of resource use and emissions. A common database for this purpose is e.g. ecoinvent (Ecoinvent 2014), but there are other commercial and national databases available. The LCA report should contain precise information on which database and datasets have been used for the LCA. Comparability of different LCAs can be improved if similar datasets are used for the background processes.

After data collection has been finished, final validation of data with technology suppliers and operators is strongly recommended to have a common agreement on all input data for the LCA. Sending the input data for validation to the experts can help to track down errors in data transfer and processing (e.g. different units or relations) which are fairly common regarding the large amount of data to be collected for the LCA. Finally, data validation will increase validity of the LCA results and consequently the trust of all stakeholders into the final results of the LCA model. Calculation of LCA results can be started before final validation which helps to track down obvious errors in input data, but public communication of LCA results should not start until all input data has been finally validated by the data providers.

4.4 Life Cycle Impact Assessment: Calculating Environmental Indicators

In general, the ISO standard for LCA does not regulate the specific indicators and related impact models that have to be chosen for a specific LCA study. Hence, the LCA practitioner can choose from a variety of different indicators and impact assessment methods which are available in the LCA community. Recently, the Joint Research Centre of the European Commission published a review of available methods for LCA impact assessment and recommended the use of selected models (Hauschild et al. 2013). For an adequate selection of environmental indicators in LCA, two aspects have to be taken into account. First, the environmental impact category should be affected by the process under study. Second, suitable data should be available to allow for a meaningful characterization of this environmental impact in the respective LCA model. In practice, the selection of indicators will also rely on practicability and data availability in each LCA study.

4.4.1 Selection of Indicators

Most LCA impact assessment models calculate “midpoint” indicators which characterize the environmental impact with a physically measurable effect model at the midpoint of the cause-effect chain, using a representative substance as reference unit (e.g. CO_2 equivalents for global warming potential). In addition, some methods also offer “endpoint” indicators which extend the modelling of the cause-effect chain towards the endpoint receiving the potential damage, i.e. the affected areas of protection (human health, ecosystems, resources). While enabling the comparison between different indicators and their relevance for a specific endpoint, the endpoint approach introduces a higher level of uncertainty in the evaluation, as the final impacts of environmental phenomena (e.g. impact of global warming on human health) are often difficult to estimate. For reasons of clarity and transparency, it is

Table 4.2 Recommended minimum set of LCA indicators for comparative LCAs of P recovery from sewage sludge

Environmental aspects	Indicators	Recommended indicators models	Comment
Resources			
Fossil fuels	Cumulative energy demand of fossil fuels	VDI 4600	LCI indicator
Nuclear fuels	Cumulative energy demand of nuclear fuels	VDI 4600	LCI indicator
Mineral resources	Mineral resource depletion	ReCiPe 2008	
Emissions			
Global warming	Global warming potential	IPCC 2014	
Acidification	Terrestrial acidification potential	ReCiPe 2008	
Eutrophication	Freshwater eutrophication potential (P)	ReCiPe 2008	
	Marine eutrophication potential (N)	ReCiPe 2008	
Human toxicity	Human toxicity potential	ReCiPe 2008	
	Total human toxicity (cancer + non-cancer)	UseTOX® 2008	Interim, high uncertainties
Ecotoxicity	Freshwater ecotoxicity potential	ReCiPe 2008	
	Terrestrial ecotoxicity potential	ReCiPe 2008	
	Total freshwater ecotoxicity	UseTOX® 2008	Consensus model, for sensitivity

Sources: VDI 4600 (VDI 2012), IPCC 2014 (IPCC 2014), ReCiPe 2008 (Goedkoop et al. 2009) using hierarchist perspective without long-term emissions, USEtox® 2008 (Rosenbaum et al. 2008)

highly recommended to report LCA results using midpoint indicators, while end-point indicators may be reported in addition if required by the study goal and scope.

For LCA studies of P recovery from sewage sludge or incineration ash, a selection of suitable midpoint indicators is recommended to reflect the most relevant environmental aspects of P recovery (Table 4.2). This list of indicators is based on previous experience of LCA studies in the field of sludge treatment and resource recovery and can serve as a minimum requirement in terms of environmental categories. Concerning the choice of indicator models, it is useful to stick to a consistent impact assessment method as far as possible (e.g. the ReCiPe model Goedkoop et al. 2009) and complement it with selected indicators of other methods, also for sensitivity analysis. Naturally, the list can be amended with selected indicators as necessary for the specific LCA study (e.g. ozone depletion, photochemical oxidant formation, land use, water footprint).

Impact assessment factors are influenced by the time horizon and other specific factors, which are reflected in ReCiPe by different sets of factors for different perspectives. Usually, the hierarchist perspective serves well for prospective LCAs and reflects a suitable timeframe for political decisions (e.g. global warming potential over 100a). The effect of long-term emissions (e.g. emissions from leaching at opencast mines or landfill sites >>100 a) can be deliberately excluded from the assessment to focus on effects within meaningful time frames.

Indicators can be normalized to a reference system, e.g. to the total environmental impact per person and year. Thus, the contribution of the P recovery system under study to the total environmental impacts of society can be quantified, indicating where the system of P recovery has significant or minor contribution to a specific area of environmental concern. Normalization factors are usually available on continental or national level and can be found in the impact assessment method (Sala et al. 2015; Sleeswijk et al. 2008).

4.5 Interpretation: Using LCA Results for Decision Support

For interpretation of the LCA results and subsequent communication towards relevant stakeholders, transparency and reproducibility of LCA methodology both play a major role for building trust in the outcomes of an LCA study. Thus, LCA reporting should include a detailed documentation and justification of the LCA methodology and definitions, the used input data, the indicator results of the impact assessment and a detailed discussion on the final interpretation of the results. This documentation also includes information on data quality and an estimate of uncertainty or variability of input data affecting the stability of the results. It is recommended to check the influence of selected definition parameters with major influence on the results (e.g. dissolved PO_4-P concentration in digested sludge or liquor for P recovery processes) in sensitivity analysis to be able to comment on the consequences of system definitions for the comparative results.

For communication of LCA results to stakeholders to support decision-making in political or municipal bodies, aggregation of indicators with endpoint methods or final weighting to end up with a single indicator result (e.g. "eco-points") has the intrinsic danger of lack of transparency for the involved stakeholders. Single-score results are attractive for decision-makers because they pretend to be scientifically derived and thus enable "unambiguous" conclusions of which process or scenario is preferable from an environmental point of view. However, weighting of indicators against each other is always based on a subjective value choice and will depend on local and regional conditions, priorities of different actors or political targets. Consequently, the weighed and aggregated result of an LCA should only be communicated with explicitly describing the weighting method and result factors. In fact, the ISO standard explicitly recommends no weighting in comparative studies ("Weighting [...] shall not be used in LCA studies intended to be used in comparative assertions intended to be disclosed to the public" ISO 14044 2006).

Therefore, it is recommended to report at least a defined set of midpoint indicators as results of the LCA and feed that into the discussion of the stakeholders about the different alternatives. In this manner, LCA can facilitate the discussion on potential environmental impacts associated with P recovery processes by giving quantitative information into the decision-making process. The LCA outcomes will not intrinsically lead to the recommendation of one process or pathway over another from an environmental perspective but can rather be seen as a defined framework in which process features can be revealed and environmental hotspots are identified. Finally, LCA does not take the decision on which process to choose based on the environmental profile but gives further that has to be correctly communicated and used by the relevant stakeholders. The responsibility for the decision and the overall evaluation of environmental relevance remains with the stakeholders.

4.6 Summary and Outlook

The present chapter discusses the specific features of LCA with relation to its application on P recovery processes from sewage sludge or incineration ash. Based on experience from the European research project P-REX, it proposes a methodological framework that can be adopted for future studies in this field to improve comparability between different LCA studies and increase the quality and validity of results. Recommendations are provided for a suitable definition of goal and scope of the LCA with regard to system boundaries, scenario selection and functional unit. Advice on data collection and validation is given to assure a high quality of input data as basis for the LCA. Finally, a defined minimum set of LCA impact indicators is proposed to improve the comparability of the results of future LCA studies and prevent the exclusive use of endpoint or single-score indicators in stakeholder communication. Using this methodological framework, future LCA studies in this field can follow a more harmonized approach which will increase trust in this tool and enable intelligent and meaningful use of its results towards an improved and more sustainable management of limited P resources.

References

Corominas L, Foley J, Guest JS, Hospido A, Larsen HF, Morera S, Shaw A (2013) Life cycle assessment applied to wastewater treatment: state of the art. Water Res 47 (15):5480–5492. S0043-1354(13)00540-X [pii] https://doi.org/10.1016/j.watres.2013.06.049

Ecoinvent (2014) Ecoinvent data v3.1, ecoinvent reports no. 1–26. Swiss Center for Life Cycle Inventories, www.ecoinvent.org

Frischknecht R, Steiner R, Jungbluth N (2008) Ökobilanzen: Methode der ökologischen Knappheit – Ökofaktoren 2006 (Life Cycle Assessment: Method of ecological scarcity – eco-factors 2006). Öbu – Netzwerk für nachhaltiges Wirtschaften, Zurich

Goedkoop MJ, Heijungs R, Huijbregts MAJ, De Schryver A, Struijs J, Van Zelm R (2009) ReCiPe 2008, a life cycle impact assessment method which comprises harmonised category indicators at the midpoint and the endpoint level; first edition report I: characterization. http://www.lcia-recipe.net

Hauschild MZ, Goedkoop M, Guinée J, Heijungs R, Huijbregts M, Jolliet O, Margni M, De Schryver A, Humbert S, Laurent A, Sala S, Pant R (2013) Identifying best existing practice for characterization modeling in life cycle impact assessment. Int J Life Cycle Assess 18(3):683–697. https://doi.org/10.1007/s11367-012-0489-5

IPCC (2014) Climate change 2014: synthesis report. Contribution of Working Groups I, II and III to the fifth assessment report of the Intergovernmental Panel on Climate Change IPCC, Geneva, Switzerland

ISO 14040 (2006) Environmental management – life cycle assessment – principles and framework. International Standardisation Organisation, Geneva

ISO 14044 (2006) Environmental management – life cycle assessment – requirements and guidelines. International Standardisation Organisation, Geneva

Klöpffer W, Grahl B (2009) Ökobilanz (LCA): Ein Leitfaden für Ausbildung und Beruf (Life Cycle Assessment: A guide for education and practice). Wiley-VCH, Weinheim

Milieu (2010) Environmental, economic and social impacts of the use of sewage sludge on land. Final report part I: overview report. Prepared for the European Commission, DG Environment under Study Contract DG ENV.G.4/ETU/2008/0076r. Brussels, Belgium

Remy C, Jossa P (2015) Comparative life cycle assessment of treatment-recovery paths (D9.2). KWB, www.p-rex.eu

Rosenbaum RK, Bachmann TM, Gold LS, Huijbregts MAJ, Jolliet O, Juraske R, Koehler A, Larsen HF, MacLeod M, Margni M, McKone TE, Payet J, Schuhmacher M, van de Meent D, Hauschild MZ (2008) USEtox-the UNEP-SETAC toxicity model: recommended characterisation factors for human toxicity and freshwater ecotoxicity in life cycle impact assessment. Int J Life Cycle Assess 13(7):532–546. https://doi.org/10.1007/s11367-008-0038-4

Sala S, Benini L, Mancini L, Pant R (2015) Integrated assessment of environmental impact of Europe in 2010: data sources and extrapolation strategies for calculating normalisation factors. Int J Life Cycle Assess 20(11):1568–1585. https://doi.org/10.1007/s11367-015-0958-8

Sleeswijk AW, van Oers LFCM, Guinée JB, Struijs J, Huijbregts MAJ (2008) Normalisation in product life cycle assessment: an LCA of the global and European economic systems in the year 2000. Sci Total Environ 390(1):227–240. https://doi.org/10.1016/j.scitotenv.2007.09.040

van Dijk KC, Lesschen JP, Oenema O (2016) Phosphorus flows and balances of the European Union Member States. Sci Total Environ 542:1078–1093. https://doi.org/10.1016/j.scitotenv.2015.08.048

VDI (2012) VDI-Richtlinie 4600: 2012-01: Kumulierter Energieaufwand – Begriffe, Berechnungsmethoden (VDI guideline 4600: 2012-01: cumulative energy demand – terms, definitions, methods of calculation). Beuth Verlag, Berlin

Yoshida H, Christensen TH, Scheutz C (2013) Life cycle assessment of sewage sludge management: a review. Waste Manag Res 31(11):1083–1101. https://doi.org/10.1177/0734242X13504446

Chapter 5
Phosphorus: Reserves, Production, and Applications

Marissa A. de Boer, Lukas Wolzak, and J. Chris Slootweg

Abstract The demand of phosphate fertilizers is growing as a result of a rising population, changing human diets resulting in the increasing (meat) consumption per capita, and an expansion in the production of biofuels. Besides the fertilizer industry, there is a steady growth of using phosphorus compounds in the chemical industry for applications in, e.g., soft drinks, pharmaceuticals, and flame retardants. To meet this growth, it is important to know if the P reserves are sufficient and what kind of processes are used to produce such phosphorus compounds. Reserves are not equally spread around the world, with three-quarters located in Morocco and The Western Sahara. Prices can be volatile, as shown in 2008 with an eightfold price increase. Moreover, the estimated time till depletion of phosphate rock differs substantially between several studies. Therefore, phosphate rock was added on the critical material list of the European Commission. An important aspect for the processing of phosphate rock is the quality of the rock, which is dependent on the ore type (sedimentary or igneous), level of radioactivity, and hazardous metal contents. The main intermediary compounds for phosphorus products are phosphoric acid and white phosphorus. About 95% of the phosphoric acid is made via the wet process: acidulation of phosphate rock to create wet phosphoric acid and the main by-products phosphogypsum and hydrogen fluoride. The purity and thus the reusability of phosphogypsum are dependent on the type of digestion process. However, at the moment, reusing phosphogypsum is not a common practice. Wet phosphoric acid can be purified via several processes. The most common processes are extraction and precipitation. Via extraction, wet phosphoric acid can be purified up to phosphoric acid comparable to that produced with the thermal process. Separation of specific compounds can be done through precipitation. Additionally, cationic impurities can be removed via precipitation, but the product will then be changed into a phosphate salt, which is unfavorable for its use in industrial applications.

Keywords Reserves · Phosphate rock · Phosphoric acid · White phosphorus

---

M. A. de Boer (✉) · L. Wolzak · J. C. Slootweg
Van't Hoff Institute for Molecular Sciences, University of Amsterdam, Amsterdam, The Netherlands
e-mail: m.a.deboer@uva.nl

H. Ohtake, S. Tsuneda (eds.), *Phosphorus Recovery and Recycling*,
https://doi.org/10.1007/978-981-10-8031-9_5

5.1 Introduction

Phosphorus (P) is an essential element for life, our current food production system, manufacturing industries, and general economic growth (Withers et al. 2015). P is a key limiting factor for plant growth and is therefore one of the three main elements (N:P:K) in fertilizers. The demand for phosphate fertilizers has grown over the last years due to the expanding production of crops, as a result of the rising population, changing human diets resulting in increased (meat) consumption per capita, and also biofuel production. However, phosphorus is a finite resource, and its global distribution is uneven; almost the entire global P demand is supplied by just four countries, namely, China, the USA, Russia, and Morocco, and with three-quarters of the reserves located in the latter (USGS 2017).

Currently, phosphorus is produced from phosphate rock (PR), which is mined and processed into valuable P-containing products using two main production processes. The first method is the production of phosphoric acid for the fabrication of fertilizers via the wet process, which accounts for 95% of the global P demand. The second process converts phosphate rock into elemental (white) phosphorus (P_4), which is then used to produce phosphorus compounds for industrial purposes, including feed additives, detergents, flame retardants, pesticides, and medicines. Both methods will be discussed in this chapter, as well as the application of phosphorus-containing products, preceded by giving a geopolitical assessment of the reserves of phosphate rock.

5.2 Reserves and Production of Phosphate Rock: A Geopolitical Assessment

In the beginning of the nineteenth century, Liebig founded a theory that stated that nutrients, such as nitrogen and phosphorus, were crucial elements to grow crops and that these nutrients were circulating between living and dead material (von Liebig 1840). During the same time, the value of guano was discovered. This phosphorus-containing matter originates from bird droppings deposited over several millennia. Guano soon became a valuable world trade material; however, this phosphate resource was limited and declined at the end of the nineteenth century (Stewart et al. 2005). Soon other sources of phosphate were discovered, most notably phosphate rock deposits that appeared to be of unlimited quantity and soon became the main resource for phosphorus. However, high-grade phosphate rock reserves take millions of years to form via the decomposition of (marine) organisms, and with the current mine production, significantly more is being consumed than is geologically replaced. Currently, it is uncertain how large the global phosphate rock reserves are, and it is thus unclear when these will be depleted. The estimated time of depletion varies greatly among different studies, ranging from 40 to more than 400 years (Van Kauwenbergh 2010). Therefore, several researchers have focused on taking stock of the current global reserves and the quality of these.

5.2.1 *Reserves*

There are three types of phosphate rock, namely, sedimentary rock, igneous rock, and metamorphic rock. Since each of these ore types differs in chemical composition, for example, the phosphate content and amount of impurities, these three types will be discussed individually, and an assessment of the quality of the reserves will be provided.

5.2.1.1 Sedimentary Rock

Sedimentary marine phosphorite is the most common phosphate rock. More than 80% of the produced phosphate originates from sedimentary rock, which is often composed of accumulated fossilized shells and aquatic animals. The most favorable deposits to exploit are frequently found as one or more thick beds with a high phosphate grade that are unconsolidated and have a uniform granular composition with a minimal structural deformation (Van Kauwenbergh 2006). Sedimentary rock contains more than 7.8% P_2O_5 and is called phosphatic sedimentary rock when it contains 7.8–19.5% phosphate and phosphorite if above 19.5% phosphate. The largest sedimentary deposits can be found in Northern Africa, China, and the USA (Table 5.1).

Sedimentary rock consists of the mineral apatite, a calcium phosphate combined with either a hydroxide, fluoride, or chloride ion: $Ca_5(PO_4, CO_3)_3(OH, F, Cl)$, with fluorapatite, $Ca_5(PO_4)_3F$, being the most common (McKelvey 1967). Sedimentary deposits exhibit a wide range of chemical compositions, including pollutants, and great variation in physical form. This results in the formation of several possible by-products during processing, such as fluorine, cadmium, uranium, and rare earth elements. The mineral apatite is sparingly soluble in water. However, the different compositions influence the solubility of the mineral. Incorporation of hydroxide, for example, results in apatite minerals with a higher solubility than apatite containing fluoride (McKelvey 1967).

Table 5.1 Composition of phosphate rock in wt % based on dry material (Diskowski and Hofmann 2000)

Phosphate rock	P_2O_5	CaO	F	SiO_2	Al_2O_3	Fe_2O_3	CO_2	SO_3
Kola (Russia)	39.1	51.5	3.4	2.0	1.2	0.7	0.0	0.0
Phalaborwa (South Africa)	38.4	52.5	2.3	1.9	0.2	0.4	1.3	0.1
Florida (USA)	31.6	47.7	3.9	9.0	1.0	1.6	3.7	1.4
Idaho (USA)	31.2	45.7	3.6	8.5	0.6	0.5	3.1	1.2
Khouribga (Morocco)	34.8	52.5	4.2	0.9	0.5	0.1	4.1	1.5
Togo (Togo)	29.7	40.3	3.9	10.2	4.4	5.3	1.6	0.1

5.2.1.2 Igneous Rock

Igneous rock is formed through the crystallization of cooling lava or magma. Phosphorus concentrations in igneous rocks often range from 0.005% to 0.4% P_2O_5, but higher values (1.5–2.0% P_2O_5) can be found in strongly alkaline, low-silica igneous rocks. Igneous rock mainly consists of carbonatites, minerals consisting of more than 50% carbonate minerals and alkaline intrusions (Bell 1989). An example of a commonly encountered alkaline intrusion is diopside ($MgCaSi_2O_6$). In addition, the minerals calcite and dolomite can also be present in igneous rock (Van Kauwenbergh 2006). These igneous deposits can provide phosphate rock with 35–40% P_2O_5 content via beneficiation methods (often grinding followed by flotation to remove quartz and calcination to remove organic impurities). Examples of crystalline fluorapatite igneous deposit regions are Canada, Brazil, Russia, and Finland.

5.2.1.3 Metamorphic Rock

Metamorphic rock deposits are often deep-lying phosphate deposits, which contain both sedimentary and igneous deposits. Metamorphic rock often contains 0.01–1.3% P_2O_5. These deposits have been subjected to high levels of pressure and heat resulting in a less porous texture, more interlocked crystals, and higher induration compared to igneous and sedimentary rock (Van Kauwenbergh 2006). As a result, exploitation of metamorphic rock deposits is not currently economically viable. Metamorphic rock deposits can be found in China and India.

5.2.1.4 Quality of the Phosphate Rock Deposits

The highest possible apatite (and thus phosphate) content is always the pursuit both for technical and economic reasons (Van Kauwenbergh 2010). Consequently, the ambition is to remove as much of the impurities in phosphate rock as possible. These impurities concern many types of different chemicals. One could distinguish three main types of unwanted impurities in phosphate rock: other non-phosphorus-bearing minerals, heavy metals, and radioactive materials. Other non-phosphorus-bearing minerals need to be removed from PR as much as possible in order to increase the phosphate grade. Heavy metals and radionuclides need to be removed due to health and environmental concerns.

5.2.1.5 Other Minerals

Due to its mineral structure, phosphate rock can contain other sparingly soluble minerals which do not contain phosphorus, such as fluorite (CaF_2), magnesite ($MgCO_3$), and barite ($BaSO_4$). Substitutes for PO_4^{3-} are CO_3^{2-}, As_2O_4, SO_2, SO_4^{2-}, and VO_4^{3-}, while the calcium ion can be substituted by Mg^{2+}, Mn^{2+}, Sr^{2+}, Pb^{2+}, Na^+, U^{3+}, Ce^{3+}, and Y^{3+-} as well as other rare earth elements (McKelvey 1967). The substitutes are not all of the same valency. When anionic substitutes cause a gain or rise in valency, this can be compensated with other cationic substitutes of opposite nature (McKelvey 1967).

The most common minerals in sedimentary phosphate rock are clay minerals, such as kaolinite ($Al_2SiO_5(OH)_4$) and sepiolite ($Mg_4Si_6O_{15}(OH)_2{\cdot}6H_2O$), quartz, and two types of carbonates, calcite ($CaCO_3$) and dolomite ($CaMg(CO_3)_2$). A replacement of PO_4^{3-} by CO_3^{2-} of 25% and of Ca^{2+} by Mg^{2+} of 10% is sometimes found (Corbridge 2000). Carbonates can lead to complications during the processing of phosphate rock, especially during froth flotation (separating the hydrophobic from the hydrophilic minerals), because of the high similarity between carbonates and phosphate minerals (McConnell 1938). A thermal process called calcination can be used to remove the carbonates, but this is economically unfavorable due to the high energy input. The Florida Institute of Phosphate Research points out that a large amount of today's phosphate reserves are left untouched because of a high dolomite concentration. Other common materials found in sedimentary rock are iron, alumina, quartz, montmorillonite, organic matter, and many other elements in trace amounts.

5.2.1.6 Heavy Metals and Radioactivity

Russia is the only country of the four main producers with mainly igneous deposits, while the others have sedimentary PR deposits (Table 5.2). Igneous deposits are mostly low in grade (less than 5% P_2O_5), but both the level of radioactivity and the hazardous metal content (e.g., cadmium, arsenic, and mercury) are low. This is also the case for the igneous deposits in Russia compared to the sedimentary deposits of Morocco, China, and the USA (Table 5.3). A more in-depth comparison of several sedimentary and igneous deposits on cadmium, uranium, and ^{226}Ra contents can be found in Table 5.4. The USA has the highest level of radioactivity, and Morocco relatively has the highest hazardous metal content, which will be discussed in more detail below.

Heavy metals, such as arsenic (As), cadmium (Cd), and uranium (U), are dangerous because of their tendency to bioaccumulate (Mehmood et al. 2009) and therefore be potentially harmful for human health (Kratz et al. 2011). There are several metallic elements present in PR with an up to 10–100 times higher concentration than in soil, such as arsenic, cadmium, mercury, lead, and uranium (Corbridge

Table 5.2 Comparing phosphate deposits of the four major countries in phosphate rock production

	Morocco	China	Russia	USA
Ore type	Sedimentary	Sedimentary	Igneous	Sedimentary
Level of radioactivity	Moderate	Low–moderate	Low	Moderate–high
Hazardous metal contents	Moderate–high	Low–moderate	Low	Moderate

Table 5.3 Comparison of heavy metals in Moroccan and Russian phosphate rock. Average concentration (mg/kg) (Corbridge 2000; Mehmood et al. 2009; Mar and Okazaki 2012)

	Type of heavy metal					
	Cr	Pb	Cd	Zn	Cu	As
Morocco	291	7	30	345	22	11
Russia	23.3	3	0.1	19	30	1

Table 5.4 Average concentration of cadmium (mg/kg), uranium (mg/kg), and ^{226}Ra (Bq/kg) in the major igneous and sedimentary deposits (Davister 1996; Cioroianu et al. 2001; European Commission 2017a; Saueia et al. 2005)

Origin	Cadmium content (mg/kg P_2O_5)			Uranium content (mg/kg)	^{226}Ra content Bq/kg
	Davister (1996)	Cioroianu et al. (2001)	European Commission (2017a)	Saueia et al. (2005)	(Saueia et al. (2005)
Igneous					
Russia (Kola)	<13	0.3	0.25	20	70
South Africa	<13	0.1	0.38		
Sedimentary					
USA	23–166	19.8–125	24–120	80–150	600–1500
Jordan	<30	12.1–28	18	80–110	800–900
Morocco	46–121	17–63	55	100–160	800–1600
Syria	52	13–46	22		
Algeria	60	42–62.6		100–120	700–850
Egypt	74				
Israel	100	81–112	61	80–140	800–1200
Tunisia	137	94	173	30–50	250–350
Togo	162	164–179	147	100–110	950–1000
Senegal (Senegal)	203	165–180.6	221	100–120	950–1100
Nauru	243				

2000). The chemical composition of PR, however, varies largely and depends on its origin. When comparing the most common toxic heavy metals chromium (Cr), lead (Pb), cadmium (Cd), zinc (Zn), copper (Cu), and arsenic (As) in PR from Moroccan and Russian origin, it is clear that the heavy metal content in average Moroccan phosphate rock (sedimentary) is significantly higher than in Russian phosphate rock (igneous) (Table 5.3). For these six heavy metals, phosphate rock in Morocco has up to 2–30 times higher concentrations than in Russian phosphate rock.

The amount of heavy metals is very variable, even within the same deposit. The presence of heavy metals in fertilizers is primarily due to impurities in phosphate rock (Schnug and Lottermoser 2013) that have not been removed during the production process. This creates the risk for these metals to accumulate in the soil and crops and enter the human food chain. However, this risk is not only influenced by the heavy metal content of the fertilizer, as soil properties and the chemical form of the metal also play an important role. An acidic environment, for example, increases the availability of heavy metals in soils for the uptake in crops. Cadmium has the highest transfer factor (cadmium uptake from soil to crop ratio) for any type of crop and is therefore the element with the highest risk of being detrimental to human health in the phosphate fertilizer field (Mehmood et al. 2009). Due to the high toxicity of cadmium, it is classified as a class 1 carcinogen (Group B1) by the International Agency for Research on Cancer and the World Health Organization (WHO) (IARC 1993; WHO 1992). If one examines the cadmium, uranium, and ^{226}Ra concentration in the several deposits, we can see in a single glance that the amounts differ notably depending on origin and are significantly higher in sedimentary phosphate rock mines than igneous phosphate rock mines (Table 5.4). The incoherency between the data of three different references for the cadmium content for the same deposits shows that it does not only differ between the mines but also between the researches. A reason can be the variability within the same deposit or different types of measurement.

The high content of cadmium in some PR deposits (Table 5.4) raised various concerns. As a result, technologies for decadmiation are currently under debate, but have not yet been employed on industrial scale, primarily due to cost issues. Besides this, the current decadmiation processes remove the cadmium from the phosphoric acid and are therefore not a solution for other products for other products, such as the fertilizers SSP, TSP, and NPKs. Due to the numerous, complex factors which can influence the bioaccumulation of cadmium and uptake by crops, it is difficult to quantify the real health risks of the use of P-containing fertilizers since it depends on a lot of factors (Cioroianu et al. 2001), resulting in controversy among researchers, policies, and fertilizer companies. Nevertheless, human exposure to cadmium resulted in an increased risk for several types of cancer and is related to kidney failure, renal failure, and bone demineralization. The diet exposure of average European citizens is close to or slightly exceeding the tolerable intake. People with special diets, like vegetarians, even exceed the tolerable diet with 200% (European Commission 2017a). Several institutes and companies have advocated limits of cadmium, such as 60 mg/kg (Fertilizers Europe), 80 mg/kg (OCP, Timac Agro), and 90 mg/kg (Grupa Azoty, Fetinagro). The European Commission is also debating on setting limits, and a maximum level will be put into practice from 2019.

Phosphate rock contains radionuclides of the uranium (U) and thorium (Th) natural decay series (Saueia et al. 2005). The radioactivity of PR ranges from 70 to 100 Bq/kg ^{226}Ra for igneous PR from Russia and 600 to 1500 Bq/kg ^{226}Ra for sedimentary Moroccan PR (Pérez 2005). On average, 90–95% of the uranium can be found in the phosphoric acid, while almost the entire ^{226}Ra content in phosphate rock ends up in the phosphogypsum (Cioroianu et al. 2001). Every year, 10,000–15,000 tons of uranium is deposited on agricultural land (Cioroianu et al. 2001). Through mining and processing of PR, radionuclides are redistributed into the environment and introduced in products and by-products from the phosphate industry. Naturally occurring radioactive materials (NORM) are found everywhere in our surroundings, and consequently we are exposed to them every day. These materials do not pose threats for human health per se. Only when these materials, such as PR, are concentrated to a certain degree, then they may pose risks to humans and environment.

The concentration of radionuclides (and their respective decay products) in by-products is technically enhanced. Therefore, these by-products are labeled as Technologically Enhanced Naturally Occurring Radioactive Materials (TENORM) (El Afifi et al. 2009). One of these by-products in the phosphate industry is phosphogypsum, which contains calcium sulfate, small amounts of silica, unreacted phosphate rock, radium, uranium, and US EPA toxic metals.

5.2.2 Geopolitics and Reserves

Currently, there is speculation on the amount of global phosphate reserves. The estimated time until depletion differs substantially between the several studies. The time left before depletion will rise according to the "*Qui quaerit, reperit*" (He who seeks, will find) principle. When the prices of these finite sources will rise, previous uneconomic reserves become economically profitable, and more reserves will be discovered. This could be compared to the oil reserves; nowadays oil is extracted from the ocean floor, which was first considered as an uneconomic/unavailable resource. Thus, the volume of phosphate rock reserves is dynamic. The precursor of the USGS (USBM) developed a classification method with which four types of PR deposits could be distinguished:

1. Current economic deposits as reserves
2. Marginally or subeconomic deposits as the reserve base
3. Uneconomic deposits, which have a reasonable potential of becoming economic in the future as resources
4. Deposits with no reasonable prospect of economic viability in the foreseeable future, as occurrences

This classification method is not universally required by law (Edixhoven et al. 2013). The lack of a universal classification method for PR deposits caused speculations on the amount of phosphorus left. An example occurred in 2010, when the reserves and resources as reported by the USGS were reassessed by the International Fertilizer Development Center (IFDC). The USGS reported in 2010 the global PR reserves as being 16,000 Mt (million tonnes), of which 5700 Mt of PR are located in Morocco. In the same year, the IFDC issued a report in which the global reserves were increased to 60,000 Mt PR with an estimation of 51,000 Mt found on Moroccan territory, which is an increase of almost four times within 1 year. In the USGS reports, reserves are defined as upgraded (beneficiated) concentrate. This term is used by the mining industry for upgraded ore that is upgraded to such an extent that it is ready to be sold on the market. The phosphate (P_2O_5) content for this beneficiated ore needs to be at least 30%. However, from the IFDC report, it does not

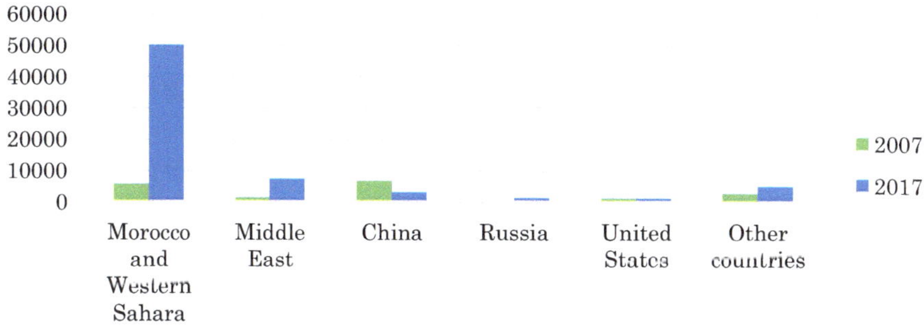

Fig. 5.1 The estimated world reserves in million tonnes according to USGS in 2007 and 2017 (USGS 2007, 2017)

become clear what factor is used to calculate this beneficiated ore back to phosphate reserves. In the report of IFDC, the term "phosphate rock" deposit is used for both unprocessed rock and beneficiated concentrates. Despite the fact that both institutes used different definitions, the USGS has adopted the reserve estimation of IFDC.

The difference in estimated reserves in Morocco can be clearly seen in Fig. 5.1, which shows the estimated world reserves in 2007 and in 2017. Besides Morocco and the Western Sahara, also the estimated reserves in the Middle East significantly increased, while the estimated reserves for China are only half of what has been estimated in 2007. In 2017, the world resources of phosphate rock are estimated around 300 billion tonnes and reserves around 68 billion tonnes, and 75% of the total estimated reserves can be found in Morocco and the Western Sahara (see Fig. 5.1) (USGS 2017).

5.2.3 *Mines in EU*

In 2010, Europe imported 7,518,000 tonnes of phosphate rock (De Ridder et al. 2012), and the remaining ~10% of the current European phosphate rock demand is produced in Europe, in Finland. The Siilinjärvi phosphate mines in Finland are currently mining phosphorus at a rate of 1,000,000 tonnes per annum and are expected to be operational until 2035. Several other phosphate deposits are located in Europe, with the largest in Turkey, and minor deposits in, for example, Belgium, Germany, Spain, France, Italy, and Greece. However, these are no longer exploited. The Terres Rouges Mine in Lorraine was the last phosphate mine in France and was closed in 1998, due to the increasing criticism of environmental damage. The UK also has significant phosphate resources in former tin mines and widely dispersed deposits, but these are not economically viable to mine. Belgian and Italian phosphate mining plans received similar criticism. Greek phosphate mines, although functional for a short time, are deemed as economically unviable due to the scattered phosphate resources and the low grade of 2% (Notholt et al. 1989).

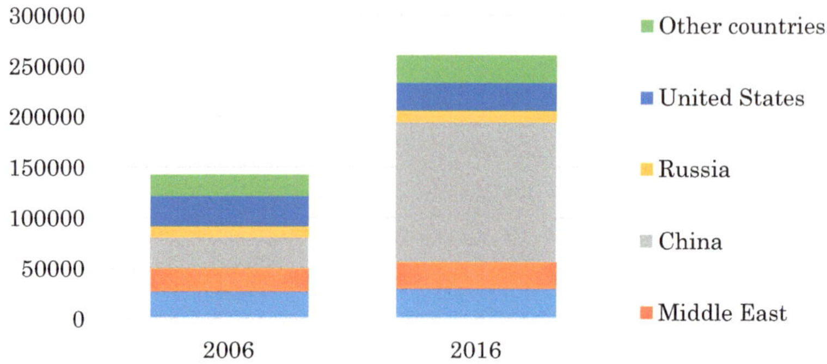

Fig. 5.2 The world production of phosphate rock in 2006 and 2016 (USGS 2017)

5.2.4 Phosphate Rock Production

The mining of phosphate rock is mainly concentrated in four countries, namely, Russia, the USA, China, and Morocco (including the Western Sahara). Together, they accounted for 79% of the global phosphate rock production in 2016 (USGS 2017). Compared to 2006, China increased their production by 450% from 30,700 to 138,000 tonnes, primarily to keep up with the growing domestic demand in agriculture. The Middle East increased their production from 22,870 in 2006 to 26,300 tonnes in 2016. Mainly due to the Arabic spring, changes in the production size occurred in the Middle East. Syria produced 3600 tonnes in 2006, but in 2016 the production came to a complete halt due to the civil war. Moreover, Tunisia lowered their production from 8400 in 2006 to 3500 tonnes in 2016. Other countries in the Middle East have increased their production, such as Egypt, which doubled their production, and Jordan increased their production from 6400 to 8300 tonnes in these 10 years (Fig. 5.2).

The world consumption of phosphate rock in 2016 was 44.5 million tonnes and is expected to increase to 48.9 million tonnes in 2020. The estimated global phosphate rock reserves contain around 68 billion tonnes of phosphate rock, which indicates that there is no urgent geological scarcity in the coming decades.

5.2.5 Price Volatility

The price of phosphate rock was stable until 2008, when suddenly a price peak occurred. The price for 1 tonne of phosphate rock was around 50 dollars before 2008, but after the price peak in 2008, prices rose over 800% (De Ridder et al. 2012). After this price peak, the price dropped but is still on average more than two

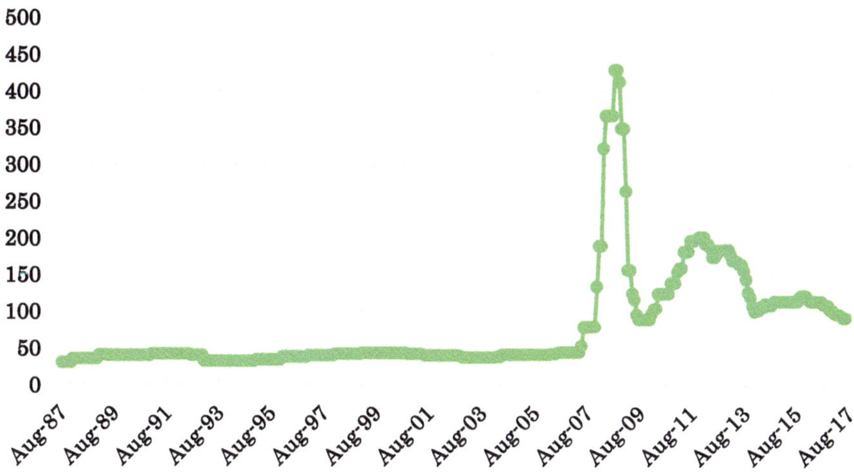

Fig. 5.3 The price for phosphate rock from August 1987 to August 2017 in dollars (Indexmundi 2017)

times higher than prior to 2008. Besides this, with increasing scarcity of P, countries are likely to implement protectionist policies to protect their reserves, which can subsequently increase the price (De Ridder et al. 2012). China, for example, has added phosphate as third on the list of most important strategic resources and protects the domestic demand via export quotas on phosphate fertilizers. In 2008, China imposed an export quota of 135% on phosphate rock, which is one of the reasons of the price peak in 2008 (Fig. 5.3).

5.2.6 European Critical Material List

The European Union is highly dependent on P import. The main suppliers of phosphate rock in 2011 of Europe were Morocco, Russia, and Algeria (De Ridder et al. 2012). According to the European Commission, the management of P is of critical importance for the development of circular economies and sustainable development. For this reason, P was added to the list of the 20 critical raw materials for Europe, where materials are rated on economic importance for the EU and the possible supply risk (European commission 2014). Phosphate rock has been considered as having a high supply risk since the high corporate concentrations in production, meaning a small amount of producers (four countries, China, Morocco, the USA, and Russia) all with a large market share.

5.3 Phosphate Processing

Following mining, phosphate rock is primarily processed to intermediary products including phosphoric acid (PA) and white phosphorus (P_4), which together form the basis of the industrial phosphorus chemistry. All the production processes from phosphate rock are given in Fig. 5.4. The reduction of phosphates to elemental phosphorus at elevated temperatures is the route to many high-grade phosphorus products (Corbridge 2000), while the production of phosphoric acid through the wet processes is closely related to the manufacturing of fertilizers. The wet process is the common treatment for the production of phosphate fertilizers and the production of technical phosphoric acid. About 95% of all the phosphoric acid is produced through the wet process, since this is the route toward readily available phosphorus-based fertilizers (Schrödter et al. 2008). In this process, strong mineral acids react with the phosphate ore, where in most cases sulfuric acid is used and on a smaller scale nitric acid or hydrochloric acid. The resulting product is wet phosphoric acid. The type of added acid in the wet process, possible purification steps, and the addition of other macronutrients (for fertilizer production) determine the final type and quality of the product.

An alternative route is the thermal process where crude phosphate, coke, and silica are heated to over 1500 °C in an electrical resistance furnace to produce elemental phosphorus (P_4), the starting material of many high-grade phosphorus compounds in food and industrial applications (Gantner et al. 2014). This process is very energy intensive, and for the production of various compounds, hazardous chemicals like chlorine are required as intermediates. On the upside, however, the product has nearly no impurities. The wet process to obtain several qualities of phosphoric acid and the thermal process to obtain white phosphorus will be both described in more detail below.

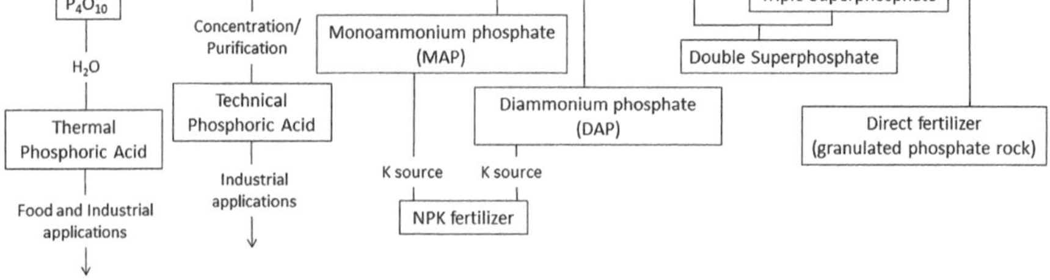

Fig. 5.4 An overview of the production processes for white phosphorus, phosphoric acid, and common fertilizers

5.3.1 *Wet Process*

The most commonly used method to produce phosphoric acid is via the wet process using sulfuric acid (H_2SO_4) for the acidulation of phosphate rock. The wet process consists of two basic steps: digestion of phosphate rock in a reactor and filtration of the created gypsum ($CaSO_4$). A third step can be added to reduce the liquid content afterward and purify the product. In the first step, fluorapatite ($Ca_5(PO_4)_3F$) is digested with sulfuric acid, resulting in phosphoric acid (H_3PO_4) and the main by-products phosphogypsum ($CaSO_4$) and hydrogen fluoride (HF).

Different types of digestion processes exist for phosphate rock, namely, the dihydrate process (DH), the hemihydrate process (HH), and combinations of the two (DH/HH and HH/DH). The most applied technology for the production of phosphoric acid, the classical dihydrate process, can be operated at large volumes. It does not have corrosion issues due to the low reaction temperature of 70–80 °C and is applicable with most grades of phosphate rock (Schrödter et al. 2008). However, the quality of the phosphogypsum ($CaSO_4\ 2H_2O$) is low, and the impurities are high, which disqualifies it as a precursor in other industries. The need for a more concentrated stream of phosphoric acid, less impurities, and reduced losses of phosphorus to phosphogypsum led to the development of the hemihydrate (HH) process. This process works at an elevated temperature of 100 °C, which gives rise to the formation of calcium sulfate hemihydrate, $CaSO_4\ 0.5H_2O$, a very pure gypsum and a higher concentration of phosphoric acid, 40–48% of P_2O_5 with a yield of 90–94% (Schrödter et al. 2008). The HH process is not often used, due to the low process yield, but a combination of the DH and HH process is used, for example, in Japan because of the advantage of the production of pure gypsum for the manufacturer of plasterboard and cement. Nevertheless, the traditional dihydrate process remains the primary dihydrate wet process to obtain phosphoric acid, with more than 90% of the production of the wet phosphoric acid (Du et al. 2014).

During the digestion step, phosphogypsum is formed: a thick slurry of solid particles of gypsum ($CaSO_4$), unreacted phosphate rock, and other impurities. The second step involves filtering out this solid phosphogypsum in a partial vacuum resulting in a 28–31 wt% P_2O_5 (50% phosphoric acid) yield. The remaining acids on the solids can be washed off and used again in the process.

The concentration of 28–32 wt% P_2O_5 is too low to be used as a fertilizer and has to be concentrated to 40–55% (or even 70% if superphosphate is the desired product). Therefore, the dihydrate process can be followed by reducing the liquid content via evaporation with submerged combustion burners and vacuum circulation evaporators resulting in merchant-grade phosphoric acid, 54% P_2O_5 (70% phosphoric acid) (Table 5.5) (Schrödter et al. 2008).

By-products of this concentration process are SiF_4 and HF. When concentrating from 30% to 55% P_2O_5, 50–60% of the hydrogen fluoride evaporates. Emissions to the atmosphere are prevented by reacting the gaseous HF with a by-product SiF_4 to form H_2SiF_6, which is a liquid at room temperature (boiling point 108.5 °C). Roughly 90–95% of the HF is captured this way, which complies with the fluorine emission limits.

Table 5.5 Composition of wet phosphoric acid in wt% (Schrödter et al. 2008)

P_2O_5	30.0
F	2.17
SO_3	2.1
Ca	0.45
Si	0.41
Mg	0.13
Fe	0.08
Al	0.04
Cr	0.022
Zn	0.019
V	0.013
Ni	0.0034
Cu	0.0027
Mn	0.0009
As	0.0007
Pb	0.00002

The by-product phosphogypsum is produced in large quantities and leads to enormous amounts of waste. For the production of 1 tonne of phosphoric acid, roughly 5 tonnes of phosphogypsum is produced, resulting in an estimated production of 100–280 Mt per year (Tayibi et al. 2009). In theory, phosphogypsum could be a source of gypsum for construction works, but the presence of trace metals, such as thorium, radium, cadmium, and uranium, prevents extensive use and turns it into a material classified as hazardous (Ayres et al. 2001). About 85% of the phosphogypsum is disposed without any treatment, which poses storage problems and provokes environmental concerns (Tayibi et al. 2009). The gypsum produced via the HH/DH and DH/HH process can be used in the cement or plaster industry. Depending on the digestion process, the gypsum formed can be a dihydrate ($CaSO_4$ $2H_2O$), a hemihydrate ($CaSO_4$ $0.5H_2O$), or a mixture of the two. Anhydrite ($CaSO_4$ without any H_2O) can be formed, but the required conditions (120–130 °C) cause excessive corrosion problems, which makes it an insufficient method for industrial processes.

5.3.2 *Purification Process of Wet Phosphoric Acid*

Many of the impurities in the phosphate rock are still present in the phosphoric acid product. The amount of impurities in the phosphoric acid and phosphogypsum is not only dependent on the type of digestion process but also on the origin of the phosphate rock. The phosphoric acid produced with phosphate rock from Russia contains 5–10 mg/L uranium compared to 120–160 mg/L in the phosphoric acid from Morocco. The amount of ^{226}radium is even negligible in the phosphoric acid

Table 5.6 Uranium and radium concentrations in phosphoric acid and phosphogypsum (Cioroianu et al. 2001)

Origin phosphate rock	Phosphoric acid		Phosphogypsum	
	Uranium mg/L	^{226}Ra Bq/L	Uranium mg/L	^{226}Ra Bq/L
Igneous rock				
Russia (Kola)	5–10	–	–	60–100
Sedimentary Rock				
Florida	120–140	40–70	10–20	500 1200
Jordan	80–100	30–50	5–10	500–1000
Morocco	120–160	30–60	10–15	600–1300
Israel	90–110	40–50	10–20	600–1200
Tunisia	40–60	40	5–8	300–400
Togo	100–110	60	10	700–1100

produced with the PR from Russia, compared to 30–60 Bq/L from Morocco (see Table 5.6). Most of the uranium is taken up by the phosphoric acid, while most of the ^{226}radium ends up in the phosphogypsum. Roughly 87–92% of the uranium ends up in the phosphoric acid and 8–13% in the phosphogypsum. In the case of ^{226}Ra, it is the other way around; 5–10% ends up in the phosphoric acid and 90–95% in the phosphogypsum.

Impurities in phosphoric acid can be partially or completely removed by extraction and precipitation. Typically, the wet phosphoric acid contains impurities such as sulfites, fluorine, silicon, and calcium. The full list and general quantities can be found in Table 5.5 (Schrödter et al. 2008). Purification is a major challenge and can be performed via precipitation, ion exchange, adsorption, membrane technologies, nano-filtration, reverse osmosis, and extraction processes (El-Asmy et al. 2008). The two most common purification processes for phosphoric acid, extraction and precipitation, will be discussed below in more detail.

5.3.2.1 Extraction

Different extraction processes have been designed to remove various impurities, ranging from the removal of specific compounds to almost complete purification. Extractive purification processes of wet phosphoric acid contain the following steps: purification of the crude wet phosphoric acid, phosphoric acid extraction, extract wash, phosphoric acid recovery, and, if necessary for the final product, a post treatment. Firstly, in a pre-purification process, arsenic, sulfate, and organic components are removed by precipitation or adsorption from the crude wet phosphoric acid (40–50% P_2O_5). The pre-purified phosphoric acid is then dissolved in an organic solvent (such as methyl isoamyl ketone and diisopropyl ether) through countercurrent distribution. While most impurities remain in the aqueous phase, the phosphoric acid ends up in the organic phase. Additionally sulfuric acid (H_2SO_4) can be used to react with cationic impurities to form sulfates, which remain in the aqueous phase when a yield of 95–98% of phosphoric acid is preferred. After treatment, the purity of phosphoric acid is comparable with the food-grade phosphoric acid produced via the thermal process.

5.3.2.2 Precipitation

Separating arsenic and to an increasing frequency cadmium (decadmiation) from wet phosphoric acid is crucial when it is used in the fertilizer industry. Arsenic can be removed by addition of Na_2S to the acid and filtration of the precipitated arsenic sulfide. Cadmium can be removed after adding complexing agents, such as dialkyl dithiophosphoric acid ester (Von Plessen and Schimmel 1987). If necessary, cationic impurities (e.g., Fe, Al, Mg, Ca) can be removed by neutralizing the acid with caustic soda, sodium hydroxide. This will however change the product in a phosphate salt solution and metal hydroxides instead of an acid, thereby limiting its use to only detergent phosphates.

5.3.2.3 Acidulation with Hydrogen Chloride or Nitric Acid

The other two types of acidulation processes of phosphate rock besides sulfuric acid are with the use of hydrochloric acid and nitric acid. The nitrophosphate process is an alternative route to form phosphoric acid with the significant advantage of the formation of the marketable calcium nitrate instead of phosphogypsum. But, due to the price difference between sulfuric acid and nitric acid, only 10% of all the fertilizers are produced via the wet process with nitric acid (Schrödter et al. 2008). Another process for the treatment of phosphate rock without the production of phosphogypsum involves HCl, the solvent extraction process (SX process), which produces calcium chloride and hydrogen fluoride as by-products via the following chemical reaction:

$$Ca_5(PO_4)_3F + 10HCl \rightarrow 3H_3PO_4 + 5CaCl_2 + HF$$

The phosphoric acid is extracted and concentrated via distillation to obtain food-grade phosphoric acid. The advantage of the HCl method is the high concentration and low impurity of the product (>50 wt% P_2O_5), but the disadvantage is the relatively high cost. The wet process (with H_2SO_4) and the SX process can be combined via mixing both slurries and extracting the phosphoric acid with an aliphatic alcohol afterward.

5.3.3 *White Phosphorus*

Elemental phosphorus has ten allotropic forms, but only two of them are industrially important, namely, white (P_4) and red phosphorus (in an amorphous network). Examples of other allotropic forms are diphosphorus (P_2), violet phosphorus (in a tube structure), and black phosphorus (a puckered sheet structure) (Corbridge 2000). White phosphorus is extremely reactive with air and is therefore stored in heated vessels (so P_4 remains a liquid) and always covered with water and on top of the water nitrogen to exclude any contact with air. Red phosphorus is less reactive and stable in air and is, for example, used as the striking surface of a matchbook. White phosphorus, P_4, accounts for 99% of the global elemental phosphorus demand with 850,000 tonnes per year and is required for the production of important organophosphorus derivatives such as PH_3, PCl_3, P_4S_{10}, P_4S_3, and P_2O_5 (Kongshaug et al. 2004). White phosphorus is the starting material of almost all high-end industrial and food products containing phosphorus, such as flame retardants and acidification of soft drinks.

The industrial manufacturing of white phosphorus is primarily done via the Wöhler process, in which phosphate rock is mixed with coke and silica and heated by electrodes. White phosphorus is commonly produced from phosphate rock (apatite) and to a lesser extent from guano, manure, and recycled phosphates. Overall, the Wöhler process proceeds according to the following equation:

$$4Ca_5(PO_4)_3F + 18SiO_2 + 30C \rightarrow 3P_4 + 30CO + 18CaSiO_3 + 2CaF_2$$

The first step in this process is to reduce the size of the rock in order to avoid blocking of the furnaces (Schipper et al. 2001). Carbon monoxide provides 85–90% of the heat needed for roasting the pellets after the sintering process, and the dust is removed through scrubbing and recycled into the process. The pellets are mixed with cokes, which act as a reducing agent, and SiO_2 pebbles, a liquid slag, which has its use in road construction and other applications. This mixture is heated to 1500 °C in an electric resistance furnace. At these temperatures, silica is a strong acid that is able to release phosphate from fluorapatite and form molten calcium silicate that binds fluoride, yielding CaF_2. In addition, some of the silica also combines with fluoride to form SiF_4 gas. In the furnace, the phosphates are reduced to diphosphorus. The furnace gas, containing the P_2, is cleaned with electrostatic precipitators to remove dust. To obtain P_4, two condensation steps, above and below the boiling point of phosphorus (280 °C), are performed to condense the P_2 vapors to solid P_4.

The production of P_4 is energy intensive with a consumption of 12.5–14 MWh per metric tonne of white phosphorus. Around 55% of the energy is lost as heat in the slag, ferrophosphorus, furnace gasses and radiation, and electrical and cooling losses (Gilmour 2014). The furnace CO gases, after P_2 removal, containing less than 0.1% of the phosphorus product, are used as fuel for power generation that is needed for the sintering process. Applications for the furnace slag are limited, since it can contain radionuclides out of the phosphate rock. The water used in the phosphorus condenser and for storage contains a substantial amount of phosphates, up to 1 g/L that can be recovered via filtration. P_4 gas (P_2 as a precursor in between) is formed along with CO, some dust, and calcium oxide. The P_4 gas leaves the furnace and is subsequently condensed.

Iron impurities in the phosphate rock are reduced in the furnace and form a separate slag: ferrophosphorus. Ferrophosphorus contains about 75% iron and 25% phosphorus with trace amounts of other metals and can be used as an additive in steel manufacturing. The radioactivity of the ferrophosphorus is constantly measured, and depending on the level of radioactivity, it can be decided if it has to be stored (temporarily or indefinitely depending on the radioactivity) or sold.

The oxidation of P_4 followed by hydrolysis to yield P_2O_5, used for the production of thermal phosphoric acid, is the most prominent application of P_4 with 70% of the production volume. This production of thermal phosphoric acid from P_4 contains the two steps:

$$P_4 + 5O_2 \rightarrow P_4O_{10}, \Delta H = -3053 \text{ kJ / mol}$$
$$P_4O_{10} + 6H_2O \rightarrow 4H_3PO_4, \Delta H = -377 \text{ kJ / mol}$$

The phosphoric acid that is obtained is very pure due to the use of elemental phosphorus as a starting material. Its H_3PO_3 content is no more than 0.1%, and it barely contains traces of impurities. Only arsenic (5–30 ppm) subsequently needs to be removed via the precipitation of arsenic sulfide by the addition of hydrogen or sodium sulfide and filtration. This is necessary, since arsenic replaces phosphorus in the phosphorus lattice resulting in an almost impossible separation on an industrial scale.

The industrial production of white phosphorus in Europe disappeared in 2012 with the bankruptcy of Thermphos B.V. The current P_4 producers worldwide are located in the USA, China, Kazakhstan, and Vietnam (Table 5.7). The producer with the biggest capacity is located in China, with a yearly capacity of 1900 kt P_4. All the P_4 produced in the USA is used for domestic consumption, while just 40% of the Chinese production is used for domestic use. China also protects their internal demand with an export tariff of 20% on white phosphorus (Gantner 2015).

Currently, white phosphorus is exclusively produced out of phosphate rock, but several companies investigated the process to produce white phosphorus out of waste products too. Thermphos had developed a method to produce P_4 out of sewage sludge ashes, but this never went into practice. In 2014, a new technology called RecoPhos has been developed to produce P_4 from sewage sludge ashes. This technology was acquired in 2016 by ICL, Israel Chemicals Ltd., a multinational manufacturing specialized in chemicals for agriculture, food, and engineered material purposes.

Table 5.7 Overview of the current producers of white phosphorus and their characteristics (Gantner 2015)

		USA	China	Kazakhstan	Vietnam
Capacity in kt P_4		80	1900	120	45
Production in kt P_4		80	750–900	80–120	38
P	PR-supply	Self-supply			
	Labor cost (monthly average in USD)	High (4900–5100)	Low (650–750)	Medium (1000–1100)	Very low (200–300)
Utilities	Coal (USD/tonne)	High (70–80)	High (105–115)	Low (60–70)	Very low (25–35)
	Electricity (cents/kWh)	Medium (6–7)	Low (4–5)	Low (4–5)	Medium (6–7)
	Transportation	Convenient	Inconvenient	Inconvenient	Convenient
Domestic consumption in kt P_4		80	750	60	k.A.
Export tariff		–	20%	–	

5.4 Industrial Applications of Phosphoric Acid and White Phosphorus

Most phosphoric acid is used for fertilizers but also in a broad range of industrial and food products (Gantner et al. 2014). Different concentrations and impurity levels of phosphoric acid are required for diverse purposes, resulting in various production methods. For fertilizers, merchant-grade phosphoric acid, with a concentration of 55% P_2O_5, 74% H_3PO_4 (phosphoric acid) is the standard. As for food purposes, food-grade phosphoric acid with an 85% concentration of H_3PO_4 is commonly used. The 85% food-grade phosphoric acid is a viscous and corrosive liquid and must therefore be stored in stainless-steel containers with a resistant inner lining. The different forms of commercial phosphoric acid and the concentrations and impurity levels in ppm are listed in Table 5.8.

The main application of wet phosphoric acid is in fertilizers. With this process, the main fertilizers used in agriculture are diammonium phosphate (DAP), monoammonium phosphate (MAP), single superphosphate (SSP) and triple superphosphate (TSP). DAP and MAP are made via the wet process and SSP via treatment of phosphate rock with sulfuric acid and TSP with phosphoric acid (see Fig. 5.4). To produce MAP and DAP, an ammonium source is added to the wet phosphoric acid and a potassium source when NPK-fertilizers are the desired final product. The nutrient ratio of the four fertilizers can be found in Table 5.9. The nutrient grade ratio of the fertilizers differs among the four fertilizers and it depends on the demand of the crop and soil type which of these fertilizers suits best.

Table 5.8 Commercial phosphoric acid is sold in different forms (Gilmour 2014)

Acid group	Concentration in P_2O_5 (%)	Concentration in H_3PO_4 (conversion factor 1.381) (%)	Impurity level in ppm
Filter acid (dihydrate process)	28	39	5.000–25.000
Fertilizer/merchant-grade acid MGA (concentrated)	42–54	74	10.000–50.000
PWA feed acid (green acid)	54–59	81	5.000–25.000
Raffinate acid	25–45	62	20.000–50.000
Technical acid	50–61.6	76	500–5.000
Food-grade acid	61.6	85	0,5–250
Cola/pharma acid	61,6	85	0,5–100
Thermal (from P_4)	61,6	85	
SEMI LCD acid	61,6	85	0,1–1
Semiconductor grade acid	61,6	85	0,01–0,1

Table 5.9 Common phosphate fertilizers and nutrient grades (Van Kauwenbergh 2006)

Fertilizer	Nutrient grade ratio, N:P:K
Diammonium phosphate (DAP)	18-46-0
Monoammonium phosphate (MAP)	10-50-0
Single superphosphate (SSP)	0-18-0
Triple superphosphate (TSP)	0-46-0

The demand for fertilizers is still emerging due to the rising population, the upcoming biofuel demand, and the changing standard of living in developing countries (Fig. 5.5). The global fertilizer P_2O_5 production steadily increased from 1973 to 2011 with 166% from 25.6 to 42.6 Mt, with a downward trend in the 1990s. This trend was mainly a result of the fall of the USSR.

Different applications for each of the phosphoric acid types exist, such as an acidifying food additive (i.e., in soft drinks), rust remover for metals, etching agents, and catalysts. An overview of all the applications of phosphoric acid can be found in Fig. 5.6 (Corbridge 2000).

Both (highly purified) merchant-grade phosphoric acid and thermal phosphoric acid derived from P_4 can be used. Which one is suitable depends on the cost and level of purity requirements. Recent numbers of industrial phosphate demand do not exist, but in the most recent research in 2015, they assume that around 38% of the industrial phosphate demand is for detergents and cleaners (Fig. 5.7). Currently there is a ban in several parts of the world for the use of phosphates in detergents to reduce the amount of phosphates ending up in surface waters and lakes, thereby resulting in eutrophication (European Commission 2017b). This ban has resulted in a decline of the use of phosphates in detergents in the last decade. Besides the application in detergents, around 23% of the phosphoric acid demand is for the production

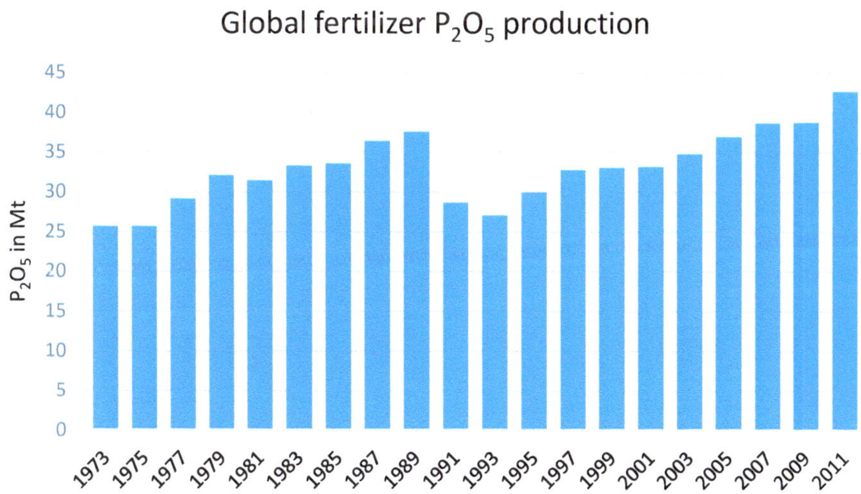

Fig. 5.5 The global fertilizer P_2O_5 production from 1973 to 2011 (IFA 2014)

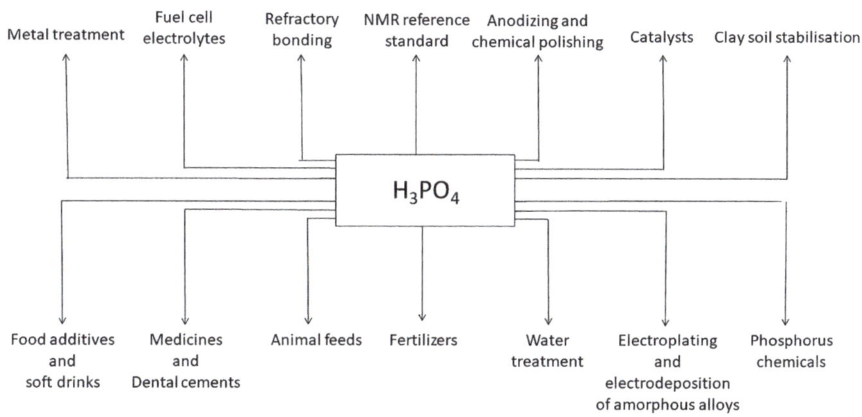

Fig. 5.6 Overview of the use of phosphoric acid

of food additives, 10% in metal treatment, 10% in water treatment, 3% in toothpaste, 2% in special fertilizers, and the other 14% used for other purposes like medicines and fuel cell electrolytes. For the segments beside detergents, a moderate growth is expected for the coming years.

The highest industrial phosphate demand in 2011 is in North America (19%, 40.000 tonnes), Western Europe (17%, 36.000 tonnes), and East Asia, particularly China (15%, 32.000 tonnes) (see Fig. 5.8).

Besides phosphoric acid, white phosphorus is a precursor for many inorganic phosphorus compounds (Fig. 5.9). White phosphorus has to be the starting material, since this cannot be prepared otherwise. White phosphorus has a small direct application as a chemical weapon, but it is mainly a precursor for several organophosphorus

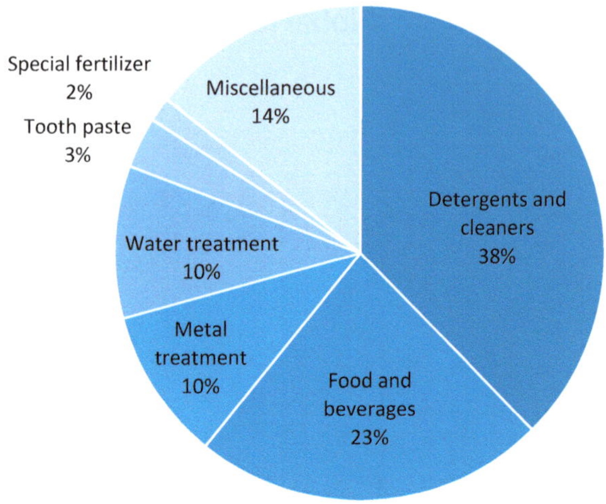

Fig. 5.7 Demand for industrial phosphates in tonne P_2O_5 by application area in 2011 (Gantner 2015)

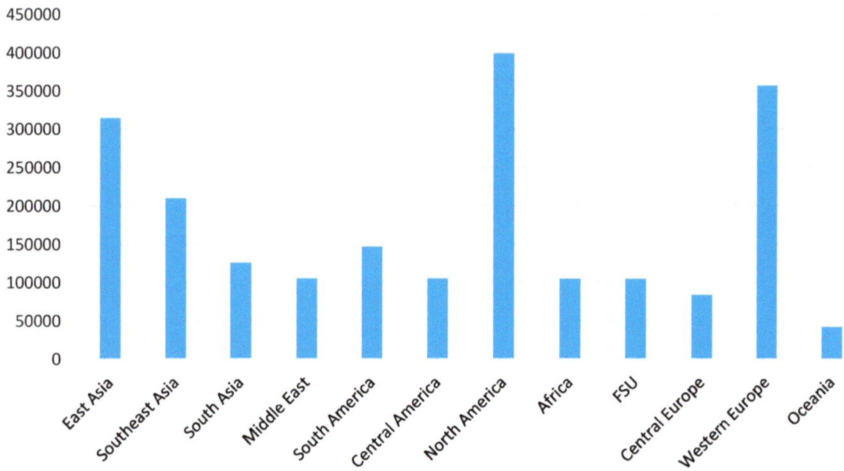

Fig. 5.8 Demand for industrial phosphates in t P_2O_5 in 2011 (Gantner 2015)

compounds. Organophosphorus compounds derived from white phosphorus are PCl_3, red P, P_xS_y, PH_3, NaH_2PO_2, and ferrophosphorus (Gantner et al. 2014). The derivative from P_4 with the largest market share is PCl_3, which, for example, is used as a derivative for making glyphosate (a pesticide). Around 1% of the global P usage is for the production of glyphosate. Examples of other phosphorus-containing products manufactured from P_4 are catalysts, battery electrolyte ($LIPF_6$), flame retardants, vitamin A, plastic additives, doping, matches, antioxidants, water treatment, and reducing agents (Gantner et al. 2014).

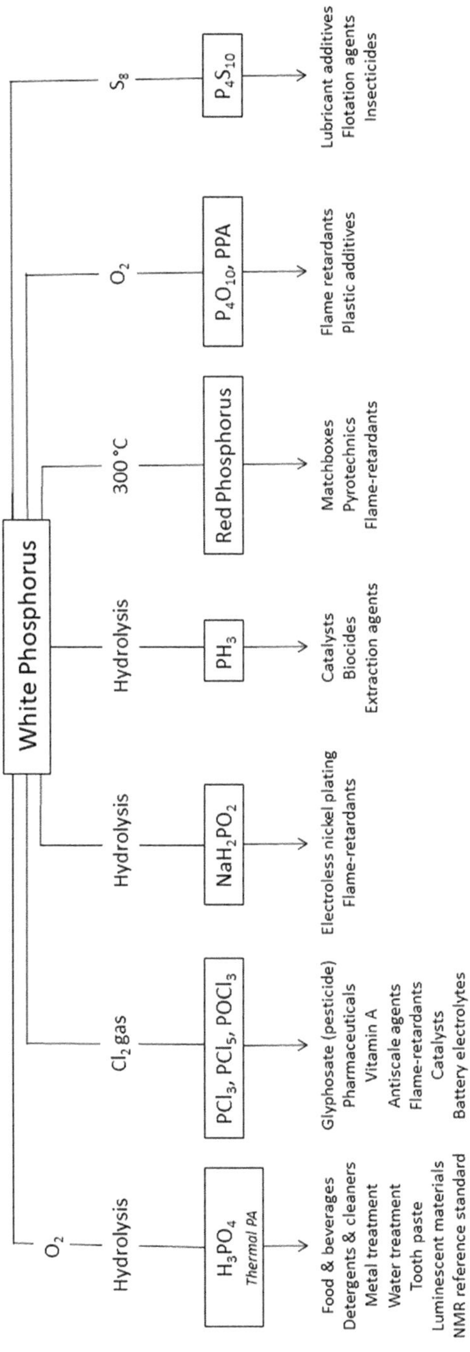

Fig. 5.9 The production routes and applications of the white phosphorus derivatives

5.5 Conclusion

The world consumption of phosphate rock in 2016 was 44.5 million tonnes and will increase to 48.9 million tonnes in 2020. The estimated global phosphate rock reserves contain in total around 68 billion tonnes of phosphate rock. This indicates that there is no urgent scarcity in the coming decades. Yet, reserves are not equally spread around the world, with three-quarters located in Morocco and the Western Sahara. Moreover, the price for phosphate rock can be volatile, and Europe is reliable for 92% on import (see Fig. 5.3); therefore, phosphate rock is added to the critical material list of the European Commission.

The reserves located in Russia mainly contain igneous ore and relatively have the lowest levels of radioactivity and hazardous metal contents. Morocco and the Western Sahara reserves, mainly sedimentary ore, have on average the highest hazardous content out of the four main phosphate rock producers. Moreover, the amount and type of other mineral structures besides apatite are important to assess the suitability for phosphate production.

Phosphate rock is mainly processed as intermediary products including phosphoric acid and white phosphorus. About 95% of the phosphoric acid is made via the wet process, which includes acidulation of phosphate rock for creating wet phosphoric acid and by-products including phosphogypsum and hydrogen fluoride. Five times more phosphogypsum as phosphoric acid is produced during this process. The purity and thus the reusability of phosphogypsum are dependent on the type of digestion process and can be as pure as necessary for reuse in the plaster industry. However, at the moment reusing phosphogypsum is not yet a common practice.

Wet phosphoric acid can be purified via several processes. The most commonly used processes are extraction and precipitation. Via extraction, phosphoric acid comparable to phosphoric acid produced through the thermal process can be obtained. Separation of specific compounds can be done through precipitation. Also, cationic impurities can be removed via precipitation, but the product will then be changed into a phosphate salt which is unfavorable for its use in many industries besides detergents. Production of white phosphorus is primarily done through the Wöhler process. The production of white phosphorus is energy intensive, but the amount of impurities is low, which is necessary for certain applications such as in the food and high-tech industry.

References

Ayres RU, Holmberg J, Andersson B (2001) Materials and the global environment: waste mining in the 21st century. MRS Bull 26:477

Bell K (1989) Carbonatites: genesis and evolution. Unwin Hyman, London

Cioroianu TM, Bunus F, Filip D, Filip G (2001) Environmental considerations on uranium and radium from phosphate fertilizers. IAEA-TECDOC, Intern. At Energy Agency, Vienna 1244:215–224

Corbridge D (2000) Phosphorus 2000. Elsevier Science B.V, Amsterdam
Davister A (1996) Studies and research on processes for the elimination of cadmium from phosphoric acid. OECD, Paris, pp 21–30
De Ridder M, De Jong S, Polchar J, Lingemann S (2012) Risks and opportunities in the global phosphate rock market. The Hague, The Hague Centre for Strategic Studies. ISBN/EAN: 978-94-91040-69-6. Rapport No 17 | 12 | 12
Diskowski H, Hofmann T (2000) Phosphorus. In: Ullmann's encyclopedia of industrial chemistry; Wiley-VCH Verlag GmbH & Co.: Weinheim, 2012. 726–745
Du H, Chen A, Yang L (2014) Research in dihydrate calcium sulphate crystallization during the production of phosphoric acid by wet process. J Mat Sci Chem Eng 2:7–11
Edixhoven JD, Gupta J, Savenije HHG (2013) Recent revisions of phosphate rock reserves and resources: reassuring or misleading? An in-depth literature review of global estimates of phosphate rock reserves and resources. Earth Syst Dyn Discuss 4:1005–1034
El Afifi EM, Hilal MA, Attallah MF, El-Reefy SA (2009) Characterization of phosphogypsum wastes associated with phosphoric acid and fertilizer production. 100, 407–412
El-Asmy AA, Serag HM, Mahdy MA, Amin MI (2008) Sep Purif Technol 61(3):287–292
European commission (2014) Report on critical raw materials for the EU. Report on the Ad hoc working group on defining critical raw materials. Ref. Ares(2015)1819503
European Commission (2017a) Cadmium. Cadmium in food. Online available at https://ec.europa.eu/food/safety/chemical_safety/contaminants/catalogue/cadmium_en, last retrieved at 04/09/2017
European Commission (2017b) EP supports ban of phosphates in consumer detergents. European Commission Press release. Online available at http://europa.eu/rapid/press-release_IP-11-1542_en.htm, last retrieved at 05/09/2017
Gantner O (2015) Ressourcenstrategische Betrachtung der Kritikalität von Phosphor. PhD thesis, Universität Augsburg
Gantner O, Schipper W, Weigand JJ (2014) Technological use of phosphorus: the non-fertilizer, non-feed and non-detergent domain. In: Scholz RW, Roy AH, Brand FS, Hellums DH, Ulrich AE (eds) Sustainable phosphorus management–a sustainable roadmap. Springer, Berlin
Gilmour R (2014) Phosphoric acid. Purification, uses, technology, and economics. CRC Press, Boca Raton
IARC (1993) Beryllium, cadmium, mercury, and exposures in the glass manufacture industry, IARC monographs on the evaluation of carcinogenic risks to human and their supplements. 58
IFA (2014) Phosphate rock and phosphoric acid statistics_detailed figures. International Fertilizer Industry Association. Online available at http://www.fertilizer.org//En/Statistics/PIT_Excel_Files.aspx, last retrieved at 07/09/2017
Indexmundi (2017) Rock phosphate monthly price – US dollars per metric ton. Aug 1987-Jun 2017. Data online available at http://www.indexmundi.com/commodities/?commodity=rock-phosphate&months=360, last retrieved at 05/09/2017
Kongshaug G, Brentnall B, Chankey K, Gregersen J, Stokka P, Persson B, Kolmeijer N, Conradsen A, Legard T, Munk H, Skauli Ø, Kiiski H, Solheim K, Brentnall B, Rauman-Aalto P (2004) Phosphate fertilizers, Ullmann's encyclopedia of industrial chemistry. Weinheim, Wiley-VCH Verlag GmbH & Co
Kratz S, Godlinski F & Schnug E (2011) Heavy metal loads to agricultural soils in Germany from the application of commercial phosphorus fertilizers and their contribution to background concentration in soils. The new uranium mining boom, pp 755–762
Mar SS, Okazaki M (2012) Investigation of Cd contents in several phosphate rock used for the production of fertilizer. Microchem J 104:17–21
McConnell D (1938) A structural investigation of the isomorphism of the apatite group. Am Mineral 54:1379–1391
McKelvey VE (1967) Phosphate deposits. United States Government Printing Office, Washington
Mehmood ST, Chaundhry MM, Tufail M, Irfan N (2009) Heavy metal pollution from phosphate rock used for the production of fertilizer in Pakistan. Microchem J 91:94–99

Notholt A, Sheldon R, Davidson D (1989) Phosphate deposits of the world. Cambridge University Press, Cambridge
Pérez JP (2005) Natural radionuclides in the fertilizer industry. Ph.D. Thesis, University of Seville (in Spanish)
Saueia CH, Mazzilli BP, Fávaro DIT (2005) Natural radioactivity in phosphogypsum and phosphate fertilizers in Brazil. J Radioanal Nucl Chem 264:445–448
Schipper W, Klapwijk B, Potjer B, Rulkens W, Temmink H, Kiestra F, Lijmbach D (2001) Phosphate recycling in the phosphorus industry. Environ Tech 22(11):1337–1445
Schnug E, Lottermoser BG (2013) Fertilizer-derived uranium and its threat to human health. Environ Sci Technol 47(6):2433–2434
Schrödter K, Bettermann G, Staffel T, Wahl F, Klein T, Hofmann T (2008) Phosphoric acid and phosphates, Ullmann's encyclopedia of industrial chemistry. Wiley-VCH Verlag GmbH & Co., Weinheim
Stewart W, Hammond L, van Kauwenbergh SJV (2005) Phosphorus as a natural resource. Phosphorus: agriculture and the environment, agronomy monograph No. 46. Madison, American Society of Agronomy, Crop Science Society of America, Soil Science Society of America
Tayibi H, Choura M, López FA, Alguacil FJ, López-Delgado AJ (2009) Environ Manag 90(8):2377–2386
USGS (2007): Mineral commodity summaries of phosphate rock 2017. Online available at https://minerals.usgs.gov/minerals/pubs/commodity/phosphate_rock/phospmcs07.pdf. Last retrieved at 28/08/2017
USGS (2017): Mineral commodity summaries of phosphate rock 2017. Online available at https://minerals.usgs.gov/minerals/pubs/commodity/phosphate_rock/mcs-2017-phosp.pdf. Last retrieved at 28/08/2017
Van Kauwenbergh SJV (2006) Fertilizer raw material resources of Africa. IFDC
Van Kauwenbergh SJV (2010) World phosphate rock reserves and resources. IFDC. Online available at: http://pdf.usaid.gov/pdf_docs/PNADW835.pdf
von Liebig J (1840) Die organische Chemie in ihrer Anwendung auf Agricultur und Physiologie (Organic Chemistry in its applications to agriculture and physiology). F. Vieweg, Braunschweig
Von Plessen H, Schimmel G (1987) Verminderung des Cadmium-Gehaltes von Rohphosphaten und Mineraldüngern. Chemie Ingenieur Technik 59(10):772–778
WHO (1992) Cadmium, environmental health criteria. World Health Organ 134:1–280
Withers PJA, Van Dijk KC, Neset TS, Nesme T, Oenema O, Rubaek GH, Schoumans OF, Smit B, Pellerin S (2015) Stewardship to tackle global phosphorus inefficiency: the case of Europe. Ambio 44(Suppl. 2):193–206

Chapter 6
Success Factors for Implementing Phosphorus Recycling Technologies

Willem Schipper

Abstract Currently available phosphate recycling technologies are compared and rated on the basis of eight criteria. Expectations and predictions regarding implementation are given. Sewage sludge ash-based technologies are identified as overall most promising in terms of recycling efficiency and compatibility with existing industries.

Keywords Phosphorus recycling efficiency · Compatibility · Success factors · Cost and benefit

6.1 Introduction

Starting in the 1990s, and especially since the seminal paper by Cordell and White (Cordell et al. 2009), phosphorus/phosphate sustainability has gained a large momentum. A lot of effort has been devoted to assessing the Earth's phosphorus (P) resources, and a discussion has ensued on if we are running out of phosphorus and when (Van Kauwenbergh 2010; Edixhoven et al. 2013; Koppelaar and Weikard 2013; Wellmer and Scholz 2016). Regardless of the urgency of the issue, it is clear that phosphates, which accumulate through geological processes and on an accompanying, slow timescale, are mined at a much faster rate than they reaccumulate (Cummins 2014; Withers et al. 2015). In view of the paramount importance of this element in feeding the world's growing population, reviewing the imbalance in the processing and usage chain of phosphorus may be viewed as a responsible course of action.

On the other hand, these phosphates, released from geological deposits on an unprecedented scale since the beginning of the last century, may cause environmental issues if they are allowed to enter the environment. More specifically this concerns eutrophication. Increasing urbanization adds a degree of complexity, as phosphates are released, e.g. in wastewater in large quantities far away from where

W. Schipper (✉)
Willem Schipper Consulting, Middelburg, The Netherlands
e-mail: willemschipper@wsconsulting.nl

H. Ohtake, S. Tsuneda (eds.), *Phosphorus Recovery and Recycling*,
https://doi.org/10.1007/978-981-10-8031-9_6

they can potentially be reused in agriculture. Preventing phosphates from freely entering the natural environment therefore makes sense, and by doing so, the possibility to recycle and reuse phosphorus presents itself.

To this effect, a large number of processes have been proposed, acting on various points of the processing, usage and waste chain, mostly but not exclusively in wastewater. An increasing number is seeing implementation on a significant scale. The merits and disadvantages of these processes are not always immediately obvious. Their position in the phosphorus chain, the details of the method and secondary effects make a comparison not entirely trivial. The designers/owners of the process, in many cases commercial enterprises, are often the primary source of information which further complicates an assessment of the advantages and disadvantages.

This article nevertheless attempts to categorize a number of these processes, looking at their advantages and disadvantages by using a list of criteria. In many cases, processes are still under development (i.e. being improved), or operational data are proprietary for obvious reasons. This severely limits the possibilities to make a definite comparison based on completely objective criteria. Nevertheless, data are available from public sources, and in addition in many cases, information, often of a nonquantitative nature, was kindly shared by the process operators. Recycling technologies were finally assessed using a standardized, semiquantitative scoring card.

6.2 Major Entry Points for P Processing in Various Processing Chain

Elsewhere in this book, sources of P are assessed and quantified in detail. For this study, the main entry points of P recovery are discussed briefly, to set the scene for assessing recovery technologies.

6.2.1 Wastewater

An obvious source of recoverable P is urban wastewater. This source of recyclable P is the major focus of P recovery research and development.

Fully grown humans are in balance with regard to P, i.e. they ingest and excrete identical amounts of phosphorus. Among many other requirements, the body needs phosphorus to maintain bone health, for essential molecules in the energy cycle (ADP/ATP) and for DNA. The P metabolism in humans is quite complicated, but in order to maintain the required level of phosphate in the body, a daily dose of phosphorus is needed. This amount is excreted and ends up in sewage. The contribution of phosphate to sewage from other sources varies by region but is nowadays quite limited in Europe, North America and Australia, since the application of phosphate bans for detergents, historically the second most important source of phosphate to the sewage system. If phosphate is discharged into the environment, severe eutrophication occurs in the form of algal blooms, which kill off much of aquatic life. In industrialized countries, and increasingly in developing countries, wastewater treatment plants (WWTPs) are therefore being implemented. The bacterial sludge at the heart of these removes about half of the phosphate in a traditional setup. An additional phosphate removal stage ("third stage") is needed. Phosphate removal is done by adding iron or (less common) aluminium salts as precipitants, or a specific setup is used – EBPR, enhanced biological phosphate removal – in which phosphate is removed sufficiently by the bacterial sludge in a special sequence. The EBPR setup is currently only found in a minority of WWTPs but is becoming increasingly popular. Recoverable P from wastewater is quite significant, estimated at 300 kt/y in the EU-28 and 150 kt/y in the USA (Buckwell and Nadeau 2016; Leip et al. 2015; Sutton et al. 2011).

6.2.1.1 Recovery at or Before the WWTP

The primary mode of P recovery currently found at WWTPs is by struvite precipitation. Struvite, NH_4MgPO_4 6 H_2O, can be precipitated from a secondary loop in the WWTP where the sludge is treated to release part of the phosphate contained therein. This is essentially limited to the EBPR setup. Various modes exist to recover EBPR-based phosphate, differing in their exact location in the WWTP system.

6.2.1.2 Recovery from Sludge

The excess sludge (which needs to be removed continuously from the WWTP to prevent build-up of mass, including P) can be treated with heat or chemicals to access the phosphate and treat it to obtain a recycled product. Even though these methods differ from processes acting within the WWTP cycle, they often use the same principles. Without the need to recycle the active sludge to the WWTP, harsher thermal and chemical methods can be applied.

6.2.1.3 Recovery from Sewage Sludge Ashes

Excess sludge poses a disposal issue. In many cases (countries/regions), it is used as a fertilizer. The merits or disadvantages of this are outside the scope of this chapter, but it can generally be said that in some countries, the direct use of sewage sludge (biosolids in the USA) is being increasingly limited (Hayward 2016). Concerns over drug residues, pathogens and heavy metals contained in the sludge have led to a restriction or ban on agricultural use of sludge in, e.g. the Netherlands, Switzerland and most German federal states. Especially in urbanized and industrialized areas, this is a major concern. Large food processors have issued their own bans on the use of sewage sludge on farms delivering to these industries (Zeit 2001; Arla 2017).

It should be mentioned that the P removal method seems to determine the plant availability of the phosphate contained in the sludge. If iron salts are applied, which is the most common method for P removal, due to the low cost involved, the solubility – even over time – of the sludge phosphate in the soil may be restricted, or even excess iron can reabsorb phosphate in the soil. Therefore recycling of phosphate via sludge could be less effective, even though the sludge itself may act as a source of organic matter and thus benefit the soil. Aluminium-precipitated and especially EBPR-based sludge are indicated to have more readily accessible phosphate for crops (Wragge 2015). It is therefore inferred that EBPR-based sludges in rural areas are relatively safe and effective for recycling (including P) and that sludges from large cities with iron-based P removal, a common scenario, are not particularly suited to direct use on fields. Sewage sludge is quite wet; typically it contains some 70–75% water. Transporting large amounts of sludge from urban areas to the surrounding agricultural areas can significantly impact the cost of wastewater treatment. For these reasons, sewage sludge incineration has been installed, e.g. across Germany, the UK, Switzerland and the Netherlands. This dries the sludge and removes the organic content, leaving 5–10% of the initial weight as ashes. If set up properly, a modest amount of energy may be recovered as steam or electricity. Due to the high moisture content of the sludge, the energy to be recovered is limited.

The ash is a dry powder with a very significant phosphate content (see Sect. 6.4.2).

The advantages of sewage ashes for large-scale industrial recycling include:

- Industry-compatible material (dry powder)
- Centralized, large scale
- Relatively high phosphate content
- Destruction of organic pollutants, drug residues and bacteria

Its disadvantages are:

- Destruction of organic matter (potentially compost)
- Destruction of ammonium nitrogen content of the sludge to elemental nitrogen or N oxides
- Relatively large investments

For sewage sludge, it is important to mono-incinerate the material to allow an ash with high P content. Co-incineration, such as seen on large industrial sites, or use of the sludge in cement works leads to dilution and lock-up of phosphate in cement or fly ash, making recovery virtually impossible. Ashes are generally low in plant-available P, even though they have seen use in agriculture.

6.2.2 Slaughter Waste

Bones in the meat industry have traditionally found their way into, e.g. feed supplements and bone china. Much the same can be said about meat and bone meal (MBM), the protein and bone mix left after rendering (melting out fat) of slaughter waste, which can be used as a cattle feed supplement. Since the BSE crisis in the UK, a part of the MBM which poses a risk of BSE infection (category I) needs to be taken out of the cattle food cycle, either by feeding it to non-ruminants or by incineration. In practice, part of the MBM category I is incinerated as biofuel, co-fed into cement and power plants. Whereas the considerable fuel value of the MBM is used, it dilutes the phosphate into cement or coal fly ash, effectively making recovery impossible. In some cases, CO_2 credits are awarded for this practice. This highlights the conflicts sometimes encountered when sustainability is being implemented.

In the UK, at the height of the BSE crisis, a number of dedicated facilities to mono-incinerate MBM were constructed. These produce a high-phosphate ash similar to phosphate rock (ore) which poses a good entry point for P reuse. Optimizing this cycle to allow maximum P and energy recovery, by implementing MBM mono-incineration outside of these limited occurrences in the UK, would have a significant impact on P sustainability. These would still allow energy recovery, as MBM is essentially a dry, high-energy fuel. Legislation enforcing P recovery from waste may cause the construction of more MBM mono-incinerators.

Generally across the EU and USA and other industrialized regions, the amount of P in cat. I MBM amounts to about 30% of the amount of P present in sewage. This is a significant amount and often underestimated in studies dealing with P sustainability. Mono-incinerated MBM give an ash with a composition close to that of rock, without cadmium. Sodium/potassium chloride and sulphate is often present as a major contaminant. Like sewage ash, the material has little plant-available P but is nevertheless used as direct fertilizer in the UK, possibly because of its K content.

6.2.3 *Manure*

In the EU alone, manure was estimated to contain 2000 kt/a of P (Buckwell and Nadeau 2016; Leip et al. 2015; Sutton et al. 2011). This clearly exceeds the amount of P in sewage and slaughter waste. However, manure, including its largely plant-accessible P content, is normally recycled directly to fields. In this sense, most of the P is kept in a cycle within agriculture. However, in regions with intensive livestock husbandry, such as Brabant (the Netherlands), Bretagne (France), Emsland (Germany) and certain areas in the USA, Canada and the UK, manure is in local surplus. Its spreading is often limited by its P content, and P application to fields essentially becomes diffuse disposal. If the local use/disposal slots are filled, transport over large distances takes place, which greatly increases disposal cost. Especially pig manure is quite wet, and hence, large volumes need to be moved around. In these areas, manure acquires a negative value, limiting local further development of intensive animal husbandry.

Technologies to extract P from manure exist (http://www.wur.nl/nl/nieuws/Kunstmest-uit-dierlijke-mest-stap-dichterbij.htm (accessed 31/8/2017)), but as manure remains a product with normally a small positive value, the implementation of such technologies in areas where it is needed critically relies on full collaboration of all farmers affected. If this is not achieved, any investment into manure treatment will lead to increased profits of all farmers, i.e. those not investing will have maximum profits, as they did not invest. This prevents any developments without government intervention. Chicken manure, which is much drier than pig manure, is incinerated in a number of plants across the UK and in the Netherlands. The ash is rich in P and K. Especially the latter is plant available, making the material essentially a PK fertilizer with notable but substandard P availability. This example shows that collaboration between farmers is possible and can be seen as a model for other sectors.

6.2.4 Other Sources

In, e.g. oil refining catalysis, polyol processing, polystyrene production, and pharmaceutical synthesis, large amounts of (poly)phosphoric acid or phosphates are used and discarded. These wastes can contain significant amounts of phosphorus. Custom-made waste processing methods can recover these P streams. These are not discussed in this chapter.

6.3 Success Factors for P Recovery Technologies

In this chapter, we assess a number of prominent P recovery and recycling technologies for their pros and cons. As only a few technologies have seen a modest breakthrough up till now, it is difficult to validate the predictive power of this assessment, but nevertheless a few trends can be identified.

6.3.1 Defining Research Needs

It is duly noted that there is no lack of research into the subject of P recovery. A scan of available literature and processes reveals that P recovery is not a subject on the cutting edge of knowledge. However relevant it is to future food supplies and food security, the transformations involved are basic chemistry. A prime goal for any research into the subject should be to prevent repetition of previous research and to allow real progress in terms of processing and product quality.

6.3.2 Defining the Envisaged Process' Output: Product or Raw Material and Its Market Potential

In many cases, P recovery involves building upon what is present in terms of P removal. Whereas this provides a good platform for further development, it does somewhat obscure the final goals for any recovery process which strives to achieve commercial success, i.e. a usable, valuable and marketable product. More specifically, in water treatment, the object for P removal is to precipitate phosphate to an essentially water-insoluble form. Since the golden standard for phosphate fertilizers is water solubility, this requirement, as, e.g. required by law in Germany for recycled phosphates (http://dipbt.bundestag.de/doc/btd/18/124/1812495.pdf (accessed 31/8/2017)), seems to be entirely incompatible with P removal. Struvite proves to be an intermediate case where it is indicated to be plant available over longer periods (Nutrient Management Solutions, WEFTEC 2016). Still, the requirements of current agriculture and P removal/recovery in wastewater treatment are seen to be at largely unreconcilable odds. Producing a material which simply presents itself because of some result of a removal process, or producing a material which has clear market potential, is an essential dichotomy at the base of any decision towards investment in research and development of a P recycling process.

6.3.3 *Pre-assessing Cost and Benefits*

When assessing cost and benefits, it always needs to be clear what stream is being assessed where in the processing chain. For instance, a technology may be favourable in terms of energy needed to recover P but lacking in P recovery efficiency. Another example is the production of white phosphorus from waste/ashes, which consumes a lot of energy compared to other recycling technologies but which needs to be compared to the production of white phosphorus from phosphate ore (P rock) which is equally energy intensive. This chapter proposes a product-based methodology which compares cost to obtain a product of comparable quality from fossil resources, compared to recycled resources. In the case of recycled fertilizers of unclarified quality, a definite assessment is not always possible.

6.3.4 *Strategies for Successful Rollout and Commercial Success*

Technologies which produce a material with a trusted and proven market profile have a clear benefit over technologies which will deliver a material for which a market needs to be developed. As will be demonstrated below, especially those technologies which require large investments will face an uncertain future if the product needs to establish its position in the market. Materials with a full functionality as a known (intermediate) product as a direct result of a P recovery technology are quite rare, as we will see, and those which can be considered as such require a number of processing steps. Typically, fertilizer products (the main application of phosphorus/phosphate) require water-soluble phosphate; at least, this is the market standard commonly encountered. Given the unit operations which are found in the wastewater chain (aimed at making phosphate insoluble), it is obvious that products which contain water-soluble phosphate are not obtained.

Any technology development should take into account the market potential of the intended output and not just consider what is technically feasible to produce. Of course, there are developments and trends visible to challenge the traditional dogma of high water solubility fertilizers and emphasize the advantages of materials providing good fertilizing efficiencies at low water solubility. This needs time and works against established industries. Markets are ripe and cost competitive, and recycled materials do not offer a premium for their "green" nature at current (Hukari et al. 2015, 2016). For this reason, we identify existing/experienced operations (such as the fertilizer industry) as having an advantage in this respect.

A number of proposed technologies require extensive reworking of the existing wastewater system (such as separative toilets) or can only be applied in a limited amount of cases, such as being compatible with only one minority type of WWTP. This obviously limits the general rollout potential of such technologies, even though local advantages may exist. High investment cost (capex) poses an economic barrier to technology implementation, especially if combined with an uncertain potential for market revenues. This is especially valid for technologies which are ready for the construction of a prototype or the first full-scale installation, currently the situation in a number of cases. As phosphate is a relatively cheap raw material, investments can be prohibitively high even if the market is clearly existent. Since we have already identified the need for a number of processing steps to obtain a readily marketable material, this generally means that development of P recovery and recycling technologies is not often a naturally evident business case. Additional benefits, e.g. for the WWTP operation, may boost overall viability (see below). Making P recovery mandatory, such as in Switzerland, also changes the playing field considerably. By extension this limitation based on capex is also valid for technologies with high operating cost (opex) due to the extensive use of chemicals, energy and the generation of waste streams which need costly disposal. In addition, societies may not accept the generation of large waste streams.

Political stimulation, driven by demands from society, can help in implementing recovery and recycling technologies by tax rebates and subsidies (incentives). Also, the implementation of legislation can drive P recovery and recycling by removing legal barriers to the sale and the use of recovered phosphates (such as the EU fertilizer directive revision) as well as by making P recovery in some form mandatory (legally binding requirements/targets), as already imposed in Switzerland (Waste Ordinance 2015) and recently adopted in Germany (http://dipbt.bundestag.de/doc/btd/18/124/1812495.pdf (accessed 31/8/2017)). Technologies presenting an overall benign profile for their energy, chemical use and waste generation, as well as having a high P recovery rate, might be in an advantageous position to be selected, e.g. by communities for local implementation.

Finally, side benefits of P recovery technologies, such as reduced operational cost of a WWTP, may strongly influence the viability of P recovery technologies, as currently proven by the so-called struvite recovery technologies (Kabbe et al. 2015; Kraus et al. 2016).

6.3.5 *Soft Issues, Societal Considerations, Legislation and Show-Stoppers*

In many cases, the WWTP and subsequent treatment are governed by a city council or water board (Germany, the Netherlands). In such cases the support of the local population can be important. For example, in the Swiss canton of Zürich, a number of documents were prepared by the AWEL agency allowing for a 95% yes vote in a referendum about sewage P sustainability. This underlines the role of local governance to make P recovery a success. Having a local, recycled phosphate available, such as WWTP-derived struvite, can be a good way to demonstrate the circular economy, by providing a hands-on example to the local community, to be used, e.g. in gardens. Such demonstrations can be very useful in convincing local and even national governments about the need to recycle phosphorus but can also slow down further development, if it is assumed the issue has been solved once and for all, whereas the technology might be characterized by, e.g. a low recovery rate.

With the subject fetching attention, a large amount of research initiatives are deployed. For these, it is imperative to keep in mind that phosphorus recycling involves basic chemical transformations, and a large number of these are both obvious and have already been studied in some form. The economic viability also needs to be kept in mind. Adding too much technological complexity will almost certainly lead to a dead end. For any new research initiative, keeping the application potential and cost in mind is imperative, as is a thorough assessment of existing literature.

Legislation is being implemented in a number of countries to stimulate or enforce P recycling (http://dipbt.bundestag.de/doc/btd/18/124/1812495.pdf (accessed 31/8/2017); Waste Ordinance 2015). This will provide a big stimulus to P recycling and likely set an example for other countries to follow. However, it needs to be made clear that P removal and P recovery are not identical. Such legislation tends to impose phosphate removal/recovery to directly yield a plant-available material, which is difficult to achieve, as discussed in Sect. 6.3.2. It is imperative to make the distinction between raw materials and products in these cases.

Finally, the success rate of phosphorus recycling is strongly influenced by the world market price of phosphate ore (P rock). This is a typical commodity, with the accompanying price fluctuations. In an economic sense, P recycling competes against an industry characterized by huge volumes and optimized technology. This gives advantages of scale and ripeness not available to new recycling technologies, posing a challenge for newcomers to become financially attractive and profitable. If the price of P rock peaks, more recycling technologies will become competitive. Since this mechanism affects all recycling technologies, it is not further regarded in this assessment.

6.4 Selected Examples: Applying the Identified Considerations

The following factors were identified to influence the success of implementation of a recycling technology:

1. The ability to generate a product with a clearly defined market potential, as a finished product or an industry-compatible raw material (primarily in/for the fertilizer industry or as fertilizer)
2. The power of the developer or seller of the technology for successful rollout (familiar with the market for products, financial backing, resources available for rollout)
3. General applicability in an existing system (such as the wastewater system); the more universally applicable the technology is, the higher the chances of success
4. A high P recovery rate, compared to the input into the system, and contaminant depletion (Note: Recovery is always assessed in terms of input to the system in question, not of the local subsystem, as the latter might be misleading and indicate overestimated recovery rates.)
5. Low or moderate investment cost
6. Low or moderate operation cost and low amounts of waste generated
7. Societal and political acceptance and the absence of legal barriers
8. Side benefits, such as reduced operational cost of a WWTP (Table 6.1)

In this chapter, we use a score card to assess the various options using these eight criteria which determine the success of a certain technology or class of technologies. As an example, the score card for direct application of sewage sludge to agriculture is as follows:

1 Known product	+
2 Rollout potential	–
3 Applicability	+
4 Recovery rate	–
5 Cost of implementation	+
6 Operating cost	+
7 Societal acceptance	–
8 Side benefits	0

In the score cards, a + means a positive effect for P recovery and a – means detrimental effects; for instance, if the cost of implementation scores a –, it means it is high. Ash-based technologies can also be suited to recover P from MBM or manure ashes. As this aspect can only be rated for these technologies, and does not allow for a full comparison based on the criteria used in the wastewater chain, it is not mentioned on the score card.

Table 6.1 Overview of phosphorus recovery from the wastewater stream facilities operating or under construction (Kabbe et al. 2015; Kabbe 2013; Schoumans et al. 2015; Ohtake and Okano 2015). A representative selection, containing all main varieties of technology, has been selected for further discussion in this chapter

Technology	Location and operator	Operational since	Recovered material or product
On-site wastewater treatment plants			
AirPrex®	MG-Neuwerk (DE), Niersverband	2009	Struvite
	Wassmannsdorf (DE), Berliner Wasserbetriebe	2010	
	Echten (NL), Drents Overijsselse Delta	2013	
	Amsterdam-West (NL), Waternet	2014	
	Uelzen (DE), SE Uelzen	2015	
	Salzgitter Nord (DE), ASG	2015	
	Wolfsburg (DE), SE Wolfsburg	2016	
	Tianjin (CN), Tianjin CEPG	2016	
ANPHOS	Land van Cuijk (NL), Aa en Maas	2011	Struvite
Crystalactor®	Nanjing (CN), Royal HaskoningDHV	2010	Struvite
EloPhos®	Lingen (DE), SE Lingen	2016	Struvite
EXTRAPHOS (Budenheim)	MZ-Mombach (DE), Wirtschaftsbetrieb Mainz	2017 (pilot)	DCP
Gifhorn	Gifhorn (DE), ASG	2007	Struvite/CaP
Hitachi-Zosen (Unitika)	Matsue (JP)	1998	Struvite
	Senboku (JP), Senboku City	2009	HAP
J-Oil	Yokohama (JP), J-Oil Mills Co.		HAP
JSA	Kawasaki (JP), Japan Synthetic Alcohol Co.	1998	HAP
KURITA	Fukuoka (JP); three plants	1997	Struvite
Kyowa Hakko	Hofu (JP), Kyowa Hakko Bio Corp.	2006	HAP
Multiform™	Yakima, WA		Struvite
	Boise, ID		
	Massey, MD, Jones Family Farms (dairy)		
	Green Bay, WI		
NASKEO	Castres (FR)	2015	Struvite

(continued)

Table 6.1 (continued)

Technology	Location and operator	Operational since	Recovered material or product
NuReSys®	Harelbeke (BE), Agristo	2008	Struvite
	2x Niewkuerke (BE), Clarebout Potatoes	2009/12	
	Waasten (BE), Clarebout Potatoes	2012	
	Geel (BE), Genzyme	2014	
	Leuven (BE), Aquafin	2013	
	Schiphol Airport (NL), Evides	2014	
	Land van Cuijk (NL), Logisticon	2015	
	Apeldoorn (NL), Vallei and Veluwe	2016	
	Braunschweig Steinhof (DE), SE|BS/AVB	2018/19	
PEARL® (OSTARA)	Tigard, OR, Clean Water Services	2009	Struvite (Crystal Green®)
	Suffolk, VA, Hampton Roads Sanit. District	2010	
	York, PA, City of York	2010	
	Hillsboro, OR Clean Water Services	2012	
	Slough (UK), Thames Water	2013	
	Saskatoon, SK, City of Saskatoon	2013	
	Madison, WI, Madison Metro. Sew. Distr.	2014	
	Burford, GA, Gwinnett County	2015	
	Amersfoort (NL), Vallei and Veluwe	2015	
	Edmonton, AB, EPCOR Water Services	2015	
	Stickney, IL, Metro. Water Recl. Chicago	2016	
	Reno, NV, Cities of Reno and Sparks	2016	
	Madrid (ES), Canal de Isabel II	2016	
	Winchester, VA, F. Winchester Service A.	2016	
PHORWater	Calahorra (ES), El Cidacos	2015 (demo)	Struvite
PHOSPAQ™	Olburgen (NL), Waterstromen	2006	Struvite
	Lomm (NL), Waterstromen	2008	
	Nottingham (UK), Severn Trent Water	2014	
	Tilburg (NL), Waterschap De Dommel	2016	
PhosphoGREEN (SUEZ)	Aaby (DK), Aarhus Water	2013	Struvite
	Marselisborg (DK), Aarhus Water	2018	
	Herning (DK), Herning Water	2016	
REPHOS® (delivered by NuReSys)	Altentreptow (DE), Remondis Aqua (dairy)	2006	Struvite

(continued)

Table 6.1 (continued)

Technology	Location and operator	Operational since	Recovered material or product
STRUVIA™	Helsingør Southcoast (DK)	2015	Struvite
Stuttgart	Offenburg (DE), AZV	2011 (demo)	Struvite (after acid leaching)
	Mobile pilot – MSE mobile Schlammentwässerungs GmbH	2015 (pilot)	
Swing	Kobe (JP), Swing Corp.	2012	Struvite
Downstream wastewater treatment plants and ashes			
Ecophos	Varna (BG), DecaPhos	2016	H_3PO_4/DCP/ MCP
	Dunkerque (FR), Ecophos	Construction	
Fertilizer industry	Various companies already apply or consider the use of secondary P sources	Tested and intended	Commercial fertilizer
MEPHREC	Nürnberg (DE), SUN	2016 (demo)	P-slag
METAWATER	Gifu (JP)	2010	HAP
	Tottori (JP)	2013	
Nippon PA	Chiba (JP), Nippon Phosphoric Acid	2009	H_3PO_4
TetraPhos®	Hamburg (DE), Remondis Aqua	2015 (pilot)	H_3PO_4
RecoPhos	ICL	Intended	P_4

6.4.1 *Struvite*

Looking at the number of struvite projects being implemented in Japan, the EU and North America, struvite can be identified as a success story. The precipitation of struvite (NH_4MgPO_4 $6H_2O$) from phosphate-rich sewage sludge water is well known and currently being practised in more than 40 full-scale installations worldwide. It is usually found in a side stream setup in WWTPs of the EBPR type with anaerobic digestion but can also be applied to leaching liquors from sewage sludge, piggery wastewater, etc., where concentrations of dissolved phosphate and ammonia are sufficiently high, so that only magnesium salts need to be added after pH adjustment to trigger struvite crystallization and precipitation.

Struvite is a known product, but not as generally applicable as traditional fertilizer, as it is not water-soluble. This can be remedied by blending it into an existing fertilizer operation as raw material, as practised today by ICL Fertilizers in Amsterdam and indicated also to happen with Ostara struvite pellets. Alternatively the material can be developed as a slow-release, next-generation fertilizer even potentially suitable as mineral phosphorus source for organic farming (Final Report on Organic Fertilizers and Soil Conditioners (II) 2016). This however requires market development and strongly depends on the physical form in which the struvite is recovered, which ranges from a sludge-contaminated mud to well-defined dewatered pellets of several mm in diameter. The dryer, purer, and more granular the material is, the higher its chances are for application as niche market fertilizer or blend-in additive.

Even though struvite is obtained in high purity from some processes, it has still been in direct contact with wastewater. Other struvites contain residual sewage sludge. Concerns about contamination of the food chain via struvite continue to be raised. The accumulation of Mg in soils after repeated struvite applications needs to be assessed. Phosphate and ammonium are used by the crops, leaving magnesium oxide in situ. Increased magnesium levels could lead to fixation of subsequently dosed phosphate.

The major limitation of struvite technologies is that they can only be applied to relatively concentrated dissolved orthophosphates in wastewater and are therefore incompatible with the majority of WWTPs removing phosphorus by chemical precipitation with iron or aluminium salts. It is therefore largely limited to EBPR-type WWTPs. This limitation to dissolved phosphates also puts a cap on the recovery rate and therefore imposes a drawback in the overall phosphorus recovery and recycling discussion targeting high recycling rates. The AirPrex technology is indicated to recover only a few % of input P to the WWTP (Heinzmann and Lengemann 2013; Technical factsheets at www.p-rex.eu n.d.). An enhanced case applied at Vallei en Veluwe WWTP, Amersfoort, Netherlands, combines WASSTRIP, LysoTherm thermal sludge disintegration and an Ostara PEARL precipitation unit to reach 45% P recovery from the WWTP input, which is estimated to be the approximate upper limit of struvite technologies. Sound results from regular operation can be expected to be balanced and calculated in 2018.

Most struvite units on WWTPs have recovery rates in between these extremes. Attention has to be explicitly drawn to the definition of recovery rate. Most technology providers refer to the orthophosphate entering their recovery systems and the precipitated P load, not the P load out of the total system input actually harvested as struvite. Precipitation rates for all struvite recovery technologies are above 85–90%, whereas the harvest rates are far below and definitely deserve efforts for optimization. With the remaining phosphate locked up in sludge solids as polyphosphates in cells, and insoluble Ca-Al-Fe phosphates, an even more extreme treatment like acid leaching instead of, e.g. pressurized sludge digestion at 140 °C (such as the aforementioned LysoTherm) is needed, with disadvantages as elaborated in the Sect. 6.4.2.2.

Obviously, the biggest advantages of struvite technologies at WWTP level are their beneficial effects of reducing overall WWTP operating cost (opex). This primarily concerns the prevention of scaling, which needs frequent and costly removal cycles, and reduced return loads for P and N in the sludge liquor recycled within the WWTP. Some technologies like AirPrex and WASSTRIP, affecting also the sludge dewaterability, provide the biggest benefits due to better sludge dewatering and therefore higher dry matter content in the sludge to be disposed of, saving on transport and processing cost. Combined with lower flocculation aid consumption, the impact on sludge dewatering and finally reduced disposal cost is of highest interest for WWTP operators. This is also proven by the market, since all up and running PEARL installations worldwide are planned to be upgraded with a WASSTRIP reactor (Nutrient Management Solutions, WEFTEC 2016). Looking at the range of pros and cons, the low P recovery rate is more than offset by the economic benefits in the form of operational benefits for the WWTP operator. This is indeed reflected by the proliferation of struvite recovery units on WWTPs. Also, the availability of local struvite provides a societal benefit. For these reasons, struvite is identified as a success story despite its limited P recovery potential.

Struvite technologies also vary very markedly in their output. Ostara technologies produce a granulated material which is easily blended into regular granular fertilizer. Other struvite operations may end up with fine material, essentially yielding a sludge which is difficult to apply on fields.

The score card for struvite is therefore as follows:

1. Known product	0
2. Rollout potential	+
3. Applicability	0/–
4. Recovery rate	–
5. Cost of implementation	0
6. Operating cost	0
7. Societal acceptance	+
8. Side benefits	++

6.4.2 *Other Sludge Processes*

6.4.2.1 Extraction by Carbon Dioxide

In the EXTRAPHOS process (Schnee and Stössel 2014), sewage sludge is sealed in a reactor with some process water, heated to 50–60 °C, and CO_2 is added under a pressure of six bar. This increases the acidity of the water, causing a fraction of the phosphate in the sludge to leach out, usually between 50% and 70%. The sludge is separated from the watery phase, which is brought down to atmospheric pressure and stripped of CO_2. This causes the dissolved phosphate to precipitate, potentially with the aid of some lime. The obtained dicalcium phosphate (DCP), which is an existing product for the company developing this technology, can be used as a fertilizer or fertilizer ingredient.

The process uses little chemicals (CO_2 is largely recycled) and has a moderately high recovery rate. Since most P at a WWTP enters the sewage sludge, the overall harvest rate at WWTP level is also expected to be moderately high. The developer already sells a comparable product, holding some promise for the market development, and the operating cost is expected to be moderate regarding the chemical consumption. However, the installation with a pressure unit is relatively expensive and offers little potential for centralization (sludge transports are expensive over large distances). The pilot operation in Mainz will allow a further assessment from 2017 onwards. The pilot operation will also reveal information about potential operational benefits in terms of sludge treatment, dewatering and volume for disposal.

The score card for CO_2 extraction is as follows:

1. Known product	+
2. Rollout potential	0
3. Applicability	0
4. Recovery rate	+
5. Cost of implementation	0
6. Operating cost	0/+
7. Societal acceptance	0
8. Side benefits	0? (still matter of investigation)

6.4.2.2 Leaching of Digested Sludge with Mineral Acids

A number of sludge treatment processes with mineral acid (as opposed to CO_2) exist. These all involve the use of relatively large amounts of acid to leach part of the phosphate from the sludge. On top of that, a form of precipitation and/or pH adjustment is needed. The modified Seaborne process, now called Gifhorn process, which has been running in a pilot at Gifhorn WWTP (Germany) for several years, combines acid leaching, heavy metal precipitation as sulphides from sludge digestion, and struvite precipitation. Similar processes have been proposed, e.g. the Stuttgart process. The setup demonstrates the trade-off between recovery rate, product quality (low contaminant level) and chemical consumption. In this case, abundant use of chemicals and a number of treatment steps are needed to achieve a satisfactory recovery rate and product quality.

Despite their general applicability, the complicated and costly nature of such processes prevents further implementation. Indeed no such process has made it past a single piloting stage, despite interest in the subject of P recovery, thus conforming the predictive power of the set of criteria used.

The score card for acidic leaching is as follows:

1. Known product	0
2. Rollout potential	+
3. Applicability	+
4. Recovery rate	0
5. Cost of implementation	–
6. Operating cost	–
7. Societal acceptance	?
8. Side benefits	–

Table 6.2 Composition ranges of common phosphate rock and sewage ashes (weight% unless otherwise stated). Component Concentration in rock Concentration in sewage ash

P_2O_5	30–38	12–23
SiO_2	1–10	5–35
CaO	42–52	10–20
Na_2O	<1	1–2
K_2O	<1	0.5–2
Al_2O_3	0.3	3–25
MgO	0.3–1	1
Fe_2O_3	0.1–1	4–23
SO_3	0.5–7	1–4
Cl	0.05	0.1
Cu mg/kg	2–15	600–1200
Zn mg/kg	20–150	1200–3000
Cd mg/kg	1–20	<10
U mg/kg	20–300	<10

6.4.3 Ash to Fertilizer or White Phosphorus

As concluded earlier, sewage sludge mono-incineration ash is an industry-compatible, dry powder with high P content. This holds some promise for large-scale P recovery. A typical composition of municipal sewage sludge ash is given in Table 6.2.

Iron and aluminium content are seen to vary strongly with the precipitation method used at the WWTP. Iron from groundwater or drinking water treatment sludge contributes to the ash in a number of cases. Especially mono-incineration of municipal sludge gives an ash high in P, as is demonstrated in Fig. 6.1. Ash-based processes generally also work on MBM and manure ashes, giving an additional positive score point.

The ash processes can be categorized into:

- Processes making phosphoric acid or superphosphates
- Processes making solid thermal fertilizers
- Processes making white phosphorus

6.4.3.1 Superphosphates: Solid Fertilizers (Rock-Based TSP/SSP with Ash Addition)

ICL Amfert fertilizers blend suitable sewage sludge ash into a larger stream of rock phosphate. The ensuing TSP (triple superphosphate) is sold as such or blended into a compound NPK fertilizer. Since sewage ash is rich in iron, and iron is not a commonly encountered or desired element in fertilizers in higher concentrations, blending in this material into a larger rock stream has its limits. The generally lower P content of sewage ashes compared to rock and the higher amount of free oxides

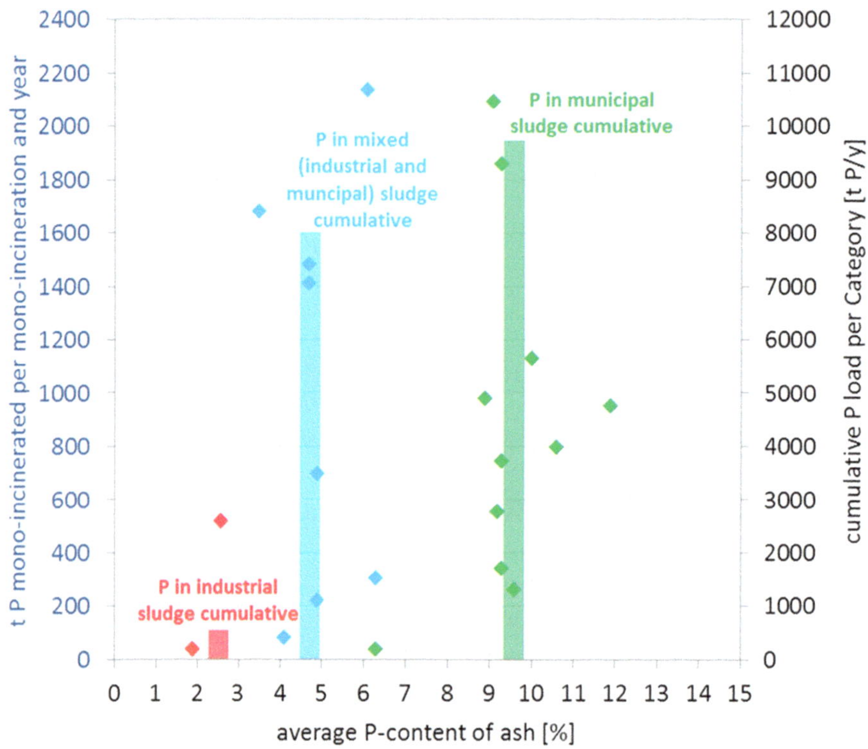

Fig. 6.1 Distribution of P content in mono-incineration ashes in Germany (Kraus et al. 2016; Krüger and Adam 2014). Legend: *left* (*red*), purely industrial sludge; *centre* (*blue*), mixed mono-incineration of industrial and municipal sludge; *right* (*green*), purely municipal sludge (courtesy P-Rex)

(Fe_2O_3, CaO) binding the phosphoric acid applied in TSP production put a limit on the amount of ash that can be blended into the P rock. ICL indicates 10% of rock input can be replaced by sewage ash; above this, technological development will be needed (Langeveld and ten Wolde 2013).

The score card for ICL Fertilizers blending operation is as follows:

1. Known product	++
2. Rollout potential	0
3. Applicability	+
4. Recovery rate	+
5. Cost of implementation	–
6. Operating cost	0
7. Societal acceptance	0
8. Side benefits	0

6.4.3.2 RecoPhos-TSP

This is a variety on the ICL model where only ash, not a rock-ash mixture, is acidulated with phosphoric acid to TSP. The output contains the sum of – partly mobilized – heavy metals in phosphoric acid (Cd, U) and sewage sludge ash (Zn, Cu). As with the blending operation described in Sect. 6.4.3.1, the process is simple, consisting essentially of a spraying and mixing unit. The product, nominally TSP, needs to establish itself. The combination of a cheap process and a more or less known product may lead to midterm success. But it has to be kept in mind that there is no heavy metal removal step in the process, limiting it to ashes already compliant to fertilizer regulation requirement for the end product. In addition, the ensuing TSP is lower in P content than the rock-derived product and still contains high concentrations of Fe and/or Al which may reabsorb P after application to soils. Superphosphates, as stand-alone product or in a blend, are also a minor product in the fertilizer spectrum, which is dominated by MAP and DAP, thus limiting the market potential of this method. This is especially true for the EU, where most sewage ashes are concentrated.

The score card for RecoPhos-TSP:

1. Known product	0/+ (potentially inferior to standard product)
2. Rollout potential	0/+
3. Applicability	+
4. Recovery rate	+
5. Cost of implementation	+
6. Operating cost	0
7. Societal acceptance	?
8. Side benefits	– [combination of heavy metals]

6.4.3.3 Ash to Phosphoric Acid by Means of Hydrochloric Acid

The Ecophos process is based on the long-known HCl digestion process (Fig. 6.2). An improved setup has been developed by Ecophos for low-grade rock but can be applied to ashes as well (De Ruiter 2014). There are several varieties, with either an internal HCl cycle, sulphuric acid entering the cycle and gypsum being purged or a variety which uses waste HCl and produces a $CaCl_2$ brine, requiring disposal, e.g. at sea. The process produces dicalcium phosphate (DCP) either as a product or as an intermediate ("second rock") to produce phosphoric acid from this and sulphuric acid. This has the advantage of yielding pure, sellable gypsum, in contrast to the classical process.

The process produces an inert iron oxide-silica waste and a heavy metal concentrate. The removal of the latter from the process by, e.g. ion exchange is still the subject of investigations and awaits final configuration. Even though the process

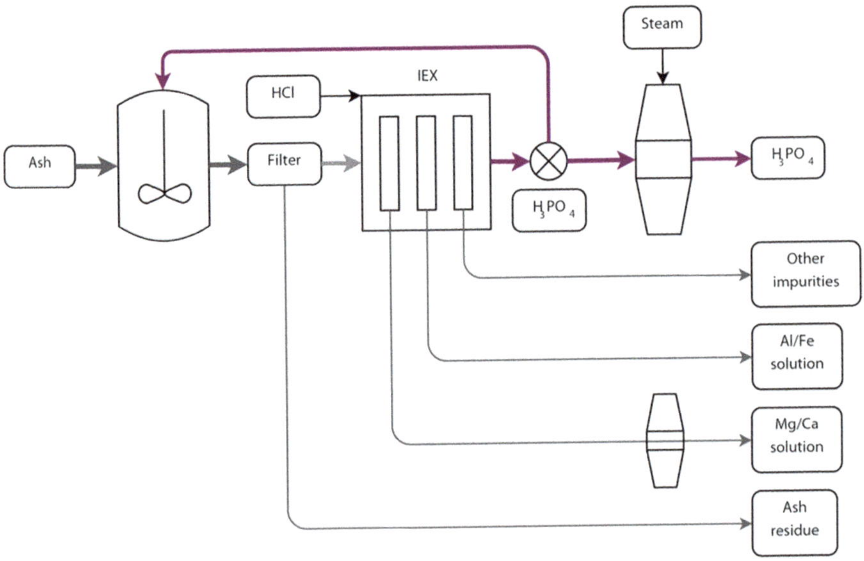

Fig. 6.2 Flow scheme of the Ecophos HCl process (Technical factsheets at www.p-rex.eu n.d.)

looks relatively complicated, the fact that the technology provider is an experienced, large supplier of feed phosphates and phosphoric acid technology/production units is expected to be an important factor for success.

The score card for Ecophos:

1. Known product	+
2. Rollout potential	++
3. Applicability	+
4. Recovery rate	+
5. Cost of implementation	–/0 (depending on variety)
6. Operating cost	–
7. Societal acceptance	+
8. Side benefits	+

6.4.3.4 Digestion of Ash with Sulphuric Acid

The Leachphos process (designed by BSH) is also based on acid digestion of ashes but here with sulphuric acid, thus essentially being a variety of the classical rock-to-phosphoric acid process (Fig. 6.3). The output can be fertilizer-grade DCP (a niche

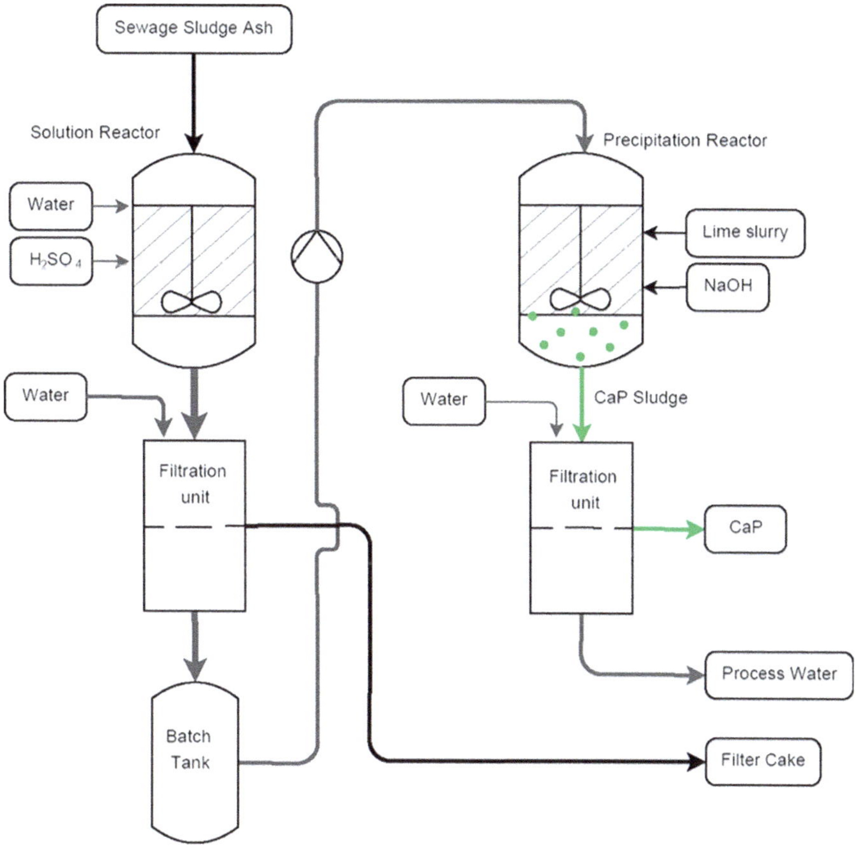

Fig. 6.3 Process scheme of the Leachphos ash leaching (Technical factsheets at www.p-rex.eu n.d.)

product) or phosphoric acid. The setup is characterized by large chemical use and generates a substantial amount of waste (filter cake) for which applications are being sought.

The combination of a nonphosphate experienced provider, potentially large waste streams and a relatively costly implementation may present a hurdle for large-scale introduction even in the case of switching from a solid mineral product towards phosphoric acid. On the other hand, in Switzerland (where the company is located), the laws imposing mandatory P recovery may give a boost. Even at increased operational cost compared to classical fertilizer operations, cost may be shared by the population via taxation, making the contribution per inhabitant quite modest (Technical factsheets at www.p-rex.eu n.d.).

The score card for Leachphos:

1. Known product	+
2. Rollout potential	−
3. Applicability	+
4. Recovery rate	+
5. Cost of implementation	−
6. Operating cost	−
7. Societal acceptance	+
8. Side benefits	0

6.4.3.5 Leaching by Phosphoric Acid Followed by Sulphuric Acid

This process, developed by Remondis Aqua, dissolves ash in phosphoric acid while using sulphuric acid as a fresh source of acid, a similar procedure as is followed in classical phosphoric acid production. It is indicated to make the precipitants from sewage works (Fe, Al) recyclable (Rak and Lebek 2016) and produce a purified phosphoric acid and gypsum as further by-product. The process is indicated by the operator to be especially suited to leaching phosphate from iron-rich ashes, whereas aluminium-rich ashes see a dissolution of the Al as well, posing a challenge in obtaining pure phosphoric acid. A minority of sewage ashes are of the aluminium-rich type, so this would limit its potential somewhat. Details were not fully available at the time of writing, making an assessment somewhat conjectural. The pilot installed in Hamburg is intended to be scaled up by Hamburg Wasser and Remondis in the coming years (Rak and Lebek 2016). The combination of a known product, little waste and a large provider (albeit not experienced in phosphoric acid) may prove to be beneficial for larger-scale implementation.

The score card for TetraPhos:

1. Known product	+
2. Rollout potential	0
3. Applicability	0
4. Recovery rate	+
5. Cost of implementation	?
6. Operating cost	?
7. Societal acceptance	+
8. Side benefits	?

6.4.4 Thermal Phosphates

6.4.4.1 AshDec

This process (now owned by Outotec) converts sewage ashes, which have little fertilizing function being largely plant unavailable, to a material with more plant-available phosphate. This is done by heating the ash to about 800 °C with an additive, driving out heavy metals via the gas phase as a side effect. Two varieties are offered: one where magnesium chloride is used and one where soda ash is the additive. The latter provides a material comparable to the now phased-out Thomas and Rhenania fertilizers but has incomplete heavy metal removal. Using magnesium chloride the metal removal is more efficient, but the output has lower P plant availability on neutral or alkaline soils. The energy use of the process is considerable, which can be remedied by linking a unit to a sewage sludge incinerator (Hermann 2013).

In the P-REX project (Herzel et al. 2016), the ashes were indeed found to have fertilizing benefits, but since they are not water-soluble, it is not easy to find a secured and trusted market for them, even though this may be developed once the material becomes available and the use, as in the case of struvite, approved for organic farming by introduction into the EC 889/2008 fertilizer regulation for organic farming (Final Report on Organic Fertilizers and Soil Conditioners (II) 2016). The combination of a large investment and an uncertain market potential for the product seems to pose a major hurdle for implementation. Indeed, the technology has been available for over a decade, but no full-scale units have been built until now.

The score card for AshDec:

1. Known product	–
2. Rollout potential	+
3. Applicability	+
4. Recovery rate	+
5. Cost of implementation	–
6. Operating cost	–/––
7. Societal acceptance	+
8. Side benefits	?

6.4.4.2 Mephrec

In this process, dried sludge is briquetted and fed into a shaft furnace with moderate amounts of air/oxygen. Upon descending, the sludge gasifies, and the remaining mineral content melts into a slag which is not unlike the AshDec output. The furnace gas, consisting of impure carbon monoxide, is intended to be sold as fuel and plays an essential role in the economics of the setup.

No information could be obtained on the energy balance with respect to the necessary drying of sludge prior to feeding into the process, which is an energy-intensive operation (usually sludge contains 70% moisture, evaporation of which eats away at the majority of the heat generated by combustion of the organic content). The process is indicated to accept ashes the briquettes. The operator indicates that selling the fuel off-gas is essential to the economic model. Should the gas prove to be unsellable or otherwise not valorizable, the project will prove to be not economically viable.

In the P-REX project, the slag was found to be only moderately effective as a fertilizer. The process, with its large investment and uncertain market for the phosphate, seems to face the same hurdles as AshDec, especially if the fuel gas proves unsellable. A large pilot is currently being operated in Nürnberg which should provide operational data from late 2017 onwards.

The score card for Mephrec:

1. Known product	–
2. Rollout potential	–
3. Applicability	+
4. Recovery rate	+
5. Cost of implementation	––
6. Operating cost	–/0
7. Societal acceptance	?
8. Side benefits	?

6.4.5 White Phosphorus

The production of white phosphorus from ashes, as once pioneered by the company Thermphos in a classical phosphorus furnace, is now being developed further by ICL, following an EU-funded project in Austria with SGL Carbon. The process heats ash to about 1500 °C by means of an inductively heated coke bed. The molten ash flows over the coke, causing the phosphate to reduce to the element (P_4) which is condensed from the gaseous output of the furnace. The CO gas generated as a by-product is used to preheat the feed, thus saving substantially on electrical power input. The phosphate-free slag received at the bottom is cooled and used as construction material or further treated to improve its value. A minor slag, a Fe-P alloy (ferrophosphorus), is formed and may decrease the product yield if formed excessively. The setup used is indicated to allow selective formation of white phosphorus even if large amounts of iron are present, due to the presence of a thin melt layer allowing an early evaporation of phosphorus before it reacts with iron. This assumption needs full-scale validation, and suppressing the formation of ferrophosphorus when iron-rich ashes are used, forming the bulk of currently available sewage ashes, is the main challenge for this technology.

Even though this process is clearly much more energy intensive than the other recovery technologies, and also has a considerable capex, comparison with the others is difficult because a much higher-value (non-fertilizer) product is obtained, which also requires a lot of energy and investment when produced in the traditional way. Elemental phosphorus has a well-defined market away from fertilizer and feed applications and is used via a number of intermediates in flame retardants, lubricant additives, crop protection agents and a range of fine chemicals. The classical process to make P_4 uses an equal amount of electrical power and should be the benchmark for the thermo-reductive RecoPhos.

The process uses ash as a powder and would not require the customary ash agglomeration stage before the actual inductive furnace, which is a major capex saving compared to the classical route. The combination of a known and valuable product and a technology developer which already uses elemental phosphorus may prove beneficial. An important driver for industries like the current process owner is given by their lack of backward integration into P_4, which could be terminated or at least reduced by operating their own P_4 production. Recent downward trends in electricity prices in several regions may make this opportunity more appealing than before.

The score card for RecoPhos-P_4 is as follows:

1. Known product	+
2. Rollout potential	+
3. Applicability	+
4. Recovery rate	+
5. Cost of implementation	−− or +[a]
6. Operating cost	− or 0[a]
7. Societal acceptance	?
8. Side benefits	Reduced dependency from global P_4 market

[a]Needs to be benchmarked against the conventional route to P_4 which has the same scores for opex and worse for capex

6.5 Conclusions

From the cases highlighted in this chapter, the following factors are concluded to be predictive for, or a barrier to, the widespread implementation of technical phosphorus recovery and recycling options:

- Low market prices for fossil P-based raw materials and products challenge the economic viability of many recovery technologies, especially when these technologies do not provide additional operational benefits and/or yield recovered material that is not directly marketable. If there is no prospect of profits, investors will spend their money in other sectors and markets. This is especially notable for technologies requiring large investments. Here realistic recovery targets implemented by a form of legislation could motivate or even enforce recovery and recycling. It is important not just to foster recovery alone. The recovered materials need to find a market. Otherwise, and as a worst-case scenario, recovered materials end up as waste and have to be disposed of as such or end up in intermediate storage for later use, in the best case (this conclusion is reflected in score card items 1, 5 and 6).

- Technologies which are developed without the direct involvement of potential users and other stakeholders tend to be more complicated than necessary and in some cases comparable to the reinvention of the wheel. Current market deployment is reflecting this, in that technologies providing operational benefits for their users (struvite, for WWTP operators) are being implemented most successfully and widely at this moment. An example is struvite, where granular material (with the physical appearance of standard, granulated fertilizer) tends to be generally sellable, whereas struvite recovered as a sludge of fine crystals is often seen to accumulate on-site without any users, as it is incompatible with fertilizer-spreading equipment and the need for further processing. Technologies which offer multiple benefits, such as cost savings, energy recovery or parallel recovery of other nutrients, may prove more successful than technologies aiming to recover P only (score card item 8).
- Technologies and recovered materials which cannot be integrated into existing infrastructure and markets have to cope with strong competition within established structures. Therefore the more varied the ways in which the recovered product can be used, the better. White phosphorus and phosphoric acid are the most promising materials, having been present in the market for over a century. But it is not only the downstream market potential that determines the vulnerability of a technology or value chain. The security of supply of the raw material is also crucial. The more versatile the technology is in terms of input material, the better. For example, a technology that can process various fossil and/or secondary P sources is less vulnerable compared to a technology depending on, for instance, sewage sludge ash alone and in worst case to the limited premium ashes rich in P and poor in Fe, Al and contaminants (score card items 2 and 3).
- Recovery rates (relative to WWTP input P) are surprisingly low for the currently most successful class of technologies, struvite recovery. In the longer run, governments and the general public may ask for more effective recovery routes. Side benefits of struvite technologies obviously play a big role in their current success. In a future scenario, struvite recovery at WWTP level and subsequent recovery of the remaining P from sludge ash could be envisaged. This highlights the sometimes complicated interplay of the identified success determining factors, in this case items 4, 7 and 8 on the score card.

From the WWTP operator's point of view, the legal requirements are the dogma to follow. If phosphorus recovery is not mandatory, implementation will be limited to the technologies providing operational benefits, as in the case of struvite recovery. But this route will remain minor in terms of tapping the P recovery potential from the wastewater stream, due to its inherent low harvesting efficiency and its limitation to WWTP operating EBPR and anaerobic digestion. Recent far-reaching legislation in Germany (2017) and Switzerland (2016) may change the playing field, as they force the wastewater chain operators to implement a much more substantial form of P recovery.

The perspective for technologies consuming a lot of chemicals and energy for P recovery on-site a WWTP no matter what P removal step is applied is rather limited. It is unlikely that operators will implement additional processes to their plants without incentives, causing a lot of additional efforts without real operational benefits. Downstream processing of sludge in some form seems more promising. This makes also sense under the light of the economy of scale. It seems less desirable to have a fertilizer production in every large WWTP only yielding small quantities of P-enriched raw materials or in best case ready-to-use fertilizers of varying composition, supply, etc. It is more likely that centralized options like P recovery from ash will be implemented on a large scale. These will ensure product homogeneity and sufficient quantities relevant for a potential customer. For them it would also be much more worthwhile to cope with the regulatory challenges connected to, i.e. the fertilizer business. It is not expected that wastewater treatment operators will become fertilizer market and regulation experts. The necessary know-how is already well covered by the various SME retailers dealing with nutrient recyclates and the large fertilizer companies. Instead of competition or invention of new markets, integration will be the key for success. Additional considerations include:

- The legal framework is tailored for existing structures and is very slow at adapting to future challenges. In relation to resource efficiency and sustainability, we are still a long way from implementing what is being discussed. For example, the upgrading of recovered material from being treated as a waste to being considered a resource or a product is proving to be a challenge with regard to legal barriers in place. The redefinition of end-of-waste criteria in the EU is a tough process but is a prerequisite to enable value chains to bridge the gap between recovery (supply) and recycling (demand) and make a circular economy really happen. Therefore, the revision of the EU Fertiliser Regulation (EC 2003/2003) is crucial to provide a level playing field for fertilizers, irrespective of whether they are produced from fossil or secondary sources. Another issue deserving consideration is the application of appropriate products for use in organic farming, for instance, by adding recovered struvite and calcined phosphates to the list of approved fertilizers in EC 889/2008, as recommended by the Expert Group for Technical Advice on Organic Production. Projects like the EU project IMPROVE-P (2013–2016) and the German project nurec4org (2017–2018) have been shaped to provide further fact-based information to facilitate stakeholder acceptance and cooperation along the relevant value chains from recovery to recycling.
- In some countries like Germany, decision makers focus only on the "highest hanging fruits" in terms of recovery and potentially recycling rates, instead of starting with the "lowest hanging fruits" and allowing technical evolution. We

observe a continuous cycle of R&D largely ignoring already viable, piloted options available on the market. Market penetration and replication will only happen with full-scale demonstrations, with the first large-scale operation requiring some form of support. Instead of broadening the range of technologies, the focus should be on setting up full-scale demonstrations of the most promising options. This should be coupled with making the most out of the existing infrastructure.
- But also downstream of WWTPs, there is still potential to improve P recovery and recycling, just by applying better or more reasonable sludge disposal or incineration logistics. As the German sewage sludge ash monitoring revealed, a big share of the German mono-incinerators burn municipal sludge mixed with industrial sludge. Dedicating these to municipal sludge exclusively may improve the ash quality and make more recycling feasible. Instead of calling for more mono-incinerators, the decision makers should foster first making the best and most out of the existing infrastructure. This would save the taxpayer a lot of money and would prevent avoidable expenditures for surplus capacities.

Looking at technologies, struvite recovery is already state-of-the-art and will spread out further around the globe, wherever there are WWTP operating EBPR and anaerobic digestion or for treatment of other liquid P-rich waste streams like pig manure. The recovery rate still bears potential for improvement, as already proven by upgrading existing first-generation facilities with various modules optimizing the sludge treatment. It seems however that it is capped at some 50%. The author sees the biggest potential where dedicated incinerators are in place or required by regulation in P recovery from ashes. They provide the highest concentrate in P and the smallest volume to be treated. Struvite recovery and recovery from ash are not competing P recovery and recycling routes; they are complementary. Actually having an efficient downstream recovery in place takes the pressure off struvite technologies to aim for optimal P recovery rates. Struvite can then optimize the WWTP for its additional benefits (less scaling and better sludge dewaterability). Where markets for struvite fail, due to product characteristics or absence of local markets, it can even be blended back into the sludge and sent to the incinerator.

On global scale, the direct application of sludge on arable land will remain an important route, since incineration is a luxury or need of industrialized countries with high population density and limited territory. Still, the trend towards more urbanization will increase the need for this route. More and more countries will impose nutrient recycling targets. Some countries, like Switzerland and Germany, can be very strict, whereas most countries will propose targets instead of requirements. This will definitely have a strong impact on technologies. The stricter the requirements, the more technologies become economically feasible.

The biggest challenge remaining is not the technology of recovery as such. More important in the coming years will be bridging the gap between recovery (supply) and recycling (demand). Here still an enabling legal framework is required to create a level playing field for all nutrients, no matter if it comes from primary (i.e. fossil) or secondary (recovery) sources. Maybe taxation will play a role in the shift from our linear to circular economy providing incentives for recycling and higher taxes on fossil-based raw materials. The revision of the European fertilizer regulation can be expected to provide a template for circularity enabling regulation.

Acknowledgement The author cordially thanks Christian Kabbe of Kompetenzzentrum Wasser for many useful discussions in shaping this manuscript.

References

Arla (2017) Arlagarden® quality assurance programme, Version 5.2, January 2017

Buckwell A, Nadeau E (2016) Nutrient Recovery and Reuse (NRR) in European agriculture. A review of the issues, opportunities and actions. RISE Foundation, Brussels

Cordell D, Drangert JO, White S (2009) Global Environ Chang 19:292–305

Cummins CC (2014) Daedalus 143(4):9–20

De Ruiter R (2014) The EcoPhos technology to close the P cycle and to safeguard the world's food chain. Abwasser–Phosphor–Dünger-Workshop inclusive Fachgespräch zum UFOPLAN-Projekt Klärschlammaschemonitoring, Berlin

Edixhoven JD, Gupta J, Savenije HHG (2013) Earth Syst Dynam Discuss 4:1005–1034

Final Report on Organic Fertilizers and Soil Conditioners (II) (2016) Expert group for technical advice on organic production, Brussels

Heinzmann B, Lengemann A (2013) Vom Betriebsproblem zum Berliner Verfahren – Phosphorrückgewinnung als Magnesiumammoniumphosphat in der Kläranlage Wassmannsdorf (From operational trouble to the Berlin P recovery process implemented at WWTP Wassmannsdorf). Proceedings 8th Klärschlammtage, Fulda, 4–6 June 2013

Hermann L (2013) Phosphate-containing waste ash process for producing mineral fertiliser. Proceedings International Fertiliser Society, 732

Herzel H, Krüger O, Hermann L, Adam C (2016) Sci Total Env 542 B:1136–1143

Hukari S, Nättorp A, Kabbe C (2015) Phosphorus recycling now! P-REX policy brief, www.p-rex.eu

Hukari S, Hermann L, Nättorp A (2016) From wastewater to fertilisers – technical overview and critical review of European legislation governing phosphorus recycling. Sci Total Environ, B 542:1127–1135

Kabbe C (2013) Sustainable sewage sludge management fostering phosphorus recovery. Bluefacts 2013:36–41

Kabbe C, Remy C, Kraus F (2015) Review of promising methods for phosphorus recovery and recycling from wastewater. In: Proceedings International Fertiliser Society, p 763

Koppelaar RHEM, Weikard H (2013) Global Environ Chang 23:1454–1466

Kraus F, Kabbe C, Remy C, Lesjean B (2016) KA Korrespondenz Abwasser, Abfall 6:528–537

Krüger O, Adam C (2014) Monitoring von Klärschlammmonoverbrennungsaschen hinsichtlich ihrer Zusammensetzung zur Ermittlung ihrer Rohstoffrückgewinnungspotentiale und zur Erstellung von Referenzmaterial für die Überwachungsanalytik (Sewage sludge ash monitoring for their material recovery potential and to produce references materials for official monitoring routines), UBA Texte 49/2014. Dessau-Roßlau

Langeveld CP ten Wolde KW (2013) Proceedings of the international fertiliser Society, 727

Leip A, Weiss F, Lesschen JP, Westhoek HJ (2015) Agric Sc 152:20–33

Nutrient Management Solutions, WEFTEC (2016) http://ostara.com/nutrient-management-solutions/ (provided by struvite operator Ostara)

Ohtake H, Okano K (2015) Development and implementation of technologies for recycling phosphorus: in secondary resources in Japan. Global Environ Res 19:49–65

Rak A, Lebek M (2016) Clean technology for P-recycling to phosphoric acid: Remondis TetraPhos®. ESPP – phosphorus stewardship in industrial applications, Brussels
Schnee R, Stössel E (2014) Extraktion von Phosphaten mit Kohlendioxid (Extraction of Phosphates with carbon dioxide), Abwasser–Phosphor–Dünger-Workshop inclusive Fachgespräch zum UFOPLAN-Projekt Klärschlammaschemonitoring, Berlin, 28–29 January 2014
Schoumans OF, Bouraoui F, Kabbe C, Oenema O, van Dijk KC (2015) Phosphorus management in Europe in a changing world. Ambio 44:180–192
Sutton MA, Howard CM, Erisman JW, Billen G, Bleeker A, Grennfelt P, van Grinsven H, Grizzetti B (eds) (2011) The European nitrogen assessment. Cambridge University Press, Cambridge
Technical factsheets at www.p-rex.eu (n.d.) Berlin Centre of Competence for Water, Berlin
Van Kauwenbergh SJ (2010) World phosphate rock reserves and resources. IFDC, Muscle Shoals
Waste Ordinance (2015) (Verordnung über die Vermeidung und die Entsorgung von Abfällen, Der Schweizerische Bundesrat. https://www.admin.ch/opc/de/classified-compilation/20141858/201607190000/814.600.pdf
Wellmer FW, Scholz RW (2016) Miner Econ. https://doi.org/10.1007/s13563-016-0094-3
Withers PJA, Elser JJ, Hilton J, Ohtake H, Schipper WJ, van Dijk KC (2015) Green Chem 17:2087–2099
Wragge V (2015) Product quality and fertiliser value of recovered products, presented at the P-Rex conference in Amsterdam, see www.p-rex.eu
Zeit (2001) Frisch und giftig auf den Tisch (fresh and toxic on the table). Die Zeit Archiv Jahrgang 2001(6):1–4

Part II
Incinerated Sludge Ash

Chapter 7
Industrial-Scale Manufacturing of Phosphoric Acid Using Sewage Sludge Ash

T. Mochiyama

Abstract Nippon Phosphoric Acid Co. Ltd. (NPA) started to use sewage sludge ash (SSA) for the industrial manufacturing of phosphoric acid (PA) in 2013. NPA purchases SSA from wastewater treatment plants (WWTP) equipped with a Bio-P process to remove phosphate (P_i) from sewage. In the NPA's manufacturing plant, SSA is blended with roller-milled P_i rock and is dissolved in concentrated sulfuric acid to generate phosphogypsum slurry. Then the slurry is filtered to separate PA from gypsum which is recycled for the manufacture of plasterboard and cement. Currently, NPA accepts a total of 1300 t/a of SSA for the manufacture of PA and gypsum. In full operation, the manufacturing plant has the capacity to accept 3000 t/a of SSA. The blend ratio of SSA with roller-milled P_i rock is currently limited to a maximum of 2.5% (<97.5% P_i rock). This is needed to guarantee the quality of both PA and by-product gypsum for customers. NPA could accept SSA from WWTP without the need to modify its existing facilities and save material costs by partially replacing P_i rock with SSA.

Keywords Sewage sludge ash · Phosphoric acid · Wet acid process · Phosphogypsum

7.1 Introduction

In Japan, only two companies produce phosphoric acid (PA) from phosphate (P_i) rock on industrial scale. Nippon Phosphoric Acid Co. Ltd. (NPA), which was established in 1967, is the principal supplier of PA for agricultural purposes. This company has produced approximately 90 kt/a of PA as P_2O_5 using approximately 250 kt (10^3 tons)/a of P_i rock. In the NPA's manufacturing process, PA has been

T. Mochiyama (✉)
Nippon Phosphoric Acid Co. Ltd, Sodegaura, Chiba, Japan
e-mail: mochiyamat@n-rinsan.co.jp

H. Ohtake, S. Tsuneda (eds.), *Phosphorus Recovery and Recycling*,
https://doi.org/10.1007/978-981-10-8031-9_7

extracted from P_i rock with sulfuric acid. Gypsum ($CaSO_4$) is generated as a by-product, whose market is an indispensable prerequisite for the economic viability of the P_i industry in Japan. NPA has used P_i rocks imported mainly from Morocco, China, and Jordan. They need to be mixed together before use to reduce the contamination of cadmium and radioactive materials, whose levels are relatively high in Morocco's P_i rock.

In 2009, NPA began a feasibility study to accept sewage sludge ash (SSA) from a wastewater treatment plant (WWTP) for the manufacture of PA. This was triggered by the surge in P_i rock price in 2007–2008. NPA needed to diversify phosphorus (P) sources, thereby minimizing the risk associated with the volatility of P_i rock price. NPA saw the possibility of using SSA as a substitute for P_i rock, because P_i-rich SSA was abundantly available from WWTP. In Japan, the total amount of P_2O_5 in SSA was estimated to be approximately 67 kt/a which is essentially equivalent to the amount of imported P_2O_5 as P_i rock. Additionally, the Ministry of Land, Infrastructure, Transport and Tourism (MLITT) encouraged NPA to use SSA as a secondary P resource for the manufacture of PA. At present, NPA purchases a total of 1300 t/a of SSA from WWTP for the manufacture of PA. In full operation, the manufacturing plant has a capacity to accept 3000 t/a of SSA. This paper describes NPA's experience to realize the industrial manufacturing of PA using SSA as a substitute for P_i rock.

7.2 Wet Acid Process

There are two different processes (i.e., dry and wet processes) for the industrial manufacturing of PA. In the dry process, PA (called thermal PA) is obtained by oxidation of elemental P that is generated by the carbothermal reduction of P_i rock in an electric furnace (Slack 1968). Since the purity of thermal PA is high, it has a wide range of industrial applications. However, the manufacture of elemental P is energy-intensive and dependent on access to low-cost, locally sourced electricity, P_i rock, and coal. Therefore, it is prohibitively difficult to generate elemental P in Japan because of energy shortage and lack of P_i rock reserve. On the other hand, in the wet process, PA (called wet PA) is extracted from P_i rock with strong acids such as hydrochloric acid, nitric acid, and sulfuric acid. Among them is sulfuric acid commonly used worldwide.

$$\underset{P_i\text{rock}}{Ca_5F(PO_4)_3} + \underset{\text{Sulfuric acid}}{5H_2SO_4} + \underset{\text{Water}}{5nH_2O} \rightarrow \underset{\text{PA}}{3H_3PO_4} + \underset{\text{Gypsum}}{5(CaSO_4 \cdot nH_2O)} + \underset{\text{Hydrofluoric acid}}{HF}$$

By-product gypsum is separated from the slurry (called phosphogypsum slurry) by filtration, while metallic impurities of P_i rock mostly move to the filtrate. Although the filtrate itself cannot meet technical-grade requirements for PA because of impurities such as metals, hydrofluoric acid, and sulfate, it is marketable as a

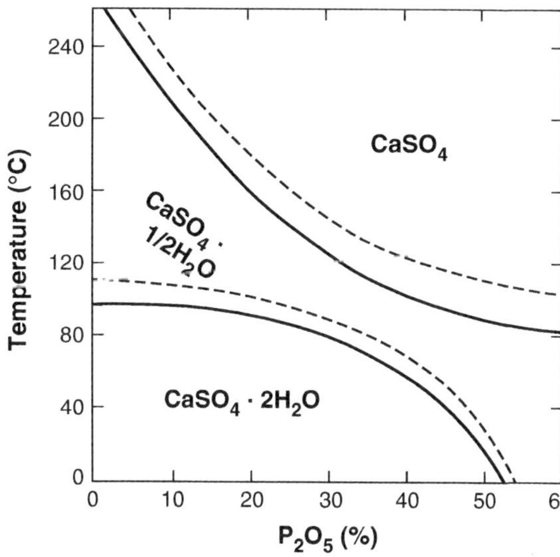

Fig. 7.1 Equilibrium diagram for the system $CaSO_4$-P_2O_5-H_2O (Slack 1968)

fertilizing material for agricultural purposes. Recovered gypsum contains unreacted P_i rock and silica which is not dissolved with sulfuric acid. Hence, the purity of gypsum relies primarily on the quality of P_i rock used as a raw material. Dewatered gypsum (called phosphogypsum) with the purity of 93–98% can be used for cement, plasterboard, and soil improvers.

Gypsum can take three different forms (i.e., anhydrous, hemihydrate, or dihydrate forms) in phosphogypsum slurry, depending on the PA content and temperature (Fig. 7.1). Based on the form of gypsum generated, wet processes can be further classified into (i) (direct or indirect) dihydrate, (ii) hemihydrate, and (iii) dihydrate-hemihydrate processes. PA-manufacturing plants located near to P_i rock mines mostly adopt either direct dihydrate or hemihydrate processes, giving the priority to the cost performance rather than the quality of gypsum. The direct dihydrate process generates gypsum with the P_2O_5 content of about 1 wt% and the purity of 90–95%. The gypsum is mostly disposed of landfilling or dumped into the ocean. By contrast, the PA-manufacturing plants operating in Japan adopt either indirect dihydrate process (Nissan "H" process) or dihydrate-hemihydrate process (Central-Prayon process). These Japanese original processes consist of multistage reaction steps for the manufacture of PA and thus are more complex than direct dihydrate processes. However, by promoting the crystal transition between hemihydrate and dihydrate forms, they can produce larger gypsum crystals with a lower P_2O_5 content compared to those from the direct dihydrate process. These processes can lower the PA content of gypsum, thereby increasing the extraction efficiency of PA from P_i rock. The generation of low-PA gypsum is critical to guaranteeing the demand from the manufacturers of cement and plasterboard and is therefore essential for the economic viability of the PA manufacturing in Japan.

7.3 Bench-Scale Tests

Although considerable attention has been paid to the potential of using SSA as a secondary P resource, the application in full-scale PA-manufacturing processes has been hampered mainly by its impurities (Table 7.1). For agricultural use, levels of heavy metals and other harmful substances in fertilizers are regulated by the Fertilizer Regulation Act in Japan. Since the PA manufactured by NPA is used as a fertilizing material, the priority target was given to minimizing the contamination of toxic heavy metals from SSA and P_i rock. Chemical analysis was carried out to examine the levels of toxic heavy metals in SSA samples collected from different WWTP (Table 7.1). Except lead, levels of heavy metals in SSA were essentially the same as those of P_i rock. Since lead moves to gypsum solids in the wet process, it could be effectively removed from phosphogypsum slurry by filtration.

Then, PA solution was prepared by varying the blend ratio between SSA and P_i rock. Table 7.2 shows the quality of diammonium phosphate (DAP) prepared using

Table 7.1 The main components of SSA and P_i rock

		SSA		P_i rock	
		Min	Max	Min	Max
P_2O_5	[wt%]	21	35	32	38
CaO	[wt%]	4	14	48	53
Al_2O_3	[wt%]	5	33	0.2	0.8
Fe_2O_3	[wt%]	2	30	0.2	0.7
MgO	[wt%]	1	6	0.2	1
SiO_2	[wt%]	22	46	2	11
As	[mg/kg]	2	44	2	20
Cd	[mg/kg]	1	20	0.1	15
Hg	[mg/kg]	0.01	2	0.01	1
Pb	[mg/kg]	20	100	1	15

SSA samples were collected from six different WWTP. P_i rock was sampled from those used in the PA-manufacturing process of NPA

Table 7.2 Characteristics of DAP generated by varying the blend ratio of SSA and P_i rock

	The SSA/P_i rock blend ratio (wt%)					Marketable DAP
	0	2.5	5.0	10.0	20.0	(wt%)
SP	46.90	47.20	46.50	46.96	45.41	46
WP	45.70	44.60	42.60	38.47	34.88	40
AN	19.10	18.10	18.80	17.50	17.18	18
Fe_2O_3	0.43	0.59	0.75	1.10	1.55	–
MgO	0.78	0.95	1.13	1.33	1.66	–
Al_2O_3	0.54	0.89	1.23	1.81	2.90	–
Total metal	1.75	2.43	3.11	4.24	6.11	–

SP citrate-soluble P_2O_5, *WP* water-soluble P_2O_5, *AN* ammoniacal nitrogen (N), Total metal: Fe_2O_3 + MgO + Al_2O_3. The standard of marketable DAP is given to SP, WP, and AN

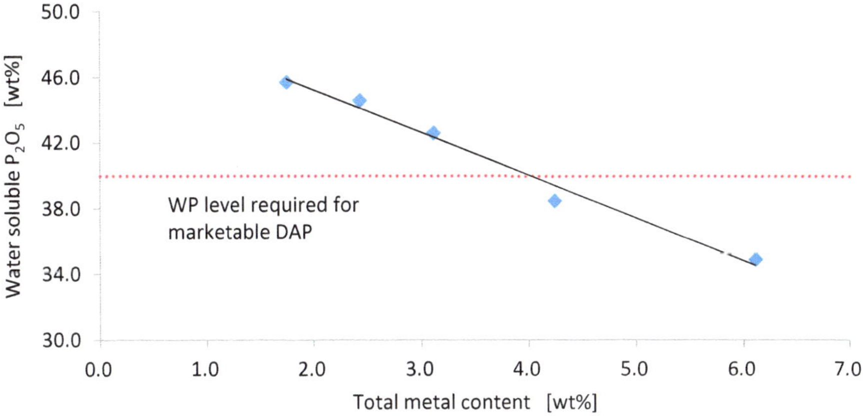

Fig. 7.2 The relationship between the water-soluble P_2O_5 and total metal contents in PA

the PA solutions. The commercially available DAP needs to guarantee the amounts of 46 wt% citrate (ammonium citrate)-soluble PA, 40 wt% water-soluble PA, and 18 wt% ammoniacal nitrogen. A remarkable reduction was observed with the water-soluble PA as the blend ratio of SSA increased up to 20%. When the blend ratio exceeded 10%, the content of water-soluble PA failed to meet the requirement for marketable DAP. This reduction was attributable to the increase in the amounts of Fe, Al, and Mg ions in PA solution.

In the literature, it has been reported that Fe^{3+}, Al^{3+}, and Mg^{2+} can form insoluble salts such as (Fe, Al)$NH_4HF_2PO_4$ and $MgNH_4PO_4$ when PA solution is neutralized by ammonia for fertilizer production (Akiyama 1980). Figure 7.2 shows the relationship between the contents of total metal ($Fe_2O_3 + Al_2O_3 + MgO$) and water-soluble PA. Obviously, the content of water-soluble PA decreased with increasing the total metal content. When the total metal content exceeded 4 wt%, the level of water-soluble PA declined below 40 wt% which is the minimal requirement for marketable DAP. Therefore, provided that the total metal content of SSA is 25 wt%, the allowable blend ratio needs to be 8% or lower.

The blend ratio between SSA and P_i rock can affect the crystal morphology of gypsum generated (Fig. 7.3). When the blend ratio exceeded 10%, gypsum showed a needle-shaped crystal structure. The needle-shaped gypsum requires more water in the kneading process, thereby consuming more energy for the manufacture of plasterboards compared to gypsum having a rhombic plate crystal structure. Consequently, the blend ratio needs to be lower than 5% to avoid the formation of needle-shaped gypsum. It has also been reported that F^- can react with Al^{3+}, thereby forming a complex AlF_n^{3-n} in PA solution (Jun et al. 1997). This complex is known to precipitate on gypsum, thereby inhibiting its crystal growth in the long axis direction. Similarly, silica is likely to react with F^- in PA solution to generate a complex SiF_6^{2-}. This reaction may competitively inhibit the formation of AlF_n^{3-n}, thus allowing the elongation of gypsum crystal. It has also been shown that the addition of

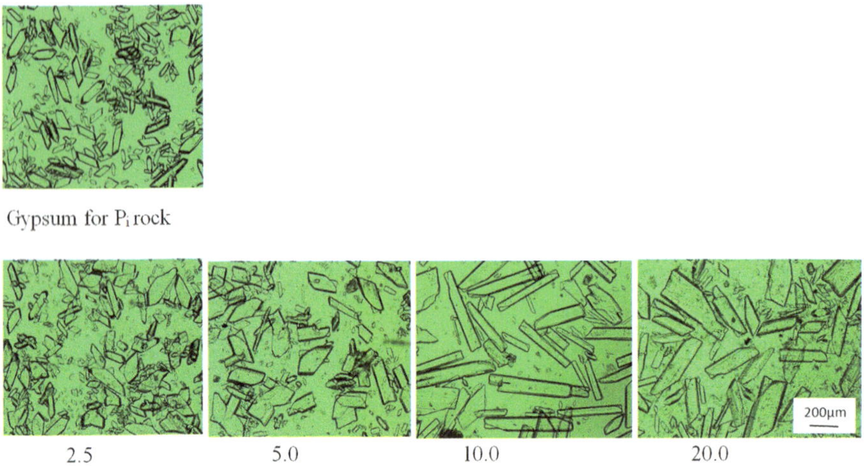

Fig. 7.3 The crystal morphology of gypsum generated from the mixture of SSA and P_i rock. The SSA/P_i rock blend ratio was varied from 0.0 to 20.0. *Bar* = 200 μm

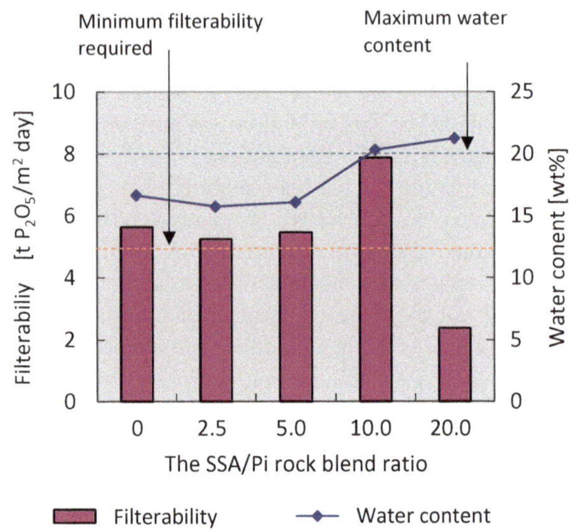

Fig. 7.4 The effect of the SSA/P_i rock blend ratio on the filterability of phosphogypsum slurry and the water content of gypsum cake

fluorine compounds such as fluorite to PA solution is likely effective in allowing gypsum to form the rhombic plate crystal structure (Mochiyama et al. 2017).

The filterability of phosphogypsum slurry is critical to the operability of downstream process in the PA-manufacturing plant. Figure 7.4 shows the dependence of slurry filterability on the SSA/P_i rock blend ratio. When the blend ratio exceeded 20%, the filterability of phosphogypsum slurry remarkably deteriorated. This was caused due to the formation of fine silica particles in the slurry.

Regarding the toxicity of DAP generated using SSA, plant cultivation tests were conducted using the leaf vegetable komatsuna (*Brassica rapa* L. var. *perviridis*) as

Fig. 7.5 The toxicity test of DAP generated from the mixture of SSA and P_i rock (5%) using the leaf vegetable komatsuna as a test plant

a test plant. As can be seen in Fig. 7.5, no significant toxicity was detected with the DAP compared to commercially available DAP samples.

Another problem regarding gypsum quality is the contamination of lead, which can move from P_i rock to gypsum in a wet acid process. Lead can react with sulfate, thereby precipitating onto gypsum in the PA-manufacturing process. The elution test of lead from gypsum was conducted using samples prepared from SSA having different lead contents (Fig. 7.6). The results showed that if the lead content of SSA is higher than 60 mg/kg, the blend ratio should be lower than 3% to prevent lead leakage which may cause health and environmental issues.

7.4 Industrial PA Manufacturing

In April 2010, NPA conducted a full-scale plant test for the manufacture of PA using 70 t of SSA at the blend ratio of 2.5%. The SSA used for the plant test fulfilled the following specifications:

- P_2O_5, 30 wt% or greater
- SiO_2, 30 wt% or lower
- Lead, 60 mg/kg or lower
- Metals (i.e., Fe_2O_3+ Al_2O_3+ MgO), 20 wt% or lower

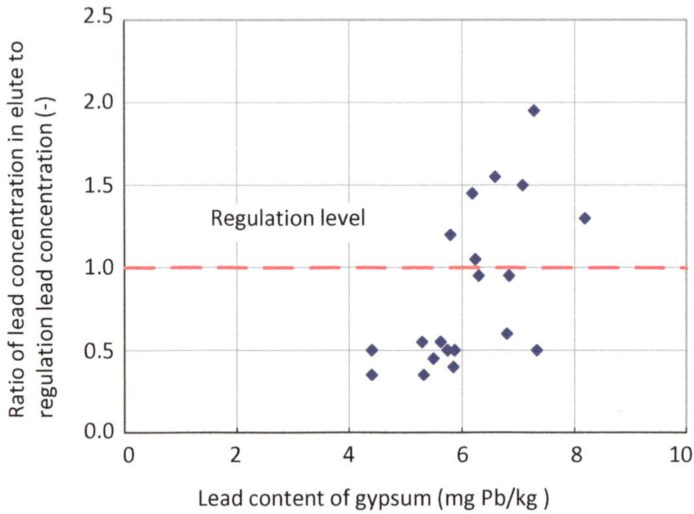

Fig. 7.6 Elution of lead from gypsum prepared from SSA having different lead contents

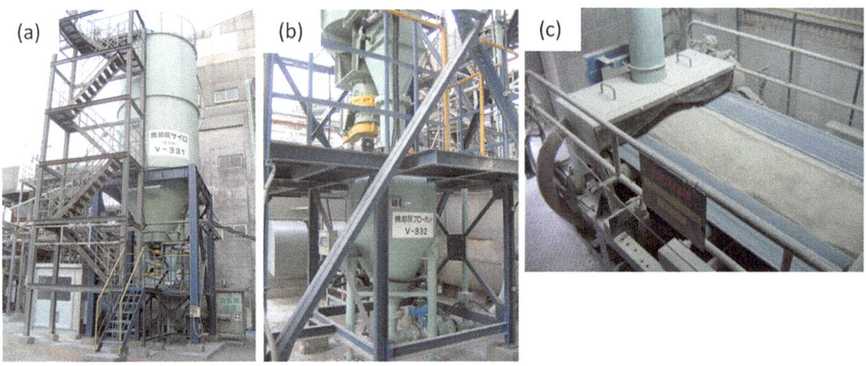

Fig. 7.7 SSA storage tank and weighing equipment implemented in the existing PA-manufacturing plant. **a** Storage tank, **b** weighing equipment, **c** unit for mixing SSA with P_i rock

The results convincingly showed that the use of SSA caused no significant problem on plant performance, product quality, and pollution prevention under the conditions tested. Based on the results, NPA decided to accept SSA for the commercial production of PA in November 2012. Since the control of blend ratio of SSA was critical to guaranteeing the product quality, SSA storage tanks and weighing equipment were newly implemented in the existing PA-manufacturing plant (Fig. 7.7). In addition, NPA acquired an industrial waste disposal business license from the government to purchase SSA from WWTP. This was absolutely needed because SSA was regarded as industrial waste by the Waste Management Law. The existing

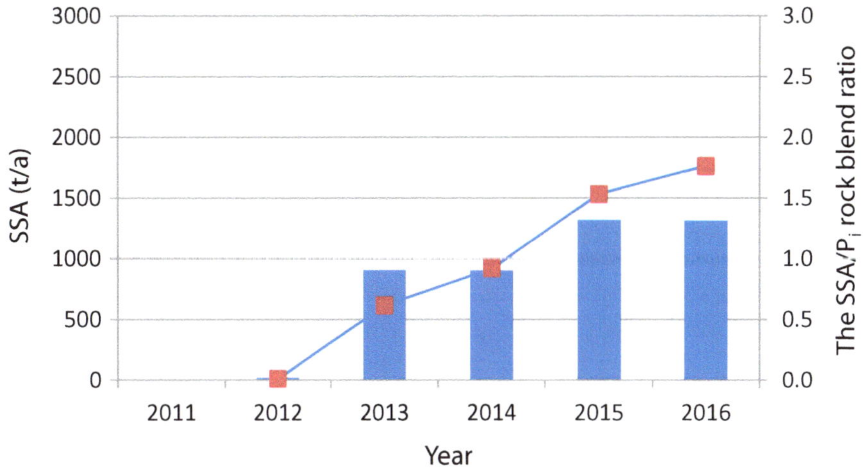

Fig. 7.8 Year-to-year changes in the amount of SSA accepted by NPA (*bar*) and the SSA/P$_i$ rock blend ratio for the manufacture of PA (broken line)

PA-manufacturing plant was therefore designated as an intermediate treatment facility for industrial waste. After a 5-month trial, the commercial production of PA using SSA commenced on April 2013 (Fig. 7.8). Currently, the blend ratio of SSA remains around 1.5% because of the shortage of SSA supply from WWTP. NPA has made efforts to develop collaborations with WWTP, advocating mutual economic benefits brought by the valorization of SSA. Although WWTP need to pay SSA shipping costs, they can not only earn revenue from selling SSA but also save sludge disposal costs.

7.5 Conclusions

NPA has realized the industrial use of SSA, which is purchased from WWTP, without modifying the existing PA-manufacturing process. Since SSA is available at costs considerably lower than P$_i$ rock, the use of SSA has allowed NPA to reduce the net cost of PA production on industrial scale. At present, SSA needs to be mixed with P$_i$ rock at a blend ratio of 2.5% or less before being subjected to the wet acid process. Although PA is readily released from SSA with sulfuric acid, it also inevitably dissolves metals from SSA, thereby contaminating PA solution. Metals such as Al, Fe, and Mg can form insoluble salts with PA in fertilizer-manufacturing processes, lowering the quality and value of fertilizer. To overcome this problem, it is necessary to develop a cost-effective technology capable of effectively removing metals from PA solution. Although various technologies have been potentially applicable to metal removal from PA solution, no technology has been successfully applied to industrial wet acid processes mainly due to economical reasons. Further study is needed to develop cost-effective technology for removing metals from PA solution.

In industrial PA-manufacturing processes, by-product gypsum is generated about 1.5 times more than P_i rock input as a raw material. Since the revenue from selling by-product gypsum is critical to the economic viability of industrial PA production, NPA makes every effort to guarantee the quality of not only PA but also gypsum to end users. Importantly, this effort can make a contribution to the improvement of resource efficiency by turning waste in a resource. In Japan, however, the wastewater treatment sector still gives the priority to P_i removal rather than P_i recycling. The precipitation of P_i with Al^{3+} or Fe^{3+} is a cost-effective, commonly used technology for removing P_i from sewage. The chemical precipitation of P_i with Al^{3+} or Fe^{3+} leads to the increase of metal contents, making the application of SSA to PA manufacturing difficult. Hence, it would be desirable to adopt biological P_i removal (known as Bio-P process) in WWTP rather than chemical P_i precipitation. In addition, to reduce the contents of silica, aluminum, and lead in SSA, it seems effective to avoid the mixing of primary and secondary sludge before incineration. Primary sludge generally contains more silica, aluminum, and heavy metals compared to secondary sludge (Kanjou and Sone 2017).

References

Akiyama T (1980) Formation and reaction of mineral constituents in compound fertilizer (in Japanese). Japan Phosphatic and Compound Fertilizers Manufactures' Association, Tokyo

Jun L, JianHua W, YunXiang Z (1997) Effects of the impurities on the habit of gypsum in wet-process phosphoric acid. Ind Eng Chem Res 36:2657–2661

Kanjou T, Sone K (2017) Reuse Recycl (in Japanese) 41(154):64–69

Mochiyama T, Morita K, Oishi Y (2017) Influence of fluorine and aluminum ion on crystal habit of gypsum in production of phosphoric acid by wet-process (in Japanese). J Soc Inorg Mater Jpn 24:3–8

Slack A (1968) Phosphoric acid. Marcel Dekker, New York

Chapter 8
Alkaline Leaching of Phosphate from Sewage Sludge Ash

Ken-ichi Sonoda

Abstract The alkaline technology has been applied to phosphate (P_i) recovery from sewage sludge ash (SSA) at wastewater treatment plants (WWTP). P_i is extracted from mono-incinerated sewage sludge with NaOH and recovered from the leachate using chemical precipitation with $Ca(OH)_2$. Approximately 30–40% of P_i could be recovered from SSA as calcium P_i while minimizing the leaching of toxic heavy metals at high pH. The recovered P product can be recycled as a fertilizing material for agriculture.

Keywords Alkaline leaching · Calcium phosphate · Sewage sludge ash · Phosphate fertilizer

8.1 Introduction

Phosphorus (P) needs to be removed from sewage to control eutrophication in natural bodies of water. The P removed from sewage ultimately ends up in sewage sludge at wastewater treatment plants (WWTP). Dewatered sewage sludge is often mono-incinerated to reduce its volume and to recover energy at WWTP. This allows phosphate (P_i) to be concentrated in the sewage sludge ash (SSA). Since SSA has a high P content, typically ranging from 20 to 30 wt% P_2O_5, it is becoming increasingly a priority target for P recovery in the wastewater treatment sector.

In Japan, about 300 incinerators are operating at WWTP. Together, they have a capacity of incinerating c. 25,000 t/year (tons per year) of dewatered sludge, generating c. 33,000 t/year of SSA. Approximately 70% of the SSA has been accepted by the recycling sector, mainly the cement industry in Japan. However, since P_i can cause a detrimental effect on the cement quality, it is becoming a nuisance issue for the cement industry to accept P_i-rich SSA from the wastewater treatment sector. Hence, it is critical to develop an alternative technology option that

K.-i. Sonoda (✉)
Business Strategy Division, METAWATER Co., Tokyo, Japan
e-mail: sonoda-ken-ichi@metawater.co.jp

H. Ohtake, S. Tsuneda (eds.), *Phosphorus Recovery and Recycling*,
https://doi.org/10.1007/978-981-10-8031-9_8

enables the valorization of ever-increasing P_i-rich SSA. In this chapter, the alkaline P leaching technology to recycle P from SSA to farmland will be described.

8.2 Alkaline P_i Leaching Technology

Figure 8.1 shows a schematic diagram of P_i recovery from SSA using the alkaline leaching technology (Yanase 2009). In the alkaline leaching process, ash is mixed with 1.0 M NaOH solution to extract P_i (PO_4^{3-}). The P_i-rich leachate is separated from the rest (de-phosphorus ash) by mechanical dewatering. P_i is recovered as calcium phosphate ($Ca_3(PO_4)_2$) by the addition of $Ca(OH)_2$ to the leachate in the precipitation process. The phenomenological reaction for P_i leaching and precipitation may be given by:

$$P_2O_5 + 6OH^- \rightarrow 2PO_4^{3-} + 3H_2O \tag{8.1}$$

$$2PO_4^{3-} + 3Ca(OH)_2 \rightarrow Ca_3(PO_4)_2 + 6OH^- \tag{8.2}$$

After the precipitation process, the slurry is subjected to mechanical dewatering to recover P_i. The high-pH rejected liquor is returned to the P_i extraction process. This is effective in saving the chemical costs for P_i recovery from SSA.

Figure 8.2 shows the full-scale process for P_i recovery from SSA using the alkaline leaching technology (Moriya 2009). Ash, which is collected by a dust collector of the incineration system, is fed to the P_i leaching tank through a buffer ash hopper. SSA is then mixed with 1.0 M NaOH solution at temperature of 50–70 °C for about 30 min of gentle mixing. NaOH is dissolved in hot water, which is prepared using waste heat energy of the exhaust gas from the mono-incinerator.

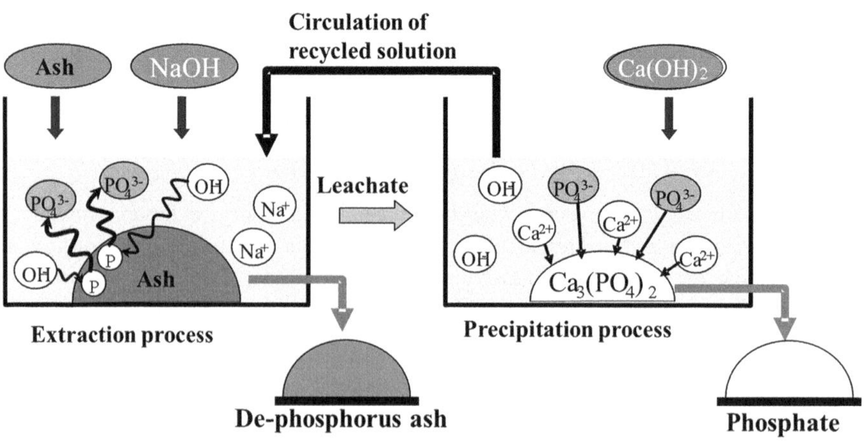

Fig. 8.1 A schematic diagram of P_i recovery from SSA using the alkaline leaching technology

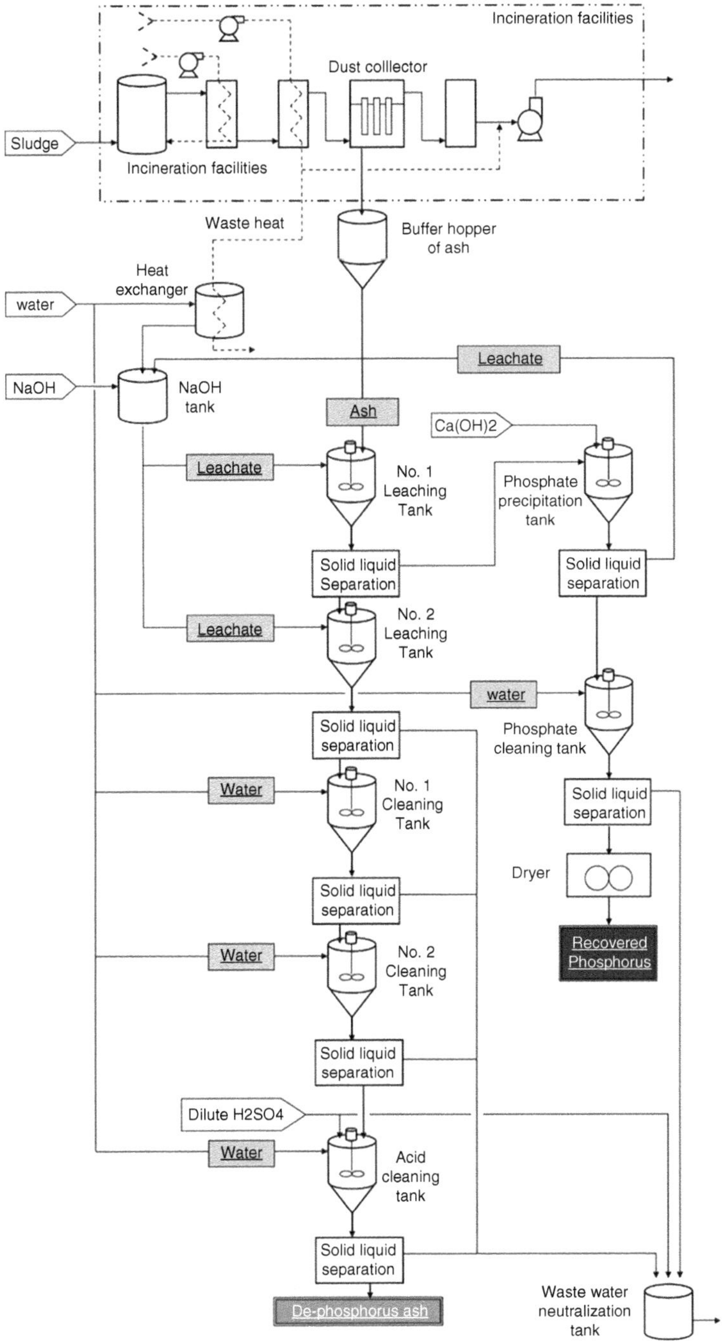

Fig. 8.2 Process flow of alkaline leaching of phosphate from SSA

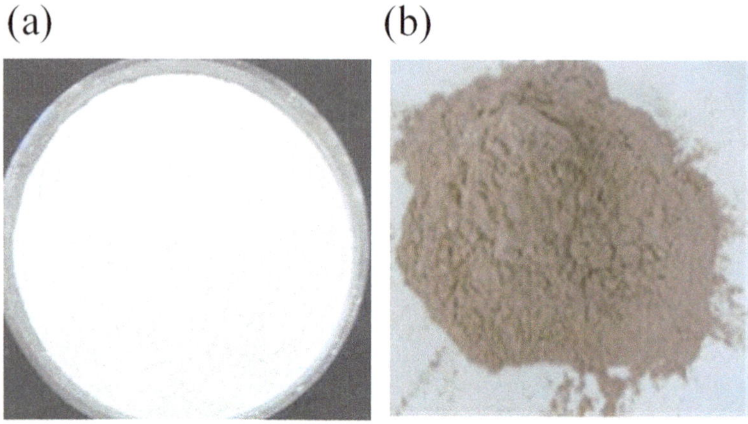

Fig. 8.3 Recovered P_i product (a) and de-phosphorus ash (b)

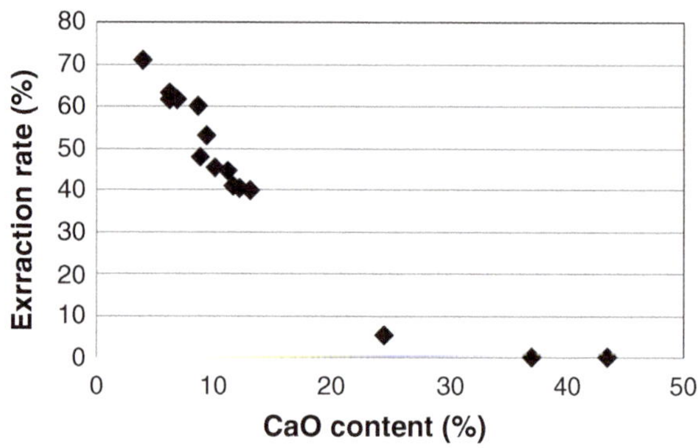

Fig. 8.4 The relationship between the CaO content of SSA and the rate of P_i leaching

Although the temperature required for P_i leaching is about 50–70 °C, the heat energy can be supplied by the waste heat from the incinerator, thereby saving the energy costs.

In the P_i leaching tank, the concentration of slurry is kept around 8–10%. In case that the P_i content in SSA is high, the P_i leaching step can be repeated to increase the amount of P_i leached. After leaching P_i, the mixture is subjected to mechanical solid-liquid separation. Then the P_i-rich liquor is sent to P_i precipitation tank where $Ca(OH)_2$ is fed at Ca/P molar ratio of 1–1.5. The mixture is slowly stirred for about 6–18 h to gradually dissolve $Ca(OH)_2$ and to generate calcium P_i as precipitates.

Calcium P_i precipitates are recovered by mechanical dewatering, washed with water, dried, and stored for shipping (Fig. 8.3a). To reduce the chemical cost, high-pH rejected water is returned to the NaOH tank for NaOH recycling. The solid material remaining after P_i leaching is washed two times with water and finally with dilute H_2SO_4 at pH of 4.5–5.5 to remove toxic heavy metals. The resulting solid (de-phosphorus ash, Fig. 8.3b), which meets the environmental quality standard for soil in Japan, can be used for cement, asphalt filler, and roadbed.

The alkaline leaching technology can recover approximately 30–60% of P_i from SSA in the form of calcium P_i. Importantly, the rate of P_i leaching from SSA is strongly dependent on the CaO content of SSA. Figure 8.4 shows the relationship between the CaO content in SSA and the rate of P_i leaching in the alkaline extraction step (Moriya 2009). There is a clear relationship between them. The alkaline P_i leaching becomes difficult when the CaO content of SSA exceeds 20%. Hence, if lime is used for sludge dewatering, the alkaline leaching technology is not a suitable option for P_i recovery from SSA.

8.3 Recovered Products

The weight of recovered P product is about 30–50% of SSA on a dry weight basis. The typical composition of recovered P product is shown in Table 8.1 (Yanase 2009). The total P content of recovered product is about 30% P_2O_5, most of which is citric acid-soluble P. The levels of toxic substances such as As, Cd, and Se are much lower than their regulation levels for by-product P_i fertilizer in Japan. Since the main component of recovered product is calcium P_i, it can also be used as a substitute for P_i rock in a variety of technical applications.

The alkaline P_i leaching process generates about 0.75 tons of de-phosphorus ash from each ton of SSA treated. De-phosphorus ash is a microporous material whose average particle size is typically about 50% smaller than that of SSA. Therefore, the specific surface area of de-phosphorus ash is about one order of magnitude larger than that of SSA. As mentioned before, de-phosphorus ash meets the environmental quality standard for soil in Japan so that it can be used for cement, asphalt filler, and roadbed. The stable sale of recovered P product and de-phosphorus ash is critical to bring economic benefits to WWTP. However, fertilizer companies do not accept recovered P product unless it brings some economic benefits to their business. The expanded use of this technology requires further reduction of the cost of chemicals required for P_i leaching and recovery.

Table 8.1 The typical composition of recovered P_i product

Main components				Toxic substances		
Total P	C-P[a]	Ca	Al	As	Cd	Se
[% P_2O_5]	[% P_2O_5]	[%CaO]	[% Al_2O_3]	[mg/kg]	[mg/kg]	[mg/kg]
29.8	29.2	45.1	4.40	21.4	1.4	6.1

[a]Citric acid-soluble P

8.4 Conclusions

The alkaline leaching technology can be applied to P_i recovery from sewage sludge ash at wastewater treatment plants. Approximately 30–40% of P_i could be recovered from sewage sludge ash as calcium phosphate while preventing toxic heavy metals from leaching at high pH. The rate of alkaline P_i leaching from SSA is strongly dependent on the CaO content of SSA. The alkaline P_i leaching technology is suited for P_i recovery from SSA having a low CaO content. Recovered P product can be used as a fertilizing material for agriculture, while the rest is usable for cement, asphalt filler, and roadbed after being washed with a dilute acid solution.

References

Moriya Y (2009) Phosphorus recovery technology from sewage sludge ash. J Resour Environ 45(6):35–39. (in Japanese)

Yanase T (2009) Phosphorus recovery and detoxification system from sewage sludge ash. J Earth Environ 40(4):72–73. (in Japanese)

Chapter 9
Phosphorus Recovery from Sewage Sludge Ash: A Case Study in Gifu, Japan

H. Nakagawa and J. Ohta

Abstract The full-scale plant for recovering P from sewage sludge ash (SSA) started operation at a wastewater treatment plant in Gifu City, Japan, in April 2010. P is released from SSA using the alkaline (NaOH) leaching technology and recovered as calcium hydroxyapatite (HAP). The full-scale plant recovers approximately 30–40% P_i from SSA and supplies about 300 t/year of by-product P_i fertilizer, named Gifu-no-daichi®, mainly to local farmers. The P_i fertilizer, Gifu-no-daichi®, is sold through the JA-Zen-Noh (National Federation of Agricultural Cooperative Associations) in Gifu City with a favorable reputation from farmers. Gifu City has been increasing the volume of the fertilizer sales year by year and considers it is critical to make the P recovery process more efficient and stable.

Keywords Sewage sludge ash · Calcium phosphate · Phosphate fertilizer

9.1 Introduction

Gifu is a city located in central Japan with a population of about 410,000. In 1934, Gifu started to construct the sewer system for the first time in Japan. The Central wastewater treatment plant (WWTP), which was the first plant implemented in Gifu, began operation in July 1937. Since then, the Gifu's sewerage system has been serving citizens and protecting the environment (Fig. 9.1). Currently, Gifu also uses the prefectural intercommunity sewerage system, which began operation in April 1991, to cover the city area at the north bank of the Kiso River. Today, 6087 ha and 2669 ha (a total of 8756 ha) of the city area are covered by the city's own and the intercommunity sewerage system, respectively. The sewer-served population of Gifu City is now 382,090, covering 92.2% of the city's total population.

Gifu is currently operating four WWTP in the sewerage areas (Fig. 9.1). Following the Central WWTP, the North and South WWTP started to treat

H. Nakagawa · J. Ohta (✉)
Gifu City Office, Gifu, Japan
e-mail: oota-jni@city.gifu.gifu.jp

H. Ohtake, S. Tsuneda (eds.), *Phosphorus Recovery and Recycling*,
https://doi.org/10.1007/978-981-10-8031-9_9

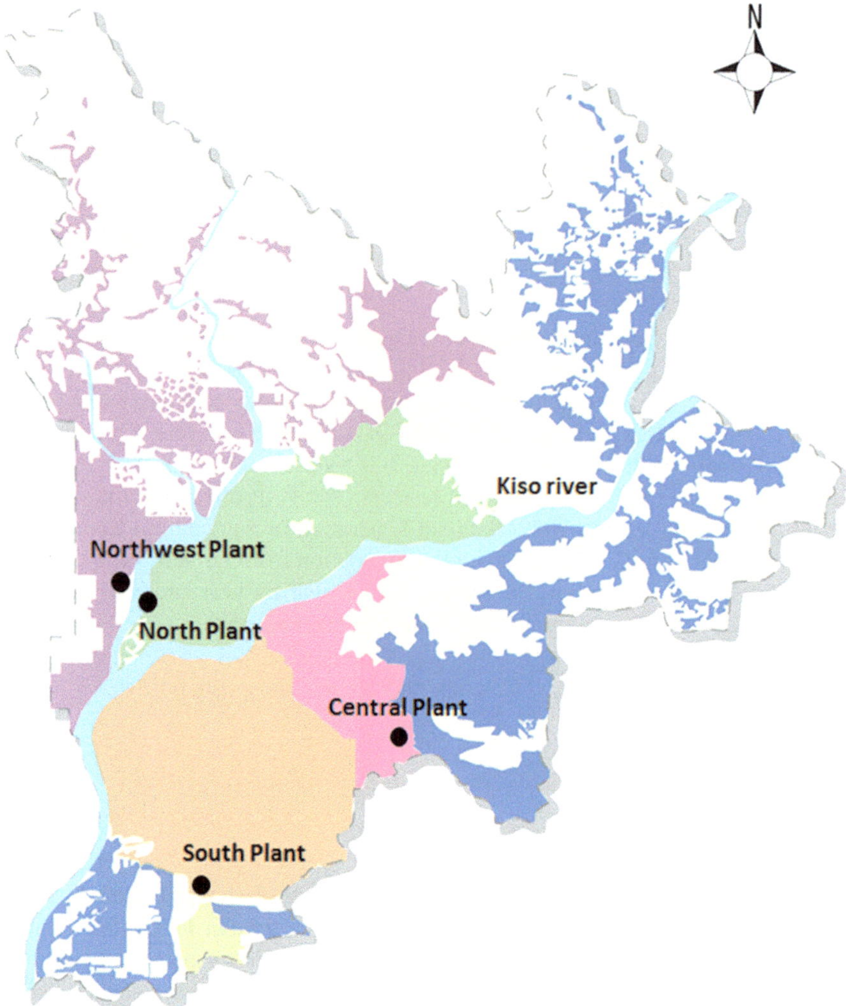

Fig. 9.1 Sewerage areas and wastewater treatment plants in Gifu City

wastewater in July 1966 and June 1973, respectively. Finally, the Northwest WWTP commenced operation in February 2004. The city's four WWTP generate a total of 29,000 t/a of dewatered sludge. The North and South WWTP have a mono-incinerator that is fed solely with dewatered sludge. The Central and Northwest WWTP transport dewatered sludge by road to the North WWTP, where it is incinerated in the mono-incinerator. Since the sewage sludge ash (SSA) of the South WWTP is also transported by road to the North WWTP, a total of 700 t/a of ash piles up at the North WWTP. No biogas production plants have been installed at the WWTP, mainly because of space shortages.

9.2 Historical Background

Since the mono-incineration of sewage sludge started in 1975, the cost reduction of ash disposal has been a priority challenge in the management of Gifu's WWTP. To address this challenge, Gifu City government decided to use SSA for the manufacture of construction bricks at the North WWTP in 1994. Although bricks could be used for surfacing sidewalks and parks in the city area, the demand for bricks gradually declined year after year because of spending cuts for public works. Accordingly, Gifu decided to cease the manufacture of bricks and alternatively to use SSA as a fertilizing material. This was motivated by the price spike of phosphate (P_i) rock in 2008. Meanwhile, Gifu City has adopted bio-P processes to upgrade the activated sludge process at the WWTP to enhance P_i removal from sewage since 1998. The P content of sludge ash has increased up to 20–35% P_2O_5, which was nearly equivalent to that of low-grade P_i rock.

Gifu City started a collaborative research project with Metawater Co. (formerly Nihon Gaishi Co.) in 2003 to develop a technology for recovering P from SSA (see Chap. 8). Then, the city participated in a 2-year development project, called the LOTUS (Lead to Outstanding Technology for Utilization of Sludge) project, which was supported by the Ministry of Land, Infrastructure, Transport and Tourism (MLITT) in 2004 (Nakagawa 2010). The feasibility study was conducted using a test plant which was installed at the North WWTP in 2005. After MLITT's project evaluation in 2006, Gifu City acquired a business license endorsed by Japan's sewerage law in 2007 and applied for legislation of the by-product P_i fertilizer in 2008. The full-scale plant was constructed in the North WWTP with the cost of approximately 0.7 billion Japanese yen (c. 7 million euros) in 2009. Finally the full-scale P recovery plant started operation in April 2010 (Fig. 9.2) (Nakagawa 2011).

9.3 P Recovery from SSA

At present, the North WWTP generates approximately 200 t/a of recovered P product (i.e., calcium hydroxyapatite, HAP) using the alkaline leaching technology. P_i is released from SSA using NaOH at 50–70 °C for 1.5 h (Fig. 9.3). The residual heat of the mono-incinerator is used for heating the NaOH solution. After repeating this several times, the leachate is obtained using a membrane-type liquid-solid separator. The alkaline leaching technology was adopted because acid leaching inevitably dissolved heavy metals, thereby contaminating recovered products. After P_i is extracted (mainly from $AlPO_4$) with NaOH, it is precipitated with $Ca(OH)_2$ from the leachate at 20–50 °C for 9 h. The NaOH solution is returned to the head of P_i-leaching reactor for recycling, thereby saving chemical costs. The recovered HAP is washed with water, dewatered, dried, granulated into pellets, and packed in a flexible container bag. The recovered P product is a white powder having no sewage odor and contains 25–30% citrate-soluble P_2O_5 (Fig. 9.4). The

Fig. 9.2 Full-scale plant for P recovery from SSA at the North WWTP of Gifu City

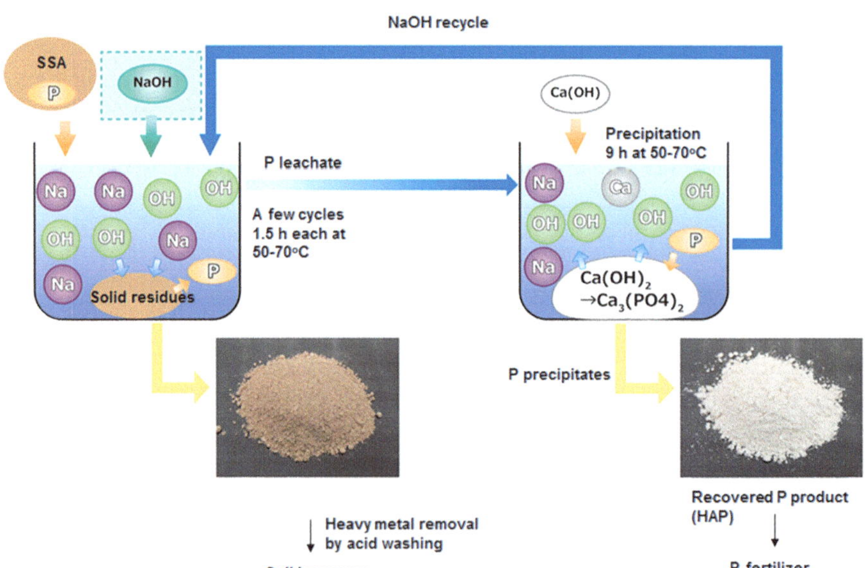

Fig. 9.3 The flow diagram of P_i recovery process at the North WWTP in Gifu City

Fig. 9.4 The recovered P product (HAP)

contents of toxic heavy metals such as As and Cd were much lower than their regulatory levels for by-product P_i fertilizer in Japan. The contents of Hg, Ni, Cr, and Pb were even low compared to the strict level for calcined sludge fertilizer. The recovered P product can be used as a slow-release by-product P_i fertilizer or a component for compound fertilizers. It is also possible to use it as a raw material for some industrial applications.

The efficacy and safety of the recovered product as P_i fertilizer were confirmed by plant cultivation tests. The recovered product, which warranted product safety and quality requirements specified by the Fertilizer Regulation Act, was registered as by-product P_i fertilizer by the Ministry of Agriculture, Forestry and Fisheries of Japan in March 2009. The P_i fertilizer is now sold to local farmers under the brand of Gifu-no-daichi® (The land of Gifu). This brand name, which was chosen from among names proposed by the Gifu office, implies that it would be consumed as a locally sourced fertilizer, thereby contributing to P recycling in the land of Gifu. The solid residues that are left after P recovery are subjected to weak-acid cleaning with a dilute H_2SO_4 solution for 5–30 min at 50–70 °C to remove heavy metals. After drying, they became brown-colored pellets having no sewage odor (Fig. 9.5). Since they meet environmental criteria for soil contamination prevention, they can be used as roadbed materials or a substitute for stone powder in asphalt filler. At present, they are also sold as a soil improver to materials companies.

Fig. 9.5 Solid residues that are left after P recovery

9.4 Fertilizer Sales

When the full-scale plant began operation, no channel was available for the sale of Gifu-no-daichi®. To create sales channels, Gifu City has conducted advertising campaigns, including free sample offer, briefing session for farmers, advertisement in a local magazine, and leaflet distribution. As a result, JA-Zen-Noh (National Federation of Agricultural Cooperative Associations) has launched the sales and marketing of Gifu-no-daichi®. Currently, Gifu-no-daichi® is sold in a 20 kg package at the JA branches in Gifu City and enjoys a good reputation from farmers (Fig. 9.6) (Ohta 2014).

Gifu City is currently making efforts to establish stable channels for the distribution and sale of recovered P products. It prioritizes the sales of Gifu-no-daichi® in the Gifu area to locally recycle P from wastewater to farmland. Gifu City began the sale of Gifu-no-daichi® by tons to fertilizer companies in July 2011 and has been increasing the volume of the fertilizer sales year by year. Hereafter, Gifu City gives the priority to making the P recovery from sludge ash more efficient and stable. The P recovery operation has brought some economic benefits to Gifu City, since it reduces the total operating cost compared with the manufacture of construction bricks. However, if compared with ash disposal as industrial waste, the P recovery cost is still high due to the cost of chemicals (mainly $NaOH$ and $Ca(OH)_2$). Further cost reduction is likely required to make P recovery from incinerated sludge ash more economically viable (Ohta 2014).

Fig. 9.6 By-product P_i fertilizer, Gifu-no-daichi®

9.5 Conclusions

The full-scale plant for recovering P from SSA started operation at the North WWTP in Gifu City in April 2010. P is released from SSA using the alkaline (NaOH) leaching technology and recovered mainly as calcium hydroxyapatite (HAP). The full-scale plant recovers approximately 30–40% P_i from SSA and supplies about 300 t/year of by-product P_i fertilizer, named Gifu-no-daichi®, mainly to local farmers. The P_i fertilizer, Gifu-no-daichi®, is sold through the JA branches in Gifu City with a favorable reputation from farmers. Gifu City is increasing the volume of the fertilizer sales year by year and considers it is critical to make the P recovery process more efficient and stable.

References

Nakagawa H (2010) Phosphorus recovery project in Gifu city. J Civil Eng 51:43–46. (in Japanese)

Nakagawa H (2011) Strategy for recovering phosphorus from sewage sludge ash in Gifu city. J Jap Soc Water Environ 34:16–20. (in Japanese)

Ohta J (2014) Phosphorus recovery from sewage sludge ash. Agric Hortic 89:436–440. (in Japanese)

Chapter 10
Urban Phosphorus Mining in the Canton of Zurich: Phosphoric Acid from Sewage Sludge Ash

L. Morf, S. Schlumberger, F. Adam, and G. Díaz Nogueira

Abstract The constitutional duty of sustainability in Switzerland requires economical use of valuable and scarce raw materials such as phosphorus. The government of Canton of Zurich recognised this as an opportunity around 10 years ago with respect to the bottlenecks in waste disposal threatening to materialise in 2015 in the existing sewage sludge disposal plan. In 2007, it already gave the Department of Public Works the assignment to design the future sewage sludge disposal so that phosphorus recovery is possible. In 2009, an evaluation of all then-known phosphorus recovery procedures as well as their integration into different sewage sludge disposal pathways showed that the procedures with P recovery from sewage sludge ash are clearly superior to P recovery from sludge and sludge water. The first milestone in the implementation was that it was possible within 6 years and with the involvement of all parties impacted to realign the existing sewage sludge disposal concept completely with respect to the new framing conditions. Since mid-2015, a new central sewage sludge treatment plant at the most optimal location in the Canton has been producing high-phosphorus ash from incinerated sewage sludge. It contains more than 90% of the phosphorus potentials of the entire potential in untreated community waste water from the Canton. By switching over the sewage sludge disposal system from an inefficient, decentralised one to an efficient, centralised system, it has been possible to cut the average sewage sludge treatment costs by more than half, including ash disposal. No modifications to the waste water treatment plants were needed. The Canton has worked with the Foundation ZAR and selected development partners on this implementation of the large-scale engineering

L. Morf (✉) · F. Adam
Public Works Department, Canton Zurich, AWEL, Office for Waste, Water, Energy and Air, Zurich, CH, Switzerland
e-mail: leo.morf@bd.zh.ch

S. Schlumberger
Foundation Development Center for Sustainable Management of Recyclable Waste and Resources (ZAR), Zuchwil, CH, Switzerland

G. Díaz Nogueira
Management Senior Advisor, Técnicas Reunidas (TR), Madrid, Spain

H. Ohtake, S. Tsuneda (eds.), *Phosphorus Recovery and Recycling*,
https://doi.org/10.1007/978-981-10-8031-9_10

of phosphorus recovery from sewage sludge ash since 2011 (Phosphorus-Mining-Project). The initial focus lay on the production of high-grade raw material for fertiliser. This led, among other risks in product sales, to settling on the production of technically pure, conventional phosphoric acid as an already established product. Currently there is development work going on using the Phos4life® procedure by Técnicas Reunidas. The attempts to this point demonstrate a P-recovery rate of >95% from the ash, a material recycling of the minerals and the separated metals as well as the use of iron as a precipitant.

Keywords Urban mining · Phosphorus mining · Sewage sludge · Waste water · Phosphorus recovery · Sewage sludge ash · Phos4life®

10.1 Introduction

Phosphorus (P) is a vitally important nutrient for humans and animals. This resource cannot be substituted for with another substance and is becoming more and more precious as a raw material. In contrast to nitrogen, for example, which is present in the air in nearly limitless quantities, P only appears in natural (geogenous) ore deposits and, after they are used, in the 'waste' (useable as anthropogenic deposit). Even other nutrients such as potassium (K), calcium (Ca), magnesium (Mg) and carbon (C) are from a current perspective no longer scarce. The recoverable geogenous phosphate ore deposits are on the other hand limited worldwide. Geopolitically they are concentrated in just a few countries, and they will only become scarcer in the foreseeable future.

Switzerland, like most other European nations, has no phosphorus deposits of its own. Explorations have shown that the quantity of phosphorus bound up in sewage sludge annually is as great as the total mineral fertiliser imported into Switzerland in which is approximately 6000 t P/a. Figure 10.1 presents a simplified version of the P-inventory in Switzerland at the start of this millennium.

Ores from these scarce geogenous sources also contain increasingly high shares of harmful substances (e.g. cadmium and uranium). These harmful substances, when used in fertiliser lead to an ever-increasing concentration of pollutants in the soil. Just how critical the situation is today with the concentration of pollutants in Switzerland can be seen in the studies commissioned by the Federal Department of Agriculture (BLW 2015; FitzGerald and Roth 2015). Figure 10.2 shows the situation in 2011–2012 with respect to the quality of fertiliser used in Swiss agriculture, for example, the cadmium pollution. This makes it apparent that just about half of all mineral fertilisers examined do not fall under the applicable limit of 22 mg Cd/kg P_2O5. At this limit the precautionary principle for soil protection can be satisfied

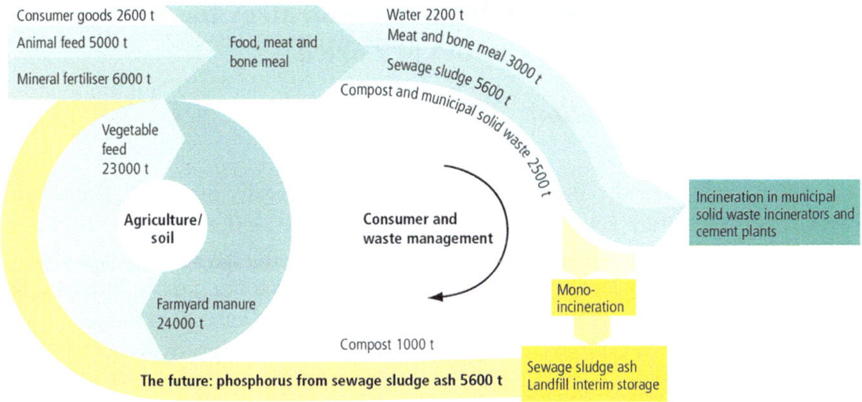

Fig. 10.1 Simplified depiction of phosphorus management in Switzerland (AWEL 2008)

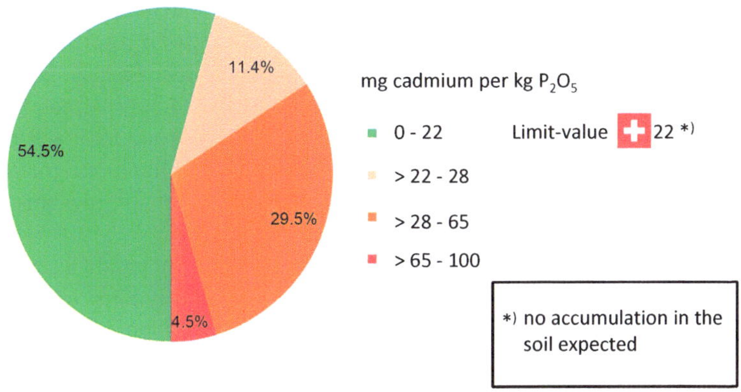

Fig. 10.2 Overview of the quality of the mineral fertiliser used in Switzerland for cadmium (BLW 2015)

so that no long-term accumulation is to be expected. The planned pathway to the limit from the European Union (EU) provides for a timed, gradual procedure for reaching a similar end value of 20 mg Cd/kgP_2O_5.

Both of these facts demand perspicacious and holistic action with respect to this highly valuable resource and its long-term use. The increasing soil pollution from mineral fertilisers with excessive heavy metals content is not sustainable. Along with phosphorus, there are other valuable resources such as iron and other metals together with the organic component, which is a valuable energy source; the best possible use should be made of all of them.

10.2 Urban Phosphorus Mining: New Strategy for Sewage Sludge Disposal in Canton of Zurich

10.2.1 Starting Situation

In Canton of Zurich, the most populous and economically strongest Canton in Switzerland, 69 public sewage treatment facilities treat 230 million m^3 of waste water annually and produce in total around 550,000 m^3 of digested sewage sludge with 6% dry substance (DS) per year. This corresponds approximately to 100,000 tons of dewatered sewage sludge (DSS) with a DS of approximately 30%. It was last possible in 2005 to deliver it directly for agricultural use. In 2003 the sewage sludge disposal plan already created a basis in Canton of Zurich for the thermal mineralisation of Zurich's sewage sludge and thereby anticipated the national ban on release that started in 2006. The precautionary principle in Environmental Protection Act was thus consistently applied in order to prevent the emission of pollutants in the direct release of this sewage sludge as a nutrient source in agriculture. The primary focus lay at that time on safety of disposal as well as utilisation of energy. The recovery of phosphorus played no role in it. As a consequence, the disposal pathway in the Canton was distributed approximately as follows: around 65% in waste-to-energy plants (WtE plants), around 9% – after being dried using fossil fuels – in cement works, ca. 25% in two smaller mono-incineration plants (with no separate deposition of ash containing P) and ca. 1% in plants outside of the Canton.

In 2006, it was recognised that the existing disposal concept would lead to bottlenecks in capacity starting in 2015. It also became increasingly apparent that phosphorus as a vital nutrient is a limited resource and the supply of low-pollutant mineral fertilisers is no longer adequately secured. Both were taken by the decision-makers in Canton of Zurich as an impetus, with respect to sewage sludge disposal to define a sewage sludge recovery strategy taking phosphorus recovery into account as the basis for the objectives of a modern, sustainable waste and resource management programme.

10.2.2 New Strategy

The government of Canton of Zurich therefore formulated the following clear limiting conditions already in 2007 in its decision to the "Implementation of sewage sludge disposal plan" (RRB 572/2007):

The limiting conditions for planning future disposal pathways derive from the decisions on phosphorus recovery and energy use. In the sense of sustainable resource management, planning of future disposal pathways, especially in the construction of new plants, must be designed so that the (later) recovery of the resource phosphorus is possible. The renewable energy trapped in sewage sludge should be used in the most economical way possible regardless of where it is treated.

The strategy is a concrete and important contribution to the urban mining strategy that is being used, defined in the plan of action for waste and resource management for the Canton of Zurich (AWEL 2010). It is based on the principles of sustainable development, the federal natural resources policy (BAFU 2006), and follows, e.g. the recommendations of the expert advisory board on the environment and the European resource strategy as well (SRU 2005).

10.3 Evaluation and Implementation of the Strategy

10.3.1 Preliminary Project and Field Observations for Selecting the Procedure

A specially assigned working group composed of representatives from all parties responsible for sewage sludge disposal in the Canton reached the conclusion in the framework of its first discussions that the greatest energy use benefit and at the same time the demand for future recovery of phosphorus can be fulfilled with a single sewage sludge treatment plant (SSTP) – designed as a sewage sludge mono-incineration plant – in the immediate vicinity of an WtE or a large waste water treatment plant (WWTP).

Parallel to the development of the preliminary project, the agency for waste, water, energy and air (AWEL) of the Canton of Zurich issued the commission for the project "Field Observation Sewage Sludge Disposal Canton Zurich" (AWEL 2009). The objective of this project was to clarify whether there are better, alternative pathways to the preferred method of sewage sludge disposal through sewage sludge mono-incineration with phosphorus recovery from the sewage sludge (incineration) ash that is produced.

In principle, there are four possible options for the use of P in the entire disposal chain (Fig. 10.3). Procedures in which the P recovery proceeds from waste water and sludge water (1 + 2), or from sludge (3) from a waste water treatment plant (WWTP). Construction on the WWTPs is necessary for these three options. The forth option is to recover the phosphorus from the ash from sewage sludge mono-incineration (4). In this case, there would be no need for structural and operational interventions in the WWTPs. The phosphorus present in the sewage sludge ash can be recovered most efficiently with respect to the total potential in the waste water input from the WWTPs.

Any alternatives would need to comply with the framing conditions of the council's decisions (see above) as well as the applicable laws, be compatible with the existing inventory of waste water treatment plants in Canton of Zurich and conform to the technical state of the art. In the first evaluation phase, all known P-recovery procedures in 2009 were assessed according to the following exclusion criteria:

- An operational start-up of the sewage sludge disposal variant with the selected P-recovery procedure must be possible according to the prescribed planning as

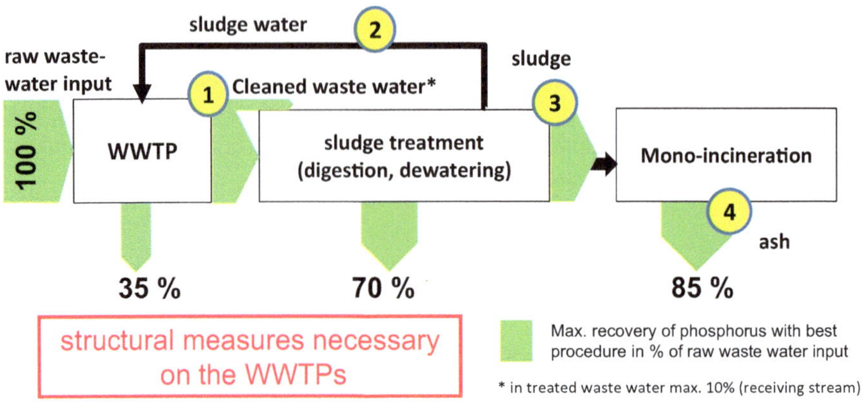

Fig. 10.3 Possible sites of maximum phosphorus recovery rates with the best procedures in % of the supply in the WWTP in waste water management and sewage sludge disposal (Source: AWEL, according to Montag 2008)

of 2015. This means that phosphorus recycling must be assured and/or secured as of 2015 (at best through later use from an interim storage facility to be constructed).

- The energy use from the sewage sludge must be secured.
- Applicable laws must be obeyed.

Only 8 of the 29 recovery procedures known from literature and practice at that time corresponded to the requirements stated above. For each of the eight selected P-recovery procedures, additional data were collected, and a detailed assessment was made with respect to the main criteria of costs, energy consumption and reclaimable quantity of phosphorus for the planned full expansion in Canton of Zurich.

The evaluation provided the following results (AWEL 2009):

Crystallisation procedures from waste water and sludge water in WWTPs yield a certain advantage with respect to costs and energy consumption at very low phosphorus recovery quotas. Procedures with recovery from sewage sludge ash from mono-incineration plants achieve the highest P-recovery rates, and the costs per quantity of phosphorus are lower than in procedures with P recovery from sludge. Procedures with P recovery from sludge are very expensive and require once again considerably more energy than procedures from incinerated sewage sludge ash (ISSA). They cannot compensate for this with the significantly higher P recovery compared to the procedures from waste water. From the insights gained at the time, it was clear that the procedure with P recovery from ISSA from mono-incineration plants was on the whole definitely superior to the alternative procedures with P recovery out of sludge and sludge water.

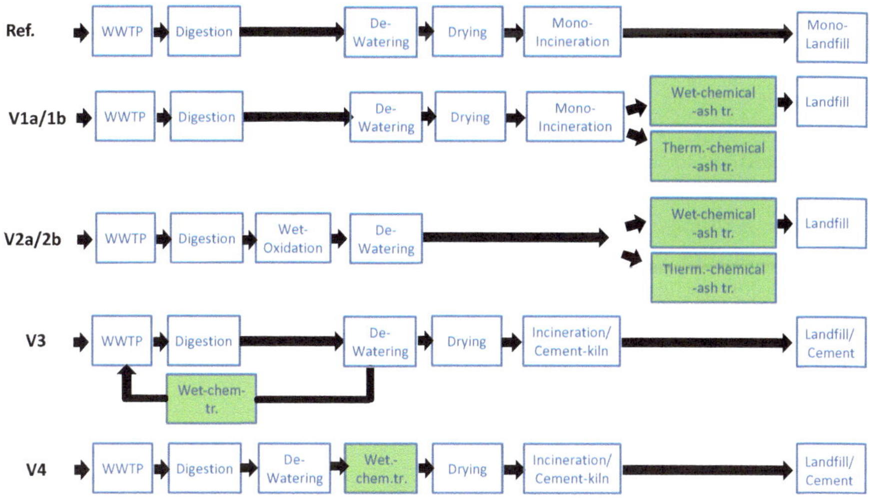

Fig. 10.4 Different variants of sewage sludge disposal chains with possible locations/procedures for P recovery compared with the reference disposal pathway of sewage sludge mono-incineration with P recovery from its ashes; system limit of process chains in each case from raw waste water input into the WWTP (AWEL 2009)

In the following the selected reference disposal pathway (Ref. = mono-incineration with interim storage of ISSA at a monofill) is compared with various alternative disposal pathways (V1a to V4: see Fig. 10.4).

The assessment was completed in an evaluation of the entire system with the system limit starting from raw waste water as input into the waste water treatment plant. For each process chain in the alternative disposal pathway, the following aspects were assumed: (a) compliance with applicable laws, (b) purification of the waste water until the required waste water input quality was achieved and (c) production of a market-ready fertiliser (raw material) or industrial raw materials and/or landfill-capable residual materials in the event that temporary interim storage of ash containing phosphorus is necessary. The same criteria were used for the assessment of the disposal pathways as were considered separately in the comparison of P-recovery procedures (see above). They were, however, supplemented with the criteria transport cost and CO_2 balance of the process chain. Thus, the three dimensions of the requirements for future sewage sludge planning were covered:

- Quantity dimension: Share of P in the waste water of Canton of Zurich, which can be recycled as fertiliser and/or industrial raw material (in %)
- Ecological dimension: Quantity of energy that can be obtained and/or must be used (in MJ/t DSS) as well as CO_2 balance (g CO_2/t DSS)
- Economic dimension: Costs/yield (in CHF/t DSS)

10.3.2 Summary of the Evaluation: Selection of Procedure

The clarifications by AWEL (AWEL 2008) led to the conclusion that the reference disposal pathway "sewage sludge mono-incineration with (future) P reclamation from the ash" with the prescribed criteria (state of the art, P recovery, energy and CO_2 requirements) best fills the criteria and evinces some additional clear advantages (low additional load on the WWTP, residual materials are compatible with landfilling, proven technology). For P recovery there are basically two variants that came into consideration: (V1a) wet-chemical processing and (V1b) thermochemical processing.

The following alternative disposal pathways were thus no longer given consideration for the implementation of the strategy:

- Wet oxidation in combination with wet-chemical or thermochemical processing out of ISSA (V2a/2b)

  This variant evinced in comparison with sewage sludge mono-incineration a considerable worse energy – and/or CO_2 balance at similar P-recycling rates and comparable costs. It had other disadvantages:

  (i) Added load on the WWTP.
  (ii) Compatibility of residual materials with landfilling in the event that the ash would need to be temporarily stored would not be assured.
  (iii) The technology is not adequately tested for community sewage sludge (disposal safety would be in danger).

- Thermal (co-)processing of sewage sludge in waste-to-energy plants (WtE) or in cement work with upstream P recovery from waste water (Var. 3). This is because it became apparent that P-recovery procedures from waste water plants evince unacceptably low in P-recycling rates.
- Disposal in the WtE plants (or cement works and/or wet oxidation) with P-recovery procedures from sludge (Var. 4). With a modest P-recycling rate, this variant was not economically feasible.

Based on these insights, the decision-makers responsible for sewage sludge disposal in Canton of Zurich reached the following decisions at the end of 2009:

- Mono-incineration of the sewage sludge which has been digested and optimally dewatered at various locations will be put into action as befits the state of the art.
- A single, centralised incineration line realised as fluidised bed combustion at the best location in the Canton in order to meet the required capacity (approx. 100,000 tDSS/a). WtE plants are still available for emergency disposal and/or revisions.

- The Canton will allocate all cantonal sewage sludge to the central plant (investment security).
- More detailed clarifications of P recovery from ISSA will be carried out.

Today (2017), around 10 years later, it can be stated that these insights are in essence still valid. Both the results of the comprehensive support initiative of the German Agency for Education and Research (BMBF) "Recycling Economy for Plant Nutrients, Especially Phosphorus" (Pinnekamp et al. 2011) and the final statement of the European research project P-REX (Kabbe et al. 2015) affirm the selection made in 2009 as expedient.

10.3.3 Selection of Location and Allocation Decision

Detailed investigation was done for a total of five possible locations in the Canton, to determine which would be the best location for a new, centralised sewage sludge mono-incineration plant. The assessment criteria are selected on the same way as for the assessment of the disposal pathways and in comparison with the P-recovery process (see Sects. 10.3.1 and 10.3.2).

The parties responsible for sewage sludge disposal in the Canton today then reached a unanimous decision in June 2010 that the location of the sewage treatment facility Werdhölzli in the middle of the city of Zurich with its approximately 400,000 residents was ecologically and economically the best solution. Decisive for the selection were the optimal possibilities for energy recycling and CO_2 reduction, since the location already has the infrastructure and sufficient potential for energy use. In the transports, the locations on the periphery of the Canton proved to be poor, as expected. The decisive factor there by is the proximity to the main accumulating point for sewage sludge from the city of Zurich itself with approximately a third of the total amount.

The City Council of Zurich then confirmed that it was taking charge of the planning and realisation of the plant. To assure planning security for the future construction and operation of the new centralised sewage sludge treatment plant in Werdhölzli starting in mid-2015, the currently valid cantonal sewage sludge disposal plan will be adapted, and all Zurich sewage sludge will be allocated anew to the central sewage sludge treatment plant for 20 years starting from mid-2015. In a consultation process in the 171 political communities as well as all WWTP joint management authorities, the draft for a new sewage sludge disposal plan met with great affirmation, thanks to a great deal of persuading, and it was approved by the government of Canton of Zurich at the beginning of August 2011 (Cantonal sewage sludge disposal plan 2015, RRB 1035/2011).

Fig. 10.5 New central sewage sludge treatment plant in Werdhölzli (Source: ERZ)

10.3.4 The New Central Sewage Sludge Recycling Plant Werdhölzli

On March 3, 2013, the people of the city of Zurich approved the building credit of around 68 million francs for the construction of the new central sewage sludge recycling plant on the property at Werdhölzli with around 93.9% AYE votes. Two and a half years after this vote, the company Entsorgung + Recycling Zurich (ERZ) as the operator of the plant for the city of Zurich opened the new central SSTP with a capacity of 100,000 tonnes of digested, dewatered sewage sludge (DSS) (Fig. 10.5).

Since August of 2015, around 85,000 tonnes of digested and dewatered sewage sludge (with dry substance of around 30%) per year from the entire Canton have already been recycled as energy. Annually around 13,400 tonnes of high-phosphorus ISSA and an annual average P content of just over 80 g P/kg ISSA accumulate (Fig. 10.6). This yields an approximate annual P quantity from the total quantity of Zurich's sewage sludge of 900 tonnes (or 0.6 kg P/resident per year).

The change of the sewage sludge disposal concept in mid-2015 resulted in the following advantages for the entire Canton:

- By using this proven, highly efficient fluidised bed technology, it is possible to assure a high availability, economical operation as well as security of disposal at all times.
- Compared to the situation before 2015, it results in a significant reduction of the average sewage sludge disposal costs for the communities of more than half, to

Fig. 10.6 ISSA containing phosphorus (Source: AWEL)

around 100 CHF/t DSS (digested, dewatered sewage sludge, with a dry substance of 30%).

- A good, positive energy balance (net heat yield in the form of natural gas for heating around 5000 apartments) could be achieved.
  - A CO_2 reduction of around 14,000 tonnes/year.
  - The Preparation of the way for the future posphorous recovery from ISSA.

With a fair, transparent transport cost compensation model, financed by the City of Zurich, there is also the assurance that differences in the transport expenses of the different suppliers form the entire Canton will be levelled out as much as possible.

10.4 Phosphorus Mining from Incinerator Sewage Sludge Ash

10.4.1 Evaluation of Optimal P-Recovery Process

Since there is as yet no procedure that is able to convert the phosphorus present in the ISSA into market-ready products on a large technical scale according to the objectives of the P-Mining-Project, the Department of Public Works of the Canton of Zurich has sought possible solutions with various partners since 2011, looking for ways to recover phosphorus from ISSA and feed it back into the economic cycle. Initial studies which focused on the production of a raw material for fertiliser showed (AWEL 2012/2013):

1. The use of phosphorus from ISSA is technically possible.
2. None of the procedures for producing fertiliser investigated at the time were suitable for immediate use in the Canton of Zurich.
3. The wet-chemical extraction procedure LEACHPHOS, however, offers an interesting potential for optimisation and will be actively refined.
4. Until a procedure is market-ready, the ISSA shall be stored in a monofill.

Since fall of 2013, the Department of Public Works for the Canton of Zurich together with the Foundation ZAR (Development Center for Sustainable Management of Recyclable Waste and Resources) has been exploring the promising optimisation potential of wet-chemical P recovery from ISSA. The LEACHPHOS process of the Swiss company BSH Umweltservice AG from Sursee has been made available to this end. It was possible to demonstrate its technical feasibility beforehand in a large-scale experiment in Bern. In the ongoing development, two goals were achieved:

- First, an increase in the quality of the extraction residue, so that it can be used as a valuable mineral resource or stored with inert material quality
- Secondly, an increase in the product quality of the phosphorus products obtained as a fertiliser or fertiliser component

These clarifications however also showed that on the one hand the market for raw materials for fertilisers both in Switzerland and in Europe as a whole is only in some degree able to integrate secondary phosphorus products and on the other hand that the creation of value in this process is very limited. Added complications are the high sales risk dependencies of individual consumers. Based on these insights, phosphoric acid production from ISSA was pursued as the most promising pathway starting in mid-2014. This pathway evinces decisive advantages over the direct production of fertiliser and/or fertiliser components:

- Phosphoric acid is a conventional trade product with good demand domestically and abroad.
- Phosphoric acid has a higher added value as a fertiliser or fertiliser component.
- It eliminates the uncertainties inherent in the currently discussed adjustment of the quality requirements for recycled phosphorus products in Switzerland and the EU (Fertiliser Ordinance) as well as time-consuming product registrations (REACH).

A comprehensive evaluation of the different options for phosphoric acid production from ISSA was carried out based on the criteria of phosphorus efficiency, reduction of material to be landfilled, product qualities, sales market, process stability, the providers' know-how, economic efficiency and potential for synergy in the Swiss systems context. The best possible technological implementation of these

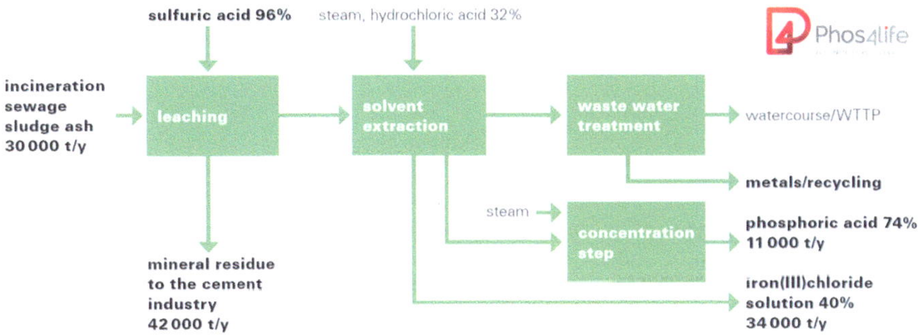

Fig. 10.7 Simplified process diagram of Phos4life® process (Source: TR)

objectives was found in solvent extraction to separate the heavy metals from the target component phosphorus. Since no tested large-scale technical procedure fulfils these criteria, the Spanish plant constructor and technology developer Técnicas Reunidas SA (TR) in Madrid was selected from the potential technology providers and development partners as the most suitable partner for collaboration, and the joint development of procedure was advanced. The goal of this development is a procedure which facilitates efficient phosphorus reclamation with simultaneous material recycling of other resources contained in sewage sludge such as iron precipitates, heavy metals and useable mineral components. The resulting products must be produced at a constant high quality while assuring high plant reliability at the same time.

10.4.2 PHOS4LIFE® Process

The development of a process for the production of technically pure phosphoric acid from ISSA began with a laboratory phase that lasted a year and a half. Once it was successfully concluded, continuous process operations were simulated in the framework of a pilot micro-plant operation, and the procedure was further optimised. In the current continuous pilot project, the insights gathered to date shall be confirmed and consolidated through March 2018.

With the new Phos4life® procedure (Fig. 10.7), it is possible to separate phosphorus and most heavy metals from sewage sludge using sulphuric acid so that the mineral leaching residue meets the highest Swiss quality requirements and material can be recycled for use in cement works. Table 10.1 shows the average annual element content of the starting material sewage sludge ash based on a 1-year sample as well as the resulting leaching residue in Columns II and III. In the first step of purification, the iron contained in the ISSAs that are added for the chemical elimination of phosphorus in the waste water treatment plant was extracted and recovered as saleable iron(III) chloride solution. Subsequently the phosphoric acid is released from the co-extracted heavy metals by means of solvent extraction. In the

Table 10.1 Typical element contents of ISSA, leaching residue and Phos4life® phosphoric acid. The limits for mineral recycling of leaching residue for Swiss cement works are shown supplementally (Source: ZAR, TR)

I	II	III	IV	V	VI
Element	Unit	Contents of ISSA	Contents of leaching residue	Maximum contents for secondary raw materials in Swiss cement clinker manufacturing plants	Phos4life® technical grade acid, 74% H_3PO_4
Al	mg/kg	30,000	3000	–	n.d.
Ca	mg/kg	130,000	190,000	–	<4
Fe	mg/kg	170,000	8000	–	<2
K	mg/kg	4000	2000	–	n.d.
Mg	mg/kg	15,000	400	–	<4
P	mg/kg	81,000	1000	–	234,000
Si	mg/kg	96,000	128,000	–	n.d.
As	mg/kg	15	2	15	<3
Cd	mg/kg	2	3	5	<1
Cl	mg/kg	160	<50	–	n.d.
Co	mg/kg	22	2	–	n.d.
Cr	mg/kg	92	23	250	<1
CrVI	mg/kg	0.05	<0.01	0.05	n.d.
Cu	mg/kg	850	47	250	<3
F	mg/kg	n.d.	n.d.	–	<10
Hg	mg/kg	0.1	<1	1	<1
Mn	mg/kg	1200	n.d.	–	<1
Ni	mg/kg	60	11	250	<1
Pb	mg/kg	90	130	250	<1
S	mg/kg	500	450,000	–	<100

(continued)

Table 10.1 (continued)

I	II	III	IV	V	VI
Element	Unit	Contents of ISSA	Contents of leaching residue	Maximum contents for secondary raw materials in Swiss cement clinker manufacturing plants	Phos4life® technical grade acid, 74% H_3PO_4
Sb	mg/kg	10	2	15	n.d.
Sn	mg/kg	60	9	–	n.d.
Tl	mg/kg	0.4	<1	–	n.d.
Zn	mg/kg	2100	180	500	<5

concentration stage, the conventional concentration of phosphoric acid of 74% H_3PO_4 is reached by vaporising the corresponding amount of water. The quality of the acid produced matches that of technically pure phosphoric acid (Table 10.1, Column V).

With the Phos4life® process, it is possible to fully meet the objectives for phosphorus recovery as well as material recycling of the mineral residue. In the future, it will be possible to recover phosphorus from ISSA with more than 95% yield and market it as a technically pure, conventional product. Iron can be recovered with a yield of >90% and used again in the waste water treatment plant for chemical phosphorus elimination. The separated heavy metals (>85%) are processed using techniques already existant in Switzerland and fed back into the materials stream. The near one hundred per cent recycling of ISSA reduces the landfill volume considerably and offers other impressive ecological advantages. With the thermal sewage sludge treatment and Phos4Life® process as central elements for sustainable phosphorus-mining strategy, it is possible, as shown in Fig. 10.8, to close important material circuits and conserve resources.

10.4.3 The Ecology of Phosphorus Mining

Worldwide phosphoric acid is primarily manufactured from raw phosphates (ores) by treating them with sulphuric acid. This raw phosphoric acid is mainly used in the manufacture of fertilisers. A part of it is refined to higher quality (technical quality TGA, food-grade quality FGA, etc.), in that harmful substances such as heavy metals, fluoride and organic materials are separated out. Today, the technology mainly used for this refining is solvent extraction, as a solvent-based liquid-liquid separating procedure.

Since the quality sources of raw phosphate that are laden with few harmful materials –especially heavy metals – are running out, the environmental pollution from

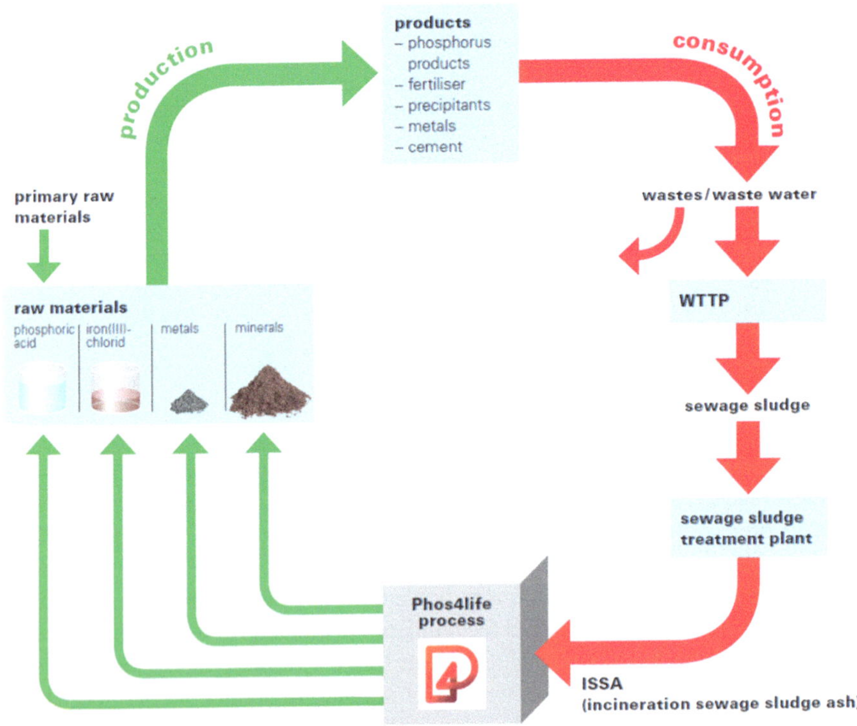

Fig. 10.8 The material cycle is closed – from ISSA to new raw material products (Source: AWEL)

fertilisers used around the world can constantly increase absent any new measures. The therewith associated increasing pollution of the soil with harmful materials coming from mineral fertiliser with excessive heavy metal content is not sustainable (see Sect. 10.2.1). The reasons for this lie on the one hand in the absence of a readiness to finance a phosphorus product that is low in harmful substances and on the other hand in the tolerance for exceeding the limits on such substances in mineral fertilisers. This practice is tolerated with the statement that the market does not offer any better products. The supply is controlled however by the demand, and so this development must be stopped by the strict compliance with established limits.

Compared to that the Phos4life® process and the phosphoric acid produced from it evince a significantly lower degree of environmental pollution. Important resources can thus be conserved more efficiently and replaced with less environmentally dangerous recyclates. Compared to the primary products, the recyclates are lower in cadmium and uranium content. By producing additional iron compounds, recycling the gypsum present in the residues in cement work and reclaiming the metals, additional environmental advantages are achieved which make a considerably lower degree of environmental pollution possible compared to the current-day practice of landfilling the ISSA. From the original waste product ISSA, high-value products are created which close the resource cycle and save valuable landfill space.

The Phos4life® process, as a sustainable and ecological process, shows a way for phosphorus to be fed back into the economic cycle free of heavy metals. Used correctly, this can also be an opportunity for our soil. As already explained in Sect. 10.2.1, in 2012 nearly half of the phosphorus fertilisers used in Switzerland exceed the permitted cadmium limit of 22 mg Cd/kg P_2O_5 in some cases considerably. Nearly 5% of the fertilisers tested in 2011–2012 showed cadmium content between 65 and 100 mg/kg P_2O_5. In another 41%, the cadmium content was in the range from 22 to 65 mg/kg P_2O_5 (BLW 2015). A telling example is the link between the cadmium content in the raw materials used to make the fertiliser based on the triple super phosphate (TSP) from phosphate rock and phosphoric acid, shown in Fig. 10.9. In the current starting scenario, typical cadmium contents of raw phosphate of approximately 100 mg/kg P_2O_5 (Kratz and Schnug 2005; Kharikov and Smetana 2000; Van Kauwenbergh 1997) and 50 mg/kg P_2O_5 (Gilmour 2013) in phosphoric acid used for fertiliser production are assumed as examples.

This results in a phosphorus fertiliser with a cadmium content of >50 mg/kg P_2O_5.With these raw materials, the objectives for fertiliser quality cannot be achieved without further technical process optimisation. This scenario thus covers approximately 11% of the fertilisers used in Switzerland. To qualitatively improve this situation, either higher-quality starting materials can be used, which in a global context however are becoming ever scarcer, or processes for separating heavy metals can be integrated into fertiliser manufacture, which will increase the cost of the products. In the second scenario, fertiliser manufacturing proceeds from the same raw phosphate with a high-quality phosphoric acid low in heavy metals, as can, for example, be obtained from ISSA using the Phos4life® procedure (see Table 10.1).

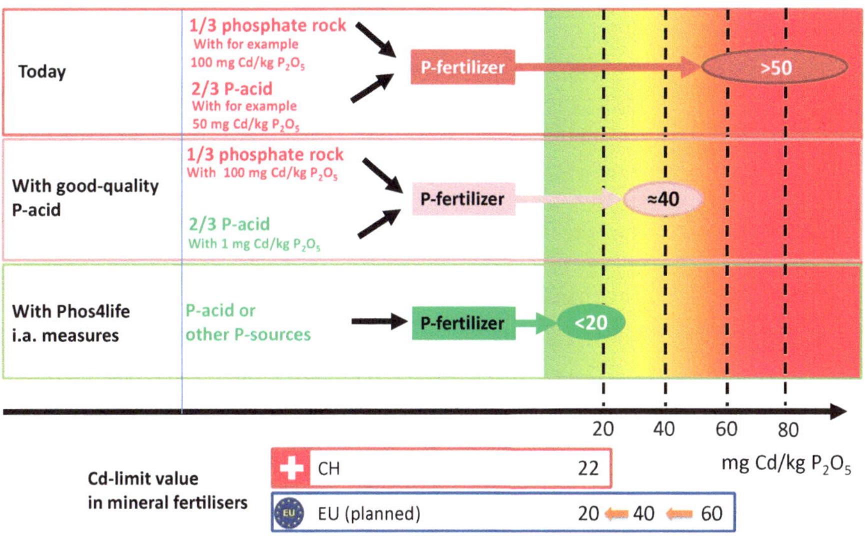

Fig. 10.9 Cadmium limits in mineral fertiliser depending on the quality of the starting materials, assessed on the currently valid Swiss limits and the planned future European limit with gradual tightening of the limit from 60 to 20 mg Cd per kg P_2O_5 within 15 years

The resulting fertiliser still evinces a cadmium content of approximately 35 mg/kg P_2O_5, which is slightly above the permitted limit. With reference to the fertilisers examined in Switzerland, this corresponds to a share of 34%, which evinced a cadmium content between 22 and 50 mg/kg P_2O_5. Only in the third scenario, in which, along with a pure phosphoric acid, phosphate rock low in heavy metals or alternative phosphorus sources are used, can the legally prescribed limit be observed. This corresponds to 55% of the Swiss mineral fertilisers tested. With a cadmium content in phosphoric acid of 1 mg/kg P_2O_5, the phosphate rock or alternative phosphorus source used may contain at most 60 mg Cd/kg P_2O_5 and still comply with the Swiss limit. Should the acid used likewise contain more cadmium, the phosphate rocks would have to be commensurately lower in cadmium. This applies analogously for the planned tightening of the limits in the EU, which aims at a reduction in the cadmium content from 60 to <20 mg/kg P_2O_5 within 15 years. This ecological achievement however also has its price, through the increased time and effort required for the procedure.

10.4.4 Costs

Completely financing the costs with revenue from the product's earnings from the Phos4life® procedure is currently not possible with the planned plant size of 15,000–30,000 t ISSA/a in Switzerland. The added expense of phosphorus mining cannot, under the strict environmental regulations in force in Switzerland, be compensated solely through product revenues. These also take into consideration the accumulated landfilling costs for ISSA. To implement a phosphorus recovery plant with approximately twice the capacity of the ISSA accumulating in the Canton of Zurich, one must reckon with added costs compared to the sewage sludge treatment of around 70 CHF/t DSS (Fig. 10.10). The total cost estimate for sewage sludge treatment (SST) and phosphorus recovery would then be at around 170 CHF/t DSS still lower than the average sewage sludge disposal costs for the Canton of Zurich before the operation of the central SSTP which were at around 210 CHF/t DSS.

If one were also to include the sewage sludge treatment steps digestion, dehydration and transport (see Fig. 10.11) so far not taken into account in the cost for the sewage sludge treatment, then the calculated additional costs of phosphorus mining would lie at around 20% compared to sewage sludge disposal as currently practised.

10.5 Conclusion and Look Ahead

In the Canton of Zurich, it was possible, together with the parties responsible for sewage sludge disposal, to make courageous and perspicacious decisions that completely realigned the existing sewage sludge disposal concept with a new strategy

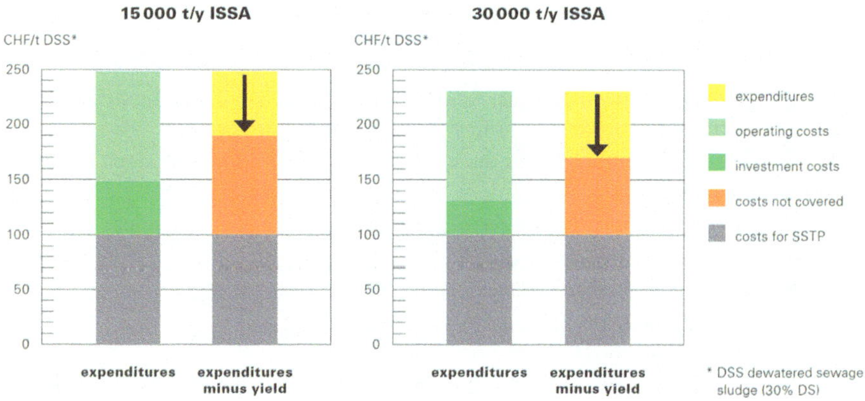

Fig. 10.10 Costs for sewage sludge treatment plant (SSTP), for the Phos4life® process and for the total phosphorus mining from sewage sludge in CHF per tonne of dewatered sewage sludge (DSS, 30% dry substance content). Two scenarios are shown with 15,000 and 30,000 t/a ISSA (Source: ZAR, AWEL)

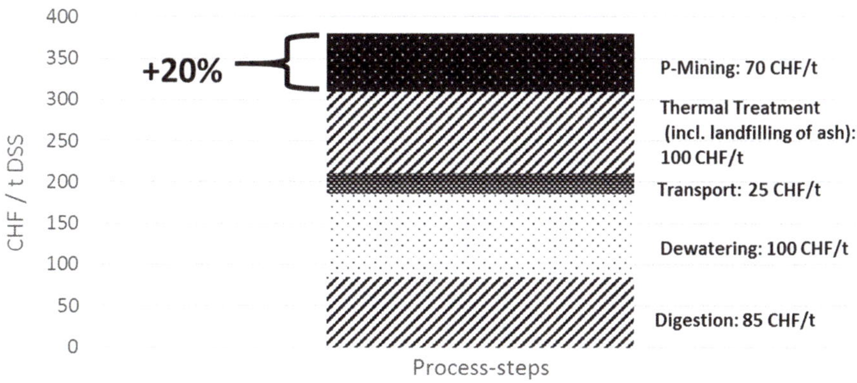

Fig. 10.11 Added costs for phosphorus mining from ISSA compared with the sum of today's total costs for sewage sludge disposal in the Canton of Zurich split into the individual process steps; thermal treatment includes the costs of landfilling the ash (Source: AWEL)

for phosphorus reclamation in fewer than 6 years. With that the Canton of Zurich continued to consistently apply the precautionary principle with respect to sewage sludge disposal as before. The high standards for surface water protection and air quality remain intact. Disposal safety is assured at all times, and the disposal costs for sewage sludge can be cut by more than half, thanks to the new central treatment plant. No structural or operational changes to the WWTP were necessary.

Around 90% of the total phosphorus potential in untreated waste water now stands ready for use in the ISSA that is produced. With the strategy now followed, which has as its goal production of phosphoric acid with the reclamation of other valuable substances, the urban mining concept defined by the plan of action for waste and resource management of the Canton of Zurich is consistently implemented.

The following next steps for the implementation of the strategy are already running or are being prepared:

- Pilot plant for the Phos4life® process (plan to be concluded by March 2018)
- Confirmation of the product qualities in the running pilot project
- More precise information about the costs for phosphorus mining once the pilot project is concluded
- Contacts with possible partners for a larger phosphorus recovery plant
- Final evaluation of the process selection once the pilot project is concluded (compare with any alternatives to P-recovery procedures from ISSA available on the market at the start of 2018)
- Evaluation of solutions for financing

References

AWEL (2008) Phosphor in Klärschlamm– Informationen zur künftigen Rückgewinnung (Phosphorus in sewage sludge – information about the future P-recycling), Baudirektion Kanton Zürich (Department of Public Works Canton Zurich), AWEL Amt für Abfall, Wasser, Energie und Luft. www.klaerschlamm.zh.ch

AWEL (2009) Synthesebericht für interessierte Fachpersonen – Zürcher Klärschlammentsorgung unter besonderer Berücksichtigung der Ressourcenaspekte (Sythesis report for interested qualified persons –sewage sludge disposal of Canton of Zurich under special consideration of resource aspects), Baudirektion Kanton Zürich (Department of Public Works Canton Zurich), AWEL Amt für Abfall, Wasser, Energie und Luft. www.klaerschlamm.zh.ch

AWEL (2010) Massnahmenplan zur Abfall- and Ressourcenwirtschaft 2011–2014, Urban Mining – Rohstoffe der Zukunft (Waste and resource management action plan 2011 to 2014, urban mining – future resources), Department of Pub-lic Works Canton Zurich. www.abfall.zh.ch

AWEL (2012/2013) Phospor- Mining im Kanton Zürich (phosphorus-mining in Canton Zurich), Projektblätter Nr.1–3. http://www.awel.zh.ch/internet/baudirektion/awel/de/abfall_rohstoffe_altlasten/rohstoffe/rohstoffe_aus_abfaellen/naehrstoffe.html#a-content

BAFU (2006) Nachhaltige Rohstoffnutzung and Abfallentsorgung, Grundlagen für die Gestaltung der zukünftigen Politik des Bundes. (Sustainable resource utilisation and waste disposal, fundamentals for the design of future national politics), Bundesamt für Umwelt, Bern (Swiss environmental protection agency, Bern)

BLW (2015) Marktkampagne Dünger 2011/2012, Kennzeichnung und Schwermetalle (Swiss fertilizer market campagne 2011/2011, identification and heavy metals)

FitzGerald R, Roth N (2015) Cadmium in mineral fertilisers – human and envi-ronmental risk update, SCAHT report for BLW

Gilmour R (2013) Phosphoric acid: purification, uses, technology and economics. CRC Press/Taylor&Francis Group, Boca Raton

Kabbe Ch et al. (2015) Phosphorus recycling – now! Building on full-scale prac-tical experiences to tap the potential in European municipal wastewater. www.p-rex.eu

Kharikov AM, Smetana VV (2000) Heavy metals and radioactivity in phos-phate fertilisers: short term detrimental effects. In: IFA Technical Conference, New Orleans, Louisiana

Kratz, S.; Schnug E. (2005) Schwermetalle in P-Dünger (Heavy metals in P-firtiliser) n. Landbauforschung Völkenrode 286 (Special Issue):37–45

Montag D (2008) Phosphorreklamation bei der Abwasserreinigung – Entwick-lung eines Verfahrens zur Integration in kommunalen Kläranlagen (Phosphorus recovery in waste water treatment –Development of a process to integrate in municipal WTTP). Dissertation ISA, Aachen, RWTH

Pinnekamp, J. et al. (2011) Resümee and Schlussfolgerungen aus den Gesamtergebnissen der Vorhaben der Förderinitiative „Kreislaufwirtschaft für Pflanzennährstoffe, insbesondere Phosporous („Phosporous-Recycling") im Rahmen der Schlusspräsentation der in der BMBF-Förderinitiative geförderten Vorhaben (Conclusions out of the total results oft he project «circular economy of plant nutrigents, especially phosphorous («phosphorous-recycling»), conference with presentation), 14. September 2011, Umweltforum Auferstehungskirche Berlin

SRU (2005) Auf dem Weg zur Europäischen Ressourcenstrategie: Orientierung durch ein Konzept für eine stoffbezogene Umweltpolitik – Stellungnahme (On the way to a European resource strategy: orientation with a concept for a substance based environmental policy), November 2005, Nr. 9, Sachverständigenrat für Umweltfragen, Berlin

Van Kauwenbergh SJ (1997) Cadmium and other minor elements in world resources of phosphate rock, Proceedings of the Fertilizer Society No. 400, London

Chapter 11
Calcination Technology for Manufacturing Mineral Fertilizer Using CaO-Enriched Sewage Sludge Ash

T. Imai

Abstract A calcination technology was developed to directly convert sewage sludge ash (SSA) to mineral fertilizer. SSA samples, having a wide range of P_2O_5 content (15.6–36.0 mass%), were collected from ten different wastewater treatment plants. $CaCO_3$ was added to SSA samples to adjust their CaO content to 45 mass%. The CaO-enriched SSA samples were then heated in an electric furnace at temperatures of 1250 or 1300 °C for 10 min. This could increase the citric acid solubility of P_2O_5 to 80–99%, regardless of the P_2O_5 content of SSA tested. In addition, approximately 75–97% of SiO_2 in SSA became soluble in 0.5 M HCl after being subjected to calcination. The high citric acid solubility of P_2O_5 and HCl solubility of SiO_2 in calcined products were attributable to the generation of silicocarnotite ($Ca_5[(SiO_4)(PO_4)](PO_4)$) and gehlenite ($Ca_2Al_2SiO_7$), respectively. The levels of toxic heavy metals such as Cr, Ni, As, Cd, and Hg in the calcined product were below their regulation levels for calcined sludge fertilizer in Japan. The calcination of CaO-enriched SSA is potentially a useful option to increase the solubility of P_2O_5 and SiO_2, thereby improving their plant availability, regardless of the P_2O_5 and SiO_2 contents of SSA.

Keywords Sewage sludge ash · Calcium carbonate · Calcination · Citric acid-soluble P_2O_5

11.1 Introduction

Sewage sludge ash (SSA) has been considered as a promising secondary phosphorous (P) resource, since it often contains high levels of P, comparable to those of phosphate (P_i) rock in the marketplace (Appleton 2002). In Japan, approximately

T. Imai (✉)
Central Research Laboratory, Taiheiyo Cement Corporation, Chiba, Japan
e-mail: Toshio_Imai@taiheiyo-cement.co.jp

H. Ohtake, S. Tsuneda (eds.), *Phosphorus Recovery and Recycling*,
https://doi.org/10.1007/978-981-10-8031-9_11

70% of dewatered sewage sludge is incinerated to reduce its volume and to recover energy at wastewater treatment plants (WWTP) (MLITT 2010). Fluidized bed incinerators are most commonly used for combusting dewatered sewage sludge. The direct application of SSA to farmland as a mineral fertilizer is prohibitively difficult, mainly because it contains toxic heavy metals. The chemical composition of SSA, in particular the content of P_2O_5, considerably varies, depending primarily on the type of sewer system (combined or separate system), the choice of flocculants for sludge dewatering, and the mode of incineration (mono- or mixed incineration). The type of P_i salts, e.g., $AlPO_4$, $FePO_4$, and $Ca_3(PO_4)_2$, in SSA is a key factor controlling the citric acid solubility of P_2O_5, thereby determining the P bioavailability when applied as P_i fertilizer (Toyama 2016). The improvement of P_2O_5 solubility is critical to the valorization of SSA for fertilizer.

Akiyama (2008) has investigated the influence of calcination on the citric acid solubility of P_2O_5 (defined as % citric acid-soluble P_2O_5 in total P_2O_5) in SSA. Using SSA with the P_2O_5 content of 25–30 mass%, he has demonstrated that the citric acid solubility of P_2O_5 could be effectively improved by mixing SSA with calcium carbonate ($CaCO_3$) at a Ca/P molar ratio of about 3.5 prior to the calcination at 1200 or 1300 °C. However, it is known that the P_2O_5 content of SSA varies in a wide range of 10–40 mass%, depending on factors such as the type of sewer system, the choice of chemical flocculants, and the mode of incineration as mentioned above. The P_2O_5 content of SSA also changes seasonally and daily (Tanioka et al. 2013). In the present study, to further investigate the effects of calcination on the P_2O_5 solubility of CaO-enriched SSA, we collected SSA samples having a wide range of P_2O_5 content (16–36 mass%) from ten different WWTP. The CaO content of SSA samples was adjusted to 45 mass% using $CaCO_3$. From a practical perspective, the mass% of CaO is an easy-to-use control parameter rather than the Ca/P molar ratio, since the content of P_2O_5 considerably varies from SSA to SSA. The CaO-enriched SSA samples were then heated in an electric furnace at temperatures of 1250 or 1300 °C for 10 min. The calcination of CaO-enriched SSA would offer a simple technology option to increase the solubility of P_2O_5 and SiO_2 of SSA and to reduce the level of toxic heavy metals for the valorization of SSA.

11.2 SSA Samples

SSA samples, designated as A–J, were collected from ten different WWTP (Table 11.1). All of the WWTP have fluidized bed reactors for the mono-incineration of dewatered sewage sludge at temperatures of 820–890 °C. The mineral composition of SSA samples was determined by X-ray fluorescence (XRF) and X-ray diffraction (XRD) analyses. As shown in Table 11.2, the main components of SSA were (mass% as oxides) SiO_2 (18–37), Al_2O_3 (11–23), Fe_2O_3 (5–10), CaO (5–12), and P_2O_5 (15–36). Together, they accounted for 80–91 mass% of SSA. The P_2O_5 content of SSA widely varied from 15 to 36 mass%, depending on the type of sewer system (A–E, combined; F–J, separate) and the choice of flocculants (A–B and F–J,

Table 11.1 SSA samples collected from ten different WWTP

Sample	Sewer system	Flocculants type	Incineration temperature (°C)
A	Combined	Polymer	830
B	Combined	Polymer	860
C	Combined	PAC	820
D	Combined	PAC	820
E	Combined	PAC	840
F	Separate	Polymer	850
G	Separate	Polymer	890
H	Separate	Polymer	820
I	Separate	Polymer	840
J	Separate	Polymer	820

Table 11.2 The chemical composition of SSA samples (mass%)

Sample	SiO_2	Al_2O_3	Fe_2O_3	CaO	MgO	SO_3	Na_2O	K_2O	P_2O_5
A	37.4	20.8	8.5	5.4	2.3	0.7	0.9	2.5	16.6
B	36.0	23.0	9.2	7.6	2.7	1.0	1.0	1.4	15.6
C	31.1	18.4	10.2	6.6	2.5	1.3	1.0	2.0	20.6
D	22.1	17.8	9.3	9.0	2.7	2.3	1.0	2.6	25.4
E	22.0	20.2	7.1	7.2	3.0	1.1	0.9	2.4	27.6
F	26.8	14.3	8.7	10.1	3.1	1.0	0.6	2.0	22.4
G	25.0	12.2	5.0	12.6	3.9	1.7	0.6	2.1	25.5
H	20.3	12.4	8.1	11.6	4.9	0.5	0.6	2.7	31.3
I	19.8	11.3	8.0	12.3	4.4	0.8	0.6	2.8	33.9
J	18.9	11.1	5.9	10.0	5.4	0.6	0.8	3.9	36.0

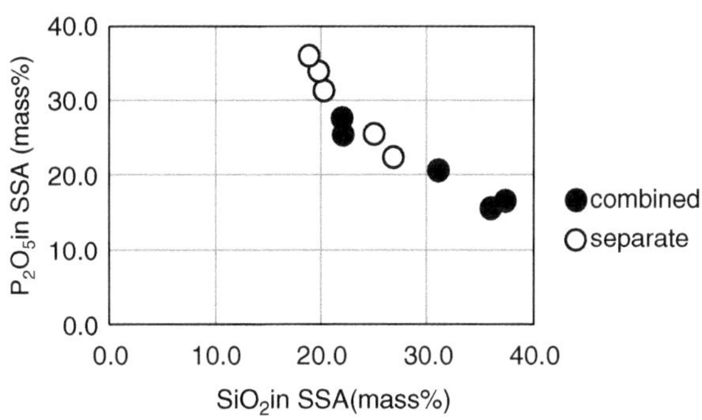

Fig. 11.1 Relationship between the contents of P_2O_5 and SiO_2 in SSA samples

high-polymer; C–E, poly-aluminum chloride (PAC)). The SiO_2 content of SSA also showed a wide variation from 18 to 37 mass%.

There was a negative correlation between the contents of SiO_2 and P_2O_5 in SSA (Fig. 11.1). SSA samples from combined sewer systems had relatively high SiO_2 and low P_2O_5, compared to those from separate sewer systems. Combined sewer

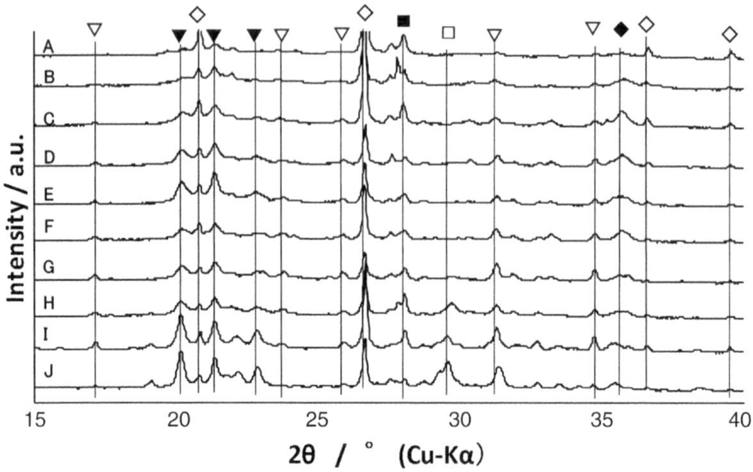

Fig. 11.2 X-ray diffraction patterns of SSA samples. Symbols: quartz, ◇; feldspar, ■; aluminum P_i ($AlPO_4$), ▼; maghemite, ◆; (($Ca_xMg_{1-x})_3(PO_4)_2$), ▽; $CaMgP_2O_7$, □

systems are designed to collect both rainwater runoff and sewage in the same pipe. This increases the levels of quartz, plagioclases, and clay minerals in SSA.

Figure 11.2 shows the X-ray diffraction patterns of SSA samples. SSA samples from combined sewer systems (A–E) showed stronger peaks corresponding to the plane (101) of quartz compared to those from separate sewer systems (F–J). Strong peaks corresponding to aluminum P_i ($AlPO_4$) and calcium magnesium P_i ($\beta(Ca_{2.589}Mg_{0.411})(PO_4)_2$ and $Ca_3Mg_3(PO_4)_4$) were clearly detected with samples from separate sewer systems (I and J). Relatively high Al_2O_3 contents were detected with SSA samples taken from WWTP using PAC as an auxiliary flocculation agent for sludge dewatering (data not shown).

There was a clear correlation between the contents of total P_2O_5 and citric acid-soluble P_2O_5 ($C\text{-}P_2O_5$) in SSA (Tables 11.2 and 11.3). The level of $C\text{-}P_2O_5$ was in the range from 6.16 (sample A) to 15.9 mass% (sample G), averaging 12.5 mass%. The $C\text{-}P_2O_5$/total P_2O_5 ratio ($C/T\text{-}P_2O_5$) ranged from 42% to 66%, suggesting that about half of P_2O_5 in SSA was potentially available for plant when applied as P_i fertilizer. Figure 11.3 represents the relationship between the peak intensity of plane (−406) for $AlPO_4$ and the $C/T\text{-}P_2O_5$ of SSA samples. Clearly, there was a negative correlation between them, suggesting that $AlPO_4$ could reduce the citric acid solubility of P_2O_5 in SSA as reported previously (Toyama 2016). By contrast, the peak intensity of plane (0210) for $\beta(Ca_{2.589}Mg_{0.411})(PO_4)_2$ showed a positive correlation with the $C\text{-}P_2O_5$ content of SSA samples (data not shown).

Table 11.4 shows the levels of toxic heavy metals in SSA and their regulation for commercial calcined sludge fertilizer by the Fertilizer Control Act. Only sample H could meet regulation levels for all the heavy metals tested. There was a clear correlation between the contents of Cr and Ni in SSA. Similar results have been

Table 11.3 The acid solubility of P_2O_5, SiO_2, MgO, and K_2O in SSA samples

Sample	C-P_2O_5 mass%	C/T-P_2O_5 %	S-SiO_2 mass%	S/T-SiO_2 %	C-MgO mass%	C/T-MgO %	C-K_2O mass%	C/T-K_2O %
A	6.16	49.5	0.95	2.1	0.81	50.0	0.36	15.5
B	9.15	66.4	5.78	14.8	0.78	53.9	0.24	42.3
C	8.20	61.6	0.87	2.1	0.65	65.7	0.15	25.0
D	13.11	54.3	1.99	7.2	1.85	70.3	0.72	48.3
E	12.84	49.9	2.10	7.2	2.14	74.1	0.66	55.5
F	13.91	61.6	1.35	4.3	2.39	79.9	0.30	32.6
G	15.91	60.8	1.10	3.7	2.89	71.0	0.44	50.0
H	15.44	49.8	0.89	4.4	2.50	39.9	0.96	45.7
I	14.90	45.7	0.70	2.7	1.55	54.6	1.27	33.8
J	15.09	42.5	0.40	1.8	1.94	56.9	1.37	71.3

C-P_2O_5, C-MgO, and C-K_2O were citric acid-soluble P_2O_5, MgO, and K_2O, respectively. S-SiO_2 is HCl-soluble SiO_2. C/T-P_2O_5, S/T-SiO_2, C/T-MgO, and C/T-K_2O are the acid solubility defined as % acid-soluble oxide in total oxide

Table 11.4 Levels of toxic heavy metals in SSA and their regulation for commercial fertilizer (mg/kg)

Sample	Cr	Ni	As	Cd	Hg	Pb
A	102	79	43	4	0.12	246
B	133	216	21	7	3.19	81
C	224	279	22	8	Tr	72
D	276	442	26	4	Tr	92
E	337	479	61	4	Tr	128
F	89	125	26	5	0.16	129
G	89	245	30	3	0.03	113
H	52	51	22	1	ND	50
I	76	102	14	8	1.98	22
J	77	67	9	8	<0.05	22
Regulation[a]	500	300	50	5	2	100

[a]Regulation for calcined sludge fertilizer

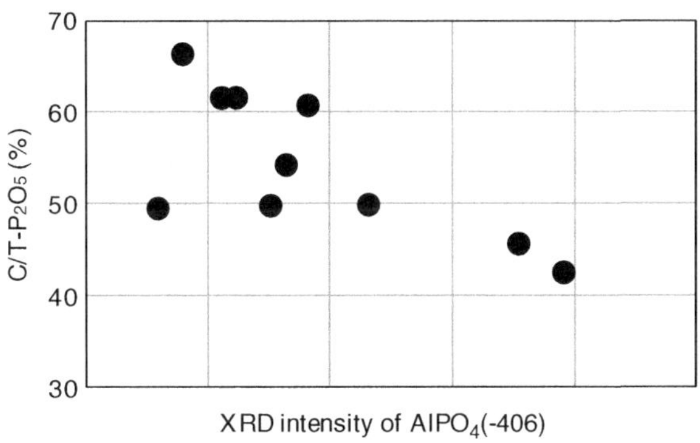

Fig. 11.3 Relationship between the C/T-P_2O_5 and XRD intensity for $AlPO_4$ (−4 0 6) in SSA

reported previously by Yazaki (1979). SSA samples collected from combined sewer systems (samples A–E) contained relatively high toxic heavy metals compared to those from separate systems (samples F–J).

11.3 Calcination Technology

SSA samples were mixed with $CaCO_3$ (325 mesh, Ube Material Industries Inc.) to adjust the CaO content to 45 mass%. The CaO level has been chosen based on a preliminary study (Imai and Miura 2015). The CaO-enriched SSA samples were pressed in a stainless steel die to make disk-like pellets (ϕ40 and 13 mm in thickness). The pellets were placed on a bisque crucible and heated in an electric furnace at 1250 or 1300 °C for 10 min. After being cooled down, the calcined product (designated as calcined fertilizer) was crushed into particles having a diameter of 600 μm or less. XRD analysis showed that all products contained gehlenite (aluminosilicate) and silicocarnotite (calcium phosphate silicate) as stable solid phases (Fig. 11.4). However, the ratio of silicocarnotite to gehlenite considerably varied, depending on the P_2O_5 content of SSA. High values were detected when calcined fertilizer was generated from high P_2O_5 SSA samples. Dicalcium silicate was also identified in calcined fertilizer generated from SSA samples with low P_2O_5 contents.

Elemental mapping of calcined fertilizer was performed using an electron probe microanalyzer (EPMA). The energy-dispersive spectroscopy (EDS) was also used to perform semiquantitative chemical analysis. Four typical regions (P_1–P_4) were

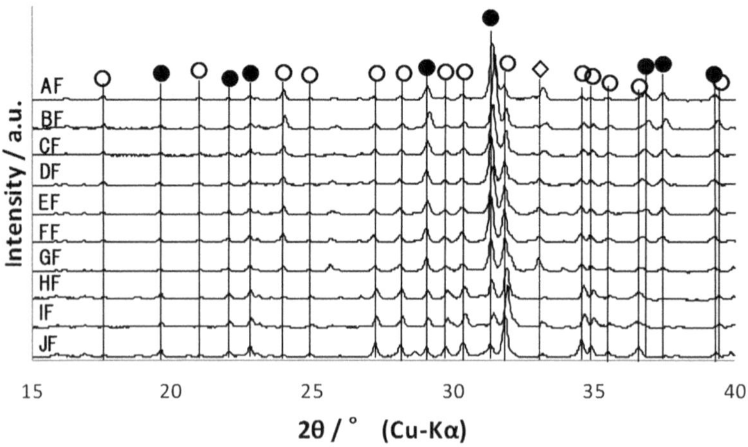

Fig. 11.4 X-ray diffraction patterns for calcined fertilizers. Symbols: gehlenite, ●; silicocarnotite, ○; di-calcium silicate, ◇. The calcined fertilizers (AF–JF) were generated from SSA samples (A–J), respectively

Table 11.5 The acid solubility of P_2O_5, SiO_2, MgO, and K_2O in calcined fertilizer

Sample	C-P_2O_5 mass%	C/T-P_2O_5 %	S-SiO_2 mass%	S/T-SiO_2 %	C-MgO mass%	C/T-MgO %	C-K_2O mass%	C/T-$_2O$ %
AF	7.31	97.5	25.80	94.2	0.95	60.5	0.99	75.6
BF	8.17	99.6	22.78	96.4	1.04	56.2	0.60	88.2
CF	9.97	97.8	22.43	89.2	1.09	77.9	0.62	66.0
DF	8.45	82.8	27.23	87.1	1.04	69.7	0.99	91.7
EF	13.00	82.5	18.60	88.8	1.16	81.7	1.21	85.8
FF	12.35	90.7	20.42	80.2	1.57	87.2	1.01	93.5
GF	13.41	79.6	18.17	75.2	1.86	77.2	1.21	87.7
HF	17.87	90.8	13.22	88.7	1.78	48.1	1.33	86.3
IF	19.54	89.0	13.75	85.7	1.01	51.0	1.32	80.0
JF	17.10	97.2	13.60	91.9	1.42	81.1	1.80	90.9

identified in calcined fertilizer (data not shown). P_1 showed high contents of P_2O_5, CaO, and SiO_2, while P_2 was characterized by high P_2O_5 and CaO but low SiO_2. P_3 was rich in SiO_2, Al_2O_3, and K_2O, while P_4 showed a needle-shaped region with high MgO and Fe_2O_3. Based on the proportion of oxides, it was suggested that P_1, P_2, P_3, and P_4 corresponded to silicocarnotite, CMP ($\beta(Ca_{2.589}Mg_{0.411})(PO_4)_2$), gehlenite, and Fe spinel, respectively.

The solubility of P_2O_5 and SiO_2 is critical to the availability of P and Si for crops when applied as mineral fertilizer. Table 11.5 summarizes the acid solubility of oxides in calcined fertilizer. The P_2O_5 content of calcined fertilizer was relatively low (T-P_2O_5 of 7–22 mass%) compared to that of SSA (15–36 mass%) because of the mass increase due to the addition of $CaCO_3$. Nevertheless, calcined fertilizer showed higher values of C/T-P_2O_5 (79–99%) compared to those of SSA samples (42–66%). This is likely due to the formation of silicocarnotite with a high citric acid solubility by the reaction of $AlPO_4$ with SiO_2 and CaO during calcination. In fact, there was a clear relationship between the contents of silicocarnotite, which was estimated from the XRD signal intensity at plane (330), and C-P_2O_5 in calcined fertilizer (Fig. 11.5).

Similarly, the S/T-SiO_2 of SSA was increased to 75–96% by calcination due to the formation of gehlenite which had a high 0.5 M HCl solubility. Gehlenite is known to be formed by the reaction of $AlPO_4$ with SiO_2 and CaO (Akiyama 2008). Gehlenite and akermanite belong to a melilite solid solution in the multicomponent oxide system consisting of SiO_2-CaO-MgO-Al_2O_3 (Schairer 1944). Since gehlenite and akermanite have different solubilities in 0.5M HCl, the composition of minor elements such as Mg, Fe, Na, and K in melilite might have an influence on the 0.5M HCl solubility of SiO_2. The citric acid-soluble K_2O in SSA was also increased by calcination, while no significant increase was detected with MgO. MgO residing in gehlenite is unlikely solubilized by calcination.

Table 11.6 shows the levels of toxic heavy metals in calcined fertilizer. Except lead (Pb), the levels of heavy metals were lower than their regulation levels for calcined sludge fertilizer. Regarding Pb, only calcined fertilizer AF exceeded the

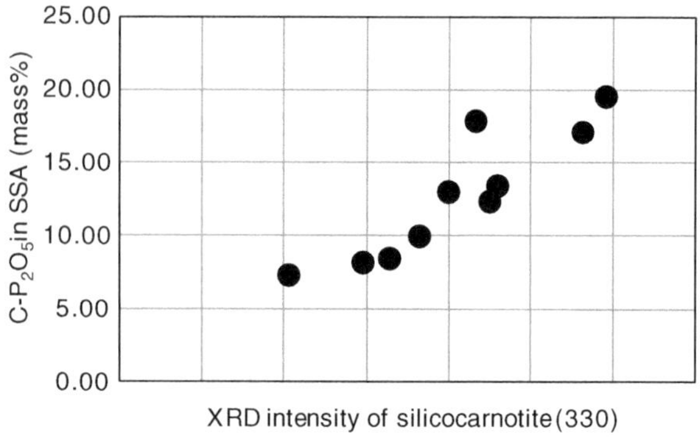

Fig. 11.5 Relationship between the content of C-P_2O_5 and XRD intensity of silicocarnotite (330) in calcined fertilizer

Table 11.6 Levels of toxic heavy metals in calcined fertilizer (mg/kg)

Sample	Cr	Ni	As	Cd	Hg	Pb
AF	127	74	27	ND	ND	134
BF	148	151	13	2.0	<0.05	73
CF	194	168	15	<1.0	ND	49
DF	379	287	19	3.0	ND	81
EF	201	161	19	3.0	ND	63
FF	90	70	22	3.0	ND	65
GF	95	131	23	3.0	ND	71
HF	37	44	10	ND	ND	38
IF	100	46	<2	<1.0	<0.05	38
JF	47	19	14	1.7	<0.05	30
Regulation[a]	500	300	50	5	2	100

[a]Regulation for calcined sludge fertilizer

regulation level (i.e., 100 mg/kg), because SSA sample A contained the exceptionally high level of lead (see Table 11.4). Cd and Hg could be readily vaporized at temperatures higher than 1200 °C. However, lead cannot be vaporized by heating SSA at 1200–1300 °C. Hence, to reduce the level of lead in calcined fertilizer, it seems necessary to increase the amount of $CaCO_3$ added to SSA. Alternatively, lead needs to be removed separately using the chloride extraction technology as described previously (Mattenberger et al. 2008; Nowak and Pessl 2009).

11.4 Conclusions

SSA samples, having a wide range of P_2O_5 content (15.6–36.0 mass%), were collected from ten different wastewater treatment plants. The CaO content of SSA samples was adjusted to 45 mass% by adding $CaCO_3$. The CaO-enriched SSA samples were then heated in an electric furnace at temperatures of 1250–1300 °C for 10 min. This increased the citric acid solubility of P_2O_5 and HCl solubility of SiO_2 of SSA samples. The levels of heavy metals such as Cr, Ni, As, Cd, and Hg of calcined products were lower than their regulation levels for calcined sludge fertilizer in Japan. Regarding Pb, only one calcined product, which was generated from a SSA sample containing exceptionally high Pb (246 mg/kg), exceeded the level of regulation. The high solubility of P_2O_5 and SiO_2 in calcined products were attributable to the generation of two solid phases consisting of silicocarnotite (Ca_5 [(SiO_4) (PO_4)] (PO_4))) and gehlenite ($Ca_2Al_2SiO_7$), respectively. The calcined fertilizer can potentially be used as a mineral fertilizer which can show the high solubility of both P_2O_5 and SiO_2 and is likely suitable for growing crops such as paddy rice, barley, wheat, and maize.

References

Akiyama T (2008) Use of sewage sludge for fertilizer. Q Issue Fertil 109:110–114. (in Japanese)

Appleton JD (2002) Local phosphate resources for sustainable development in sub-Saharan Africa. Brit Geol Surv Rep. CR/02/121/N. 67

Imai T, Miura K (2015) A calcination technology for converting sewage incineration ash to mineral fertilizer. J Res Taiheiyo Cem Corp 169:51–61. (in Japanese)

Mattenberger H, Fraissler G, Brunner T, Herk P, Hermann L, Obernberger I (2008) Sewage sludge ash to phosphorus fertilizer: variables influencing heavy metal removal during thermochemical treatment. Waste Manag 28:2709–2722

MLITT (2010) A manual for P recovery and recycling in the wastewater treatment sector, pp 15–16. (in Japanese)

Nowak B, Pessl A, Aschenbrenner P, Mattenberger H, Friebert A, Rechberger H, Hermann L Winter F (2009) Thermal processing for heavy metal removal from municipal solid waste fly ash. Proc Europ Combust Meeting. p 1–6

Schairer JF (1944) Some aspects of the melting and crystallization of rock-forming minerals. J Mineral Soc Am 29(3 and 4):75–91

Tanioka N, Imai T Toda M (2013) Seasonal changes in phosphorus mass balance in a wastewater treatment plant. Proc. 34th Ann meeting of Jap Waste Manag Assoc. p 248–250

Toyama T (2016) Development of a technology for recovering P from sewage sludge ash. Recover Reuse 39(150):65–69. (in Japanese)

Yazaki J (1979) Heavy metals in sewage sludge. Sewag Sludge, Jap Soc Soil Sci and Plant Nutr, Haku-yusha. pp 67–82. (in Japanese)

Chapter 12
Phosphorus Recovery from Sewage Sludge by High-Temperature Thermochemical Process (KUBOTA Process)

Fumiki Hosho

Abstract With the increase in world population causing a rise in food demand, demand for nutrients such as phosphorus, potash, and nitrogen is on an upward trend. To maintain sustainable agricultural production, it is essential in many areas of the world to ensure stable procurement and to utilize these nutrients more efficiently. It is known that phosphorus is concentrated in sewage sludge through the cycle of consumption of food, human living, and sewage treatment systems. Hence, discussions have been held and studies done on ways of recovering phosphorus from sewage sludge and how to utilize it as a fertilizer. On the other hand, in addition to phosphorus, sewage sludge also includes micropollutants and heavy metals. How to separate heavy metals from phosphorus and decompose organic matter are important from the viewpoint of effectively and safely using phosphorus and preventing groundwater contamination. With the above background, we have developed a thermochemical process (melting process) under high temperature (around 1300 °C) that recovers 90% of phosphorus from sewage sludge and converts it into slag. Organic components are utilized as fuel in the furnace and decomposed. Heavy metals are vaporized in the furnace and collected by a gas treatment system. More than 90% of the phosphorus in the slag is citric acid-soluble phosphorus, indicating that plants can utilize the recovered phosphorus in the slag. We have also confirmed the effectiveness and safety of the slag as a fertilizer with a plant cultivation test. In this chapter, we describe the performance of a technology developed to recover phosphorus from sewage sludge based on the results of laboratory and pilot plant tests. We also describe the properties of the slag as a fertilizer based on the results of chemical analysis and the plant cultivation test.

Keywords Melting process · Nutrient · Phosphorus recovery · Sewage sludge · Thermochemical process

F. Hosho (✉)
KUBOTA Corporation, Osaka, Japan

H. Ohtake, S. Tsuneda (eds.), *Phosphorus Recovery and Recycling*,
https://doi.org/10.1007/978-981-10-8031-9_12

12.1 Introduction

Phosphorus is a substance that is necessary for living organisms and is not substitutable. With the increasing global population, the importance of phosphorus has grown larger and more critical. Sewage sludge is rich in phosphorus discharged from agriculture, daily life, and sewage systems. Direct application to farmland is an effective option of recycling nutrients such as phosphorus and nitrogen contained in sewage sludge. However, direct application is prohibited in some countries because of the risks of contamination of soil and groundwater with micropollutants (persistent organic pollutants (POPs), pathogenic bacteria, antibiotics, etc.) and heavy metals. Establishing a system that can use phosphorus safely and sustainably is a future challenge.

Three tasks exist to effectively apply the phosphorus in sewage sludge as a safe fertilizer resource for the future:

1. Decomposition of micropollutants
2. Separation of heavy metals
3. Recovery of phosphorus in a form that can be effectively consumed by plants

With our unique high-temperature thermochemical process, we took up the challenge of solving these problems. The technology has been installed in more than 30 projects from the 1970s to treat various wastes such as sewage sludge and solid waste residues from sorting plants or municipal solid waste incineration plants and for environmental remediation of illegal dumping sites. Especially in the field of sewage sludge treatment, the technology is implemented in 11 projects with capacities that range from 1.7 to 35 t-DS (dry solid)/day (28,000–600,000 population equivalent (p.e.), assuming sludge load 60 g-DS/capita/day). This robust technology backed by historical use is redefined here as a technology for phosphorus recovery, creating renewed value in phosphorus recycling.

12.2 Process Description

12.2.1 Recovery Principle

KUBOTA melting process is a thermochemical technology for separation and concentration of materials under high-temperature atmosphere (Hosho and Yoshioka 2016; Goto et al. 2016; Kamata et al. 2015). Lab tests using an electric furnace are effective for understanding the behavior of elements in the process. Table 12.1 shows a component analysis of sewage sludge ash samples. Figure 12.1 shows the results of a laboratory test that simulates the behavior of phosphorus and heavy metals in sewage sludge in the high temperature range of 1250–1350 °C. Sewage sludge consists of ash matter and combustible matter. Therefore the lab test samples are a mixture of sewage sludge ash and activated carbon, an alternative to the carbon in combustible matter.

Table 12.1 Major components of sewage sludge ash

Component	Minerals (%-dry)				Heavy metals (mg/kg-dry)	
SiO_2	CaO	Al_2O_3	FeO	P_2O_5	Zn	Pb
25.2	7.2	12.9	13.4	22.9	5600	180

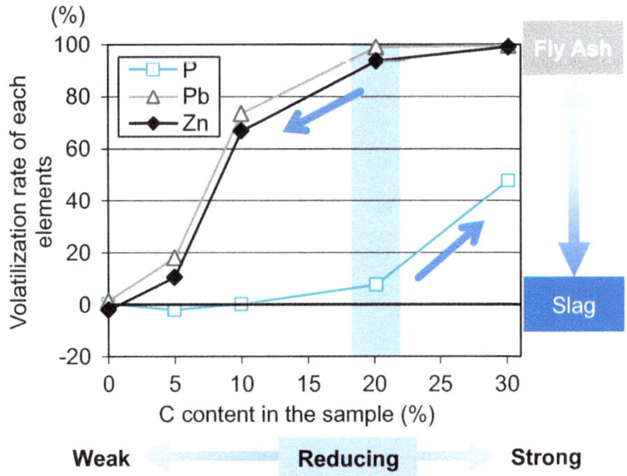

Fig. 12.1 Behavior of phosphorus and heavy metals in sewage sludge ash with varying the carbon content. Sample: the mixture of sewage sludge ash and activated carbon. Test conditions: 1350 °C, in the air, and 15 min

Along with the increasing carbon content in the sample, atmospheric reduction in the vicinity of the sample becomes stronger because of the combustion of carbon. At the 20%-carbon content, the volatilization rate of heavy metals is higher than 90%. On the other hand, the volatilization rate of phosphorus is less than 10%. Hence, heavy metals can be separated from phosphorus by controlling the reducing atmosphere in the vicinity of sludge in the furnace.

In the lower carbon condition, the volatilization rate of heavy metals decreases such that heavy metals remain in the slag. In the higher carbon condition, the volatilization rate of phosphorus increases such that phosphorus tends to move to the fly ash. The content of combustible matter in sewage sludge is 80–85% (dry wt%) for raw sludge and 50–70% for digested sludge. Controlling the atmosphere can be achieved by controlling the air supply to the vicinity of the sludge in actual sludge treatment.

12.2.2 Reactor

The KUBOTA Surface Melting Furnace (KSMF), the core reactor, is illustrated in Fig. 12.2. This reactor has the following features:

1. *A wide variety of wastes*

The KSMF can handle a wide variety of wastes whether combustible, incombustible, or filled with moisture. Until now, the KSMF has been applied to recycling

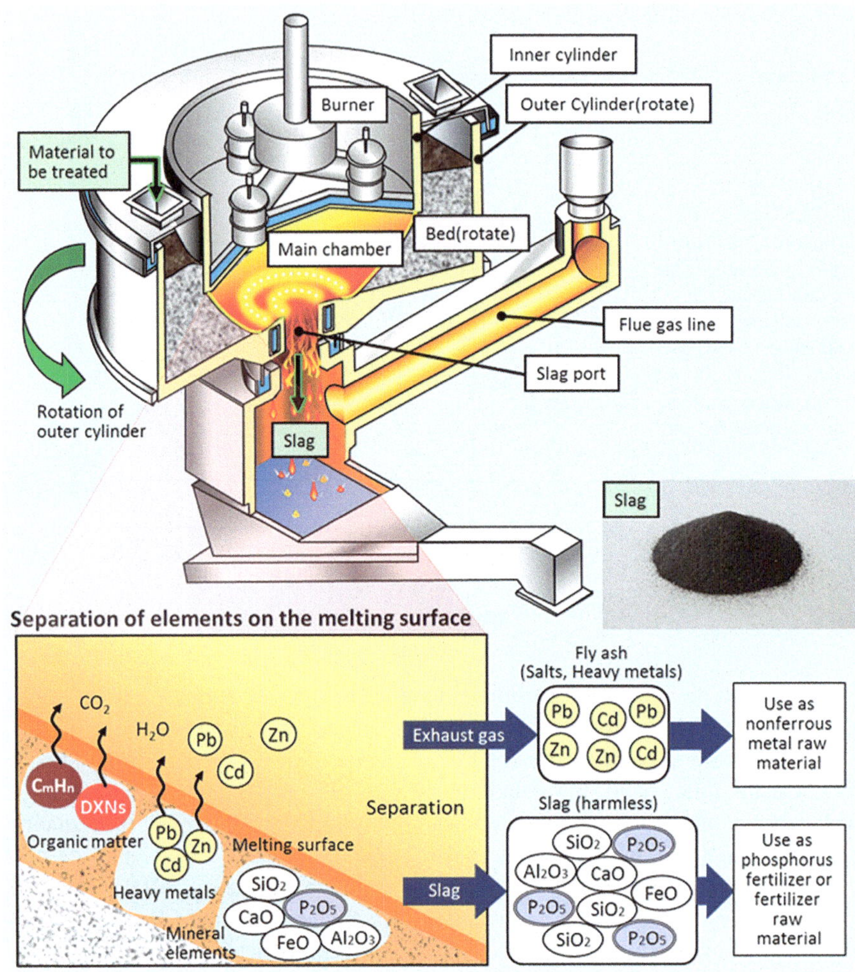

Fig. 12.2 Schematic diagram of the KSMF and conceptual diagram of the separation of elements on the melting surface

wastes such as municipal or industrial waste incineration residues, sewage sludge, industrial sludge, waste plastics, char, and reclaimed illegal dump sites (mixed soil, ash, oil, plastics, and metals).

2. *Continuous and stable feed of material into the furnace*

KSMF has a double cylindrical structure. The outer cylinder can rotate. The space between the inner and outer cylinders is filled with the material to be treated, which is continuously fed into the furnace along the entire circumference of the inner cylinder by the rotation of the outer cylinder (about 1 rph). Continuous and stable feed of material enables to keep stable atmosphere in the furnace such as temperature.

3. *Unique combustion mode*

The space near the inner cylinder is the combustion area. The ash component melts and flows down through the slag port at the center of the furnace floor. Because operating temperature in the furnace ranges from 1250 to 1350 °C, organic matter in the sewage sludge burns and is decomposed. Silicon, calcium, phosphorus, and iron tend to remain in the slag (slag or metal phase). Heavy metals such as lead, zinc, and cadmium tend to volatilize from the melting surface. Phosphorus is thus recovered in the slag, while heavy metals are separated to the gas phase. Combustible matter can work as fuel in the furnace. In case the material being treated has a low heat value of more than 10 MJ/kg (wet), the operating temperature can be maintained by the combustion heat of the material itself without additional fossil fuel during stable operation (self-sustaining combustion). The conversion efficiency from nitrogen to N_2O is very low in the furnace temperature range (1250–1350 °C) so that N_2O hardly occurs.

4. *Separation of exhaust gas from slag in the secondary chamber*

Both slag and exhaust gas pass through the slag port. Slag drops down to the water bath under the secondary chamber. Exhaust gas including volatilized elements then goes to the flue line. Thus, slag and volatilized materials are separated. The slag then rapidly cools in the water bath and becomes granulated. The granulated slag is then transported by the slag conveyer.

12.2.3 Sewage Sludge Treatment System

The process flow diagram of the sewage sludge treatment system is shown in Fig. 12.3. Dewatered sewage sludge is dried and supplied to the melting furnace. The respective pre-drying targets necessary to maintain self-sustaining combustion in the furnace are 85–90%-DS for digested sludge and 70–80%-DS for raw sludge.

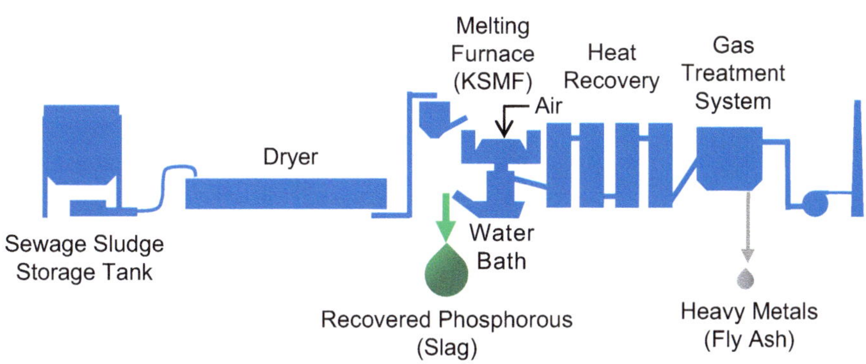

Fig. 12.3 Process flow diagram of sewage sludge treatment system

Table 12.2 Outline of the pilot plant

Capacity	**3.6 t-DS/day (60,000 p.e.[a])**
Furnace size	Inner cylinder
	Diameter: 1.8 m
Input	Dried sludge

[a]Assuming 60 g-DS/day/person

The heat recovery system recovers heat from the exhaust gas. The recovered heat is then used as a heat source for the sludge dryer and air preheater. This gas treatment system neutralizes acidic gases and separates fly ash, which includes heavy metals, and neutralized salts from the exhaust gas.

12.3 Pilot Plant Test

The lab test shown in Sect. 12.2.1 explains the potential of the melting process in recovering phosphorus from slag and in separating heavy metals in fly ash. In this section, the potential of the KSMF on phosphorus recovery from sewage sludge will be discussed based on the KSMF pilot plant test. The effectiveness of phosphorus recovered in the slag will also be discussed based on the slag characteristics.

12.3.1 Pilot Plant

An outline of the pilot plant is shown in Table 12.2, and the facade and process flow diagram are shown in Fig. 12.4. Sewage sludge is dried and transported to the plant and supplied to the furnace. The slag produced by the furnace is rapidly cooled in a water tank so it becomes granulated. The heavy metals volatilized in the furnace are condensed and solidified into the fly ash, which is collected by the bag filter. Acidic gas neutralized salt generated by slaked lime blown in before the bag filter is also collected by the bag filter.

12.3.2 Test Results

Table 12.3 shows the characteristics of the sample sludge and slag. Phosphorus in the slag is almost the same level as in the sludge ash matter, and the heavy metal content in the slag is less than 10% in the sludge ash matter. Figure 12.5 shows the volatilization rates of phosphorus and heavy metals in the pilot plant test. The volatilization rate of heavy metals is more than or equal to 95%, with the volatilization rate of phosphorus less than 5%.

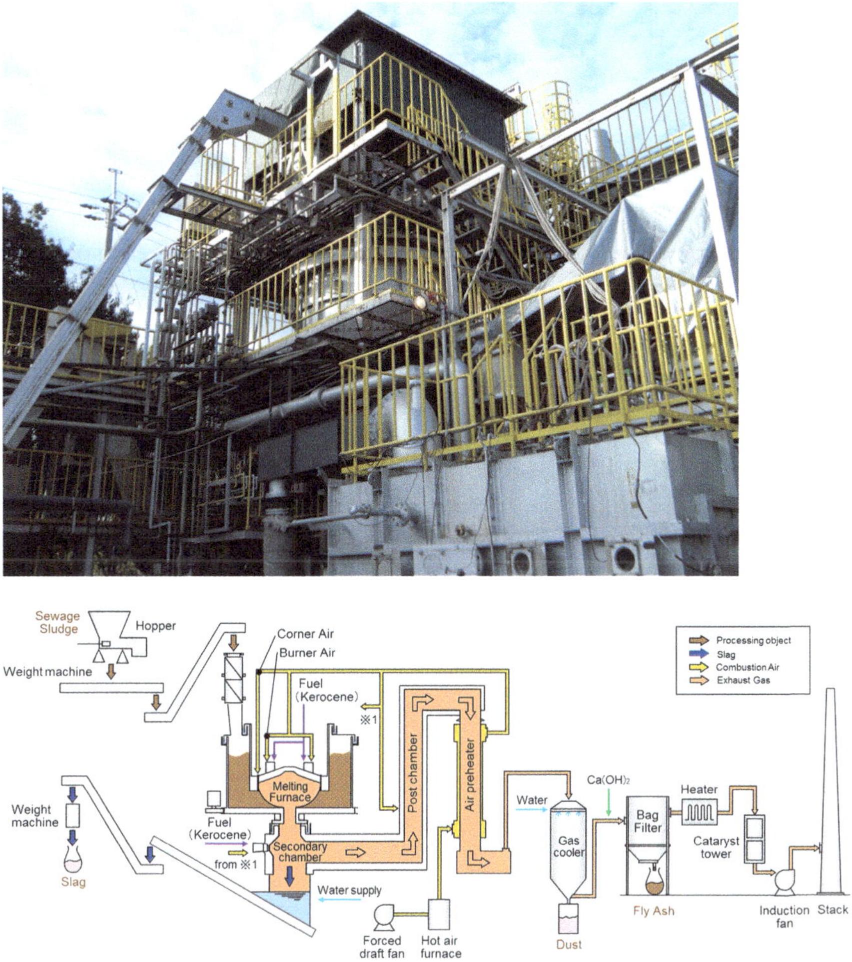

Fig. 12.4 The pilot plant: (**a**) the plant facade and (**b**) the process flow diagram (Osaka, Japan)

Table 12.3 Components of sample sludge and slag

		Sludge (ash)	Slag
T-P_2O_5	% (dry)	16.3	17.3
C-P_2O_5	% (dry)	–	16.9
Cd	mg/kg (dry)	12.0	<1
Pb	mg/kg (dry)	259	16
Zn	mg/kg (dry)	4217	250

Sludge (ash): components converted to concentration in ash matter

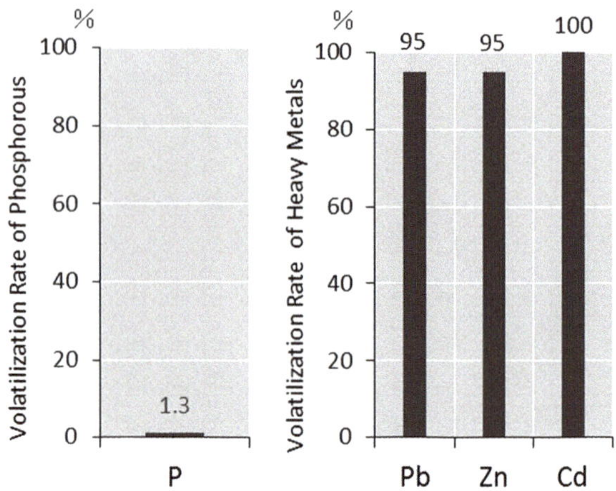

Fig. 12.5 Volatilization rate of phosphorous and heavy metals

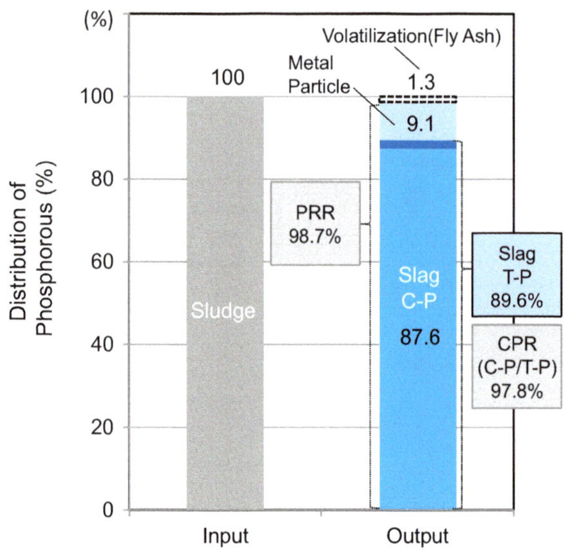

Fig. 12.6 Mass balance of phosphorus

The phenomenon that phosphorus in sewage sludge can be separated from heavy metals could be confirmed by the KSMF pilot plant test as well as laboratory test results.

Figure 12.6 shows the mass balance of phosphorus. Citric acid-soluble phosphorus (C-P) is an index indicating the effectiveness of phosphorus fertilizer. From the viewpoint of plant availability of phosphorus in the slag, it is important to increase the citric acid-soluble phosphorus rate (CPR) in the total phosphorus (T-P), which is equal to C-P/T-P. For an input of phosphorus, 89.6% (T-P) was recovered in the slag and 87.6% was C-P. As a result, the CPR of the slag was 97.8%. The operation of the pilot plant demonstrated a high phosphorus recovery rate (PRR) and the recovery of phosphorus with high C-P.

Table 12.4 Test conditions and results of plant cultivation test

Phosphorus			Blank	Superphosphate	Slag (<0.3 mm)
Amount		g-P_2O_5/pot	0	1.19	1.19
Growing conditions					
Breed: Kinumusume (rice)					
Soil: sandy soil					
Pot size: 1/2000 a					
Fertilizer condition on N and K					
N: 2.38 g-N/pot					
K: 1.04 g-K_2O/pot					
[Effectiveness]		g (dry)/plant	4.0	10.2	11.0
Dry matter yield					
Unpolished rice					
Stems and leaves		g (dry)/plant	12.2	30.5	29.9
[Safety]	Cd	Mg/kg (dry)	N.D. (<1)	N.D. (<1)	N.D. (<1)
Heavy metals in unpolished rice	Pb	Mg/kg (dry)	N.D. (<1)	N.D. (<1)	N.D. (<1)

12.4 Bioavailability and Safety

A plant cultivation test was conducted to evaluate the effectiveness and safety of the slag as fertilizer compared with commercial phosphorus fertilizers. The pot experiment (pot size: 1/2000 a) was done on paddy rice, a major cereal. Regarding the phosphorus fertilizer condition, slag (milled to less than 0.3 mm), superphosphate (commercial fertilizer), and no phosphorus condition (control) were compared. Nitrogen and potassium fertilization conditions were the same in all test plots.

To eliminate the effects of fertilization history, soil with low phosphorus content was used as the cultivation soil. The results of the cultivation test are shown in Table 12.4. Dry matter yield of unpolished rice, stems, and leaves shows that the slag has similar plant availability to commercial phosphorus fertilizer. In addition, the heavy metal contents of the unpolished rice were identical to commercial fertilizer; no heavy metals were detected from the rice in the plot using the slag. It is considered that the slag has a similar level of safety to commercial fertilizer in terms of heavy metals.

12.5 Conclusion

Phosphorus is an essential element for maintaining sustainable food production. Utilizing and recycling this resource more efficiently are important issues for the future. The sewage sludge melting technology described in this chapter can separate phosphorus from toxic heavy metals in sewage sludge and recovers the phosphorus into slag at a high recovery rate and in an effective form for plants. This technology also has the potential to simultaneously use heat contained in sewage sludge in treating the sludge and recovering phosphorus resources. Through the application of our technology, we would like to contribute to the development of a sustainable phosphorus resource recycling system (Fig. 12.7).

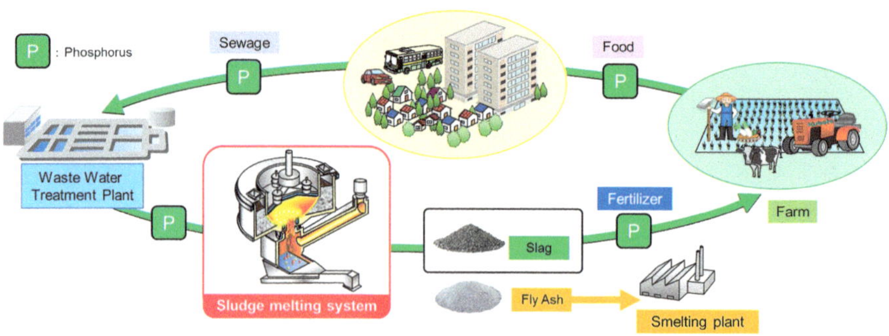

Fig. 12.7 Sustainable phosphorus recycling society with sludge melting system

References

Goto K Kamata Y Hosho F (2016) Study on upgrading of the melting process as a resource recovery technology. Environ Conserv Eng (45–6):306–312

Hosho F Yoshioka Y (2016) Thermochemical phosphorus recovery from sewage sludge and effectiveness of recovered phosphorous as fertilizer. 9th i-CIPEC

Kamata Y Yoshioka Y Hosho F (2015) Elemental separation of phosphorus, cesium using a melting furnace. Kubota Tech Rep (49):68–73

Chapter 13
Phosphorus Extraction from Sewage Sludge Ash by the CO_2 Blowing Method

Takeshi Toyama

Abstract Sewage sludge ash (SSA) often contains high levels of phosphorus (P) and has been considered as one of important secondary P resources. However, SSA also contains toxic heavy metals such as Cd, As, and Hg which can cause contamination problems in the recovery of P from SSA. Alkaline earth metal phosphate salts, which are present in SSA, can be preferentially dissolved by blowing CO_2 into the aqueous suspension of SSA. The dissolution mechanism involves the formation of soluble hydrogen carbonate by the CO_2 blowing which allows alkaline earth metal phosphate salts to form bicarbonate salts having high solubility in water. This chapter describes the selective extraction of phosphate from SSA using the CO_2 blowing method.

Keywords Alkaline earth metal phosphate · CO_2 blowing · Selective dissolution · Hydroxyapatite

13.1 Introduction

The raw material for phosphoric acid, which is widely used for chemical fertilizer, livestock feed, and various industrial products, is phosphate (P_i) rock. The global distribution of P_i rock is uneven, and its reserve is concentrated in a limited number of countries. Japan is most entirely dependent on imported P_i rock and is, therefore, vulnerable to the price volatility in the global market. For example, in the summer of 2008, China raised the tariff on P_i rock and restricted the export of phosphorus (P) products. This caused the price surge of P_i in Japan and greatly affected not only the industrial production but also food supply (Cordell and White 2014).

Municipal sewage contains relatively high levels of P_i which eventually ends up in sewage sludge at wastewater treatment plants (WWTP). In Japan, more than 70%

T. Toyama (✉)
Department of Materials and Applied Chemistry, College of Science and Technology, Nihon University, Tokyo, Japan
e-mail: touyama.takeshi@nihon-u.ac.jp

H. Ohtake, S. Tsuneda (eds.), *Phosphorus Recovery and Recycling*,
https://doi.org/10.1007/978-981-10-8031-9_13

of sewage sludge is incinerated at WWTP. Sewage sludge ash (SSA) from mono-incinerators contains high levels of P_i and has been considered as one of the most important secondary P resources (Shiroyama et al. 2015; Takahashi et al. 2004). However, SSA also contains toxic heavy metals which are harmful to plants when applied to farmland (Soma et al. 1989; Chatterjee and Dube 2005). Although various technologies are potentially available for P_i recovery from SSA, their implementation has been hampered mainly by economic reasons. It is, therefore, critical to develop a simple, low-cost technology for recovering P_i from SSA.

The author of this chapter has been conducting research on the morphological control of calcium carbonate and calcium phosphate crystals. The morphological control is a technique to increase the added value of materials by controlling the particle size, shape, and distribution. The basis of this method is the dissolution-deposition reaction, and it is critical to prepare a highly concentrated salt solution. Recently, the author has developed the CO_2 blowing method which is effective for dissolving alkaline earth metal phosphate salts. This chapter describes the selective extraction of P_i from SSA using the CO_2 blowing method.

13.2 The CO_2 Blowing Method

Hydroxyapatite ($Ca_{10}(PO_4)_6(OH)_2$, HAp), which is a main component of P_i rock, is insoluble in water and chemically stable. On the other hand, alkaline earth metal carbonates such as $CaCO_3$ is dissolved in water by the CO_2 blowing and form highly soluble bicarbonate salts (Langmuir 1968; Segnit et al. 1962). HAp is decomposed into soluble $Ca(HCO_3)_2$ and H_3PO_4 by blowing CO_2 into an aqueous suspension of HAp according to the formula (13.1) (Toyama et al. 2012).

$$Ca_{10}\left(PO_4\right)_6\left(OH\right)_2 + 20CO_2 + 18H_2O \rightarrow 10Ca\left(HCO_3\right)_2 + 6H_3PO_4 \qquad (13.1)$$

Figure 13.1 exemplifies the time-course of HAp decomposition after the start of CO_2 blowing into an aqueous suspension of HAp. No significant release of Ca^{2+} from HAp is observed in deionized water without CO_2 blowing. However, the concentration of Ca^{2+} increases soon after the start of CO_2 blowing and reached about 80 mg/L in 1 h. Thus, it is possible to prepare a highly concentrated calcium phosphate solution from insoluble HAp by simply blowing CO_2 gas to the aqueous suspension.

The potential of the CO_2 blowing method was further examined using various phosphate salts (Fig. 13.2). The P_i release from Na_3PO and K_3PO_4, which are alkaline metal phosphate salts, was not enhanced by CO_2 blowing. P_i salts such as $Fe_3(PO_4)_2$ and $AlPO_4$, which were insoluble in water, showed no significant P_i release even after 30-min CO_2 blowing. On the other hand, alkaline earth metal phosphates such as $Mg_3(PO_4)_2$ and $Ca_3(PO_4)_2$ showed a distinct release of P_i by CO_2 blowing as in the case of HAp. This is attributable to the fact that alkaline earth

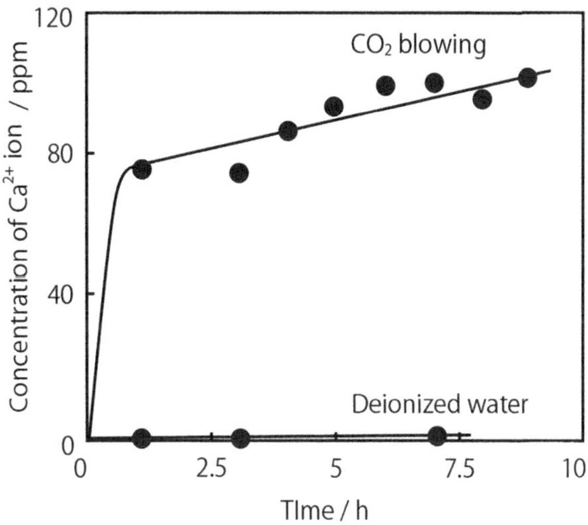

Fig. 13.1 The release of Ca^{2+} from HAp by CO_2 blowing

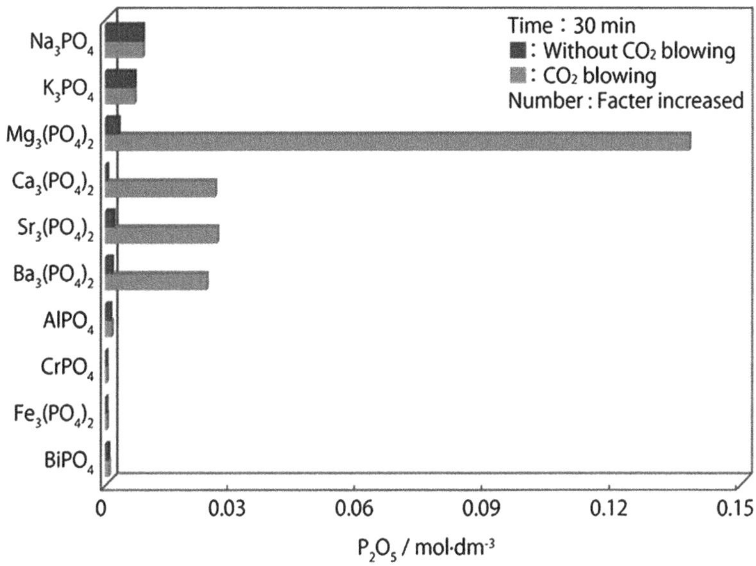

Fig. 13.2 P_i release from various P_i salts by CO_2 blowing

metal phosphate salts can react with CO_2 to form water-soluble bicarbonate salts. Thus, the CO_2 blowing method is able to dissolve alkaline earth metal phosphate salts selectively from a mixture of P_i salts.

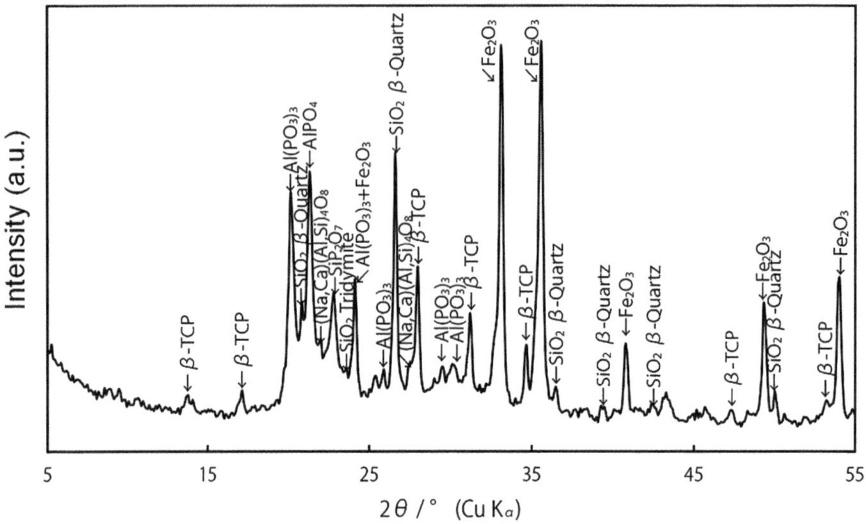

Fig. 13.3 X-ray diffraction pattern of a SSA sample

13.3 Morphologies of P_i Salts in SSA

SSA often contains P_i at levels similar to those of mined P_i rock. When extracting P_i from SSA using the CO_2 blowing method, it is necessary to know the morphologies of P_i salts in SSA. Ozaki et al. (2005) have analyzed the morphologies of P_i salts in SSA in detail using energy dispersive X-ray (EDX) spectroscopy and have detected various P_i salts in SSA. Figure 13.3 shows the X-ray diffraction pattern of a SSA sample taken from WWTP in Japan. The SSA sample, which was brownish in color, exhibited a strong diffraction peak corresponding to Fe_2O_3 in the X-ray diffraction pattern. In addition to the peaks corresponding to quartz (SiO_2), the peaks of P_i salts such as $Al(PO_3)_3$, $AlPO_4$, SiP_2O_7, and β-$Ca_3(PO_4)_2$ were clearly detectable. Among them is β-$Ca_3(PO_4)_2$ that can be efficiently dissolved by the CO_2 blowing method as shown in Fig. 13.2.

13.4 P_i Release from SSA

To examine the performance of CO_2 blowing on the release of P_i from SSA, SSA samples were suspended in deionized water at a concentration of 0.5 mass%. Then, the CO_2 gas was blown into the suspension at a rate of 1 L/min for 60 min at room temperature. Then, the suspension was filtered with a glass filter to obtain the leachate. Figure 13.4 shows the time-course data on P_i release from SSA by CO_2 blowing. The data were also compared with those obtained with HNO_3 treatment. The P_i concentration reached approximately 10 mg/L soon after suspending SSA in

Fig. 13.4 The release of PO_4^{3-} form SSA

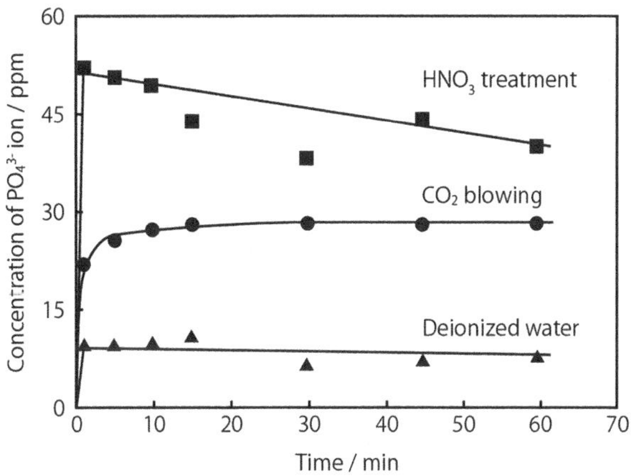

deionized water. However, there was no significant increase in P_i concentration in deionized water without CO_2 blowing. By contrast, the concentration of P_i rapidly increased up to 50 mg/L when HNO_3 was added to the suspension of SSA. Then, the P_i concentration gradually decreased down to 40 mg/L at 60 min. The P_i concentration of deionized water increased up to 30 mg/L by CO_2 blowing. Importantly, pH was about 5.3 even 60 min after the start of CO_2 blowing. The level of P_i release from SSA by CO_2 blowing was about 70–80% of that observed with HNO_3 treatment.

The time-course data on the release of Na^+, Mg^{2+}, and Ca^{2+} from SSA were shown in Fig. 13.5. The release of these cations was low in deionized water without CO_2 blowing. HNO_3 could enhance the release of Na^+, Mg^{2+}, and Ca^{2+} from SSA. Interestingly, there was no significant difference in the release of Na^+ from SSA in deionized water with and without CO_2 blowing (Fig. 13.5a). This suggests that alkaline metal P_i salts are not dissolved by CO_2 blowing. By contrast, Mg^{2+} and Ca^{2+} were significantly released by CO_2 blowing (Fig. 13.5b, c), suggesting that alkaline earth metal P_i salts could be dissolved by the CO_2 blowing method. Figure 13.6 shows the time-course data on the release of Fe ions (mainly Fe^{3+}) from SSA. The release of Fe ions was remarkable by HNO_3 treatment, but the elution of Fe ions was not enhanced by CO_2 blowing. This is likely attributable to the fact that trivalent Fe^{3+} ion is difficult to form bicarbonate salts.

Table 13.1 summarizes the data on the release of various ions from SSA. The release of P_i (PO_4^{3-}) by CO_2 blowing was about 75% of that observed with HNO_3 treatment. However, the release of K^+ and Na^+ was not enhanced by CO_2 blowing. As shown in Fig. 13.2, aluminum phosphate salts ($AlPO_4$) in SSA was not efficiently dissolved by CO_2 blowing. By contrast, the release of alkaline earth metal ions was greatly enhanced by CO_2 blowing. Hence, β-$Ca_3(PO_4)_2$ in SSA could be selectively dissolved by the CO_2 blowing method. Importantly, though heavy metal ions such as Fe^{3+}, Cu^{2+}, and Mn^{2+} can be eluted by HNO_3, their release from SSA was

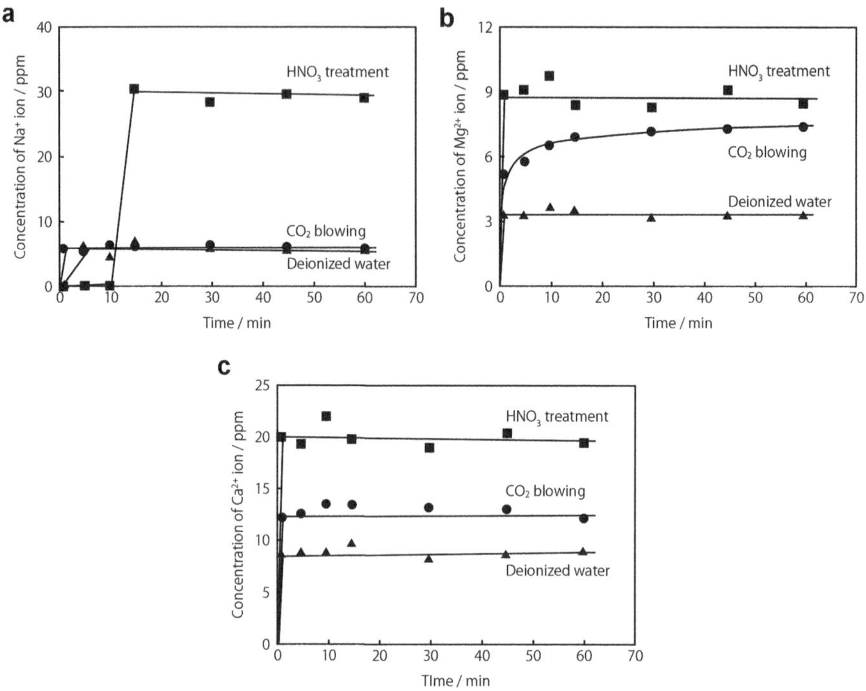

Fig. 13.5 The release of Na^+ (A), Mg^{2+} (B), and Ca^{2+} (C) by CO_2 blowing

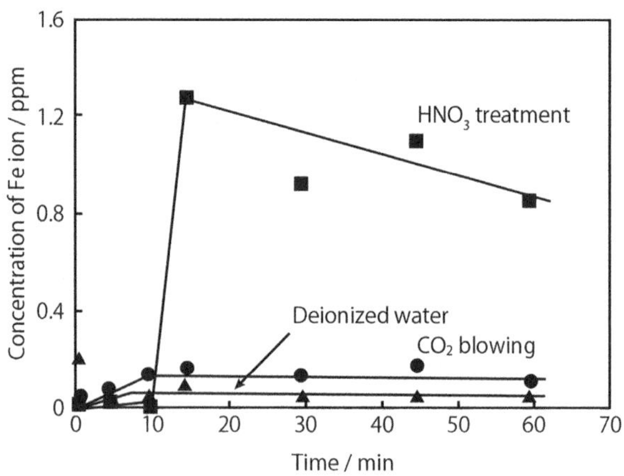

Fig. 13.6 The release of Fe ions (Fe^{2+} plus Fe^{3+}) by CO_2 blowing

Table 13.1 The release of various ions from SSA (Unit: ppm)

Component	HNO_3 treatment	CO_2 blowing	Deionized water
PO_4^{3-}	40	30	7
Mg^{2+}	9	8	3
Ca^{2+}	20	12	8
Na^{2+}	30	6	6
K^{+}	24	4	3
Fe^{3+}	1.3	0.2	0.1
Cu^{2+}	2.0	0.3	N.D.
Mn^{2+}	2.0	N.D.	0.3

insignificant by CO_2 blowing. Consequently, it is likely possible to release P_i from SSA by CO_2 blowing, while minimizing the contamination by toxic heavy metals.

13.5 Conclusions

SSA often contains a large amount of P_i and can serve as a promising secondary P resource. The CO_2 blowing method is potentially a simple, low-cost technology for releasing P_i from SSA, since it requires only water and CO_2 gas for releasing P_i from SSA. Needless to say, CO_2 is unlimitedly available from a sludge incinerator in WWTP. The CO_2 blowing method can release P_i from SSA at room temperature and without needing to adjust pH to low or high values. Further study is needed to examine the performance and cost effectiveness of this technology at a full scale.

References

Chatterjee C, Dube BK (2005) J Plant Nutr 28(10):1811–1820
Cordell D, White S (2014) Annu Rev Environ Resour 39:161–188
Langmuir D (1968) Geochim Cosmochim Acta 32:835–851
Ozaki M, Yamashita H, Miyamoto A (2005) Tech Note Natl Inst Land Infrastruct Manag 263:59–64
Segnit ER, Holland HD, Biscardi CJ (1962) Geochim Cosmochim Acta 26:1301–1331
Shiroyama H, Matsuo M, Yarime M (2015) Glob Environ Res 19(1):67–76
Soma M, Tanaka A, Seyama H, Ogura H, Kazuhara Y (1989) Int J Environ Anal Chem 36:103–109
Takahashi M, Sato K, Onari Y, Kato S, Enjyoji H (2004) Trans Mater Res Soc Jpn 29(5):2149–2152
Toyama T, Nakajima H, Kojima Y, Nobuyuki N (2012) Phosphorus Res Bull 26:91–94

Chapter 14
The Ecophos Process: Highest Quality Market Products Out of Low-Grade Phosphate Rock and Sewage Sludge Ash

Mohamed Takhim, Marc Sonveaux, and Rob de Ruiter

Abstract Ecophos s.a. has developed a unique modular process for the valorization of low-grade phosphate rock and/or various alternative P resources such as sewage sludge ash on the basis of soft digestion by hydrochloric acid or phosphoric acid. The process is extremely flexible and is, by the modular setup, capable of using several types of raw materials and producing a variety of products (fertilizer-, feed-, and food-grade phosphoric acid (PA), animal feed (DCP and MCP), and liquid NPK, PK, and NP fertilizers). The process has economic and ecological advantages over conventional industrial processes and those in development for valorization of sewage sludge ash, since it is simple, stable, and easy to control without needing expensive chemicals, raw materials, and equipment. The performance has already been tested in industrial plants at Bulgaria, Syria, and Peru as well as pilot- and lab-scale installations. Uptime longer than 7800 h/a is easily reached and the yield on P_2O_5 is 90% or higher. Furthermore the process can use excess HCl in the manufacture of products such as isocyanate, caustic soda, or SOP (Sulfate of Potassium). The energy balance is more positive than competing processes, since PA of high concentration (>42%) can be obtained without evaporation. The process can generate uranium (U)-free fertilizers, while conventional fertilizers generally contain 300–500 mg U/kg P_2O_5. Main by-products include high-purity $CaCl_2$ (as solution, flakes, or prills), radiation-free gypsum, silicate filter residue, and Fe/Al-chlorides. By applying different modules, most of the by-products can be split into sellable products, thereby minimizing final waste.

Keywords Ecophos process · Low-grade phosphate rock · Sewage sludge ash · Phosphoric acid

M. Takhim · M. Sonveaux
Ecophos s.a., Louvain-la-Neuve, Belgium

R. de Ruiter (✉)
[RBC]2 Consultancy, Kloetinge, The Netherlands

H. Ohtake, S. Tsuneda (eds.), *Phosphorus Recovery and Recycling*,
https://doi.org/10.1007/978-981-10-8031-9_14

14.1 Introduction

Phosphorus (P) is, together with many other elements such as oxygen, hydrogen, carbon, nitrogen, potassium, and various micronutrients, essential for living organisms including microorganisms, plants, animals, and humans on earth. Every cell of a living organism contains P which is irreplaceable with any other element. Phosphate (P_i) found in soil is taken up by plants. Plants are then consumed by animals that use P in biological building blocks and cellular energy metabolism. At the end of a life, P_i is returned back to nature. This natural P cycle has been in balance for many years. However, humans have influenced the P cycle mainly by intensive agriculture. While synthetic fertilizer is inefficiently used in agriculture, P is lost from soil by runoff and erosion on a global scale. This challenge needs to be addressed for feeding the ever-growing world population.

P_i-rich rocks have been found in limited geographical areas on the planet. The industry has come up with a solution by synthesizing fertilizers from these natural resources, thereby returning P_i back to farmland to feed the world population. However, today, this solution is increasingly challenged. High-grade P_i reserves which have been used by the fertilizer industry are in dangerous decline. What is more, the fertilizer-manufacturing processes consume lots of energy and produce tons of waste. Furthermore, approximately 45% of the world's high-quality P_i rock reserves is located in one specific region, North Africa. Europe's food supply is already depending on the P import from this region. Another 45% of the P_i rock reserves can be found in Asia, mainly China and Kazakhstan, but it is not readily available for export. The USA accounts for about 6% but has ceased the export since 1997. According to Fertilizers Europe, Europe imports around 1.75 Million tons of P annually in the form of P_i rock, fertilizers, phosphoric acid (PA), and feed phosphates. The situation for other countries with no substantial P_i rock resources is similar.

Nevertheless, hundreds of thousands of tons of P are lost annually in Europe alone. The main waste streams containing P are animal manure, sewage sludge, and bone meal. However, a large portion of the waste streams are disposed of into the environment without being efficiently recycled. For example, flow analyses from France show that 50% of the total P used there is lost – around 20% in wastewater, the same through erosion and leaching, and about 10% in the form of food waste and other biowastes (Consultative communication on the Sustainable Use of Phosphorus (European Commission) Brussels 2013, COM 2013). Intensive animal production is concentrated in specific areas close to ports and major population centers with available labor and expertise. This concentration has led to an oversupply of manure in these regions with a gradual build-up of P_i in soil and increased risks of water pollution. In addition, urbanization contributes to disrupt the P cycle by aggregating P-containing sewage sludge and food waste, few of which is returned to farmland. "Waste of today will generate food for tomorrow" is the slogan for the cradle-to-cradle concept of chemist Michael Braungart and architect William McDonough. The European Committee also advocates "urban mining" to find ways for valorizing wastes and to make them the mines for the future. Recycling of urban P_i resources is an essential issue for the future of the cycle of P and ultimately for the cycle of life. Once P is dispersed into the natural environment, it is practically difficult to recover for recycling.

14.2 Ecophos Philosophy

The fastest way to expand the global P availability may be to find ways to economically process the low-grade P_i rock. The United States Geological Survey (2017) estimates the world resources of P_i rock are more than 300 billion tons, while the reserves which can be economically extracted with proven technology and current economics are estimated to be about 68 billion tons (USGS 2017). However, P_i rock is generally contaminated with toxic substances such as cadmium (Cd) and radioactive uranium (U). Conventional technologies cannot completely remove these toxic substances in fertilizer-manufacturing processes. Certain plants such as sunflower, colza, and tobacco can accumulate relatively high concentrations of cadmium. In terms of health impact, the EU Risk Assessment Report on cadmium (European Commission Joint Research Center 2007) found that the major risk of cadmium is kidney damage through food consumption and smoking. The radioactive uranium ends up in fertilizers in the range of 300–500 mg U/kg P_2O_5, posing the contamination risk of food. Ecophos believes that fertilizers should be free from such toxic substances as heavy metals, radioactive elements, and other contaminants. It is critical to apply the right technology for valorizing "urban mines" and low-grade P_i rock as high-quality, contaminant-free P_i resources for the manufacture of fertilizer.

Producing high-purity P sources has another advantage. High-purity PA makes it easy to supply tailor-made fertilizers suited for the needs of the specific soil. Ecophos believes that the key concept of the fertilizer industry will shift from a volume-based standard product approach to a quality-based specific composition approach. The specifics of the soil and consumer needs will determine not only what macronutrients are needed but also the requirement for specific micronutrients. Precision agriculture in terms of fertilizer composition and timing of fertilization will become more and more important. Ecophos believes that future fertilizers will be manufactured using high-purity macro- and micronutrients. The Ecophos technology enables this approach for supplying high-quality P-containing building blocks from a variety of resources, regardless of high- or low-grade P_i rock or secondary P resources from "urban mines."

Regarding "urban mines," it is logical to valorize waste streams that have the highest P content and are most readily available. In this perspective, bone meal and sewage sludge have the highest potential. Animal manure is also an option, but its use is logistically more difficult. In Europe, more than 10 million tons of sewage sludge (dry mass) arises every year (Table 14.1). One fifth of this quantity is from Germany where about 47% of sewage sludge is used in farming and landscaping, and the remaining 53% is incinerated (45% mono-incinerated (220 kt (kilo tons)

Table 14.1 Estimates for sewage sludge and sewage sludge ash production in Europe, USA, and Japan

Sewage sludge and mono-incineration (%)					
	Sludge (DS)	Ash		Ash (mono)	As P_2O_5
	kt/a	kt/a	% mono-inc.	kt/a	kt/a
Europe	8330	3165	20	633	107
Germany	2450	931	28	260	44
Austria	245	93	31	28	4
Switzerland	203	77	47	36	6
USA	8200	3116	22	685	116
Japan	3000	1140	50	570	96
Total	22,428	8522		2214	376

ash/a) and 55% co-incinerated). In Germany alone, approximately 24 kt of meat and bone meal (MBM) and 55 kt of sewage sludge are generated annually. Germany is one of the leading countries to study valorization of these waste streams. However, according to the conclusion of the PhoBe-Study (Pinnekamp et al. 2011), today no single process to recycle P from waste streams is economically viable. Since the occurrence of BSE crisis MBM is classified into the three categories. The category 2 and 3 materials are allowed to be used as animal feed or fertilizers, whereas the category 1 (high risk) material must be incinerated. Unfortunately this material is mostly incinerated in cement kilns or in power plants as a high calorific substitute. This simply means that proteins and phosphorus are lost forever. If they are mono-incinerated, the ash would be a high-quality P resource (35–40% P_2O_5, no heavy metals and no uranium).

Several processes are known to recover P from waste streams (Sartorius et al. 2011 and see Chap. 1). P can be recovered from (i) sludge liquor, (ii) digested sludge, or (iii) sludge ash. Although chemical precipitation (mostly as calcium phosphate or struvite) from sludge liquor is a relatively easy option. However, the main disadvantage is that the overall potential of P recovery from sewage is low, normally between 50% and 60%. Another issue almost never mentioned is the fact that excess Mg (e.g., via struvite) in the soil could hamper the uptake of Ca, K, and other essential cations by crops (Hammond and White 2005). As the sludge contains organics (e.g., antibiotics, pathogens, and pharmaceuticals), the recovered P products are often contaminated with these organics. Depending on the specific process, heavy metal contamination can also occur. By contrast, sludge incineration has advantages such as energy integration and high mercury removal. Organic pollutants such as PCBs, dioxins, hormones, and POPs are also decomposed.

The Ecophos philosophy is simple and straightforward: any risk concerning organic contaminants in fertilizers, animal feed, and food must be avoided. Therefore, incineration is a must in the Ecophos's view. Furthermore, it is highly prioritized to design a technology that leads to the high-purity P-containing building blocks that are identical to today's market products and that can be easily used as a P source to manufacture specifically designed fertilizers in the new fertilizer economy. The type of P_i-precipitates should be carefully selected with respect to the crop accessibility (the water- and citric acid-solubility). Moreover, the market for P fertilizers in Germany is only 10%, while the majority are NPK fertilizers. It is therefore economically preferable to aim for NPK fertilizer production out of waste materials. This calls for a flexible, economically and ecologically feasible, new technology that is able to produce several types of P compounds that can be used either as such or in an NPK formulation out of low-grade P_i rock, MBM, and sewage sludge ash.

14.3 The Ecophos Process

To understand the advantages of the Ecophos process, it is useful to give a short introduction about the basics of conventional industrial processes for the manufacture of phosphoric acid. Conventionally, calcium apatite ($Ca_{10}(PO_4)_6(OH)_2$) in P_i Rock has been made accessible for crops using a wet acid process. Sulfuric acid is most commonly used for the wet acid process (Fig. 14.1). Drawback of this process is that the acid also solubilizes heavy metals such as Cd and Pb as well as radioactive elements such as radium (Ra) and U, causing their contamination of the end products. Radium is the most problematic radioactive element as it generates Radon which causes lung cancer. Radium is concentrated in the by-product phosphogypsum of the conventional wet processes.

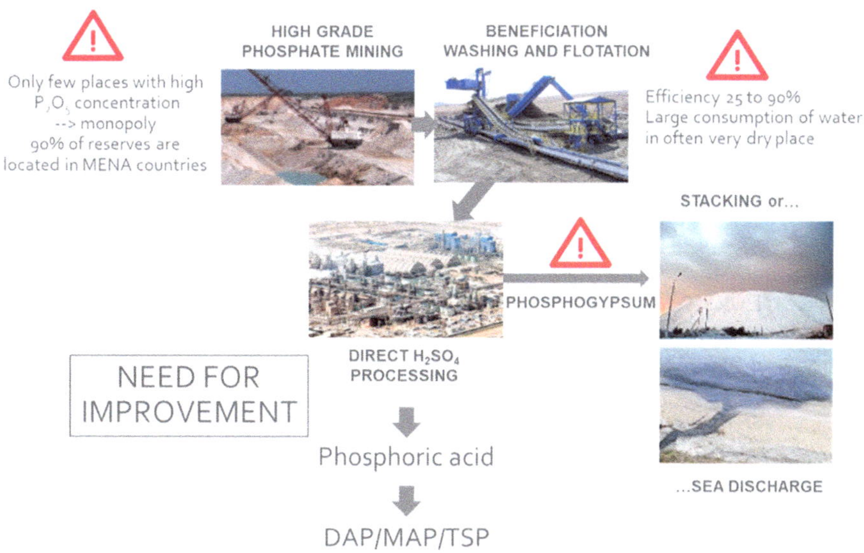

Fig. 14.1 Conventional process for the production of phosphates

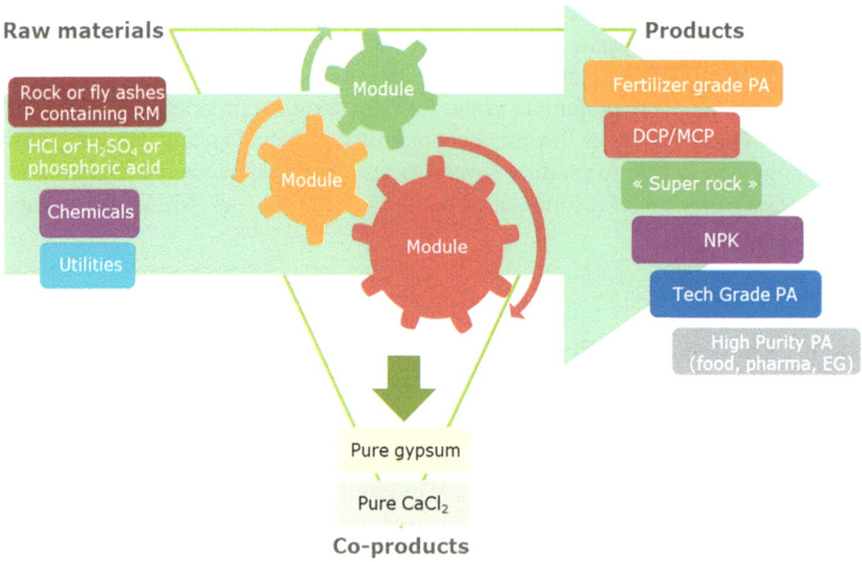

Fig. 14.2 Raw materials, products, and coproduct scheme of the modular Ecophos approach

P_i rock is obtained mostly by open-cast mining. This type of mining requires large areas of land not only for mining itself but also for spoil heaps and clay settling ponds. The quantities of total solid waste produced are so large that about 21.8 tons of mine wastes and 6.5 tons of tailings are produced for each ton of phosphoric acid (PA) produced. The PA-manufacturing plants also generate large quantities of a by-product called phosphogypsum. In some countries phosphogypsum is stored at large stacks due to regulation of radioactivity levels or because the alternative (natural gypsum and flue gas gypsum) are more competitive. In a few countries such as Brazil and China, however, it is increasingly used in the construction and agriculture sectors. It should be noted that natural radioactivity levels in P_i rock can differ widely, depending on the geology of the mine. The conventional mining is also a water- and energy-intensive process.

Taking these into account, Ecophos has developed and patented technologies for:

- The production of dicalcium phosphate (DCP) and/or phosphoric acid using a wide variety of phosphate resources including low-grade P_i rock and sewage sludge ash, regardless of their quality and P content
- The conversion of low-grade phosphoric acid such as green phosphoric acid into high-purity products such as technical-grade, feed-grade, food-grade, and electronical-grade phosphoric acid
- The production of phosphate specialties from a variety of raw materials

Ecophos has developed the processes as modular units that can be combined to generate target products using a variety of raw materials as P sources (Fig. 14.2). The philosophy of the Ecophos processes is basically very simple and therefore

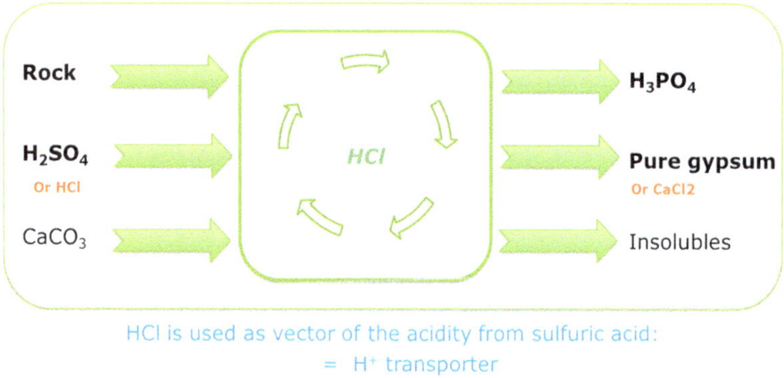

Fig. 14.3 Simple block scheme for the production of DCP and/or phosphoric acid

extremely effective. Firstly, the raw materials are digested in such a way that phosphate is mostly released from the matrix. After this first digestion, insoluble solid residues are separated from the leachate. During the second stage of the process, the phosphate is purified from the leachate either by the chemical precipitation of DCP or by the ion exchange technology, depending on the leaching agent (HCl or H_3PO_4) used in the first step (Fig. 14.3).

14.4 The HCl-Process for the Production of DCP and/or PA

The input P_i-containing raw materials react with dilute HCl in the Module 1A (Fig. 14.4), generating a phosphate solution containing dissolved $Ca(H_2PO_4)_2$, H_3PO_4, and $CaCl_2$. Cd and Mg are dissolved together with the phosphate. A liquid/solid separation step is performed to remove the solid residue that contains most of the impurities – Al, Fe, organics, radioactive elements, and fluorides present in the P_i rock or the sewage sludge ash. The liquor obtained from the Module 1A reacts with a calcium source in the Module 1B. Dicalcium phosphate (DCP) is precipitated. Then, the second liquid/solid separation step is performed. The solid phase consists of DCP, while the liquid phase is a $CaCl_2$ solution containing heavy metals and magnesium. This $CaCl_2$ solution is purified by precipitation of heavy metals and Mg at high pH (not shown in Fig. 14.4). Very pure $CaCl_2$ solution at a concentration of 15–18% is obtained. In the optional Module 4, the purified $CaCl_2$ reacts with sulfuric acid, resulting in the formation of a slurry of dihydrate gypsum in dilute HCl solution. $CaCl_2$ can be generated prior to the Module 4, and a makeup of HCl is recycled to the Module 1A to maintain equilibrium in water and impurity balances. The gypsum is separated from HCl before HCl is returned back to the Module 1A for digestion. Gypsum which is very pure and free of radioactive elements and phosphate is a coproduct of this process. $CaCl_2$ can either be purged to the sea, or sold as a solution, or after vaporization brought to the market as flakes or prills for deicing, dedusting,

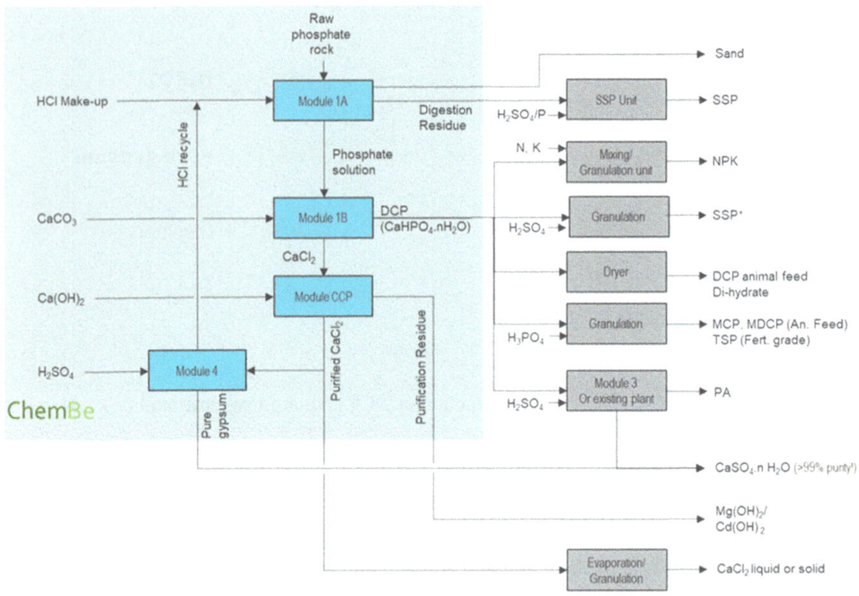

Fig. 14.4 More extended scheme for the production phosphate products by the Ecophos method

or oil-drilling applications. The main input of this process is either sulfuric acid or hydrochloric acid, depending on whether the Module 4 is used.

Similarly, in the Module 3, which is an optional module, the DCP resulting from Module 1B can be sold on the market (animal feed) or digested in sulfuric acid. Phosphoric acid is produced in a rather high concentration (>42% P_2O_5) and pure gypsum is obtained as a coproduct. The phosphoric acid is then concentrated to the merchant-grade (52% or higher). Furthermore Ecophos offers efficient and economical processes to generate NPK fertilizers, MKP, MAP (monoammonium phosphate), DAP (diammonium phosphate), STTP (sodium tripolyphosphate), MCP (monocalcium phosphate), MDCP (monodicalcium phosphate), MgP (magnesium phosphate), SSP (single super phosphate), TSP (triple super phosphate), and many others (Fig. 14.4). Other projects relevant to the HCl process are summarized in Table 14.2.

14.5 The Phosphoric Acid Process for the Production of H_3PO_4

Especially for local applications, Ecophos designed a process based on phosphoric acid digestion and ion exchange. This process (see Fig. 14.5) can be applied to downstream of sewage sludge incineration and is economically feasible from SSA-input volumes of around 10 kt SSA/a or larger.

Table 14.2 Shows reference projects of the HCl approach

			Module and unit							
Project	Country	Year	1A/6	1B	CCP	Dry DCP	3	4	$CaCl_2$ C/G	SSP
Decaphos	Bulgaria	2006	o	o		o				
UCCI	Syria	2010	o	o	o	o		o		
Quimpac	Peru	2014	o	o	o	o	o			
Eurochem	Kazakhstan	2014	o	o	o	o		o	o	
LLNP	Namibia	2015	o	o			o	o	o	
Technophos	Bulgaria	2016	o	o	o		o	o	o	
Aliphos	France	2017	o	o	o	o				
Evergrow	Egypt	2018	o	o	o	o	o		o	o
EGIL	India	2019	o	o	o	o				o
Everphos	Egypt	2020	o	o	o	o		o	o	o

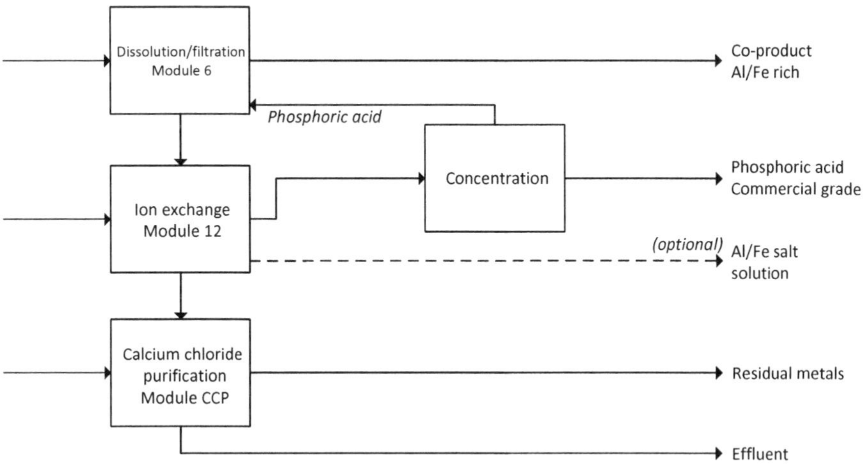

Fig. 14.5 Schematic representation of the Ecophos process based on phosphoric acid digestion

In this approach, the fly ashes are digested by phosphoric acid. The insoluble solid residues are separated by a press filter from the cation-loaded phosphate solution. In the next stage, the cations (Al, Fe, Ca, Mg, and heavy metals) are captured and exchanged with protons by an ion exchange unit to obtain purified phosphoric acid with a concentration of approximately 25% P_2O_5. This solution is partially recycled for digestion in the first stage of this process. The ion exchange resin is regenerated by hydrochloric acid. The final product is concentrated by evaporation to reach the preferred concentration for the market. All products and by-products specifications comply with or exceed market quality requirements.

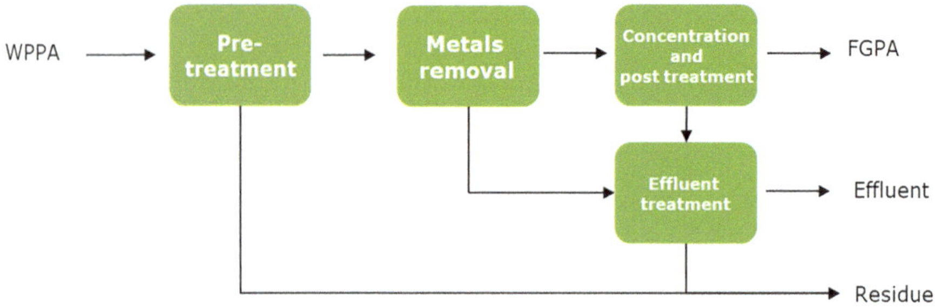

Fig. 14.6 Schematic representation of the process for the purification of phosphoric acid

14.6 Process for the Purification of Phosphoric Acid

The Ecophos process can also convert fertilizer-grade PA from the Module 3 to technical- and food-grade PA in three steps (Fig. 14.6):

Step 1: pre-purification to remove gypsum, arsenic, and cadmium
Step 2: removal of heavy metals, anions, and color by different techniques
Step 3: concentration of PA to 62% P_2O_5 by evaporation

The effluent is treated to remove the metals and impurities as solid cake. All Ecophos processes are stable and easy to control. Uptimes longer than 7800 h/a or longer are easily reached, and the P_2O_5 yields of 90% or higher can be guaranteed. This modular approach is flexible in terms of raw materials, products, and by-products and offers a new insight to fertilizer chemistry to shift from volume-based standards to quality-based dedicated products.

14.7 Conclusion

Ecophos changes one's mind about phosphorus and the circular economy thereof. The several patented processes enlarge the usable amount of phosphorus that is essential for life, in an economically and ecologically feasible way. The Ecophos process can generate the high-quality P products from low-grade P_i rock and untapped secondary P resources, thereby making a great contribution to realizing a circular economy for future generations.

References

Consultative communication on the Sustainable Use of Phosphorus (European Commission), Brussels 2013, COM (2013) 517

European Commission Joint Research Center (2007) EU risk assessment report on cadmium metal part II-human risk

Hammond, White (2005) Final project report entrust 675382.006, "Is Struvite a Valuable Phosphate Source for Agriculture?", Warwick HRI

Pinnekamp et al (2011) Abschlussbericht: "Phosphorrecycling – Ökologische unde wirtschaftliche Bewertung verschiedener Verfahren und Entwicklung eines strategischen Vertungskonzepts für Deutschland"

Sartorius et al (2011) Phosphorus recovery from waste water – state-of-the-art and future potential. Fraunhofer Institute for Systems and Innovation Research, Karlsruhe

USGS (2017) Mineral commodity summaries of phosphate rock 2017. Online available at https://minerals.usgs.gov/minerals/pubs/commodity/phosphate_rock/mcs-2017-phosp.pdf, last retrieved at 28/08/2017

Chapter 15
Outotec (AshDec®) Process for P Fertilizers from Sludge Ash

Ludwig Hermann and Tanja Schaaf

Abstract Outotec offers the AshDec® process by which inorganic calcined phosphates (thermophosphates) are produced from phosphate-rich ashes remaining from incineration or gasification of sewage sludge, animal by-products, poultry litter, and other nutrient-rich organic waste. In the thermochemical process, solid, potassium and/or sodium-based, alkaline compounds admixed to the ash decompose at a temperature of >900 °C and react with the ash-borne phosphates to form bioavailable (ammonium citrate-soluble) alkaline phosphate compounds. Simultaneously, the toxic arsenic, cadmium, and lead compounds become gaseous and evaporate from the reactor bed. As soon as the process gas is being cooled, the particles condensate and are captured and removed in an electrostatic precipitator as metal concentrate. The process produces a P or PK fertilizer with relevant mass fractions of silicates, sodium, and trace elements. Phosphates are released and taken up by crops when root exudates decompose the Ca-K/Na-PO_4 compounds preventing losses of P in solution if water-soluble P fertilizers are used. The recently explored, partial replacement of sodium sulfates by potassium phosphates avoids high sodium concentrations and leads to a PK 16-7 + 4S fertilizer with >25% total macro nutrient content and >10% sodium/potassium silicates that may enhance crop resilience. In a recent (published 2016) report, the Expert Group for Technical Advice on Organic Production (EGTOP) came to the conclusion to recommend calcined phosphates and struvite for organic production.

Keywords P recovery and recycling · Calcined phosphates · Heavy metal removal · Citrate-soluble phosphates · Organic farming · Root controlled release fertilizer

L. Hermann (✉) · T. Schaaf
Outotec GmbH & Co. KG, Oberursel, Germany
e-mail: l.hermann@proman.pro

H. Ohtake, S. Tsuneda (eds.), *Phosphorus Recovery and Recycling*,
https://doi.org/10.1007/978-981-10-8031-9_15

15.1 Introduction

The development of Outotec's AshDec process started in 2004 when the Federal Institute for Materials Research and Testing (Berlin) and ASH DEC Umwelt AG (Vienna) shared know-how and joined forces to investigate the potential of a thermochemical process for recovery and recycling of phosphorus. Together with other scientific and industrial partners, they filed the proposal for the joint European R&D project "SUSAN" (2005–2008) aiming at the development of a circular, sustainable, and safe management option for municipal sewage sludge based on sludge incineration and thermal treatment of ash (Adam et al., 2008). Sludge ash is a promising feedstock for phosphorus recycling because effective incineration completely destroys pathogens and organic pollutants, an ongoing concern of organic fertilizers produced from sewage sludge. For subsequent application of phosphorus recovery and recycling techniques, it is required to incinerate the sludges in facilities where only municipal sewage sludge and other P-rich organic waste fractions ("green" bin, chicken litter, manure, animal by-products) are combusted and a minimum P_2O_5 concentration of 15% is achieved. Co-incineration of sewage sludge with waste containing less P dilutes the ash resulting in low P concentration which makes it less attractive for recovery for technical, environmental, and commercial reasons. The solid residue from mono-incineration of sewage sludge is ash (SSA) with a phosphorus content in the range of 15–25% P_2O_5. Chicken litter ash contains similar P and relevant K concentrations, and ash from the combustion of animal by-products contains even higher mass fractions of P. Sewage sludge ash mainly consists of phosphate, calcium, silicon, iron, and aluminum compounds. Toxic heavy metals, even if present in minor fractions, should be removed to avoid their accumulation in soils and potential transfer to the food and feed chain. Particularly sewage sludge ashes often contain heavy metal compounds above the legal limits for agricultural use stipulated in the corresponding fertilizer ordinances. Furthermore, phosphorus in ash usually exhibits a low bioavailability – hampering the fertilizing efficiency when used on cropland. Therefore, a second step is required after mono-incineration to (1) reduce the heavy metal concentrations below the relevant limits and (2) to transfer phosphorus into mineral phases with high P bioavailability.

15.2 Process Evolution in Response to the Regulatory Framework

When AshDec and the Federal Institute for Materials Research and Testing started the SUSAN project in 2005, two countries had launched initiatives for phosphorus recycling: Germany and Switzerland. In both countries national fertilizer and chemical product regulations with very low metal limits (including the micronutrients copper and zinc) were in force, as shown in the following table (Table 15.1):

Table 15.1 Heavy metal limits for (recycled) fertilizers in Germany and Switzerland in 2008 (Bundesministerium für Umwelt…; Schweizerische Eidgenossenschaft, Der Bundesrat…)

	Unit	As	Cd	Cu	Hg	Ni	Pb	Zn
Germany	mg/kg	40	4[a]	70	1	80	150	1000
Switzerland	mg/kg		1	100	1	30	120	400

[a]Referring to a P_2O_5 concentration of 20% in the product

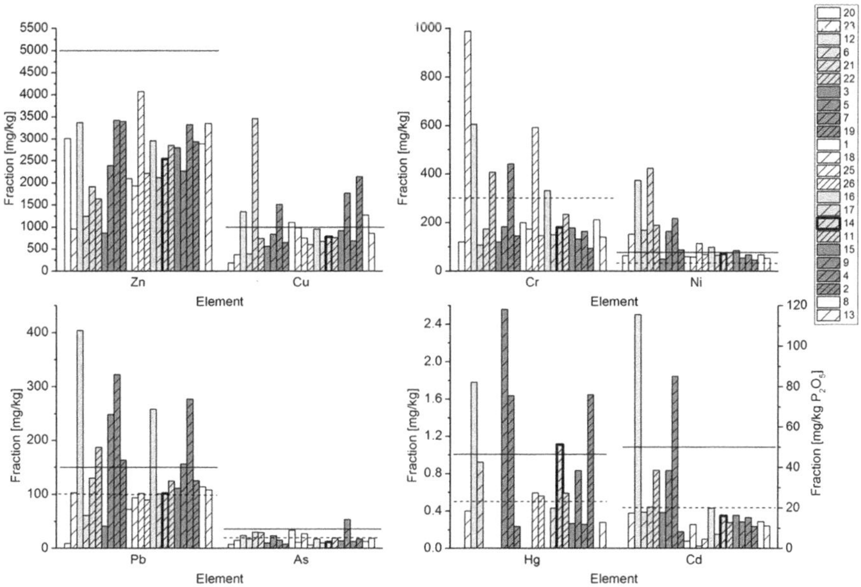

Fig. 15.1 Mean mass fractions of Zn, Cu, Cr, Ni, Pb, As, Hg, and Cd in the SSA of 24 German mono-incineration facilities. The solid horizontal lines represent the German limit values from 2009, and the dashed lines represent the concentrations above which the German fertilizer regulation requires labeling (Herzel et al. 2016)

Comparing heavy metal concentration in ashes and the legal limits for (recycled) products used on cropland – in Germany in effect until December 2008 and in Switzerland until December 2017 – makes clear that removal of heavy metals was a priority (Fig. 15.1). Since copper concentration had to be reduced by more than 90%, from about 1000 to 70 ppm, only chlorine additives in relevant concentrations could achieve the removal.

For the ease of handling, process efficiency, and product quality, solid chlorine compounds were selected – calcium and magnesium chlorides. The task was accomplished by admixing enough calcium or magnesium chlorides to achieve 10–15% Cl in the blend and heating the material to 1000 °C for 15–20 min. The use of $CaCl_2$ produced chlorapatite as a main phosphorus phase in the product, whereas the magnesium of $MgCl_2$ reacted to calcium-magnesium-phosphate compounds like stanfieldite and farringtonite. The bioavailability of chlorapatite is low as we know from phosphate rock. The magnesium-bearing phosphate phases were bioavailable as long as the soil was below pH 7. In alkaline soils, hardly any phosphorus was available for crops (Adam et al. 2008).

After 2008, with the enforcement of the new fertilizer regulation in Germany and Swiss authorities indicating that more accessible limit values may be introduced for mineral recycled fertilizers, new treatment options became viable. Whereas the toxic heavy metals without a nutritional function – arsenic, cadmium, and lead – can be removed from the ash in a reducing atmosphere, additives may be selected by their capacity of enhancing the bioavailability of phosphorus. Soda ash (Na_2CO_3) has been known for almost a century for making phosphate rock plant available after thermal treatment (Werner 1981). Consequently, Outotec invested in new research and test works investigating the effect of adding soda ash to the sludge ash before thermal treatment. Indeed, reactions took place as expected, and plant-available phosphate compounds could be produced. However, two drops of bitterness remained: (1) soda ash is quite expensive and has a high environmental footprint and (2) the reactions are not as complete at technical scale as at lab scale. Research for alternative reaction partners led to sodium sulfate as appropriate reactant achieving similar results in terms of consumption and bioavailability of phosphate compounds in the product. In addition, sodium sulfate is available as by-product or waste from cleaning of sulfur-rich gases. Additional test work proved that sodium sulfates could be mixed with potassium sulfates facilitating a higher nutrient concentration in the P-K + S fertilizer.

15.3 Process Description

Compared to other phosphate recovery technologies, processes which have ash as a starting material share the benefit of relatively low mass flows because the phosphate (P_2O_5) concentration in sludge ash is largely in the same order as in phosphate rock concentrate – 15–25% in ash vs. 25–36% in phosphate rock concentrate. With a P recovery rate of 97% of the phosphorus eliminated at a wastewater treatment plant, the process is unrivaled in terms of P recovery efficiency. Secondary waste streams do not exceed 3% of the manufacturing plant inflow. Contrary to competitive technologies, ashes from all current wastewater treatment processes qualify for the treatment by the AshDec process, regardless of their chemical characteristics and presence of impurities such as iron and aluminum.

The core process encompasses feeding hot ash to a rotating reactor (rotary kiln) where it is mixed with potassium and/or sodium compounds and a reducing agent (preferably dry sewage sludge) and treated at 900–950 °C for 15–20 min. Preferably natural gas is used as a clean fuel. In the flow sheet (see Fig. 15.3 in the Sect. 15.5) predominantly considered in this description, ash is fed hot due to being taken directly from the adjacent incineration reactor. In a greenfield flow sheet (not shown), a fluidized bed reactor-based preheating step partly compensates the additional energy consumption due to ash being fed at ambient temperature. It allows full energy recovery by counterflow preheating of the feedstock. Both options facilitate a short residence time and a compact rotary kiln with small temperature gradients. If the ash treatment plant is located adjacent to an incineration plant, the air pollution control of the main plant can be used for off-gas cleaning. In this case, only one additional electrostatic precipitator is needed for condensing and capturing the heavy metals removed from the ash. A simplified process flow sheet is shown in Fig. 15.3, in the manufacturing plant section.

The main consumables are alkaline compounds containing potassium and/or sodium. Potassium and sodium ions replace calcium ions in the phosphates and form citrate-soluble Ca-K/Na-PO_4 compounds. The level of heavy metal removal is a function of the reactor atmosphere, ash characteristics, and the additive mix. Potassium and sodium sulfates are the preferred additives due to their relatively low cost and high reactivity. A relevant part of the sodium sulfates can be recycled from the air pollution control system of sludge incineration, depending on the sulfate concentration in the sewage sludge. If gaseous sulfur compounds are removed from the off-gas using sodium hydroxides or sodium bicarbonates, sodium sulfate can be recovered and used.

Application of the product as a phosphate fertilizer leaves a smaller environmental footprint than using either conventional mineral fertilizers or recycled organic fertilizers. In comparison to conventional mineral fertilizers, concentrations of cadmium and uranium are 80–99% lower. In comparison to organic fertilizers, no risk of transfer of organic pollutants to the food and feed chain exists. Other benefits include (1) the absence of free acids enabling blends with urea and organic fertilizers without undesired reactions and (2) phosphate supply on demand of crops due to releasing phosphorus only in presence of crop root exudates avoiding P losses by runoff, leaching, and fixation. Other potentially beneficial characteristics, such as the relevant silica content, will be further investigated during a new R&D project supposed to start in September 2017.

15.4 Key Process Data (Brownfield, Adjacent to Incinerator)

The model case for exhibiting the key process data is based on the ash of one typical European sludge incineration plant processing about 100,000 annual tons of digested and dewatered municipal sewage sludge with about 30% dry matter and 46% ash. The case is modeled after the state-of-the-art sewage sludge incineration plant combusting the sludge of about 1.4 million citizens discharging wastewater into 72 sewage plants of the Swiss federal state (Kanton) Zürich.

The sludge is transported by truck to the incinerator located at the "Werdhölzli" sewage plant in Zürich city (Fig. 15.2). It converts the remaining energy (after anaerobic digestion and biogas production) to 8 MWh/year electricity used for operating the plant and about 32 MWh/year heat used for district heating. The

Fig. 15.2 Zürich Werdhölzli municipal sewage sludge incineration plant (Source Outotec)

sludge incineration plant does not use any fossil fuels (except for start-up) (Outotec…). About 46% ash (inorganic mass fraction) in the dry sludge corresponds to 13,800 tons per year or 1725 kg per hour fed at 850–900 °C to the AshDec plant. The ash fraction in municipal sewage sludge varies between 30% and 46%. The Zürich sludge has been fully digested upstream of incineration leading to a low organic and corresponding high inorganic mass fraction.

Table 15.2 shows the absolute input and output data for the modeled AshDec plant as if built and operated adjacent to the sludge incinerator and the same flows referred to 1 ton of P_2O_5. The data has been modeled after ash treatment in a pilot plant (semi-industrial), thermodynamic calculations (HSC Chemistry), as well as process and product assessment by the P-REX consortium.

15.5 Manufacturing Plant Concept

A smaller (<3 t/h) renewable phosphate fertilizer manufacturing should be built adjacent to a sludge incinerator. Hot ash is separated from the gas funnel in a cyclone positioned between the FB reactor and the boiler and fed directly to the AshDec plant that includes the following sections: rotary kiln, additive preheater, electrostatic precipitator (ESP) and bag filter, and granulator/dryer. A simplified flow sheet is shown in Fig. 15.3.

Table 15.2 AshDec model case input and output flows of materials, utilities, emissions, and waste (Joint EU Research Project P-REX 2012–2015…)

	Amount	Unit	Reference to product	Unit
Input streams				
Sludge hot ash	1725	kg/h	5,08	t/t P_2O_5
New sodium sulfate Na_2SO_4	346,59	kg/h	1,02	t/t P_2O_5
Recycled sodium sulfate Na_2SO_4	293,01	kg/h	0,86	t/t P_2O_5
Sludge hot ash	1725	kg/h	5,08	t/t P_2O_5
New sodium sulfate Na_2SO_4	346,59	kg/h	1,02	t/t P_2O_5
Recycled sodium sulfate Na_2SO_4	293,01	kg/h	0,86	t/t P_2O_5
Dry sludge – reducing agent	293,01	kg/h	0,86	t/t P_2O_5
$Ca(OH)_2$	236,5	kg/h	0,70	t/t P_2O_5
Water	26,2	kg/h	0,08	t/t P_2O_5
Natural gas	472	kg/h	1,39	t/t P_2O_5
Electricity consumed	0,674	MWh/h	1,99	MWh/t P_2O_5
Air	0,138	MWh/h	0,41	MWh/t P_2O_5
Output streams				
Calcined fertilizer	2311	kg/h	6,81	t/t P_2O_5
Waste	43,3	kg/h	0,13	t/t P_2O_5
Total off-gas	4261,5	Nm^3/h	15,48	t/t P_2O_5
Total off-gas	5252,2	kg/h	15,48	t/t P_2O_5
Off-gas composition				
H_2O	573,9	kg/h	1,69	t/t P_2O_5
N_2	3473,6	kg/h	10,24	t/t P_2O_5
O_2	813,2	kg/h	2,40	t/t P_2O_5
CO_2	396,4	kg/h	1,17	t/t P_2O_5

15.5.1 Rotary Kiln Including Venturi/Cyclone Preheating System

The rotary kiln facilitates mixing and chemical reaction of phosphates and alkaline compounds (e.g., Na_2SO_4) to produce a phosphate compound called rhenanite ($Ca\text{-}Na\text{-}PO_4$), a traditional and extensively tested phosphate fertilizer with a high fertilizer value (Werner 1981). Alternatively, a mixture of K_2SO_4 and Na_2SO_4 is used as additive directly producing a PK fertilizer with $Ca\text{-}K/Na\text{-}PO_4$ as a main phosphorus-bearing phase.

The rotary kiln is heated to 900–950 °C by a natural gas burner, hot ash, and coal dust or dry sewage sludge as a reducing agent. The reducing atmosphere in the reactor facilitates heavy metal evaporation and the formation of the bioavailable P compounds in a reaction between the alkaline additives and the phosphates. The hot ash is fed to the rotary kiln. K_2SO_4 and/or Na_2SO_4 are preheated in the upstream venturi/cyclone system and admixed in a molar ratio of 1.4:1 (K_2O/Na_2O to P_2O_5). Coal dust or dry sewage sludge is directly fed to the rotary kiln that is loaded with comparatively high material loads (>15%) to facilitate reactions in the material bed.

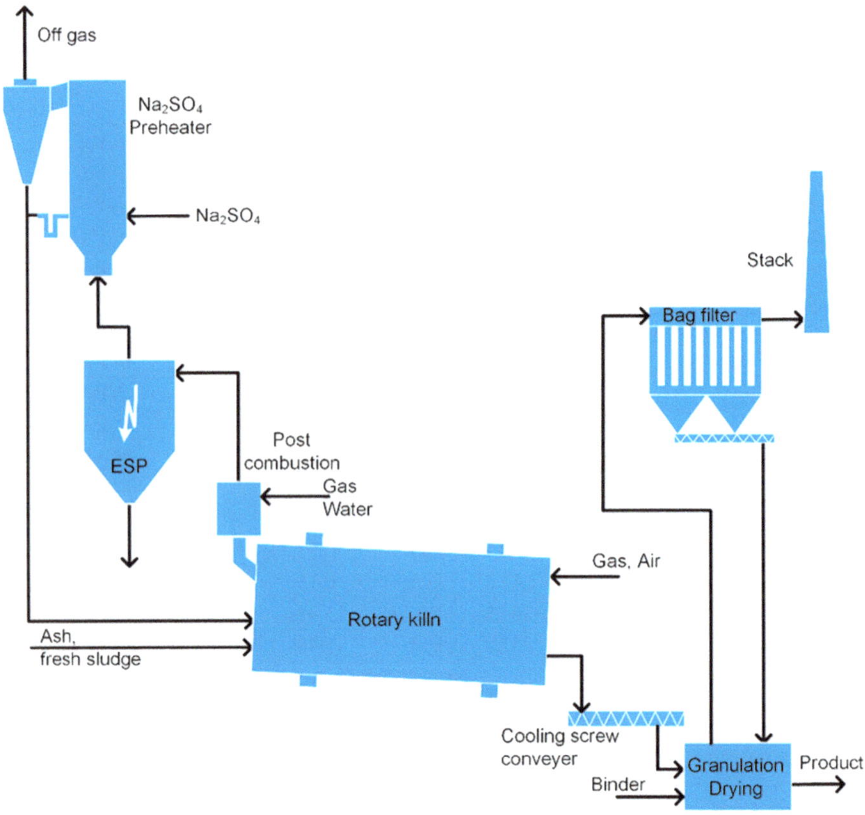

Fig. 15.3 Simplified AshDec flow sheet (Source Outotec)

In the reactor, the coal/dry sewage sludge decomposes releasing CO. Further downstream in the post-combustion chamber, air is blown in and the CO is combusted, providing additional heat. The material residence time in the reactor is about 15–20 min. The product leaves the kiln to a screw cooler equipped with a cooling circuit. Downstream of the screw cooler, the material is analyzed for its P concentration and selected heavy metal concentrations online. After the online analysis, the semifinished product is fed to the product finishing section including a granulator, dryer, baghouse filter, and stack. Optionally, an additional feeding system for admixture of a chlorine compound (e.g., $MgCl_2$) can be installed for further removal of metals such as copper and zinc. Significantly enhanced heavy metal removal, however, may be at the expense of the fertilizer solubility if the heavy metal removal and the phosphate solubilizing processes are not spatially separated.

After leaving the post-combustion chamber, the process gas is cooled in a quench (not shown) with fresh air and/or water. In an electrostatic precipitator, dust particles and particularly the recondensed heavy metals are removed from the process gas. The combustion gas is funneled to a venturi, where it is mixed with the potassium and/or sodium sulfate additives fed in counterflow, and to a cyclone, where solids are separated from the off-gas. After the cyclone, the off-gas is directly fed to the air pollution control system of the adjacent sludge incinerator.

15.5.2 Mixer-Granulator System (Possibly Located Elsewhere), Bag Filter, and Stack

Producing a fertilizer requires adjustment of nutrient concentrations within the tolerances required by the fertilizer regulation and producing round PK fertilizer granules with a corn-band of 3–5 mm appropriate for spreading with the equipment in the market. The finishing section may be integrated to the ash treatment plant or be part of a fertilizer production or blending plant. If it is part of the AshDec plant, its central device is a mixer-granulator with a rotating case and mixing tools rotating in the contrary direction. At this stage, additional nutrient-carrying compounds can be admixed to the semifinished product.

The semifinished phosphate fertilizer is online analyzed for its nutrient and heavy metal concentration and fed to the granulator through a rotating valve. Simultaneously, a standard phosphate (e.g., phosphoric acid), potassium (e.g., KCl), or nitrogen carrier (e.g., urea) is fed from a hopper to the granulator through a rotating valve. KCl or urea is stored in a silo or a bunker, acid in a tank. Alkaline bleed water from the sludge incineration scrubbers, containing most of the sulfur from the combusted sewage sludge, may be fed to the mixer as a binder and sulfur carrier.

Granules of different corn sizes are produced in the granulator. The granules are conveyed to a dryer, where excess humidity is removed via evaporation. The dryer is heated by the excess energy from the product cooling screws. Downstream of the dryer, the granules are fed to a sieve machine where oversized granules are separated and fed to a mill and undersized granules and fines are fed back to the mixer-granulator. Downstream of the mill, the fines are fed back to the mixer-granulator through a conveyor. The gaseous dryer emissions are fed to a bag filter and released to the air through the stack.

15.6 Manufacturing Plant Concept for Greenfield Operations

All other conditions unchanged, a greenfield AshDec plant would be equipped with a CFB heater to cope with the higher energy demand due to the ash fed at ambient temperature and a full air pollution control system to release clean off-gas in compliance with the legal and environmental requirements.

The function of the CFB heater equipped with a natural gas burner is to preheat the cold feedstock to a temperature below the ash sintering point, around 950 °C. It enables the recovery of energy from the combustion gas by sequential counterflow venturi/cyclone steps to transfer the sensible energy from the off-gas to the feedstock. Similar to the brownfield plant concept, the CFB heater avoids the dilution of the process gases in the rotary reactor and allows to design a compact rotary kiln with low-temperature gradients. In addition, the heater would provide the conditions for a separate heavy metal removal step by adding some chlorines or using chlorine containing off-gas. The process gas from the rotary kiln, after cooling and particle removal in the ESP, could be fed as secondary combustion air to the CFB reactor. If the process gas from the reactor is fed to the CFB heater, only one hot gas stream has to be purified for release to the air.

The CFB heater off-gas may be funneled to one or two venturi/cyclone stages for preheating the feedstock before entering the CFB heater by the sensible heat of the off-gas. By preheating the feedstock, the off-gas loses enough heat to be fed to another electrostatic precipitator and a scrubber/bag filter in the final cleaning stage. The final design depends on the local conditions and the degree of internal looping of compounds that can be reused as additives.

15.7 Product

The process in its most advanced adaptation produces a PK + S 16-7 + 4 fertilizer increasing the total macronutrient concentration to >25% and reducing nonfunctional compounds in the product (Table 15.3). The AshDec process and the fertilizer product are protected by a joint Outotec/Federal Institute of Material Research and Testing patent (Adam et al. 2015).

Table 15.3 Chemical analysis of typical sewage sludge ash and AshDec products from the same ash

	Unit	Sludge ash	AshDec 100% Na	AshDec 50/50 Na/K
Ash	kg		3	3
Na_2SO_4	kg		1,4	0,7
K_2SO_4	kg		0,0	0,8
Dry sludge	kg		0,75	0,75
N	%	–	–	–
P_2O_5	%	20,2	15,8	16,7
K_2O	%	1,7	1,2	7,4
S	%	0,8	3,8	3,5
Si	%	14,3	10,5	12,0
Mg	%	1,7	1,3	1,4
Ca	%	11,1	8,5	9,6
Na	%	0,6	11,0	6,8
Al	%	7,0	5,0	5,6
Fe	%	5,7	5,7	6,2
P_{nac} solubility	%	31	75	75

Phosphate release from the AshDec product is induced by slightly acidic excretions of the plant roots and microbial soil activity, a property that is required in accordance with organic farming principles. The fertilizer does not contain acids and can be universally blended without undesired chemical reactions. In addition, the product contains secondary and trace nutrients like potassium, sulfur, magnesium, and zinc (Adam et al. 2008; Werner 1981; Kabbe 2015) and 10–12% of sodium and potassium silicates. None of these compounds are found in conventional phosphate fertilizers, if not admixed intentionally and at an additional cost. An extended literature research and personal discussions with experts unveiled the following facts: numerous authors (Matichenkov and Bocharnikova 2012; McCune 1981; Moreira et al. 2002; Penha and Keith 2012; Sandim et al. 2008; Viana and Vasconcelos 2008; Guntzer et al. 2012) have investigated the effects of (reactive) silicates in fertilizers reporting positive results: (1) silicates increase the strength of the cell walls and outer skins of crops, observed in rice and sugar cane; (2) cell wall and outer skin strength improves the resilience of the plant against pests and diseases; (3) silicates may reduce phosphate fixation in high Fe/Al soils; and (4) silicates may enhance the uptake of micronutrients (e.g., Zn) and prevent the uptake of toxic elements (e.g., Cd). The multinational authors refer to different products, including thermophosphates produced at 1300 °C from phosphate rock and silicon-containing additives.

Greenhouse and field tests have provided robust evidence that on average European soils these products have similar effects on yield and P uptake as conventional mineral fertilizers like diammonium phosphate (DAP)/monoammonium phosphate (MAP) and triple super phosphate (TSP)/single super phosphate (SSP). Biomonitoring has proven the absence of toxic effects, and chemical analyses have shown that cadmium and uranium concentrations are by an order of magnitude lower than in most rock-based phosphates.

15.8 Conclusions and Outlook

Outotec is continuously working on improvements of the AshDec® process, predominantly with the Federal Institute for Materials Research and Testing in Berlin. The main objective is to increase the acceptability of the products under the prevailing market conditions. This has been partly achieved by increasing the overall nutrient concentrations in the product by replacing a half of the sodium sulfate additive by potassium sulfate.

In contrast to some competitors and swinging preferences of potential customers, Outotec is not departing from the original product philosophy – producing a non-water-soluble, nonacidic, calcined phosphate with a superior recovery rate and a minimum of secondary waste. The recommendation for AshDec products to be positive listed for use in organic farming expressed by the Expert Group for Technical Advice on Organic Production (EGTOP) on behalf of the Directorate-General for Agriculture and Rural Development of the European Commission

(Expert Group for Technical Advice on Organic Production (EGTOP) 2016) is supporting the tenacity and opening a global niche market potential.

In the next step, Outotec will investigate the nonobvious benefits of the product in subtropical and tropical soil/crop systems where the product characteristics may have a more accentuated impact. Since the efficiency of conventional mineral fertilizers leaves a lot to be desired and some struvites indicating the potential of recycled mineral fertilizers, there may be yield and efficiency benefits yet to be discovered. The critical questions are:

1. To what extent do AshDec fertilizers comply with the requirements of the NextGen fertilizer initiative of Prof. Susanne Schmidt and her team of Australian researchers, mostly based at The University of Queensland (UQ), who aim at producing fertilizers with nutrient release and plant nutrient uptake curves synchronized (Schmidt et al. 2016; Paungfoo-Lonhienne et al. 2013)?
2. Which are real benefits of the sodium and potassium silicates present in the product and reported to improve the pest resilience of crops and reduce the phosphorus fixation in highly weathered, Fe-/Al-rich soils?

The AshDec team and plant nutrition and soil scientists from various universities will closely cooperate to figure out the potential of calcined phosphates which goes much beyond the benefits of recycling: increased efficiency, no acidification, no accumulation of pollutants, and – possibly – better pest resilience of crops.

References

Adam C. www.bam.de. 15 02 2017. [Online]. Available: urn:nbn:de:kobv:b43-391607. Accessed 12 07 2017

Adam C, Peplinski B, Michaelis M, Kley G, Simon F-G (2008) Thermochemical treatment of sewage sludge ashes for phosphorus recovery. Waste Manag 1122–1128

Adam C, Hermann L, Stemann J (2015) Production of citrate soluble phosphates by calcination of secondary phosphate sources with a sodium-sulfuric compound. Germany, Europe Patent EP2015/063062

Bundesministerium für Umwelt, Naturschutz und Reaktorsicherheit im Einvernehmen mit dem Bundesministerium für Verbraucherschutz, Ernährung und Landwirtschaft, Verordnung über das Inverkehrbringen von Düngemitteln, Bodenhilfsstoffen, Kultursubstraten und Pflanzenhilfsmitteln (Düngemittelverordnung – DüMV). Bundesrecht, Bundesrepublik Deutschland, Bonn, 26.11.2003

Expert Group for Technical Advice on Organic Production (EGTOP) (2016) Final report on organic fertilizers and soil conditioners (II). European Commission, Directorate-General For Agriculture And Rural Development, Brussels

Guntzer F, Keller C, Meunier J (2012) Benefits of plant silicon for crops: a review. Agron Sustain Dev., Springer Verlag, vol. INRA 2012 32(1):201–213

Herzel H, Krüger O, Hermann L, Adam C (2016) Sewage sludge ash – a promising secondary phosphorus source for fertilizer production. Sci Total Environ 542:1136–1143. https://doi.org/10.1016/j.scitotenv.2015.08.059

Joint EU Research Project P-REX 2012–2015, "P-REX," 05 04 2016. [Online]. Available: www.p-rex.eu

Kabbe C (2015) Sustainable sewage sludge management fostering phosphorus recovery and energy efficiency. Kompetenzzentrum Wasser Berlin gGmbH, Berlin

Matichenkov V, Bocharnikova E (2012) Influence of plant associations on the silicon cycle in the soil-plant ecosystem. Appl Ecol Environ Res 10(4):547–560

McCune DL (1981) Fertilizers for tropical and subtropical agriculture. The Fertilizer Society of London, London

Moreira A, Malavolta E, Cardoso Moraes LA (2002) Eficiencia de fontes e doses de fósforo na alfaalfa e controsema cultivadas em Latosolo Amarelo. Pesqui Agropecuaria Bras 37(10):1459–1466

Outotec. www.outotec.com. 2016. [Online]. Available: http://www.outotec.com/company/media/news/2016/sustainable-sewage-sludge-incineration-for-zurich-canton/. Accessed 12 07 2017

Paungfoo-Lonhienne C, Schmidt S, Webb R, Lonhienne TG (2013) Rhizophagy-a new dimension of plant-microbe interactions. In: de Bruijn FJ (ed) Molecular microbial ecology of the rhizosphere, Vols. Hoboken, New Jersey, pp 1199–1207. https://doi.org/10.1002/9781118297674.ch115

Penha A, Keith S (2012) Rio Verde minerals announces positive agronomic tests on thermophosphate produced from its Sapucaia target, Fosfatar phosphate project. Rio Verde Minerals, Toronto

Sandim A, Vian JD, Silva AB, Camili EC, Silva PRA, Wolf MJ (2008) ASSESSMENT OF PRODUCTIVITY cultivar RB 83-5486 CANA-DE-AÇÚCAR (Saccharum officiarum) IN DIFFERENT SOURCES FOSFATADAS ADUBAÇÃO BASE IN THE CAMPO GRANDE – MS., FertBio 2008, Campo Grande

Schmidt S, Laycock B, Luckman P, Redding M, Batstone D, Hugenholtz P (2016) NEXTGEN FERTILIZERS – nutrient stewardship for environmentally sustainable agriculture and food security. The University of Queensland, Brisbane

Schweizerische Eidgenossenschaft, Der Bundesrat, Verordnung zur Reduktion von Risiken beim Umgang mit bestimmten besonders gefährlichen Stoffen, Zubereitungen und Gegenständen. Der Schweizerische Bundesrat, Bern 18. Mai 2005 (Stand am 20. Juni 2017)

Viana E, Vasconcelos ACF (2008) Producao de alface adubada com termophosphato e adubos organicos (Thermophosphate and organic fertilization in lettuce yield). Rev Cienc Agron Fortaleza 39(02):217–224

Werner W (1981) Der Rhenania Dünger – Monographie über Herstellung, Eigenschaften und Wirkung der Rhenania-Dünger. Verlag M. & H. Schaper, Hannover

Chapter 16
Phosphorus Recovery into Fertilizers and Industrial Products by ICL in Europe

Kees Langeveld

Abstract Phosphate recycling is an important issue, since phosphate is a finite resource which is essential to food security. The phosphate used in the fertilizer industry, which now solely comes from mining, has to be replaced with so-called secondary phosphates. At ICL Fertilizers, trials have been conducted to investigate the potential implementation of these sources of secondary phosphates into the fertilizer production. Extensive pilot-scale testing and several plant-scale tests have yielded promising results for the use of sewage sludge ash, meat and bone meal ash and struvite. The main issue remaining is the legislation for the use of these sources, as they are currently regarded as waste. Struvite is also suspected to be able to contain contaminants such as pathogens and pharmaceuticals, encapsulated in its crystals. Therefore, further research on this topic is necessary. In the draft of the new Fertilizer Regulations of the European Commission, maximum values for heavy metal content in fertilizer are discussed in greater detail. The first results show that products produced from sewage sludge ash (the big quantity of secondary phosphates) meet some of these demands; however some limits are put (without doing a risk assessment) too low and could block the use of these secondary phosphates in fertilizers. Since heavy metal content in struvite and meat and bone meal ash is low, no problems are expected. The use of secondary phosphate in fertilizer production yields great opportunities; however in parallel ICL is piloting other processes for production of industrial products (elemental phosphorous P_4 and food-grade phosphoric acid). The P_4 route is via the thermal RecoPhos process (inductive heating of ashes and evaporation, cleaning and condensation of the P_4) where no waste what so ever is created, only products with a positive market value. The food-grade phosphoric acid route is via the Tenova process, where ashes are treated with by-product HCl to produce phosphoric acid, which is then purified in several extraction stages. In the coming years, the pilot results will show the economic feasibility of these processes for which ICL has its own captive use in industrial applications. In this way ICL will try to turn the development of a circular economy from a threat into an opportunity.

K. Langeveld (✉)
ICL, Amsterdam, The Netherlands
e-mail: Kees.Langeveld@icl-group.com

H. Ohtake, S. Tsuneda (eds.), *Phosphorus Recovery and Recycling*,
https://doi.org/10.1007/978-981-10-8031-9_16

Keywords ICL · Secondary phosphate · Sewage sludge ash · Meat and bone meal ash · Wood ash · Struvite · Mineral fertilizer production · RecoPhos process · Tenova process

16.1 Introduction

For some other finite resources, like oil, it is possible to find alternative sources. For phosphorus this is not the case as this is a chemical element (Heffer et al. 2006). Therefore, usage has to be cut to make the reserves last longer. Still, this will not make phosphorus an infinite resource (Van Vuuren et al. 2010). Closing the phosphorus cycle by recovering and recycling will be required if phosphorus famine is to be prevented (Gilbert 2009).

Phosphorus is disposed of in human excreta, used detergents and food and industrial waste. This stream enters the sewage systems and offers an opportunity to recover it as it accumulates in the sewage sludge at wastewater treatment plants. The sewage sludge can be processed in many different ways to recover the phosphorus (Schick et al. 2009). These can be summarized into three main categories: the watery sludge, dewatered sludge and sewage sludge ash. At wastewater treatment plants, struvite, which is essentially magnesium ammonium phosphate, can also be formed by crystallization and precipitation. This feedstock also contains a high level of phosphorus and can be regarded as a secondary phosphate source (Jaffer et al. 2002).

Next to the wastewater treatment plants, other sources exist (Schipper et al. 2001). Since the ban on the use of meat and bone meal as animal feed due to the outbreak of BSE, it is classified as a waste material (Yamamoto et al. 2006). The meat and bone meal is incinerated, rendering it harmless, but this also renders it useless for its traditional uses. The phosphate content in this is even higher than sewage sludge ash and it contains less contaminants.

The nutrient availability for plants of the phosphate is imperative for it to be used in a fertilizer (Cabeza et al. 2011). The solubility in neutral ammonium citrate and water, which with phosphate rock is realized by acidulating, is important. The processing for this will be discussed later in this chapter. The main issue regarding the use of secondary phosphates is legislation, as the streams currently being tested are regarded as waste. Also, some contaminants are present in selected sources. In sewage sludge ash, for example, a relatively high amount of heavy metals are present. This does not, however, have to be an issue.

16.1.1 ICL Fertilizers' Position

ICL Fertilizers runs several fertilizer production units in different parts of the world with two in Europe (the Netherlands and Germany) with a combined capacity of 800,000 t/a product and a use of 250,000 t/a rock phosphate. All of them are based

Fig. 16.1 ICL Fertilizers' production plant in Amsterdam

on the attack of phosphate rock with sulphuric acid, phosphoric acid or combinations of the two (secondary attack) after which potassium chloride (MOP) or potassium sulphate (SOP) or trace elements (Cu, Mg, Mn, Mo, Zn, etc.) can be added to make different forms of PKs and on top of that ammonium sulphate to produce NPKs. These processes are very suitable for the recycling of secondary phosphates (contrary to other NPK processes) without any safety issue.

At ICL Fertilizers in Amsterdam (Fig. 16.1), a lot of research and testing has been done regarding the use of secondary phosphates in the fertilizer production. Extensive pilot-plant scale tests (Ten Wolde 2012) have been done regarding the use of struvite as well as that of sewage sludge ash and meat and bone meal ash on plant-scale tests.

As a phosphate fertilizer producer with an own supply of phosphate in the Israeli desert and a mine in China, it could be perceived as odd to be researching the use of secondary phosphates. The vision of ICL Fertilizers is that sustainability is important and the environment is to be taken care of. 'Closing the loop' on phosphorus could elongate the use of the phosphate mines and improve the distribution of phosphorus on a worldwide scale. The fact that ICL's European plants are in countries with excess phosphate adds to this philosophy: 'Using a part of all recycled phosphate in factories that export a major part of their products to countries with a deficit on phosphate helps solving the existing phosphate surplus in The Netherlands and Germany'.

16.1.2 Value Chain Agreement

In the Netherlands, ICL Fertilizers Europe is taking part in the so-called value chain agreement (covenant), initiated by the Nutrient Platform in 2011. Together with 19 other parties, the ambition is to create a sustainable market where reusable phosphate streams will be returned to the cycle in an environmentally friendly way. ICL Fertilizers Dutch ambition is to base its entire fertilizer production in the Netherlands on secondary phosphates by 2025 (Dutch Nutrient Platform 2011). Since the phosphate issue is not merely a Dutch issue but a European one, the expansion to a European Union-based platform was desirable. In 2013, at the 1st European Sustainable Phosphate Conference in Brussels, the European Phosphorus Platform was launched, with over 150 participants. This is an important development to move forward and could improve Europe's competitive position and avoid potential geopolitical tensions (European Phosphorus Platform 2013).

16.2 Secondary Phosphate Sources

Different sources are available within the European Union. These can be categorized in three main categories. These are manure and litter, phosphate-rich ashes and struvite. The main differences between these sources are the solubility of phosphate and the contaminants it can contain. Also the physical form is an important factor that differs. These facts impact the way the secondary phosphates can be employed in the fertilizer value chain. In this section, the properties of the sources will be discussed.

16.2.1 Manure and Litter

Several countries, such as the Netherlands, with intensive livestock agriculture have a surplus of animal manure and poultry litter. Since these contain phosphates, research is being done to be able to use these as a raw material for fertilizer production. Untreated manure contains organic components and water and is of a low nutrient content. For industrial applications it therefore needs to be dewatered and incinerated (Schipper et al. 2001). These ashes could be employed as raw material for the fertilizer production in secondary attack units in the same way as sewage sludge ashes.

Other techniques currently being researched is the pyrolysis of manure (Azuara et al. 2013) and the gasification of chicken litter (Kaikake et al. 2009). The products from these processes could also be implemented in the production of phosphate fertilizer. Since incineration of manure destroys the organic material and the nitrogen, an alternative route for treatment of manure by digestion (energy production), taking out the N as ammonium sulphate and the P as calcium phosphate (both products to be used as secondary raw materials in industrial NPK production), the remaining organic fraction has a good value for the farmer to be used on his fields without over-fertilization with N and P. One of the projects researching this option is the SYSTEMIC project led by Wageningen University in which ICL is a partner.

16.2.2 Ashes from Mono-incineration

The phosphate-rich ashes are a product of mono-incineration of a phosphate-rich stream, such as from wastewater treatment plants or meat and bone meal from rendering factories. Due to the incineration of phosphate-rich streams separately from phosphate-poor streams, relatively high phosphorus content can be achieved in the ash.

16.2.2.1 Sewage Sludge Ash

At wastewater treatment plants, surface water-polluting phosphate is precipitated using iron or aluminium compounds forming sludge. The Dutch sewage system contained about 32 kt P_2O_5 per year in 1998 (Schipper et al. 2001). Recovery can be done directly from the sludge, from dewatered sludge or incinerated sewage sludge ash. Wet or dewatered sludge is not suitable for the traditional industrial processing, as these streams contain water and could contain organic compounds, viruses, medicine and other contaminants. These can be rendered harmless by incinerating the sludge.

This yields sewage sludge ash. Dewatered sewage sludge is incinerated in dedicated furnaces in order to prevent more contaminants from other streams such as industrial waste to contaminate the sludge, further diluting the phosphate content (European Commission 2000). As this is the form from which over 90% can be recovered, this is the most interesting, and it could be possible to integrate it into existing infrastructure (Cornel and Schaum 2009). Another advantage of incineration of the sewage sludge is that it has a caloric value, so it yields energy on incineration.

The main problem with sewage sludge ash (SSA) is the content of heavy metals, iron and aluminium. These hinder the regular processing, which will be discussed later. Since the flocculants for sewage sludge vary, several different analyses are shown in Tables 16.1 and 16.2. A sewage sludge ash sample is depicted in Fig. 16.2.

16.2.2.2 Meat and Bone Meal Ash

Prior to 2001, meat and bone meal (MBM) was primarily used in animal feed, but in 2001 this was banned in Europe due to the fact that MBM was suspected to be the cause of the mad cow disease outbreak (Yamamoto et al. 2006). This caused large-scale waste problems which were solved by incinerating the MBM and thus creating meat and bone meal ash (MBMA) (Cascarosa et al. 2012). This meat and bone meal ash is very similar to regular phosphate rock in terms of chemical composition. Also

Table 16.1 Analyses of the main components of sewage sludge ash (SSA), meat and bone meal ash (MBMA) and phosphate rock

Total wt%	SSA 1	SSA 2	SSA 3	SSA 4	MBMA	Wood ash	Phosphate rock
P_2O_5	15,20	20,44	18,90	17,80	25,50	4,8	30,96
CaO	18,80	20,59	11,50	18,60	37,40	13,5	47,50
SO_4	5,30	4,50	1,60	3,00	6,40	6,2	2,70
K_2O	1,30	1,66	0,80	1,20	2,20	14,8	0,70
MgO	2,30	2,74	1,19	2,90	0,99	8,1	0,40
Al_2O_3	6,28	9,39	9,44	9,20	1,74	12,9	0,11
Fe_2O_3	12,08	5,82	3,05	5,60	0,99	9,5	0,17

Table 16.2 Heavy metal content analyses of sewage sludge ash (SSA), meat and bone meal ash (MBMA) and phosphate rock

ppm	SSA 1	SSA 2	SSA 3	SSA 4	MBMA	Wood ash	Phosphate rock
As	20,1	19,9	9,4	9,0	8,1	0,4	17,8
Cd	2,1	1,0	2,2	<0,1	1,7	<0,2	25,9
Cr	115,5	124,5	25,0	79,5	18,1	1,0	53,0
Cu	760,3	1146,0	404,0	749,6	365,0	1,5	13,5
Ni	44,6	49,6	17,8	37,7	7,8	1,1	30,7
Mn	871,6	825,5	3070,0	719,4	207,0	11,7	6,7
Pb	273,0	254,0	157,6	84,4	82,4	2,2	< 0,1
Zn	3053,0	2139,0	876,0	1624,0	209,0	–	260,2

Fig. 16.2 Sewage sludge ash

the content of contaminants is very low, as can be seen in Tables 16.1 and 16.2. The product received at ICL Fertilizers for testing can be seen in Fig. 16.3.

16.2.2.3 Wood Ashes

Several initiatives are taking place to incinerate clean waste wood as a biofuel. The ashes coming from this incineration are fairly pure and contain phosphates and potash as valuable nutrient, however in a not-available form for plants. They can, however, be transformed by secondary attack into soluble fertilizers. The results of component analysis can be seen in Tables 16.1 and 16.2. It should be remarked that nutrient content varies a lot over different types of wood.

Fig. 16.3 Meat and bone meal ash

16.2.3 Struvite

At wastewater treatment plants, struvite crystallization is a widely used technique to remove phosphorus, ammonium from digested sludge liquors (Martí et al. 2010) by adding a source of soluble Mg in the proper stoichiometric ratio. The main driver for this is not the nutrient recovery, but the improvement of process conditions (less scaling, better filterability) (Jaffer et al. 2002). Struvite is sometimes also referred to as MAP (magnesium ammonium phosphate) and consists of these ions in a molar ratio of 1:1:1. The reaction that takes place is shown in Eq. (16.1):

$$Mg^{2+} + NH_4^{+} + PO_4^{3+} + 6H_2O \rightarrow MgNH_4PO_4.6H_2O \qquad (16.1)$$

Since struvite is not incinerated, a certain fear exists that it could contain pathogens, pharmaceuticals, hormones and other contaminants encapsulated in the crystals (Decrey et al. 2011). Research on struvite precipitated from urine has shown that 98% of the hormones and pharmaceuticals remained in the filtrate and only a small fraction of the heavy metals remained in the struvite. Drying has shown to be effective to inactivate viruses that could be present in the struvite. Heating struvite to a temperature higher than 40–55 °C causes it to decompose, releasing gaseous ammonia (Bhuiyan et al. 2008). Therefore, drying should be done in a controlled fashion, and further research is needed in this field.

Struvite can be obtained from several different processes. The struvite samples that have been tested at ICL Fertilizers for the use in fertilizer production were obtained from the Anphos and Airprex processes (Figs. 16.4 and 16.5). Analyses from these struvite are shown in Tables 16.3 and 16.4, with a phosphate rock analysis for comparison.

16.3 Processing in Mineral Phosphate Fertilizer Production

The main two types of secondary phosphates which can be used in the production of phosphate fertilizer are ashes (from mono-incineration such as meat and bone meal ash, wood ash, sewage sludge ash) and struvite. At ICL Fertilizers, these have been extensively tested on a pilot scale, and some have also been tested on plant scale. The results of these tests will be discussed in this section, as well as the technical implications it has on the current infrastructure at the production location in Amsterdam.

16.3.1 Ashes from Mono-incineration

As well in sewage sludge ash as in wood and meat and bone meal ash, the phosphate is not soluble in water or neutral ammonium citrate (NAC) and thus not plant-available. With phosphate rock, acidulating the rock with either sulphuric or

Fig. 16.4 Struvite from the Anphos process as received in bulk

Fig. 16.5 Struvite crystals from the Airprex process (STOWA 2012)

Table 16.3 Analyses of the main components of Airprex and Anphos struvite and phosphate rock

wt% (TQL)	Struvite Airprex	Struvite Anphos	Phosphate rock
Total P_2O_5	19,8	14,7	30,96
NH_4	3,8	2,6	0,00
CaO	0,8	2,3	47,50
SO_4	0,1	0,5	2,70
MgO	10,2	7,6	0,40
Al_2O_3	0,1	< 0,1	0,11
Fe_2O_3	1,5	0,3	0,17
Moisture	14,9	21,9	1,8

Table 16.4 Analyses of the heavy metal content of Airprex and Anphos struvite and phosphate rock

ppm	Struvite Airprex	Struvite Anphos	Phosphate rock
As	<0,1	0,0	17,8
Cd	<0,1	0,0	25,9
Cr	11,1	0,0	53,0
Cu	31,6	2,0	13,5
Hg	<0,1	0,0	< 0,1
Ni	3,6	0,0	30,7
Mn	<0,1	12,0	6,7
Pb	652	0,0	0,0
Zn	85,9	11,0	260,2

phosphoric acid will yield single or triple superphosphate, respectively. These fertilizers have a typical water and water + NAC solubility of 90–92% and 95–97%. The reaction equations of the acidulation to single and triple superphosphate are shown in Eqs. (16.2) and (16.3) (Kongshaug et al. 2000):

$$2Ca_5(PO_4)_3F + 7H_2SO_4 \leftrightarrow 3Ca(H_2PO_4)_2 + 7CaSO_4 + 2HF \tag{16.2}$$

$$2Ca_5(PO_4)_3F + 7H_3PO_4 \leftrightarrow 5Ca(H_2PO_4)_2 + 2HF \tag{16.3}$$

Phosphate from ashes is not present in the form of apatite ($Ca_5(PO_4)_3(F,Cl,OH)$), but in complexes with iron, calcium or aluminium. Due to this different form and the presence of other contaminants such as heavy metals which could react with the acid, a regular acidulation mixture does not yield the physical and chemical results required. Another important difference between the apatite and these ashes is that the apatite contains fluorides, chlorides, hydroxides and carbonates. During acidulation, these are released into gaseous form such as HF, SiF_4, HCl and CO_2. In order to create sufficient surface area during the acidulation process, which delivers a product that is softer and better processable, some additives have to be mixed with the ash prior to acidulation. These ashes are much finer than regular phosphate rock.

Therefore, milling is not necessary. This does impact the handling and storage. Therefore, storage in silos and direct input in the mixers are desirable.

16.3.1.1 Meat and Bone Meal Ash

Meat and bone meal ash shows most similarities with apatite; the phosphate is mainly present as calcium phosphate. This reacts with acid in a similar way as apatite, as can be seen by comparing Eqs. (16.2) and (16.3) with (16.4) and (16.5):

$$Ca_3(PO_4)_2 + 3H_2SO_4 \leftrightarrow 3CaSO_4 + 2H_3PO_4 \quad (16.4)$$

$$Ca_3(PO_4)_2 + 4H_3PO_4 \leftrightarrow 3Ca(H_2PO_4)_2 \quad (16.5)$$

This suggests that regular acidulation with a slightly different acid concentration would be possible. Also, since the meat and bone meal ashes have a phosphate content that almost reaches the concentration found in phosphate rock, it can be mixed with phosphate rock to achieve regular products. During extensive tests on pilot scale at ICL Fertilizers, it has shown that this is the case. Mixing with regular phosphate rock was possible, achieving a high water solubility and neutral ammonium citrate solubility yield. The physical properties of the produced superphosphate are also similar to that of phosphate rock.

Besides a mixed acidulation with phosphate rock, acidulation of pure MBMA was also possible and yielded good results with regard to the chemical properties. The physical properties of the acidulated MBMA did impact the processability, and at this moment, a mixture with phosphate rock is preferred.

16.3.1.2 Sewage Sludge Ash

As shown in Tables 16.1 and 16.2, the levels of aluminium, iron and heavy metals are much higher than that in phosphate rock or meat and bone meal ash. Next to the fact that it is possible for the phosphate to be present in complexes with the iron and aluminium, it is also possible for the acid to react with these:

$$2Ca_5(PO_4)_3F + 7H_2SO_4 \leftrightarrow 3Ca(H_2PO_4)_2 + 7CaSO_4 + 2HF \quad (16.6)$$

$$2Ca_5(PO_4)_3F + 7H_3PO_4 \leftrightarrow 5Ca(H_2PO_4)_2 + 2HF \quad (16.7)$$

Acidulation of different sewage sludge ashes has been tested at ICL Fertilizers. This resulted in a large spread in the resulting products. Since the contaminant levels are high in the sewage sludge ash, mixtures with phosphate rock showed only negative effect. Acidulation using phosphoric acid also did not yield good results, as the product did not coagulate fully, which was not processable. Therefore, only acidulation using sulphuric acid was further tested.

Three sewage sludge ashes with a different iron content were acidulated and monitored in time, since the water solubility in acidulated phosphate rock normally increases in time as the reaction continues. This increase in time was not noticed with the sewage sludge ashes. The water solubility of the reaction product is a clear function of the iron and aluminium content of the ash. The solubility in neutral ammonium citrate of these products shows the same dependency, be it that in general Al-based ashes are reducing recoveries less than Fe-based ashes.

Another important factor for employing sewage sludge ash in the production of phosphate fertilizer is its physical properties. Regular acidulation mixtures showed a sticky product, which was not processable. Removing part of the water and thus increasing the acid concentration showed positive effect on this. The iron-based sewage sludge ashes delivered less processable products, regardless of the used acid concentration. The acidulation has also been tested at ICL Fertilizers on plant scale, which resulted in good acidulation of 10–14 tons sewage sludge ash (7 wt% Fe_2O_3) per hour. No optimized acidulation mixture has been found yet for the acidulation on plant scale, since tests need to be continued. The resulting product did show good physical and chemical properties.

To check the granulation properties, several different recipes were used. Six PK- and four P-fertilizer granulations were executed in the initial trials. The conclusion found from these results is that the optimum for using sewage sludge ashes is making a specific mixture of acidulated high reactive Israeli rock phosphate with acidulated sewage sludge ash in the granulator. This caused the granulation to yield a good granule size distribution and a proper nutrient content.

During the granulation, the temperature was over 10 °C higher than regular phosphate rock granulation. In the PK granulation, this could be attributed to the exothermic reaction taking place between the free acid and potash as shown in Eq. (16.8). During granulation some hydrochloric acid fumes were noticeable. However, this temperature rise is also present in the P-only fertilizer, of which the temperature rise cannot be ascribed to this reaction (Schultz et al. 2000). In the used setup, it was impossible to determine the cause of this:

$$KCl + H_2SO_4 \rightarrow KHSO_4 + HCl \tag{16.8}$$

Besides this difference in temperature, the granulation process itself is more sensitive. More water is needed for the granulation to start, and the granulation is more prone to spontaneous over-granulation than is the case with comparable non-SSA containing fertilizers. When granulating the mixtures, the P_2O_5 water-soluble yield was not proportional to the yield in both components. This suggests that there could be a reaction taking place during granulation, which could also attribute to the temperature increase.

Fig. 16.6 PK 8-27+7MgO from plant-scale test with Airprex struvite

16.3.2 Struvite

Unlike the ashes from mono-incineration, the phosphate in struvite is readily neutral ammonium citrate soluble. Therefore, the acidulation step does not have to be carried out on this feedstock which simplifies the processing of it. Several plant-scale tests have been carried out with struvite as a secondary phosphate. Struvite obtained by the Airprex process from a wastewater treatment plant was very well usable and could be added to a maximum of 20% of the total granulation input. The moisture content of the used struvite appeared to be the limiting factor. As the struvite used in the plant-scale test contained between 15% and 20% total moisture (so including crystal water), water and steam addition to the granulation drum had to be limited, and smearing occurred on several points.

The main differences in the obtained products using struvite are a decrease in heavy metal content and water-soluble phosphate and an increase in pH. Emission measurements for fluorine, phosphorus and chloride were also performed during the granulation of a PK 8-27+7MgO, which indicated lower emissions when struvite was added. Phosphate and fluorine emissions to wastewater decreased, indicating a positive impact on the environment. The granules produced during the plant-scale test are shown in Fig. 16.6.

16.4 Future Perspective

16.4.1 Implementation in ICL Fertilizers

A proposal for a Euro 2 million investment has been approved (supported by a local Dutch subsidy of Euro 0.5 million) at ICL Fertilizers in Amsterdam in order to store and process sewage sludge and meat and bone meal ashes in a more professional way and to enable us to reach the target of 100% replacement by 2025.

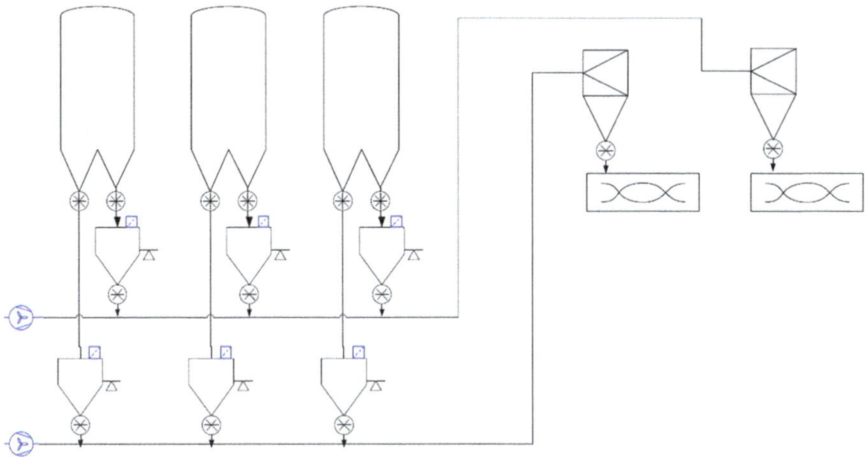

Fig. 16.7 Flow sheet for the storage, mixing and transporting secondary phosphates to the mixer for acidulation at ICL Fertilizers

The expansion will entail several silos with dosing units and transportation systems directly into the mixers for acidulation. An important ability is to achieve a constant flow in the necessary composition of the components. It should also be possible to mix milled phosphate rock from the regular process with, for example, meat and bone meal ash from this system at the mixer input. The pre-engineering led to a proposal for three silos with gravimetrical dosing and pneumatic transportation to the mixers. This way, a flow of 15 t/h can be achieved to each of the mixers. A flow sheet for this system is shown in Fig. 16.7. Apart from this development, ICL is in a process of discussions and trials for its German plant in Ludwigshafen to apply the same concept. However, approval processes go a bit slower in Germany. At the moment a license was obtained for using a limited daily quantity of ashes in the process.

16.5 Challenges and Issues

16.5.1 Legislation

As sewage sludge ash, struvite and meat and bone meal ash are currently regarded as waste. It is imperative that these streams will not be regarded as waste in the future, thus making nutrient recovery from these streams more practicable. The new EU Fertilizer Regulations which are under discussion at the time of writing have to accommodate the use of ashes and struvite as a raw material for production of EU fertilizers in an environmental acceptable way. Missing the opportunity that some conventional fertilizer plants can be a big help in creating a circular economy for phosphate would be very sad and contradicting the new sustainability views.

MBMA processing does not pose any threat to the environment and ecology, but even reduces emissions because of the lack of fluoride and other gas-forming substances. When regarding the processing of SSA, the emissions are also lowered due to the absence of gas-forming substances. During a plant-scale test with struvite addition to the granulation, the emissions of fluorine and phosphate showed lower figures than without struvite in PK production. However another trial with struvite had to be stopped due to smell issues from co-crystallized organic material. It proves each material source needs careful checking before processing.

With regard to struvite, further research is needed to prove that no pathogens, pharmaceuticals or other hazardous contaminants remain present; it is imperative that no risks are carried over to the fertilizer. Use of struvite as a raw material in processes like ICL's fertilizer process will reduce the chances of contaminants to almost zero, since the struvite will be exposed to a very acidic environment and high temperature in the dryer, which will kill all viruses and bacteria.

The cadmium content for both MBMA and SSA is far lower than that of most phosphate rock. This results in lower cadmium levels in the final products. The SSA, however, does contain some elevated levels of trace elements like copper and zinc compared to the phosphate rock. There should be room in the legislation for these beneficial heavy metals that are needed in plant growth.

16.6 Conclusions for the Fertilizer Application

The use of secondary phosphates in the mineral fertilizer industry yields great opportunities. Many different sources are possible, which could guarantee security of supply and keep the market healthy with regard to competition. This also contributes to a healthy phosphate balance, since this way countries with a surplus of phosphate could remove it from the cycle and export it through fertilizer to countries which have a phosphate deficiency.

Technically, it is already possible to replace a great deal of phosphate rock with secondary phosphates from struvite and mono-incineration ashes. However, legislation and safety issues still exist. The classification of these products as 'waste' obstructs their current employment on an industrial scale. It is imperative that the legislation issues are addressed as quickly as possible and high hopes are set to the JRC working group called STRUBIAS to come with practical proposals by the end of 2018. The European Sustainable Phosphorus Platform emphasizes this and could contribute to these issues.

Regarding struvite, it is important that further legislation is developed on the contaminants it could contain. Following this, the struvite could then easily be used in the production of phosphate fertilizer as the phosphate it contains is already soluble and thus plant-available. The processability of struvite does vary, as odour emissions and moisture content vary. These are issues that should be kept in mind, and every struvite source is therefore to be tested and reviewed individually.

As sewage sludge ashes have a high content of heavy metals, this could become an issue with regard to accumulation in the soil where the fertilizer is applied. The products from sewage sludge ash do meet the set limits from the European Commission's WG3 meetings which were based on scientific arguments. However, political battles are now taking place to introduce stricter limits and restrict the recycling potential. As was seen in trial experiments, sewage sludge ashes with a high content of iron did not acidulate as well as the ashes with a lower concentration of iron. If the sewage sludge streams and applied flocculants could be managed in a better fashion, it could be possible to achieve even better results regarding phosphate recovery from sewage sludge ashes.

Meat and bone meal ashes are the best applicable at the moment, as these show most similarities with regular phosphate rock. Therefore, no real issues exist with the implementation of this feedstock into the production of mineral phosphate fertilizer. However a big volume of the meat and bone meal is now being used in cement and concrete production, using only the caloric value of the material and wasting the phosphate in a building material. Hopefully new legislation like in Switzerland and Germany will forbid this kind of applications and will be driving more into phosphate recycling.

16.7 Recycling of Ashes in Industrial Applications

ICL is a big consumer of elemental phosphorus (P_4) and food-grade phosphoric acid in its downstream activities in Europe and the USA (lubricant additives, flame retardants, food and feed salts, fire extinguisher products). Instead of putting all the eggs in one basket for recycling P (the fertilizer route), we are in a process to develop two other processes especially for P_4 and food- grade phosphoric acid.

16.7.1 RecoPhos Thermal Process

This is a process for which ICL obtained the patents in March 2016 and is using sewage sludge ashes as raw material. The ash is mixed with carbon and heated under reducing circumstances by inductive heating to 1600 °C. At this temperature, P_4 is released as a gas which, after cleaning out contaminants, is condensed to pure solid P_4 (Fig. 16.8). The by-products from this process are:

- *Slag*, containing immobilized metals and silicate, which can be treated in a new process to produce cement. If not treated, the slag can be used for road construction.
- *Ferrophos*, a mix of elemental iron and phosphorus used as an additive in the steel industry.
- *Syngas*, mainly CO that can be used in the process to preheat the ashes.

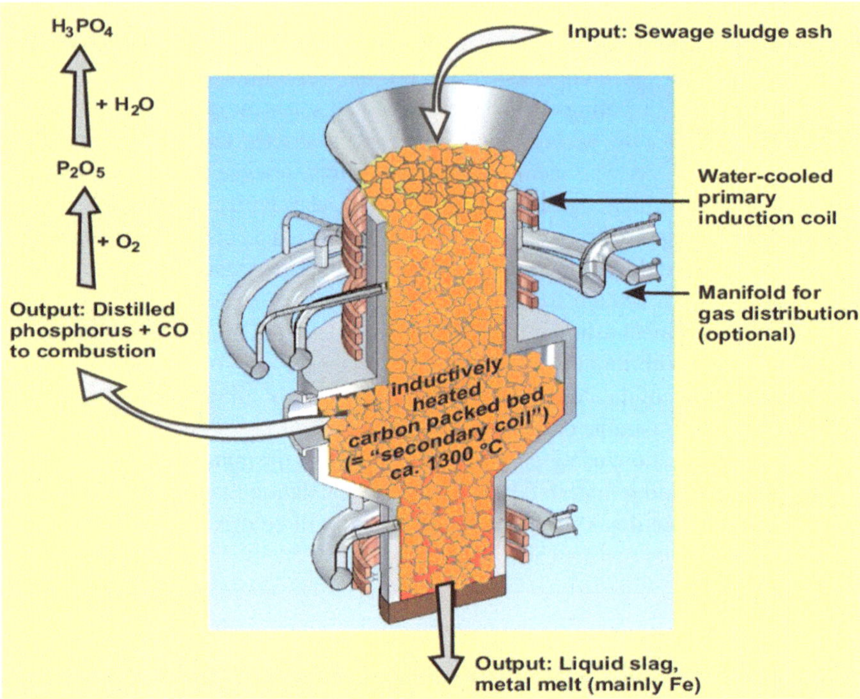

Fig. 16.8 Schematic picture of the RecoPhos thermal process

ICL is preparing the engineering for a pilot plant, to be built in Terneuzen, the Netherlands. Since the investment of the pilot will be major (about Euro 20 million), subsidies are sought as well as partners that want to participate financially in the pilot (ash suppliers, cement companies, etc.). The first commercial unit will be built in Bitterfeld Germany, where big amounts of ash are available and use of the P_4 in the chemical plant of ICL producing fire retardants is secured. RecoPhos is a very interesting process with a potential offtake for sewage sludge ashes of over 400,000 t/a and one of the few processes not generating any waste product but only sellable final products.

Apart from applications in Europe and the USA, there is a great interest for this technology from Southeast Asia where big cities are trying to find outlets for their ashes and RecoPhos (either producing P_4 or alternatively technical-grade phosphoric acid) could be the solution.

16.7.2 Tenova Process for Food-Grade Phosphoric Acid

A different approach for dealing with sewage sludge ashes: ash is treated with by-product hydrochloric acid (HCl), creating a phosphoric acid solution. This solution is purified by several stages of extraction and finally evaporated to commercial concentrations (75 or 83% H_3PO_4). This concept is very interesting in industrial areas where both ash and HCl are available as waste products. The principle was tested on lab scale and feasibility studies are now being made. The next step will be the construction of a continuous pilot plant. Also here partners are sought (HCl suppliers, ash suppliers) as well as subsidy providers (EU, local).

References

Azuara M, Kersten SRA, Kootstra AMJ (2013) Recycling phosphor us by fast pyrolysis of pig manure: concentration and extraction of phosphorus combined with formation of value-added pyrolysis products. Biomass Bioenergy 49:171–180. https://doi.org/10.1016/j.biombioe.2012.12.010

Bhuiyan MIH, Mavinic DS, Koch FA (2008) Thermal decomposition of struvite and its phase transition. Chemosphere 70(8):1347–1356. https://doi.org/10.1016/j.chemosphere.2007.09.056

Cabeza R, Steingrobe B, Römer W, Claassen N (2011) Effectiveness of recycled P products as P fertilizers, as evaluated in pot experiments. Nutr Cycl Agroecosyst 91(2):173–184. https://doi.org/10.1007/s10705-011-9454-0

Cascarosa E, Gea G, Arauzo J (2012) Thermochemical processing of meat and bone meal: a review. Renew Sust Energ Rev 16(1):942–957. https://doi.org/10.1016/j.rser.2011.09.015

Cornel P, Schaum C (2009) Phosphorus recovery from wastewater: needs, technologies and costs. Water Sci Technol: J Int Assoc Water Pollut Res 59(6):1069–1076. https://doi.org/10.2166/wst.2009.045

Decrey L, Udert KM, Tilley E, Pecson BM, Kohn T (2011) Fate of the pathogen indicators phage ΦX174 and Ascaris suum eggs during the production of struvite fertilizer from source-separated urine. Water Res 45(16):4960–4972. https://doi.org/10.1016/j.watres.2011.06.042

Dutch Nutrient Platform (2011) Phosphate value chain agreement. The Nutrient Platform

European Commission (2000) Disposal and recycling routes for sewage sludge (October 1999)

European Phosphorus Platform (2013) Press release ESPC13. The Nutrient Platform, Brussels

Gilbert N (2009) The disappearing nutrient. Nature 461(October):716–718. https://doi.org/10.1016/j.amjmed.2010.04.007

Heffer P, Prud'homme MPR, Muirheid B, Isherwood KF (2006) Phosphorus fertilisation: issues and outlook. In: Proceedings International Fertiliser Society, York, p 586

Jaffer Y, Clark TA, Pearce P, Parsons SA (2002) Potential phosphorus recovery by struvite formation. Water Res 36:1834–1842

Kaikake K, Sekito T, Dote Y (2009) Phosphate recovery from phosphorus-rich solution obtained from chicken manure incineration ash. Waste Manag (New York, NY) 29(3):1084–1088. https://doi.org/10.1016/j.wasman.2008.09.008

Kongshaug G, Brentnall BA, Chaney K, Gregersen JH, Stokka P, Persson B, … Skauli Ø (2000) Phosphate fertilizers. In: Ullmann's encyclopedia of industrial chemistry. https://doi.org/10.1002/14356007.a19_421

Martí N, Pastor L, Bouzas A, Ferrer J, Seco A (2010) Phosphorus recovery by struvite crystallization in WWTPs: influence of the sludge treatment line operation. Water Res 44(7):2371–2379. https://doi.org/10.1016/j.watres.2009.12.043

Schick J, Kratz S, Adam C, Schnug E (2009) Techniques for P-recovery from waste water and sewage sludge and fertilizer quality of P-recycling products. In: Phosphorus recycling and good agricultural management practice

Schipper WJ, Klapwijk A, Potjer B, Rulkens WH, Temmink BG, Kiestra FD, Lijmbach AC (2001) Phosphate recycling in the phosphorus industry. Environ Technol 22(11):1337–1345. https://doi.org/10.1080/09593330.2001.9619173

Schultz H, Bauer G, Schachl E, Hagedorn F, Schmittinger P (2000) Potassium compounds. In: Ullmann's encyclopedia of industrial chemistry https://doi.org/10.1002/14356007.a22_039

STOWA (2012) Struviet productie door middel van het airprex proces, 2012th–27th edn. Stichting Toegepast Onderzoek Waterbeheer, Amersfoort

Ten Wolde KW (2012) Phosphate fertilizer from Sewage Sludge Ash. University of Twente

Van Vuuren DP, Bouwman AF, Beusen AHW (2010) Phosphorus demand for the 1970–2100 period: a scenario analysis of resource depletion. Glob Environ Chang 20(3):428–439. https://doi.org/10.1016/j.gloenvcha.2010.04.004

Yamamoto T, Tsutsui T, Nonaka T, Kobayashi S, Nishiguchi A, Yamane I (2006) A quantitative assessment of the risk of exposure to bovine spongiform encephalopathy via meat-and-bone meal in Japan. Prev Vet Med 75(3–4):221–238. https://doi.org/10.1016/j.prevetmed.2006.03.003

Part III
Sewage Sludge and Night Soil

Chapter 17
Struvite Recovery from Digested Sewage Sludge

Daisuke Koga

Abstract Struvite recovery is an ideal way to simultaneously recover phosphate (P_i) and ammonia from digested sludge. A full-scale plant to recover struvite directly from digested sludge was implemented at Higashinada Wastewater Treatment Plant (WWTP) having the sewage treatment capacity of 241,500 m^3/day in Kobe City. The struvite recovery plant has a capacity of treating digested sludge of 239 m^3/day, which is equivalent to a quarter of digested sludge generated at the WWTP. On average, it can recover approximately 40 and 90% of total P and soluble P_i, respectively, from digested sludge. The struvite recovery could reduce the volume of dewatered sludge by 3.3% on average and prevent struvite-scaling problems in the sludge treatment process. Recovered struvite has been registered as a chemical fertilizer and distributed to the Kobe area through fertilizer companies.

Keywords Struvite recovery · Sludge dewatering · Anaerobic digestion · Chemical fertilizer

17.1 Introduction

Many wastewater treatment plants (WWTP) use anaerobic digestion to reduce the weight and volume of sewage sludge and to recover energy from sludge biomass. After anaerobic digestion, high-phosphate (P_i) liquor (filtrate or centrate) from sludge dewatering is typically returned to the sewage treatment process (Kato and Muroya 2005). This unavoidably increases the load of P_i as well as ammonia to the sewage treatment process, making the nutrient removal more difficult (Abe and Muro 1995). Struvite ($MgNH_4PO_4{\cdot}6H_2O$) recovery is an ideal way to simultaneously reduce the internal load of P_i and ammonia and to recycle them as fertilizing materials (Shimamura and Mizuoka 2007).

D. Koga (✉)
Engineering and Development Division, Swing Corporation, Tokyo, Japan
e-mail: koga.daisuke@swing-w.com

H. Ohtake, S. Tsuneda (eds.), *Phosphorus Recovery and Recycling*,
https://doi.org/10.1007/978-981-10-8031-9_17

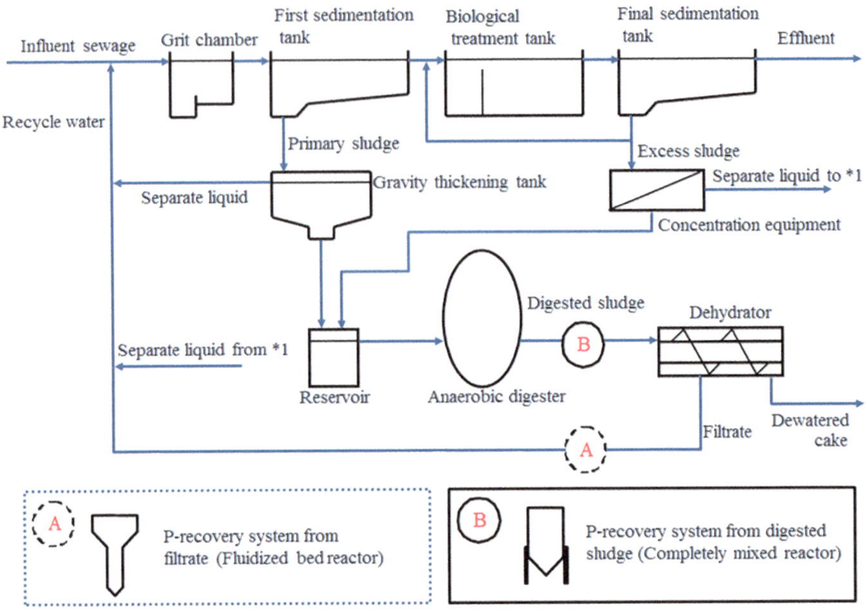

Fig. 17.1 The process flow of a conventional WWTP and the spots most relevant to struvite recovery (*A* and *B*)

Figure 17.1 shows two spots most relevant to struvite recovery in a conventional WWTP. Struvite can be recovered either from liquor (filtrate or centrate) or directly from digested sludge. At the spot A, struvite can be easily recovered from the filtrate (or centrate) because of the low suspended solid (SS) content and viscosity compared to those of digested sludge. Fluidized bed reactors are commonly used to recover struvite from the liquid phase of digested sludge (Shimamura and Ishikawa 2009). On the other hand, direct struvite recovery from digested sludge at the spot B has several advantages over the recovery from the liquor at the spot A (Koga and Hagino 2015). Digested sludge typically contains soluble P_i accounting for 15–30% of total P (T-P), though its content considerably varies depending on sludge treatment processes (Fig. 17.2). In addition, digested sludge also contains struvite particles, typically accounting for 5–15% of the dry weight, which are generated during anaerobic sludge digestion. Since both soluble P_i and struvite particles are recoverable at the spot B, it can be expected to recover more P from digested sludge than from the liquid phase alone. However, because of the high SS content and viscosity, it is necessary to use a complete-mix reactor for recovering struvite from digested sludge (Koga and Hagino 2016).

Additionally, there are potential economic benefits which can be brought by the struvite recovery directly from digested sludge (Fig. 17.3):

- P is recovered as a valuable resource which can be used as a slow-P_i-release fertilizer.

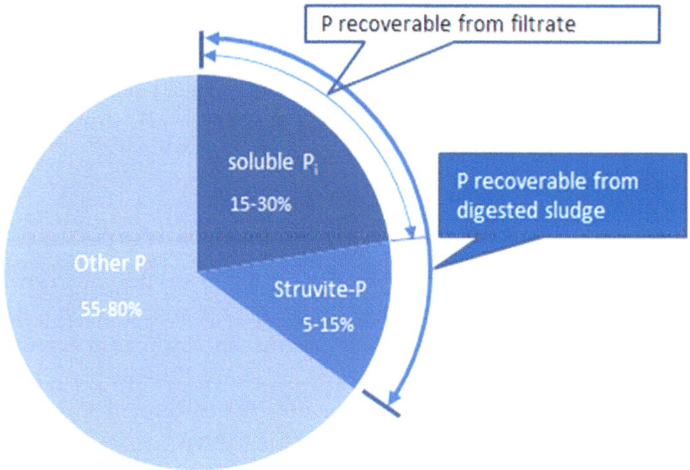

Fig. 17.2 The percentage of recoverable P in digested sludge

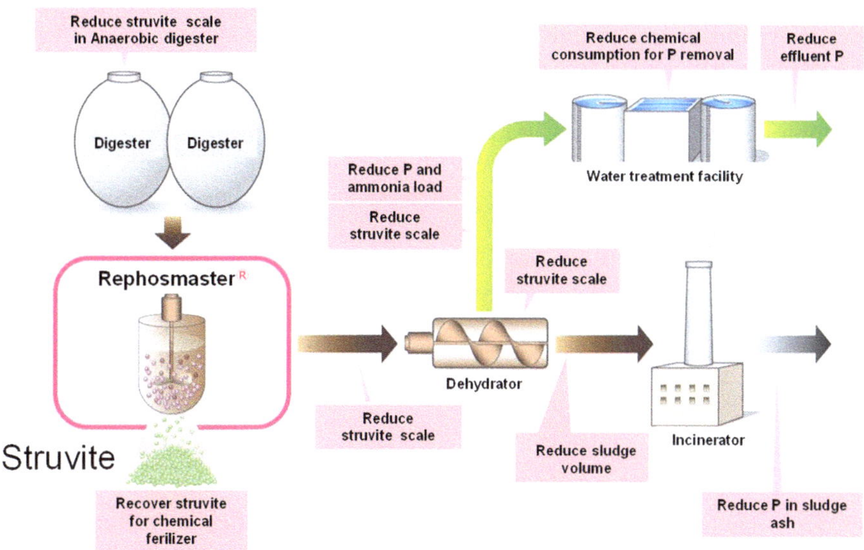

Fig. 17.3 Possible benefits given by implementing struvite recovery in a WWTP

- Sludge is stabilized, which makes plant operation stable.
- The prevention of struvite-scaling problems in pumps, pipes, and dewatering facilities enables the reduction of chemical costs.
- The recovery of struvite particles from digested sludge reduces the volume of dewatered sludge, leading to the cost reduction of sludge disposal.

17.2 Struvite Recovery Technology

Anaerobically digested sludge contains high levels of P_i and ammonia. When magnesium ions are added to the digested sludge at alkaline pH, struvite crystals are formed according to the reaction (17.1):

$$Mg^{2+} + NH_4^{+} + HPO_4^{2-} + OH^{-} + 5H_2O \rightarrow MgNH_4PO_4 \cdot 6H_2O \qquad (17.1)$$

Figure 17.4 exemplifies the process flow diagram of struvite recovery after sludge digestion. Sludge from anaerobic digesters is passed through a fine screen and fed to the crystallization reactor. Here, magnesium hydroxide is provided as a magnesium source to promote a crystallization reaction. Struvite particles are separated from the reaction mixture by centrifugation. A small part of separated struvite particles is recycled as seed crystals to the reactor. Struvite crystals are then sent to the scrubber at a frequency of several times per day. Recovered struvite is washed with water, dried, and stored for shipping. On the other hand, the rest (treated sludge) is dewatered and incinerated by a mono-incinerator. The whole of the struvite recovery process can be automatically controlled.

Typically, this process can recover more than 80% of soluble P_i from digested sludge. Approximately 30% of T-P can be recovered from digested sludge, if it is derived from a biological P removal process. The volume of dewatered sludge can typically be reduced by 3.3% mainly due to the removal of struvite particles generated during sludge digestion. The purity of recovered struvite is 80% or higher. Figure 17.5 shows a schematic view of the struvite recovery reactor (Rephosmaster®). It is a completely mixed-type reactor. The mechanical agitation enhances the contact of HPO_4^{2-}, NH_4^{+}, and Mg^{2+} with struvite seed crystals, promoting the crystallization reaction in viscous digested sludge. Without the homogeneous mixing, fine struvite particles are formed in the reactor, making the separation of struvite from digested sludge difficult.

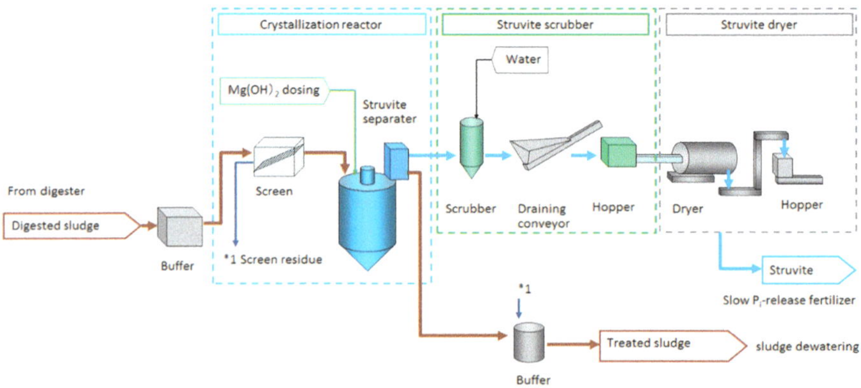

Fig. 17.4 Struvite recovery process

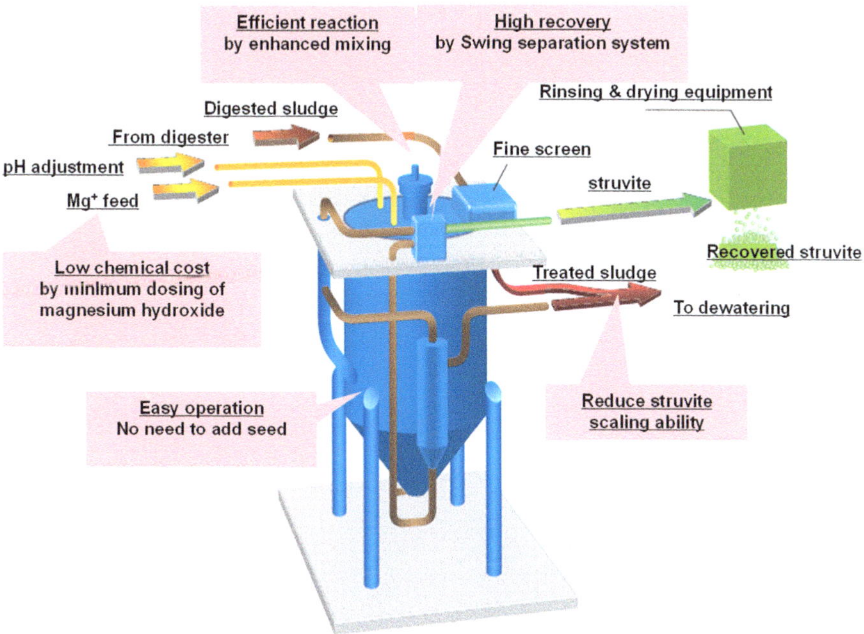

Fig. 17.5 Closeup of the reactor for struvite crystallization (Rephosmaster®)

The centrifugal struvite separation system (Swing separation system) can separate seed crystals from digested sludge and keep the concentration of seed crystals at a high level in the reactor. The particle size of seed crystals can also be kept nearly constant by the mechanical mixing, since it enhances the collision among seed crystals. The constant concentration and size of struvite particles provide a necessary surface area for struvite crystal growth without feeding seeds from outside. As a result, the system can deliver stable performance for the struvite recovery. Struvite crystals can be drawn from the reactor several times per day without the need to stop operation for harvesting. Recovered struvite products are uniform, granular crystals with a diameter of 100–700 μm (Fig. 17.6).

17.3 Case Study

The full-scale demonstration of struvite recovery was conducted during the period from August 2013 to March 2014 as a joint project commissioned by the National Institute for Land and Infrastructure Management. A full-scale plant was set up at the Higashinada WWTP having the treatment capacity of 241,500 m^3/day in Kobe City (Fig. 17.7). It has a capacity of treating digested sludge of 239 m^3/day,

Fig. 17.6 Recovered struvite

Fig. 17.7 The full-scale plant for struvite recovery plant at the Higashinada WWTP in Kobe, Japan

equivalent to a quarter of digested sludge generated at the WWTP. It was expected to recover struvite of 360 kg/day.

Figure 17.8 shows the change of P_i and T-P concentrations in digested sludge before and after struvite recovery. The concentration of T-P in digested sludge varied from 542 to 605 mg/L, averaging 581 mg/L, before struvite recovery. Struvite recovery could reduce the T-P concentration by 37% on average. It also removed approximately 90% of soluble P_i from digested sludge. The T-P/P_i removal ratio,

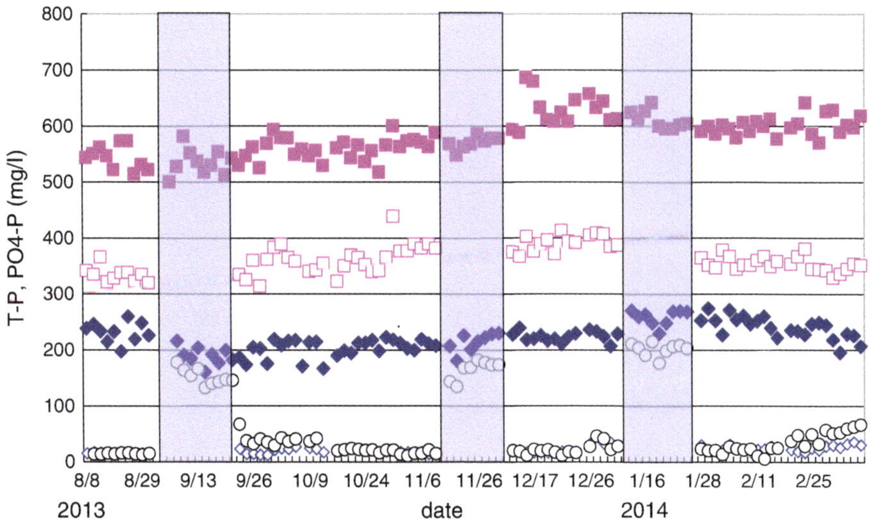

Fig. 17.8 Changes in T-P and P_i in the digested sludge before and after struvite recovery. Symbols: T-P (■) and soluble Pi (◆) in the digested sludge before struvite recovery, T-P (□) and soluble Pi (◇) in the digested sludge after struvite recovery, and T-P (O) in the filtrate of digested sludge after struvite removal. Vertical bars indicate the period of halting struvite recovery

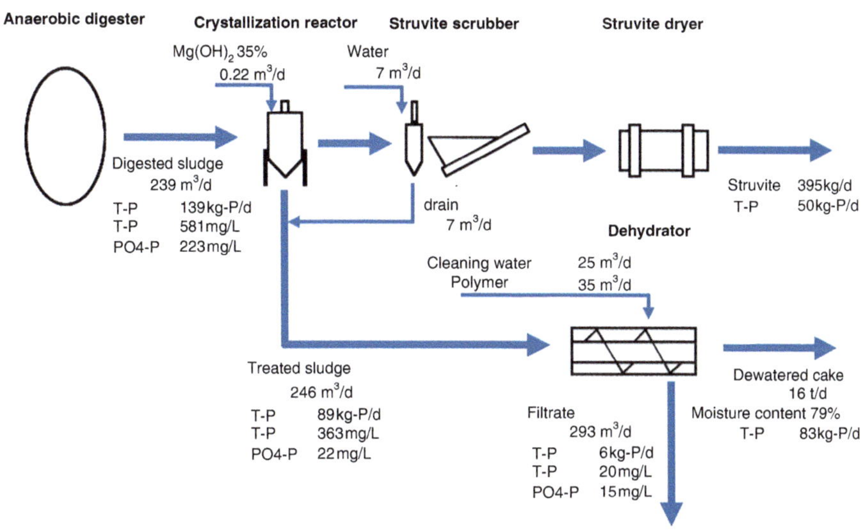

Fig. 17.9 P mass balance in the struvite recovery process

$(T\text{-}P_{in} - T\text{-}P_{out})/(P_{i\ in} - P_{i\ out})$, varied from 80% to 160%, averaging 111%. This suggested that not only soluble P_i but also struvite particles which had been formed in the sludge digester were recovered by this process. The P mass balance in the struvite recovery process is shown in Fig. 17.9.

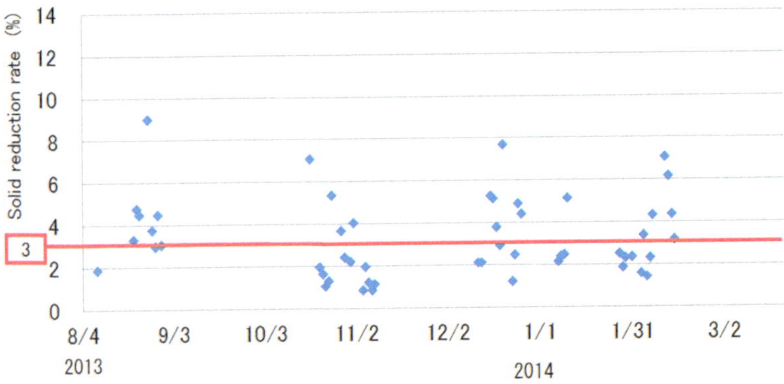

Fig. 17.10 Percent reduction of dewatered sludge weight due to struvite recovery

Fig. 17.11 The effect of struvite recovery on the scale buildup in pipe lines

Figure 17.10 shows the changes in % reduction of sludge weight due to struvite recovery. Although the data considerably fluctuated, the dry weight of digested sludge showed about 3.3% reduction on average after struvite recovery. Taking into account the high disposal costs, this reduction can be considered to be economically beneficial. To examine the effect of struvite recovery on scale buildup in pipes, stainless pipes (80A SUS) were placed in the digested sludge stream with and without struvite recovery (Fig. 17.11). Clearly, the buildup of struvite scale was prevented by struvite recovery. As a result, the screw-press dehydrators for sludge dewatering could be stably operated without causing clogging problems after the start of struvite recovery.

Table 17.1 shows the levels of major nutrients and hazardous elements in recovered struvite. All the products satisfied the specifications required for chemical fertilizers in Japan. Consequently, the recovered struvite was approved by the Ministry of Agriculture, Forestry and Fisheries of Japan as a chemical fertilizer in April 2014, allowing the Higashinada WWTP to officially distribute struvite recovered from sewage as a chemical fertilizer. Recovered struvite is shipped to fertilizer-manufacturing plants where it is commercialized as a slow P_i-release fertilizer.

Table 17.1 Levels of major nutrients and contaminants of recovered struvite

Components	Contents (%)	Content per 1% of main component ((1)+(2))	Result	Official specifications for chemical fertilizer
Nutrients				
Ammonia nitrogen (1)	4.65		>	≥1
Citric acid-soluble P_2O_5 (2)	24.56		>	≥1
Citric acid-soluble magnesium (MgO)	13.68		–	–
(1) + (2)	29.21		>	≥10 in total
Hazardous component				
Arsenic	<0.0001	<0.0000034	<	0.002
Cadmium	<0.0001	<0.0000034	<	0.000075
Mercury	0.000002	0.000000068	<	0.00005
Nickel	0.0009	0.0000308	<	0.005
Chromium	0.0016	0.0000548	<	0.05
Lead	0.0011	0.0000377	<	0.003
Ammonium thiocyanate	<0.05	<0.0017	<	0.005
Nitrous acid	<0.01	<0.00034	<	0.02
Biuret nitrogen	<0.05	<0.0017	<	0.01
Sulfamic acid (NH_2SO_3H)	<0.05	<0.0017	<	0.005
Titanium	0.028	0.00096	<	0.02

(Average: $N = 5$)

Fig. 17.12 Photos of struvite pellet fertilizer and crop cultivation

Figure 17.12 shows a commercialized fertilizer and a crop (sweet corn) cultivated using the fertilizer. Thus, P could be captured from sewage and recycled to farmland, thereby contributing to the closing loop of P flow in the Kobe area.

17.4 Conclusions

The full-scale plant to recover struvite directly from digested sludge was implemented at the Higashinada WWTP having the sewage treatment capacity of 241,500 m^3/day in Kobe City. The full-scale plant has a capacity of treating digested sludge of 239 m^3/day, equivalent to a quarter of digested sludge generated at the WWTP. On average, it could recover 40 and 90% of T-P and soluble P_i, respectively, from digested sludge at the WWTP. The struvite recovery could reduce the weight of dewatered sludge by 3.3% on average and effectively prevent struvite-scaling problems in pipes, pumps, and sludge dewatering facilities. The recovered struvite has been registered as a chemical fertilizer and distributed to the Kobe area through fertilizer companies.

References

Abe S, Muro S (1995) The grade-up of wastewater treatment plants in Fukuoka by the development of struvite recovery technology. J Jpn Sew Works Assoc 32, 389 (in Japanese)
Kato K, Muroya N (2005) Study on phosphorus balance in a municipal wastewater treatment plant. J Ind Water 98:1152–1161. (in Japanese)
Koga D, Hagino T (2015) Technology for phosphorus removal and recovery from digested sewage sludge. Ebara Eng Rev 249
Koga D, Hagino T (2016) Technology for high efficiency phosphorus removal and recovery from digested sewage sludge. Environ Sol Technol 15(3) (in Japanese)
Shimamura K, Ishikawa H (2009) A phosphorus recovery process from digested sewage sludge. Environ Sol Technol 8 (1) (in Japanese)
Shimamura K, Mizuoka A (2007) Development of a novel process for recovering phosphorus from anaerobically digested sludge using a crystallization reactor. Ebara Eng Rev (215)

Chapter 18
Phosphorus Recovery from Night Soil and Johkasou Sludge

Yoshio Okuno, Yasuyuki Nakata, and Masaaki Isse

Abstract In Japan, sanitation for 26% of the population is covered by decentralized treatment facilities called Johkasou and night soil treatment plants (NSTPs). The former is installed to treat black water from small communities or individual households. Johkasou is a general term for compact on-site wastewater treatment unit and/or facility and is applicable to a population of several to several thousands, depending on the installation condition. The latter is installed to treat mainly night soil (human feces) coming from 6% out of the 26% population that uses decentralized treatment facilities. Since the sludge extracted from Johkasou is also treated in NSTPs, they play a key role in the Japanese sanitation system. As part of a social sustainability policy, the "Plan of Sludge Resource Recycling Treatment Center" (SRRTC) was enacted as a bylaw in 1997. It demands that NSTPs be furnished with facilities for resource and/or energy recovery from organic wastes including night soil and Johkasou sludge. Facilities implemented by this plan are categorized as sludge resource recycling treatment centers. This chapter describes the first SRRTC project in which a chemical precipitation process was applied to the recovery of phosphorus as calcium phosphates from night soil and Johkasou sludge.

Keywords Night soil · Johkasou sludge · SRRTC · Phosphorus recovery · CaP precipitation

18.1 Introduction

Phosphorus (P) is an essential element for human growth, and it is the only nutritional food element for which Japan depends completely on importation. About 170 kt P/a, which was equivalent to 75% of the import, was consumed in the

Y. Okuno (✉) · Y. Nakata · M. Isse
Hitachi Zosen Corporation, Environment Business Headquarters, Osaka-shi, Osaka, Japan
e-mail: okuno_yo@hitachizosen.co.jp

H. Ohtake, S. Tsuneda (eds.), *Phosphorus Recovery and Recycling*,
https://doi.org/10.1007/978-981-10-8031-9_18

Japanese fertilizer and livestock feed industries in 2014. As the world population grows, the global agricultural and livestock industries will increase their demand for phosphorus. This may pose a serious risk to Japan's P availability (Phosphorus Atlas Research Institute 2015). The key to mitigating the P shortage risk would be the phosphorus recovery from phosphorus-rich wastes such as steelmaking slag and organic wastes including sewage sludge, night soil, and food waste. The phosphorus recovery from steelmaking slag and sewage sludge is discussed elsewhere (see Chap. 1).

The traditional and most economical method for recycling organic wastes is composting. However, as the demand of composts varies both seasonally and regionally, composting is not always a successful option in Japan. From a resource recycling perspective, user-friendly products for agriculture may be chemical compounds without heavy metal contamination. Chemical precipitation technologies have been considered as practical means of obtaining recycleable P products. Typical P products from chemical precipitation processes are struvite and calcium phosphate compounds (CaPs). However, the applicability of these technologies depends on coexisting substances, in particular ammonium nitrogen. The struvite precipitation method is applicable to ammonium and phosphate (P_i)-rich waste streams such as digested sewage sludge and animal manure, while the CaP precipitation method is suited to ammonium-free streams.

Human waste sanitation in Japan is based, in principle, on centralized systems for both black water and solid waste. Regarding black water, about 74% of the population is served by the sewer system, whereas the remainder, who lives in remote and/or geographically isolated areas, depends on decentralized systems such as compact on-site wastewater treatment tanks (called Johkasou) installed in individual households. Community wastewater treatment plants for apartment houses and small towns are also categorized as Johkasou. About 6% of the population living in individual households without access to centralized or Johkasou systems depends on specialized systems, that is, a combination of truck collection and disposal at night soil treatment plants (NSTPs). The excess sludge extracted from Johkasou is also treated by the NSTP systems. Therefore, sanitation for 26% of the population directly or indirectly depended on NSTPs in Japan in 2015 (Japanese Waste Treatment 2014), which means that phosphorus recovery from night soil and Johkasou sludge is worthy of investigation.

Because the "Plan of Sludge Resource Recycling Treatment Center" (SRRTC) was enacted in 1997 as a bylaw, in order to enhance energy and resource savings required by sustainability policies, newly built and renovated NSTPs need to incorporate facilities for energy production and/or resource recovery. NSTPs equipped with these facilities are categorized as sludge resource recycling treatment centers and can receive financial support from the national government. Since the plan was enacted 20 years ago, Japan has accelerated the development of resource recovery technologies such as composting, phosphorus recovery, and biogas cogeneration.

This chapter describes an outline of phosphorous recovery from night soil and Johkasou sludge using the CaP precipitation method in SRRTCs. The struvite precipitation method for night soil treatment is described elsewhere (Japan Waste Research Foundation 2010, 2012).

18.2 Phosphorus Material Flow

According to a study of phosphorus material flow in Japan for 2006, out of 60 kt P/a from all domestic drainage, 5 kt P/a and 2 kt P/a were drained to NSTPs as night soil and Johkasou sludge, respectively. Phosphorus in the effluent from NSTPs was only 20 t P/a. That is, almost 6 kt P/a ended up in sludge at NSTPs. The NSTP sludge consists of both biomass and chemical sludge because night soil must be treated through a combination of biological and physicochemical processes to cope with multiple pollutants (Tsurumaki et al. 2008). A more recent study showed that Johkasou and NSTPs served for about 35 million people, treating approximately 4600 t P/a in 2013 (Phosphorus Atlas Research Institute 2015).

18.3 Phosphorus Recovery Using Precipitation Methods

The form of phosphorus that is recovered by precipitation methods is mainly inorganic phosphate (P_i). The top priority of NSTPs, which need to discharge the final effluent to public water bodies, is given to the decomposition of highly concentrated organic matter using a biological treatment process. Therefore, it is unavoidable that most of the insoluble phosphorus and a portion of the phosphate end up in the biological sludge. In addition, conventional NSTPs remove the remaining phosphate by a chemical coagulation process, where phosphate is insolubilized as chemical sludge. Precipitation methods can be applied, instead of chemical coagulation methods, to recover phosphorus in more user-friendly forms from the waste streams.

There are mainly two methods for generating user-friendly products from night soil and Johkasou sludge: CaP precipitation and struvite precipitation. Phosphorus recovery processes using these precipitation methods can be placed upstream or downstream of biological processes. At the upstream stage of a biological treatment process, high concentrations of ammonium and phosphate ions are present, and, therefore, the struvite precipitation method is preferable for P recovery. On the other hand, the CaP precipitation method is more rational for the recovery of P after biological processes, because essentially no ammonium ion remains.

The CaP precipitation reaction is given as:

$$10Ca^{2+} + 6PO_4^{3-} + 2OH^- \rightarrow Ca_{10}(PO_4)_6(OH)_2$$

Among typical CaP products, hydroxyapatite is the most stable compound having the low solubility in water. With sufficient Ca^{2+} at appropriate pH, CaP precipitation occurs on seed CaP particles. Once precipitation starts, the CaP particles grow and/or are split into fine particles which can serve in series as a seed. Therefore, it is not necessary to feed CaP particles continuously to the reaction mixture under appropriate operating conditions. CaP particles larger than around 400 μm can be recovered from the mixture as a product.

18.4 Phosphorus Recovery in SRRTC

18.4.1 Outline of the SRRTC

Name	Senboku-shi Sludge Resource Recycling Treatment Center
Location	Senboku-shi, Akita-ken, Japan
Plant capacity	60 kL/day (night soil, 35 kL/day plus Johkasou sludge, 25 kL/day)
Process	Biological high-load denitrification MBR with CaP precipitation process
	Schematic flow diagram is shown in Fig. 18.1

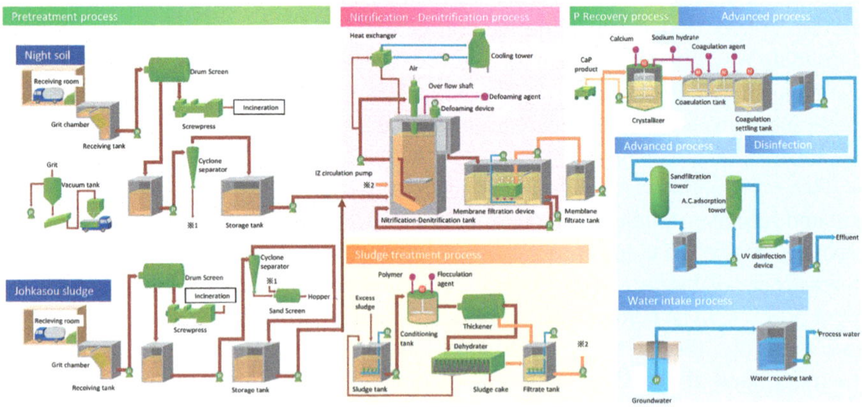

Fig. 18.1 Schematic flow diagram of the Senboku-shi Sludge Resource Recycling Treatment Center

18.4.2 Process Description

18.4.2.1 Pretreatment Process

After grit and debris separation, night soil and Johkasou sludge were stored separately for a maximum of 3 days. During the storage, both were well mixed to ensure homogenization. Debris is incinerated after dewatering by screw press.

18.4.2.2 Biological Treatment Process

Pretreated night soil and Johkasou sludge were independently fed to the membrane bioreactor (MBR) where both nitrification and denitrification occurred. The aeration of the MBR was performed using the IZ jet aeration system (Fig. 18.2) which could supply enough oxygen required for a high rate of nitrification at the upper part of the tank. Thanks to a unique tank configuration and the control of feeding rate of night soil and Johkasou sludge, the lower part of the tank could be kept anaerobic for promoting denitrification. The mixture of the nitrification-denitrification tank was then filtered using an immersion-type microfiltration (MF) membrane.

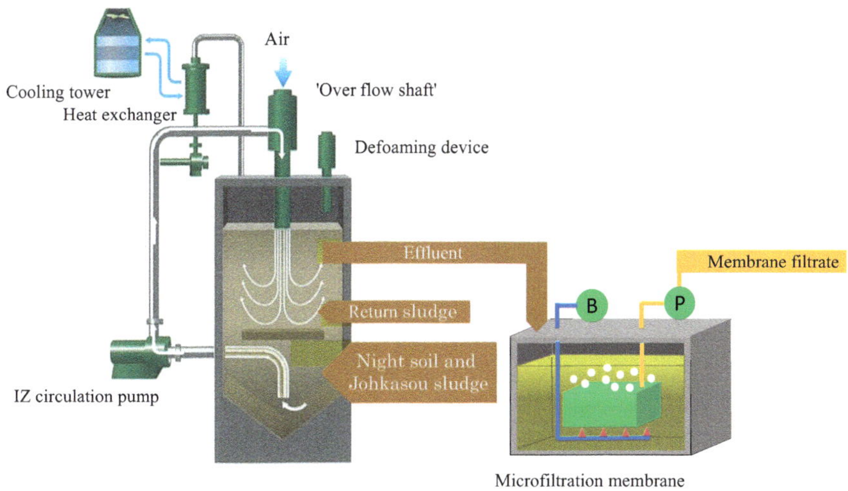

Fig. 18.2 IZ jet aeration system

18.4.2.3 Phosphorus Recovery Process

The filtrate was fed to a crystallizer where calcium chloride was dosed under controlled pH. The crystallizer consists of two zones: the continuously stirred crystallization zone in the center and the CaP particle settling zone in the outer region (Fig. 18.3 and Appendix).

The mixture in the crystallization zone moved to the settling zone where CaP particles settled and returned to the crystallization zone by gravity. The treated water overflowed from the reactor. CaP particles accumulated at the bottom of the settling zone were withdrawn using a drain pump and poured into a drainer bag. Finally, CaP particles were stored in the bag as a product. A woven container bag was used as the drainer bag in this process.

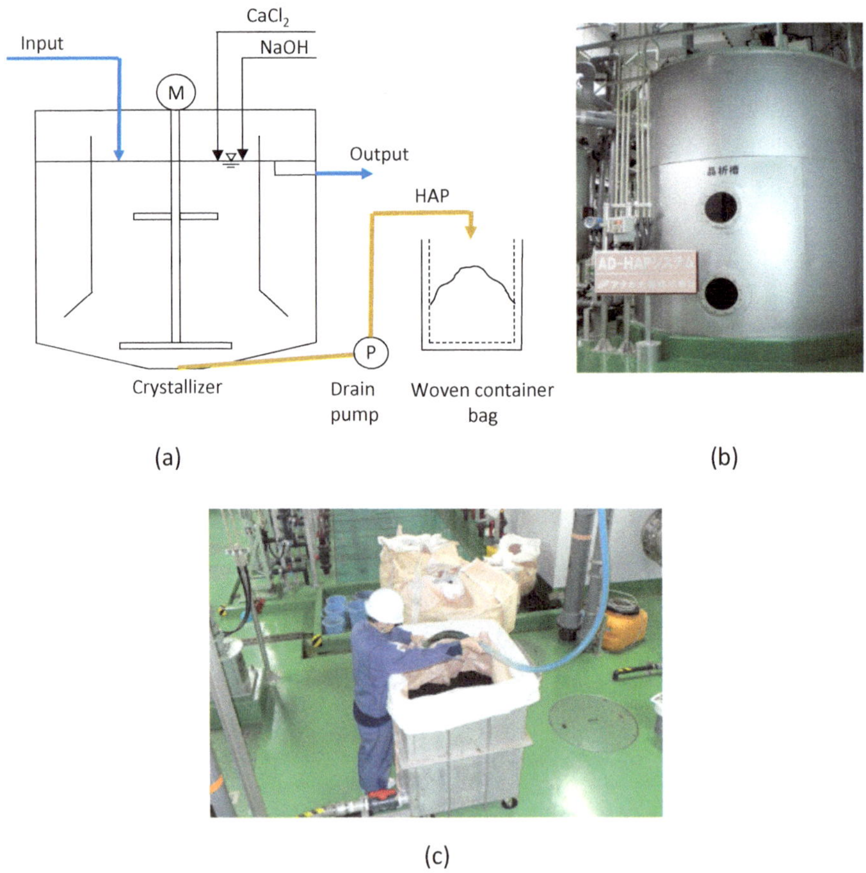

Fig. 18.3 The P recovery process. (**a**) Schematic flow of the precipitation process, (**b**) crystallizer, and (**c**) woven container bag

18.4.2.4 Advanced Treatment Process

The effluent of the phosphorus recovery process was discharged to a river after an advanced treatment process consisting of coagulation, filtration, activated carbon (AC) adsorption, and ultraviolet (UV) disinfection. The effluent quality met the regulation criteria specified by the local government.

18.5 Performance

18.5.1 Phosphorus Balance

18.5.1.1 Water Analysis

Table 18.1 shows the flow rate and water quality data at the process points specified in Fig. 18.4. The flow rates of night soil and Johkasou sludge were 18.6 and 24.0 m^3/day,

Table 18.1 Material balance of the facility

Item		BOD_5	COD_{Mn}	SS	T-N	T-P	PO_4-P	Flow rate	P	
Unit		mg/L	mg/L	mg/L	mg/L	mg/L	mg/L	m^3/day	kg/day	%
P1	Night soil	4300	8400	6800	2500	280	250	18.6	5.21	59.1%
P2	Johkasou sludge	3600	8200	9300	760	150	87	24.0	3.60	40.9%
P3	Filtrate	7	210	22	37	9.7	7.3	27.3[a]	0.26	3.0%
P4	Sludge cake	–	–	16.1%	–	2.39%dry	–	1.75[a]	6.73	76.5%
P5	Excess sludge	–	–	18,000	–	390	220	15.9	6.20	70.4%
Pin	Input in crystallizer	1	130	<1	16	43	41	66.8	2.87	32.6%
Ptank	Crystallizer	–	–	17,000	–	2400	6.9	–	–	–
CaP	CaP	–	–	–	–	15.5%dry	–	–	2.08	23.5%
Pout	Output from crystallizer	1	110	2	11	12	8.9	66.8	0.80	9.1%
P6	Coagulation sludge	–	–	11,000	–	210	210	3.8	0.80	9.1%
P7	Advanced process treated water	<1	4	<1	5.8	0.04	0.03	63.0	0.00	0.0%

[a]Calculated from mass balance

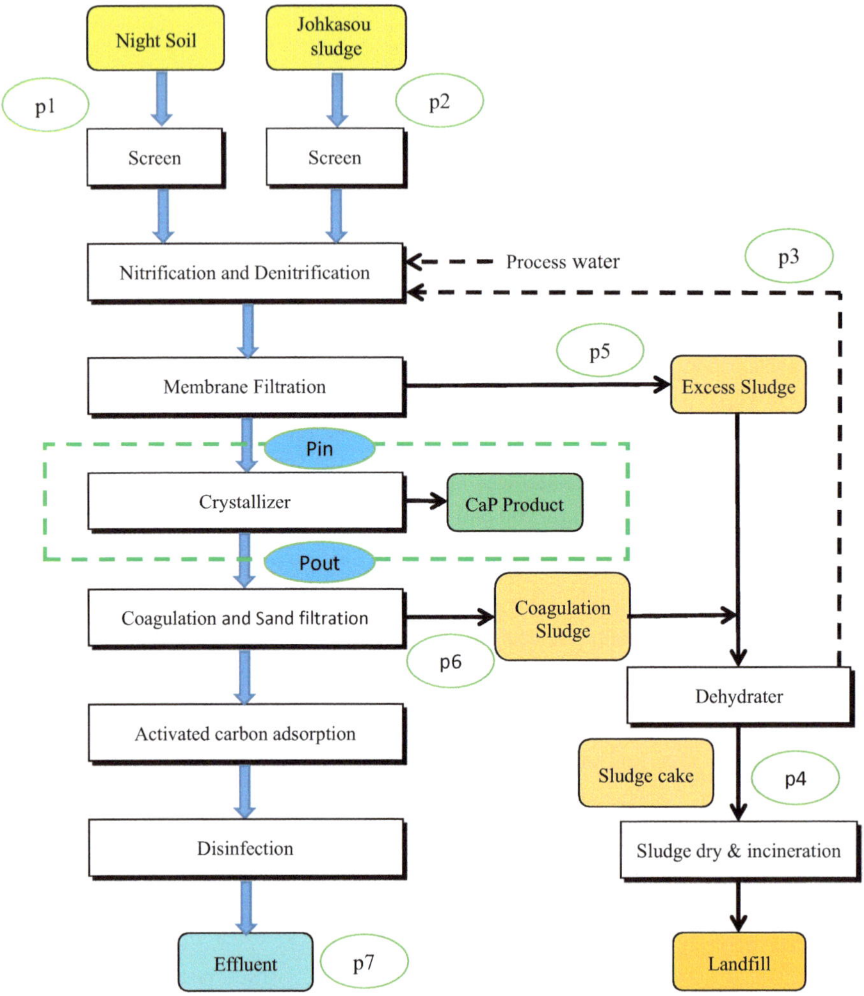

Fig. 18.4 Flow and sampling points of the facility

respectively. Thus a total of 42.6 m^3/day was fed into the STRRC. BOD_5, CODMn, T-N, T-P, and PO_4-P were effectively decreased during the biological treatment process:

BOD_5	4300–3600 mg/L to 1.0 mg/L
CODMn	8200–8400 mg/L to 130 mg/L
T-N	760–2500 mg/L to 16 mg/L
T-P	150–280 mg/L to 43 mg/L
PO_4-P	87–250 mg/L to 41 mg/L

18.5.1.2 Phosphorus Recovery Rate in the SRRTC

The material balance on phosphorus was calculated using the data shown in Table 18.1. Overall, 23.5% of input phosphorus was recovered, while 76.5% ended up in sludge. The phosphorus recovery rate of the crystallizer was 72.5%.

18.5.1.3 Summary of Operation Data

The operation data collected from January 2013 to December 2015 are shown in Fig. 18.5. The night soil ratio (night soil/(night soil + Johkasou sludge)) in the feed varied from 0.45 to 0.7. The peak ratio occurred in March, August, and January when the local population temporarily increased because of consecutive holidays. The feed rate of night soil and Johkasou sludge varied seasonally with a peak in summer. The influent PO_4-P concentration of the crystallizer varied depending mainly on the night soil ratio.

18.5.2 Properties of the Product

18.5.2.1 Component Analysis

The total phosphorus (T-P) content in the recovered product ranged from 15.4 to 19.9 (%) on a dry weight basis. This is comparable to the theoretical T-P content of hydroxyapatite ($Ca_{10}(PO_4)_6(OH)_2$) (18.5%). The citic acid-soluble P content of the product ranged from 27.4 to 37.8 (%) (Table 18.2) (Japan Waste Research Foundation 2010, 2012). The heavy metal contents were much lower than the levels specified by the regulations on sludge fertilizer in the Fertilizer Control Law, Japan.

18.5.2.2 Elution Test

The test of heavy metal elution from recovered products was conducted according to the method described previously [*]. The heavy metal concentrations in the eluate were less than one-tenth of the levels specified by the regulations on "Verification concerning industrial wastes containing metals" (the ordinance of the Prime minister's office, Japan) (Table 18.3) (Japan Waste Research Foundation 2010, 2012; Ministry of the Environment 1973).

18.5.2.3 Appearance and Size of Products

The product was brawny particles (Fig. 18.6a, b). The coloring was likely due to endocrine compounds or bile pigments. The particle size distribution is shown in Fig. 18.6c. The median diameter was 370 μm.

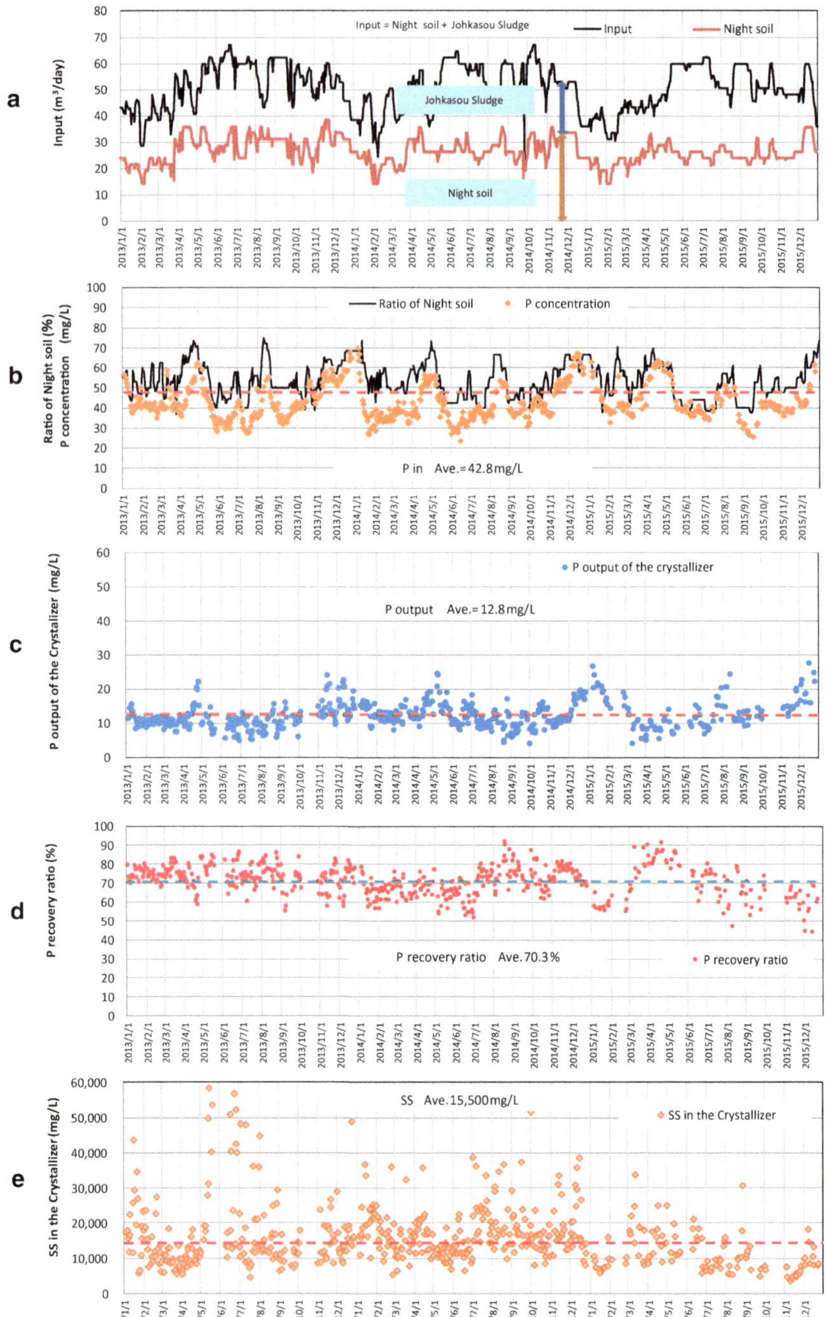

Fig. 18.5 Summary of operation. (**a**) The total and night soil feed rates to the SRRTC, (**b**) relationships between the night soil ratio (% night soil in the total feed) and the inflow PO_4-P concentration of the crystallizer, (**c**) the effluent PO_4-P concentration of the crystallizer, (**d**) the phosphorus recovery rate, and (**e**) the suspended solid (SS) concentration in the crystallizer

Table 18.2 Component analysis results

Item	Unit	2010			2012	Remarks
TS	%	72.2	74.1	71.3	41.8	
VS	%	12.1	10.4	10.7	15.6	
C	% (dry)	4.23	4.44	3.47	5.05	
H	% (dry)	0.94	0.74	0.83	0.77	
N	% (dry)	0.42	0.13	0.28	0.45	
Phosphorus	% (dry)	19.9	18.6	15.4	15.5	18.5[a]
Calcium	% (dry)	30.5	31.3	30.9	29.9	39.9[a]
Thiocyanate	% (dry)	<0.05	<0.05	<0.05	<0.05	
Biuret nitrogen	% (dry)	<0.05	<0.05	<0.05	<0.05	
Sulfamic acid	% (dry)	<0.05	<0.05	<0.05	<0.05	
Citric-soluble phosphate	% (dry)	37.8	31.4	27.4	34.5	
Soluble phosphate	% (dry)	7.31	5.73	7.39	17.3	
Water-soluble phosphate	% (dry)	0.32	0.24	0.18	0.11	
Total phosphoric acid	% (dry)	45.5	42.7	35.1	35.4	
Arsenic	mg/kg (dry)	4.4	4.7	4.7	4.0	50[b]
Cadmium	mg/kg (dry)	0.8	0.6	0.8	1.2	5[b]
Mercury	mg/kg (dry)	<0.5.	<0.5	<0.5	<0.5	300[b]
Nickel	mg/kg (dry)	4.6	3.7	2.6	2.5	500[b]
Chrome	mg/kg (dry)	<5	<5	<5	6	100[b]
Lead	mg/kg (dry)	1.1	0.5	0.6	2.3	
Potassium	mg/kg (dry)	460	380	550	370	
Magnesium	mg/kg (dry)	5400	5000	4300	5200	
Zinc	mg/kg (dry)	240	250	230	270	
Copper	mg/kg (dry)	55	41	74	100	
Titanium	mg/kg (dry)	14	7.7	6.7	28	

[a]Theoretical value
[b]The tolerance levels specified by regulations on sludge fertilizer in the Fertilizer Regulation Act, in Japan

Table 18.3 Elution test results of the product

Item	Unit	2010			2012	Tolerance[a]
Alkyl mercury compound	mg/L	ND	ND	ND	ND	ND
Mercury compound	mg/L	<0.0005	<0.0005	<0.0005	<0.0005	0.005
Cadmium compound	mg/L	<0.01	<0.01	<0.01	<0.01	0.3
Lead compound	mg/L	<0.05	<0.05	<0.05	<0.05	0.3
Organic phosphorous compound	mg/L	<0.1	<0.1	<0.1	<0.1	1
Hexavalent chromium compound	mg/L	<0.05	<0.05	<0.05	<0.05	1.5
Arsenic compound	mg/L	0.007	0.007	0.011	0.006	0.3
Cyanogen compound	mg/L	<0.1	<0.1	<0.1	<0.1	1
Polychlorinated biphenyl	mg/L	<0.0005	<0.0005	<0.0005	<0.0005	0.003
Thiuram	mg/L	<0.006	<0.006	<0.006	<0.006	0.06
Simazine	mg/L	<0.003	<0.003	<0.003	<0.003	0.03
Thiobencarb	mg/L	<0.02	<0.02	<0.02	<0.02	0.2
Benzene	mg/L	<0.001	<0.001	<0.001	<0.001	0.1
Selenium compound	mg/L	<0.005	<0.005	<0.005	<0.005	0.3

[a]Ordinance of the Prime Minister's Office on Standards for Verification concerning Industrial Wastes containing Metals, etc. Japan (1973)

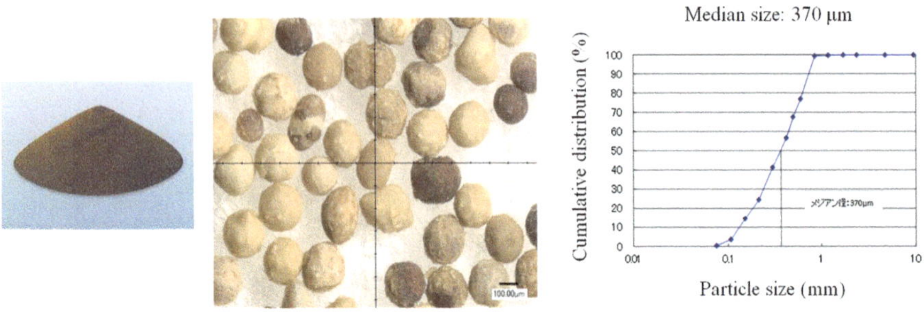

(a) CaP product (b) CaP product (× 200) (c) The particle size of CaP product (an example)

Fig. 18.6 The recovered product. (**a**) Photograph, (**b**) micrograph, and (**c**) the typical particle size distribution

Table 18.4 Fertilizer potential of recovered P product

(a) Relative agronomic efficiency						
P_i fertilizer	No.	P_2O_5 added (mg/pot)	Live weight (g/pot)			
			No.1	No.2	No.3	No.4
The product	I	300	37.4	41.7	61.8	84.4
	II	600	45.9	49.2	86.8	117.1
Fused magnesium phosphate	III	300	28.1	29.3	50.9	69.7
	IV	600	35.2	48.6	65.1	100.2
Calcium superphosphate	V	300	31.1	41.6	56.8	83.8
Control	VI	0	7.7	6.3	9.3	14.0
(b) Phosphate uptake by bok choy						
P_i fertilzier	No.	P_2O_5 added (mg/pot)	Phosphoric acid absorption (mg/pot)			
			No.1	No.2	No.3	No.4
The product	I	300	14.6	17.1	25.2	33.4
	II	600	20.8	24.2	42.6	64.1
Fused magnesium phosphate	III	300	8.7	9.9	17.1	22.4
	IV	600	13.6	20.4	26.7	37.7
Calcium superphosphate	V	300	10.0	15.1	18.9	27.3
Control	VI	0	2.4	1.9	2.2	4.0

18.5.2.4 Fertilizer Potential

The fertilizer potential of the product was compared with commercial phosphate fertilizers based on two items: (1) pot trials for relative agronomic efficiency and (2) phosphorus uptake by bok choy (Table 18.4 and Fig. 18.7) (Japan Waste Research Foundation 2010, 2012). The reference commercial fertilizers were calcium superphosphate and fused magnesium phosphate.

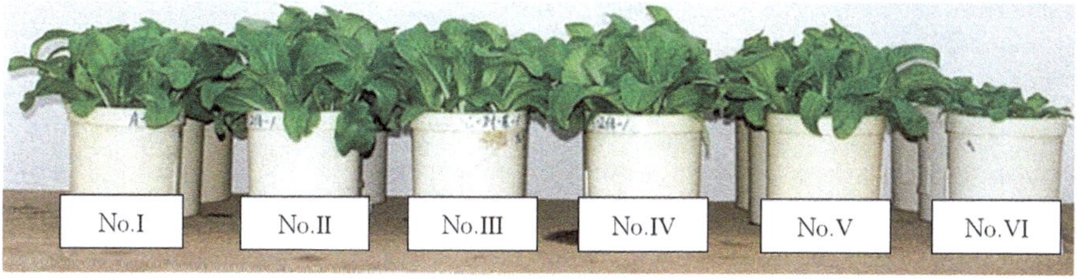

Fig. 18.7 Plant growth pot testing

18.6 Product Property

The composition of recovered products was analyzed using X-ray fluorescence and X-ray diffraction technologies. Recovered CaP samples were obtained from the Senboku-shi SRRTC. Reagent-grade apatite was used as reference. To examine the influence of heating on the composition of recovered products, the samples were dried by two different methods: heating at 100 °C or vaccum drying at room temperature (about 20 °C) for 2 days. X-ray fluorescence analysis was performed using EDX-700HS (SHIMAZU), while X-ray diffraction analysis was carried out by RINT2000 (RIGAKU).

As expected, the major elements of the samples were Ca and P (Table 18.5). Sight differences were detected with the Mg and Na contents between the recovered products and reagent-grade apatite. The recovered products contained Mg which is derived from human excreta, whereas the reference contained Na in place of Mg. Overall, the composition of the recovered products was similar to that of reagent-grade apatite.

18.6.1 Crystal Structure

X-ray diffraction analysis showed a significant difference between the recovered products (samples A and B) and the reference (Fig. 18.8). Many spike-like peaks were detected with the reference (reagent-grade apatite). Among them are two peaks that were also detected with the products. The recovered products showed several broad peaks. There was no significant difference between samples A and B, indicating that the crystal structure of the recovered products was not influenced by the two different drying methods. These results suggested that the recovered products were amorphous calcium hydroxyapatite.

Table 18.5 The components of recovered product

	Elements (%)							
	Ca	P	S	Si	Mg	Na	K	Sr
Sample A	68.35	27.06	2.489	0.916	0.645	–	0.158	0.092
Sample B	68.55	26.31	2.432	1.264	0.718	–	0.253	0.105
Reference	68.61	27.32	0.628	1.682	–	1.602	0.158	0.006

Sample A, heat-dried product; sample B, vacuum-dried product; reference, reagent-grade apatite (powder state)

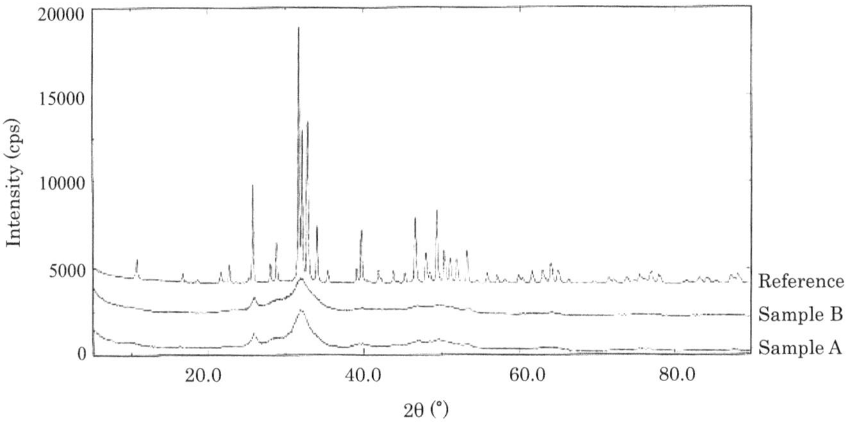

Fig. 18.8 X-ray diffraction patterns of the recovered products

18.7 Conclusions

The Senboku-shi Sludge Resource Recycling Treatment Center (SRRTC) recovers phosphorus from night soil and Johkasou sludge using the CaP precipitation method. The average phosphate recovery rate was 70.3%. The P content of recovered products varied from 15.5 to 19.9 (%) on a dry weight basis. The critic acid-soluble phosphate in the product ranged from 27.4 to 37.8 (%). Contamination by heavy metals was negligible. The recovered product showed the fertilizer potential similar to those of commercial phosphorus fertilizers. Although the number of SRRTCs equipped with phosphorus recovery facilities is expected to increase, the amount of night soil is predicted to decrease mainly because of the falling population in rural areas. To promote phosphorus recovery from night soil and Johkasou sludge, SRRTCs may need to accept phosphorus-rich waste such as livestock manure and food waste.

Appendix: Computational Fluid Dynamics of the Crystallizer

Computational fluid analysis was carried out to visualize the water movement in the continuously stirred crystallizer.

Conditions

The conditions for the computer simulation are shown in Table 18.6, and the modeling of crystallizer is shown in Fig. 18.9b.

Table 18.6 Analysis conditions

Item	Contents
Crystallizer	Diameter, 2.6 mφ; height (water depth), 2.31 m; capacity, 11.7 m^3
Fluid property	Density, 998.2 kg/m^3
Inlet and outlet	Flow rate, 112.3 m^3/day; pipe size, 65 A; outlet trough length, 0.472 m
Stirrer	Upper impeller (four paddles)
	Paddle diameter, 0.55 m; paddle width, 0.075 m; the number of paddles, 4
	Rotation speed, 41.7 rpm
	Lower impeller (four paddles)
	Paddle diameter, 0.7 m; paddle width, 0.1 m; the number of paddles, 4
	Rotation speed, 41.7 rpm (frequency, 35 Hz)

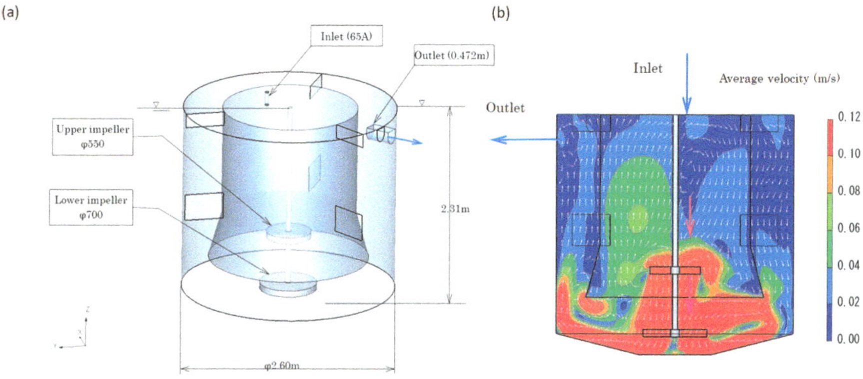

Fig. 18.9 (**a**) Simulation model of the crystallizer and (**b**) simulated velocity distribution at 35 Hz (41.7 rpm)

Software

The software, SCRYU/Tetra (Software Cradle Co., Ltd.), was used for the fluid movement analysis. The steady-state simulation was carried out using a Reynolds average-type turbulence model RANS (Reynolds-averaged Navier-Stokes) simulation.

Result of Simulation

The simulation results of the velocity distribution in the crystallizer are shown in Fig. 18.9b. The particle settling velocity calculated by the Allen's law is shown in Table 18.7.

In the crystallization zone, the vertical gyrate movement of 0.0–0.08 m/s was observed over the upper paddle and streamline directions were a mix of upward and downward. The synthesis velocity at the bottom of crystallization zone was more

Table 18.7 Particle settling velocity calculated by Allen law

Particle diameter	(μm)	100	200	300	400	500	600
Settling velocity	(m/s)	0.0200	0.0399	0.0599	0.0799	0.0999	0.1198

Particle density, 3.16 × kg/m^3 and liquid density, 1.0 × 10^3 kg/m^3

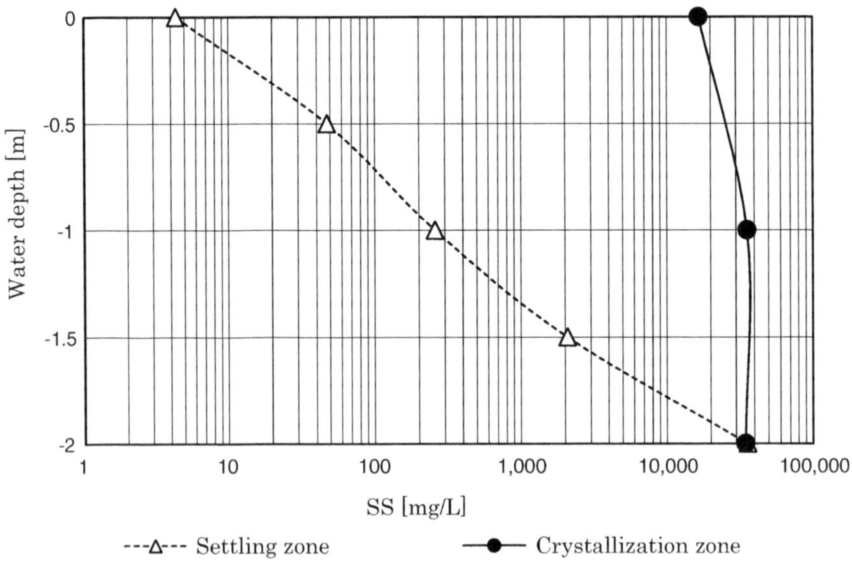

Fig. 18.10 SS distribution of crystallizer at 35 Hz (41.7 rpm)

than 0.1 m/s, and the streamline direction was roughly from the center to the periphery. This means small particles less than 300 μm could be suspended in the upper paddle zone, while the possibility of particle accumulation on the bottom of crystallizer was small. There was some potentiality of accumulation of particles more than 600 μm on the bottom of the crystallizer. In the settling zone, the streamline direction was a mix of upward and downward. The upward stream of 0.01 m/s or higher was observed near the surface. This means small potentiality of washout of particles less than 100 μm.

Figure 18.10 shows the distribution of suspended solid (SS) in the crystallization zone and settling zone (Ministry of the Environment 1973). The concentration gradient of SS was clear in the settling zone. The SS concentration was only 4.4 mg/L at the surface layer of the settling zone. On the other hand, the vertical change in the concentration of SS in the crystallization zone was relatively small. These results suggested that a gravitational particle separation was possible while avoiding the influence of stirred mixture in the crystallizer.

References

Funaishi K et al (2010) About collection and utilization of the phosphorus by the apatite method in night soil treatment plant. J Jpn Waste Manag Assoc 63(295)

Japan Waste Research Foundation (2010) The study on phosphorus recovery from sludge of night soil/Johkasou and phosphorus recycling (Summary), Japanese Fiscal Year 2010, Japan

Japan Waste Research Foundation (2012) The study on phosphorus recovery from sludge of night soil/Johkasou and phosphorus recycling (Summary), Japanese Fiscal Year 2012, Japan

Japanese Waste Treatment (2014) Ministry of the Environment Government of Japan

Ministry of the Environment (1973) Assey methods of metals contained in the industrial waste

Phosphorus Atlas Research Institute (2015) Fiscal year 2015 research report. Waseda University, Tokyo

Tsurumaki M et al (2008) Research on fundamental conditions for building the recycling system of phosphorus resources. Environ Syst Res 36, Japan

Chapter 19
The Stuttgart Process (Germany)

Carsten Meyer, Volker Preyl, Heidrun Steinmetz, Werner Maier, Ralph-Edgar Mohn, and Harald Schönberger

Abstract The Stuttgart Process for nutrient recovery aims to produce struvite as fertilizer from digested sewage sludge from wastewater treatment plants (WWTP) with chemical phosphorus removal. This chapter deals with the detailed description of the experiences with a pilot-scale test plant and its process operation, the latest process optimizations, as well as operational performance data, i.e., phosphorus recovery rates, recyclate product quality, required operational supplements, and costs. The results show that depending on the chemicals used for phosphorus elimination and on the process boundary conditions (especially pH value for dissolving phosphorus from the sewage sludge), different amounts of phosphorus can be recovered. With acidic leaching at pH of approximately 3, it is possible to gain recovery rates of more than 65% as struvite with high purity and very low contents of heavy metals and recalcitrant organic compounds. Additional operating costs for the Stuttgarter process would increase wastewater feed of about 0.15 €/m^3.

Keywords Phosphorus recovery · Nutrient recovery · Digested sludge · Struvite · Magnesium ammonium phosphate (MAP) · Recyclate quality · Costs

---

C. Meyer · V. Preyl · H. Schönberger
Institute for Sanitary Engineering, Water Quality and Solid Waste Management (ISWA), University of Stuttgart, Stuttgart, Germany
e-mail: carsten.meyer@iswa.uni-stuttgart.de

H. Steinmetz (✉)
Institute for Resource Efficient Wastewater Technology, University of Kaiserslautern, Kaiserslautern, Germany
e-mail: heidrun.steinmetz@bauing.uni-kl.de

W. Maier
iat – Ingenieurberatung GmbH, Stuttgart, Germany
e-mail: werner.maier@iat-stuttgart.de

R.-E. Mohn
Abwasserzweckverband Raum Offenburg, Offenburg, Germany
e-mail: Ralph-Edgar.Mohn@AZV-Offenburg.de

H. Ohtake, S. Tsuneda (eds.), *Phosphorus Recovery and Recycling*,
https://doi.org/10.1007/978-981-10-8031-9_19

19.1 Introduction

The amendment of the sewage sludge ordinance passed the German parliament in June 2017, stating that within 12–15 years the agricultural use of sewage sludge from wastewater treatment plants (WWTPs) larger than 100,000 p.e. (population equivalent) and 50,000 p.e., respectively, will be forbidden. In order to not completely disrupt the natural phosphorus cycle, it will become mandatory that phosphorus has to be recovered from sewage sludge or from sewage sludge ash (after mono-incineration) if the phosphorus content of the sewage sludge exceeds 20 g/kg DM (dry matter).

The most common way of P removal in Germany is chemical phosphate precipitation with iron salts and/or aluminum salts. Thus, the largest proportion of phosphorus in the sewage sludge is chemically bound as iron or aluminum phosphates which are known not to have good fertilizer quality. As only about 25% of the sewage sludge is mono-incinerated in Germany, phosphorus recovery from sludge is an important option. Against the background of the majority of WWTPs using chemical P elimination in Germany, the Stuttgart Process was developed. The basic idea is that it can be easily integrated into the existent treatment process without any need to change the type of P removal.

Thus, the Stuttgart Process aims to recover phosphorus from digested sewage sludge produced on WWTPs with chemical phosphate precipitation in the form of struvite (magnesium ammonium phosphate, MAP). The development of the Stuttgart Process by the University of Stuttgart and the engineering company iat-Ingenieurberatung GmbH started at lab scale at the beginning of the 2000s, followed by various pilot-scale tests of 1 m^3-size reactors which demonstrated the technical feasibility of the process (Weidelener 2009). Since 2011, the Stuttgart Process has been studied in a long-term, large-scale test plant at the WWTP Offenburg. Since then, the nutrient recovery technology and the test plant have undergone various process optimizations to achieve a high-quality struvite fertilizer product.

Besides the pilot plant at WWTP Offenburg, a mobile and redesigned (from batch mode to semicontinuous mode) version of the Stuttgart Process has been tested at various WWTP, taking differing sludge properties into account.

19.2 Process Technology

Sewage sludge from WWTPs with chemical phosphorus removal contains high amounts of phosphorus in the form of iron and/or aluminum phosphates which are not very useful as fertilizer. After anaerobic treatment of such sludge, the digestate contains, in contrast to digested sludge from plants with enhanced biological phosphorus removal, just a small fraction of dissolved phosphate (HPO_4^{2-}, $H_2PO_4^{-}$). In order to gain a valuable product comparable to mineral fertilizers like triple

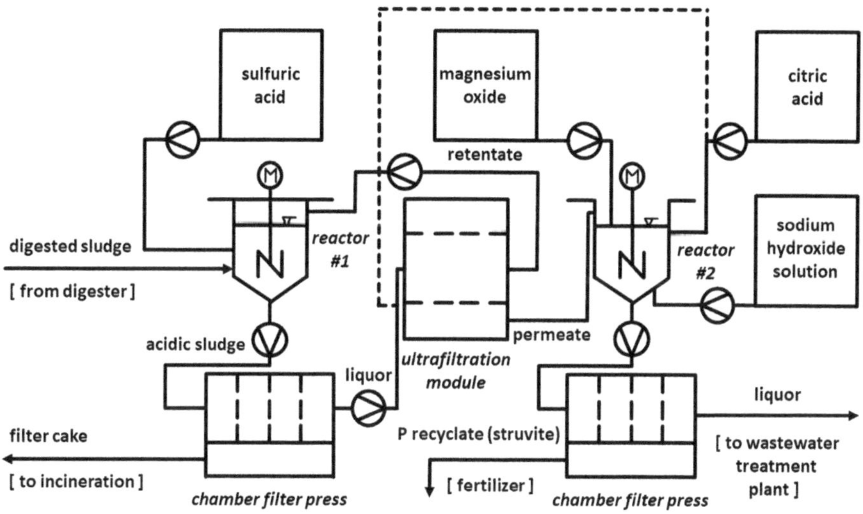

Fig. 19.1 Schematic of the Stuttgart Process

superphosphate, the solid-phase metal phosphates has to be dissolved first, so that the dissolved phosphate concentration increases. Afterwards, the dissolved phosphate is precipitated into a valuable P-rich compound. Within the Stuttgart Process, these two steps consist basically of:

1. Acidic leaching of solid-phase metal phosphates from anaerobically stabilized sewage sludge in a first reactor.
2. Precipitation of struvite (MAP) in a second reactor.

Figure 19.1 shows a simplified schematic of the Stuttgart Process. In reactor #1, solid-phase metal phosphates are dissolved through acidic leaching by sulfuric acid (H_2SO_4). A part of the phosphate which is incorporated into cell biomass is also being dissolved, though presumably not in large quantities (Weidelener 2009). The acidic suspension is separated into solids and liquid by a chamber filter press or a centrifuge. The filter cake is phosphate depleted, while the filtrate is phosphate rich (Meyer et al. 2015).

After this solid/liquid separation step, the filtrate still contains small particles which influence the recyclate quality negatively. Thus, a further step of particle elimination has been integrated in order to purify the filtrate. Therefore, *either* a settling tank *or* an ultrafiltration unit is used. Ultrafiltration leads to a significantly higher-quality product compared to sedimentation or no second separation step at all. As the acidic retentate still contains approximately 11% of the phosphorus, it can be refed to reactor #1 so that no phosphorus will be "lost" and less sulfuric acid will be needed for the dissolution process (Meyer et al. 2015).

In reactor #2 (precipitation reactor), struvite precipitation, crystallization, and agglomeration take place. To avoid metal and/or heavy metal transfer into the struvite product (recyclate), citric acid ($C_6H_8O_7$) as a metal chelating agent is added prior to precipitation. After adding magnesium oxide (MgO) as Mg source for precipitation, the pH is raised to 8.0 using sodium hydroxide (NaOH). Under these conditions, precipitation of struvite (MAP) takes place. The precipitated struvite can be harvested using either sieves or a chamber filter press. The liquid phase (liquor) is fed back to the WWTP's activated sludge tank (Meyer et al. 2015). It must be ensured that both the leaching reactor and the precipitation reactor are well-mixed.

19.3 Large-Scale Nutrient Recovery Pilot Plant

Weidelener (2009) demonstrated within semi-technical scale experiments (1 m^3) that up to 60% of the phosphorus contained in digested sludge could be recovered via the Stuttgart Process, but the amount is strongly dependent on the process conditions and characteristics of the sludge. Nevertheless, questions concerning upscaling had been left open from his experiments. Therefore, in 2011 a large-scale pilot plant for nutrient recovery with the Stuttgart Process has been constructed at the municipal wastewater treatment plant (160,000 p.e.) in the City of Offenburg, Germany, treating a partial stream of the WWTP's digested sludge, equivalent to approximately 5,000 p.e. The particular processes, i.e., acidic leaching, metal complexation, struvite precipitation, and sludge/recyclate dewatering, have been continuously optimized. Figure 19.2 shows the two reactors, for phosphate leaching and for struvite precipitation. In front, the sodium hydroxide storage tank can be seen.

Fig. 19.2 Parts of the pilot plant at WWTP Offenburg, Germany: reactor #1 (acidic leaching), reactor #2 (metal complexation and struvite precipitation), sodium hydroxide storage tank; back left, settling tank; back right, citric acid tank

19.3.1 Operation of the Pilot Plant

The nutrient recovery pilot plant at WWTP Offenburg is operated in batch mode. The various process steps and their characteristics are described in following subsections.

19.3.1.1 Acidic Leaching of Phosphate

Phosphorus in digested sludge from municipal WWTPs using chemical phosphorus elimination exists in three different forms: (1) bound as metal phosphates, (2) organic-bound phosphate, and (3) dissolved phosphate. Acidification causes primarily the conversion of metal phosphates to (dissolved) hydrogen phosphates which are then available for the subsequent struvite precipitation.

Previous lab-scale experiments with sewage sludge from the WWTP Offenburg show the relevance of the pH value for the dissolution process. The results indicate that iron phosphate starts to dissolve already at pH of approximately 5, while aluminum phosphate seems to be stable down to pH 4.0 or below. Calcium and magnesium phosphates start to dissolve much earlier at pH of approximately 6. In all cases, the phosphate concentration in the solution increases with decreasing pH values. These lab experiments indicate clearly that the composition of the solid-phase phosphate components is decisive for the dissolution process. Furthermore, the pH value influences the degree of dissolution significantly.

Semi-technical scale experiments confirm these findings. Table 19.1 demonstrates the results of dissolving rates reached during operation of the pilot-scale plant. The arithmetic average ($n = 38$) of the total P concentration of the Offenburg digested sludge reached 1,127 mg/L. The dry matter content was 2.6–3.2% (w/w). The amount of dissolved phosphate in the sludge after acidic leaching underlies large variations even at similar pH values (Table 19.1). This suggests that the composition of the digested sludge is quite variable under real conditions.

It can be clearly seen that the type of metal salts (iron and/or aluminum salts) used for chemical phosphorus removal has a huge impact on the dissolving rate. Secondly, the pH is decisive for the leaching. But even for similar leaching pH values, significant differences in the phosphorus dissolving rate could be observed. The varying amount of present calcium and magnesium phosphates which dissolve

Table 19.1 Phosphorus dissolving rates after acidic leaching of digested sewage sludge from WWTP Offenburg

	Phosphorus dissolving rate	
Leaching pH	Iron salts used for chemical P removal	Aluminum salts used for chemical P removal
pH > 4.5	36–55%	28–49%
3.5 ≤ pH ≤ 4.5	43–70%	24–47%
pH < 3.5	~80%	~50%

much earlier than iron and aluminum phosphates might play a role for the wide range of phosphorus dissolving rates.

The P dissolving rate of the Stuttgart Process is defined by the following equation, indicating how much phosphorus of the raw digested sludge is released by acidic leaching.

$$\text{P dissolving rate}[\%] = \frac{PO_4\text{–}P\left[\frac{mg}{L}\right](\text{acidified dig. sludge}) - PO_4\text{–}P\left[\frac{mg}{L}\right](\text{raw dig. sludge})}{P_{total}\left[\frac{mg}{L}\right](\text{raw dig. sludge}) - PO_4\text{–}P\left[\frac{mg}{L}\right](\text{raw dig. sludge})} * 100$$

Lower pH allows for higher release of phosphorus and, therefore, a higher recovery rate. By comparing the P dissolving rates of digested sludge from WWTPs with phosphate precipitation by aluminum salts and with phosphate precipitation by iron salts, it is evident that generally higher P dissolving (and recovery) rates are reached when iron salts are used. Most of the iron phosphates species are more easily dissolved by the use of sulfuric acid than aluminum phosphates. Thus, the type of phosphorus elimination, the wastewater characteristics and the digesting process influence the phosphorus recovery process and recovery rate significantly.

It has to be considered that the lower the chosen leaching pH, the higher the amount of required chemicals is, not only sulfuric acid for the leaching process itself but also citric acid for the complexation of more dissolved metals and sodium hydroxide for the adjustment of the precipitation pH. However, for the leaching process a pH between 3 and 5 appears reasonable, both in terms of phosphorus yield and operating resources.

19.3.1.2 Dewatering of Acidified Sludge

For dewatering the acidified sludge and obtaining the phosphate-rich fluid phase (filtrate), a chamber filter press and an inorganic flocculant (polymer) are used at the WWTP Offenburg. In general, the effects of polymers on sludge dewatering are strongly dependent on the individual boundary conditions. In the case of acidification, effects of the low pH might influence the flocculation process. Thus, pre-experiments are recommended to find out which flocculant is most suitable for the given situation.

19.3.1.3 Struvite Precipitation

In the last process step, the dissolved phosphate reacts with cations and is thus converted into a solid recyclate (product). The goal is to make specifically ammonium and magnesium ions react with phosphate ions in order to produce nearly pure struvite. However, depending on the process conditions of the prior treatment steps, competing reactions may lead to co-precipitation of non-struvite phosphate compounds. This accounts especially for iron or aluminum phosphate, if no proper measures are taken. Thus, the whole struvite precipitation process needs the following sub-steps: (1) complexation, (2) addition of magnesium, and (3) pH increase.

Complexation

During the acidic leaching process, metal phosphates and other metal compounds dissociate. Hence, the phosphate-rich filtrate also contains metal ions, mostly Fe, Al, Ca, and Mg, and, depending on the leaching pH, even heavy metal ions such as Cu, Zn, Pb, Ni, Cd, and Cr. In order to obtain a high-quality struvite product (magnesium ammonium phosphate), the (co)precipitation of these metals must be avoided. Therefore, citric acid is added which forms a stable complex with the metals and stays in solution (Meyer et al. 2018).

A complete complexation is indicated by an unchanged concentration of metals in the filtrate after the precipitation, except for magnesium as a matter of course. The effect of complexation depends on the stoichiometric ratio of citric acid ($C_6H_8O_7$) and the sum of the predominant cations Fe, Al, Ca, and Mg. A satisfactory complexation has occurred if metal concentrations in the filtrate measured before and after pH increase show similar values. More details concerning the effect of complexation are described elsewhere (Meyer et al. 2018).

Addition of Magnesium

Struvite consists of magnesium, ammonium, and phosphate with the theoretical stoichiometric ratio of 1:1:1. Based on the available phosphate in the digested sludge filtrate, ammonium occurs in an over-stoichiometric ratio, while not sufficient magnesium is available for struvite formation, so that it has to be added externally. In practice, the entire amount of magnesium dosed will not react with phosphate and ammonium to form struvite, particularly when magnesium oxide (MgO) in powder form is used, as it is the case in the Offenburg plant. In our case, the optimal ratio between magnesium and phosphate in the filtrate turned out to be 1.5:1. Under these conditions phosphate precipitation rates of 95% are achieved. The addition of MgO has to be clocked, e.g., every 60 s with a dosing pulse of 5 s. However, it must be ensured that prior to the next dosing pulse all MgO has been dissolved. After finishing the dosing process, a reaction time of 30 min should be met for complete dissolution (Meyer et al. 2018).

pH Increase (Precipitation)

In order to trigger the struvite precipitation reaction, the pH of the acidic filtrate has to be raised. Although the dosage of MgO leads to an increase of the pH, this is not sufficient to reach the target pH of approximately 8.0 which is needed for struvite precipitation. The use of sodium hydroxide (NaOH) leads to satisfactory rates of magnesium precipitation (mainly as struvite) of approximately 95%. At a pH of 8.5, it is possible to achieve higher precipitation rates of approximately 99%, but due to the increased ammonium/ammonia equilibrium shift, more ammonia gas is produced. This should be avoided with respect to work safety and environmental protection (Meyer et al. 2018).

Due to high struvite precipitation rates, generally, very little phosphorus is found in the reject water stream (refeed to WWTP). Eighty to ninety percent of the filtrate's or permeate's NH_4-N is refed to the WWTP. However, compared to a situation without nutrient recovery, the reject water contains between 10% and 20% less NH_4-N, taking the additional release of NH_4-N, due to the acidic treatment, into account (Meyer et al. 2015).

The struvite crystals of the Offenburg nutrient recovery plant show the typical "coffin lid" shape and are of the size of approximately 100 μm (Fig. 19.3).

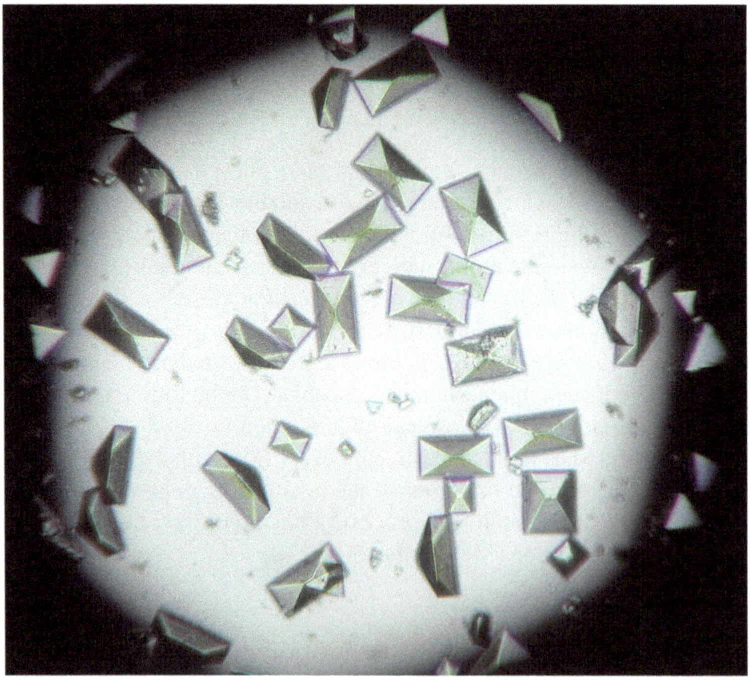

Fig. 19.3 Microscopic image of harvested struvite crystals (size approximately 100 μm) of the nutrient recovery plant in Offenburg

19.3.1.4 Recyclate/Struvite Harvesting

By using a chamber filter press, an efficient recyclate product recovery is possible without losing small struvite crystals. The filter cake produced (without using additives) is compact and easily released from the press (Meyer et al. 2018).

19.3.2 Performance Data

19.3.2.1 Phosphorus Recovery Rates and Recyclate Yields

Meyer et al. (2015) described that the phosphorus (P) recovery rate can be calculated by the following equation which indicates how much phosphorus is transferred from the raw digested sludge into the struvite product. Therefore, the PO_4-P concentration of the filtrate or permeate (UF) before and after precipitation as well as the total P concentration of the raw digested sludge have to be known. This approach has the advantage of avoiding the direct and more difficult measurement of the phosphorus content of the solid phase, i.e., the struvite recyclate.

$$\mathrm{P\,recovery\,rate}\left[\%\right]=\frac{\left(\mathrm{PO_4\!-\!P}\left[\frac{\mathrm{mg}}{\mathrm{L}}\right]\left(\begin{matrix}\mathrm{before}\\ \mathrm{precip.}\end{matrix}\right)-\mathrm{PO_4\!-\!P}\left[\frac{\mathrm{mg}}{\mathrm{L}}\right]\left(\begin{matrix}\mathrm{after}\\ \mathrm{precip.}\end{matrix}\right)\right)\times V\left[L\right]\left(\begin{matrix}\mathrm{supernatant}\\ \mathrm{or\,permeate}\end{matrix}\right)}{\mathrm{P_{total}}\left[\frac{\mathrm{mg}}{\mathrm{L}}\right]\left(\mathrm{raw\,dig.sludge}\right)\times V\left[L\right]\left(\mathrm{raw\,dig.sludge}\right)}*100$$

The system boundary for the calculation of the P recovery rate is the nutrient recovery process (Fig. 19.1). In the event that it is necessary to compare the P recovery rate to the phosphorus influent load of the WWTP, the aforementioned recovery rate has to be multiplied by the phosphorus elimination efficiency of the WWTP which is, e.g., in Germany, usually 90–95% (Meyer et al. 2018).

The P recovery rate depends strongly on the specific conditions like precipitants (aluminum or iron salts) used for phosphorus removal or the leaching pH. Table 19.2 demonstrates the large range of the P recovery rate over the long-term test of the Stuttgart Process at WWTP Offenburg due to the mentioned changing boundary conditions and comprising all technical enhancements during the research phase. Generally, the P recovery rate depends strongly on the P dissolving rate (see Sect. 19.3.1) which indicates that the pH increase and precipitation steps of the process are not the limiting factors of the Offenburg nutrient recovery plant. In the course of

Table 19.2 Phosphorus recovery rates (Meyer et al. 2018)

Leaching pH	P recovery rate
pH > 4.5	25–46%
3.5 ≤ pH ≤ 4.5	21–58%
pH < 3.5	33–67%

Table 19.3 Specific phosphorus yields (Meyer et al. 2018)

	Specific product yield		
Leaching pH	kg P[a]/m^3 DS	kg P_2O_5[b]/m^3 DS	kg $MgNH_4PO_4(6{\bullet}H_2O)$[c]/m^3 DS
pH > 4.5	0.3–0.5	0.7–1.1	2.2–3.6
3.5 ≤ pH ≤ 4.5	0.3–0.6	0.7–1.4	2.2–4.3
pH < 3.5	0.3–0.5	0.7–1.1	2.2–3.6

DS digested sludge
[a]Phosphorus calculated as P
[b]Phosphorus pentoxide
[c]Magnesium ammonium phosphate hexahydrate (struvite)

Table 19.4 Specific consumption of chemicals for the nutrient recovery process (Meyer et al. 2018)

Chemical	Purity	Specific amount required
Sulfuric acid	78% (w/v)	1.7–5.7 L/m^3 DS
Citric acid	50% (w/v)	3.9–11.5 L/m^3 DS
Magnesium oxide	95% (w/v)	0.8–1.7 kg/m^3 DS
Sodium hydroxide	20% (w/v)[a]	5.0–25.1 L/m^3 DS (2.1–10.3 L/m^3 DS)[b]
Flocculant (stock solution)	0.4–0.6% (w/w)	3.0–5.7 L/m^3 (DS)

DS = digested sludge
[a]Solidification point at approx. −25 °C, note that NaOH of higher/lower concentrations may have a significantly higher solidification temperature
[b]When ultrafiltration is applied (see Sect. 19.2)

increasing operational experience and technical enhancement, higher P recovery rates were obtained.

Table 19.3 indicates that the achievable specific phosphorus yield (e.g., kg P per m^3 digested sludge (DS)) depends mainly on the P dissolving rate, respectively, on the P leaching.

19.3.2.2 Consumption of Chemicals

As described above, the recovery rate strongly depends on the process conditions (leaching pH, stoichiometric value of magnesium ions for complexation) which are controlled by the addition of chemicals. Table 19.4 shows the range of required chemicals for differing leaching conditions and treatment options.

Besides the pH value, the integration of an ultrafiltration step (see above) is relevant for the required quantity of sodium hydroxide. It is much less in the case of ultrafiltration treatment, while the amount of sulfuric acid, citric acid, and magnesium oxide is nearly not influenced by the ultrafiltration. This can be explained by the reduction of the buffer capacity through ultrafiltration.

Table 19.5 Nutrient contents of the recyclate (struvite)

Medium	Total organic content	Phosphorus as P_2O_5	Total nitrogen
Filtrate	6.3% (w/w)	19.8% (w/w)	1.9% (w/w)
Permeate (UF)	0.40–0.55% (w/w)	25.3–27.6% (w/w)	5.5% (w/w)

19.3.3 Recyclate Quality

19.3.3.1 Nutrients

Table 19.5 indicates the positive influence of the ultrafiltration on the product quality. With ultrafiltration, the composition of the recyclate comes closer to that of pure struvite which contains 28.9% phosphorus as P_2O_5 and 5.7% nitrogen, while the recyclate from the experiments without ultrafiltration contains significant amounts of organic matter and lower amounts of phosphorus and nitrogen.

19.3.3.2 Metals

The transfer of metals from the digested sludge into the filtrate/permeate and into the struvite product depends on the metal type and the process setup. While Al and Fe are redissolved to a high extent during acidification, the complexation step is decisive for avoiding a later precipitation of iron or aluminum phosphates. Concentrations of critical heavy metals like Cd, Cr, Ni, and Pb are generally low, but differences could be found between the approaches with and without ultrafiltration. Depending on the matrix used for struvite precipitation – permeate or filtrate – only between 0.04% and 3.5% (w/w) of these metals contained in the digested sludge are found in the struvite products. A large percentage of the metals are kept in the filter cake and are safely removed from the process.

The concentrations of the heavy metals Cd, Ni, and Pb in the struvite products fall well below the limits of the German Fertilizer Directive; this was also proven for Cr(VI), thallium, and arsenic. The results correspond to the findings of Uysal et al. (2010) for As, Cr, Ni, Fe, Zn, and Hg. As heavy metals are released at pH values smaller than 4 (Cohen 2009; Stark et al. 2006), the chosen pH in the acidification step is crucial for keeping the heavy metals in the sludge.

19.3.3.3 Recalcitrant Organic Compounds (ROCs)

The following ROCs were analyzed: antiepileptic drug (carbamazepine), synthetic musks (galaxolide (HHCB), tonalide (AHTN), and OTNE (patchouli ethanone)), flame retardants (TCEP, tris-(chloroethyl)-phosphate; TCPP, tris-(chloropropyl)-phosphate), and nonylphenol (used for tenside production, on the priority list of EU water framework directive). Generally, the transfer of the analyzed ROCs from the digested sludge into the struvite is extremely low.

Table 19.6 Specific yearly costs of the Stuttgart Process (Meyer et al. 2018)

OPEX (power, chemicals)	20 EUR/kg P
P load (1.8 g P/cap/day)	0.657 kg P/year
P load in digested sludge (90% P elimination)	0.9 × 0.657 = 0.590 kg P/year
Amount of recovered P (~50% recovery rate)	0.30 kg P/cap/year
Costs for recovered P	0.30 kg P/cap/year × 20 EUR/kg P = 6 EUR/cap/year
Increase of wastewater treatment fees for 40 m^3/cap/year	(6 EUR/cap/year)/(40 m^3/cap/year) = 0.15 EUR/m^3

Depending on the filtrate purification method (ultrafiltration or filtration by bench filter press), only between 0.0007% and 0.1919% (w/w) (0.0011–6.28 μg/kg) of the ROCs contained in the digested sludge were found in the struvite products. Of the analyzed ROCs, nonylphenol showed the highest transfer potential. Depending on the polarity, i.e., the sorption tendency, of the substance, between 52% (w/w) (carbamazepine) and 0.10% (w/w) (OTNE) were transferred into the reject water stream. Similar to the findings for metals, most of the ROCs are adsorbed to the filter cake and thus are discharged from the process. More details have been described elsewhere (Meyer et al. 2018). Uysal et al. (2010) found the same for PCBs from anaerobic digester sludge not being transferred into struvite.

19.3.4 Cost Analysis

The specific operational expenditures (OPEX) for the current operation of the pilot plant account for approximately 20 EUR per kilogram phosphorus (elemental) recovered. OPEX include chemicals and power.

It is to be assumed that the specific OPEX will be less for new and larger plants due to an optimized and more effective design and lower delivery costs for the required chemicals. Nevertheless, a simple calculation (Table 19.6) of the specific yearly costs demonstrates that the technology is affordable in general (Meyer et al. 2018). In the near future, also a detailed life-cycle assessment of the Stuttgart Process will be carried out.

References

Cohen Y (2009) Phosphorus dissolution from ash of incinerated sewage sludge and animal carcasses using sulphuric acid. Environ Technol 30(11):1215–1226

Meyer C, Preyl V, Steinmetz H (2015) High quality MAP production from digested sewage sludge. Oral presentation and proceedings of the IWA specialist conference on nutrient removal and recovery: moving innovation into practice, Gdańsk, Poland, 18.-21.05.2015

Meyer C, Preyl V, Steinmetz H, Maier W, Mohn R-E, Schönberger H, Piersson T (2018) The Stuttgart Process. In: Schaum Chr (ed) Phosphorus: polluter and resource of the future:

removal and recovery from wastewater. IWA Publishing. ISBN13: 9781780408354, eISBN: 9781780408361, in process (estimated publication date: 15/02/2018)
P-REX (www.p-rex.eu) (2015) Stuttgart process sludge leaching: technical fact sheet (pdf) n.p.; http://p-rex.eu/uploads/media/PREX_Factsheet_STUTTGART.pdf; 15 Feb. 2015. Web. 07 Feb. 2017
Stark K, Plaza E, Hultman B (2006) Phosphorus release from ash, dried sludge and sludge residue from supercritical water oxidation by acid or base. Chemosphere 62(5):827–832
Uysal A, Yilmazel YD, Demirer GN (2010) The determination of fertilizer quality of the formed struvite from effluent of a sewage sludge anaerobic digester. J Hazard Mater 181:248–254
Weidelener A (2009) Phosphorrückgewinnung aus kommunalem Klärschlamm als Magnesium-Ammonium-Phosphat (MAP). Fakultät Bau- und Umweltingenieurwissenschaft der Universität Stuttgart. Bericht: Prof. Dr.-Ing. Heidrun Steinmetz, Prof. Dr.-Ing. Jörg Londong, Stuttgarter Berichte zur Siedlungswasserwirtschaft. Band 202

Chapter 20
Phosphorus Recovery from Wet Sewage Sludge Using CO_2

R. Schncc and E. Opitz

Abstract Germany introduced a legislation to require the recovery of phosphorus from municipal wastewater in 2017. The ExtraPhos® process from chemical company Budenheim is emerging as one of the country's promising technology options in this area. Budenheim began developing the phosphorus recovery process "ExtraPhos®" in 2010, forming part of a current wave of investigations in the country. Now the company's process is emerging as one of the promising options for helping meet the recovery target.

Keywords ExtraPhos · Sewage sludge · Carbon dioxide

20.1 Introduction

As essential component of all living beings and plants, phosphorus cannot be replaced by any other substances. Phosphorus is a key limiting factor for plant growth. Over 80% of phosphate rock mined worldwide ends up in fertilizers. Phosphate supply and price are thus strongly linked to food security.

Phosphate rock is a nonrenewable, fossil resource. Estimates of how long these resources may last are variable. They are concentrated in a few countries/regions, and Europe is almost entirely dependent on imports posing geopolitical supply security risks and vulnerability to price fluctuations. Alternative sources of phosphorus therefore need to be found. Recovery from sewage sludge is a promising approach as it contains nutrients, like phosphates, which are valuable for agriculture.

Now one might assume that applying the treated sewage sludge (e.g., after digestion or composting) directly onto the fields would be the easiest way of fertilizing and so reusing phosphates. However, this method is increasingly rejected by authorities, farmers, or the food industry because of concerns about contaminants, as well as logistic obstacles.

R. Schnee (✉) · E. Opitz
Chemische Fabrik Budenheim KG, Innovation P-Recovery, Budenheim, Germany

H. Ohtake, S. Tsuneda (eds.), *Phosphorus Recovery and Recycling*,
https://doi.org/10.1007/978-981-10-8031-9_20

Sewage sludge is the "pollutant sink" of the sewage plant. Many pollutants and contaminants are extracted from the sewage water and concentrated in the sludge, meaning that it contains not only valuable components (organic carbon and nutrients) but also pollutants such as heavy metals, pharmaceutical residues, persistent chemicals, or pathogens. In some countries in Europe and under certain conditions, sewage sludge can also directly be used as a fertilizer on fields. In Germany, at present, about 30% of the sewage sludge generated is still used in agriculture (Wiechmann et al. 2013). Stricter limits will however lead to a significant decrease of this quantity (Fertilizer Ordinance 2015).

20.2 Description of the Budenheim ExtraPhos® Process

For the extraction of the phosphates, only carbon dioxide is used, which is run through the process in a cycle. The process can be roughly subdivided into three process stages. These are carbon dioxide extraction, solid/liquid separation, and phosphate precipitation.

For the extraction, the sewage sludge/water suspension is aerated with carbon dioxide under pressure of approx. 10 bars. With this treatment, carbon dioxide becomes carbonic acid in the sewage liquor, the pH decreases to a value of between 4.5 and 5.5, and a part of the phosphates bound to the sewage sludge matrix are dissolved.

In the following solid/liquid separation, the sewage sludge particles are separated from the liquid phase, using state-of-the-art sewage sludge dewatering coagulants. The remaining, dewatered sewage sludge can be used for incineration or further recycling. The carbon dioxide used to lower the pH value comes back out of the liquor when pressure is released (off-gassing), captured, condensed, and recycled back for reuse in the process.

The sludge liquor is introduced into the third process stage, phosphate precipitation, in which the dissolved phosphates are precipitated as dicalcium phosphate (DCP). In order to accelerate the precipitation process, a small quantity of lime (calcium hydroxide) is added, so pH increases, whereas most of the calcium required is already present in the sewage sludge liquor. After separation, drying, and granulation, the dicalcium phosphate is used as fertilizer.

20.3 Realization and Implementation in Sewage Plants

The Budenheim ExtraPhos® Process[1] can be easily implemented and integrated into sewage plants in between fermentation and dewatering according to Fig. 20.1. The Budenheim ExtraPhos® process is suitable for all sewage plants, with or without

[1] Dr. Rainer Schnee is the joint project leader for the work at Budenheim with colleague Eva Opitz.

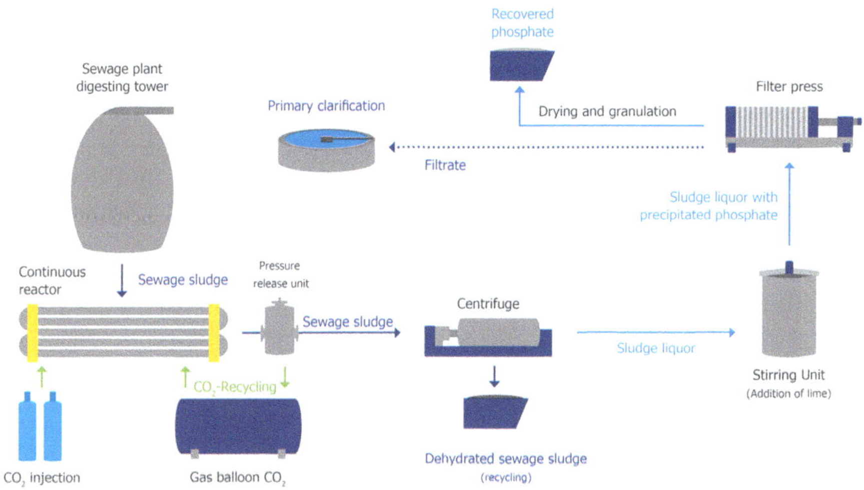

Fig. 20.1 Process implementation

anaerobic sludge digestion. Its successful implementation is also regardless of the type of phosphate removal; both chemical and biological approaches can be pursued, because part of the phosphorus precipitated with iron or aluminum chemicals can be remobilized by the carbon dioxide treatment. This means complex and expensive process changes within the sewage plant are unnecessary.

20.4 Properties of the Recovered Fertilizer Product

The product generated in the Budenheim ExtraPhos® process is dicalcium phosphate (DCP), initially as an aqueous suspension. The solids are separated by filtration and a relatively viscous phosphate suspension remains. The final process stage is drying to phosphate powder. Depending on its intended purpose, this powder can be further processed. For direct use in agriculture, for example, it would generally be granulated in order to enable application using farmers' existing equipment. In plant availability trials at the University Bonn, the product from the Budenheim ExtraPhos® process has been rated comparable to conventional phosphate fertilizers. Currently, the use of the recovered phosphates is only planned for agriculture.

20.5 Feasibility Study and Eco-Efficiency

As the Budenheim ExtraPhos® process operates in the wet sewage sludge phase, a fine or granulated phosphate fertilizer can be produced directly at the sewage plant. This can be used as a fertilizer in the surrounding region, thus saving transport costs.

The dewatered phosphate-depleted sewage sludge can be used in various valorization routes, e.g., co-incineration in the cement industry. Detailed cost and economic efficiency calculations will be made with the pilot plant in the Mainz-Mombach sewage plant within 2017 and 2018. Initial calculations carried out so far are based on the test plants and suggest that the Budenheim ExtraPhos® process is potentially competitive compared to other recovery processes. In particular, the low consumption of chemicals is important, because the carbon dioxide, used to lower the pH value, can be recovered, compressed, and used again for extraction. Furthermore, the Budenheim ExtraPhos® process runs under ambient temperature. An additional use of thermal energy is not required.

In general, coagulant chemicals already used in the sewage plants can be used to dehydrate the sewage sludge. This has a positive effect on the investment costs for phosphate recycling because no additional storage tanks are needed.

20.6 Conclusions

ExtraPhos® is suitable for all sewage treatment plants, independent if anaerobic sludge digestion is in place or not. More than 90% of wastewater treatment plants in Germany also remove phosphate from the wastewater (Wiechmann et al. 2013), and ExtraPhos® can be used whichever type of phosphate removal is in place. Recovery of phosphate in the form of struvite is currently seen as one of the most viable options, but this is only possible where biological phosphate removal is used. ExtraPhos® can be used also where chemical removal is in place, since the precipitated phosphorus can still be remobilized by the carbon dioxide process. This has the additional benefit that there is almost no need for changes to the overall wastewater treatment process and also the coagulation chemicals needed for ExtraPhos® can be provided by facilities which are already present on the site.

The recycled phosphate, dicalcium phosphate, is already an approved fertilizer. The advantage of dicalcium phosphate is that it is a single fertilizer. For fertilizer companies, it is an easy-to-handle raw material. The dicalcium phosphate provides a bioavailability similar to struvite and regular phosphate fertilizers.

The costs may look promising, but the ExtraPhos® process is not likely to deliver an end product costing less than one based on raw mineral phosphate. However, the new German regulations will contribute to the growth of a market that is intended to signal a shift away from reliance on raw phosphates, in common with wider recovery and recycling efforts generally (Sewage Sludge Ordinance 2017).

The investment in the ExtraPhos® process represents a contribution by Budenheim to the goal of sustainability in the way phosphates are used. At the same time, the company is a phosphate chemical specialist for high-end applications in food, pharma, and industrial applications and is used to delivering end products that meet customer needs.

References

Fertilizer Ordinance (2015) Fertilizer ordinance of 27 May 2015. Federal Law Gazette, p 886 (in German)
Sewage Sludge Ordinance (2017) Sewage sludge ordinance of 27. September 2017. Federal Law Gazette, p 3465 (in German)
Wiechmann B, Dienemann C, Kabbe C, Brandt S, Vogel I, Roskosch A (2013) Sewage sludge management in Germany (in German)

Chapter 21
Effect of Iron on Phosphate Recovery from Sewage Sludge

Leon Korving, Mark Van Loosdrecht, and Philipp Wilfert

Abstract Iron is omnipresent in sewage treatment systems. It can unintentionally be present because of, e.g., groundwater seepage into sewers, or it is intentionally added for odor and corrosion control, phosphate removal, or prevention of hydrogen sulfide emissions. The strong affinity of iron for phosphate has advantages for efficient removal of phosphate from sewage, but it is also often considered a disadvantage for phosphate recovery. For instance, the strong affinity between iron and phosphate may reduce recovery efficiencies via struvite precipitation or for some phosphate recovery methods from ash. On the other hand, iron may also have positive effects on phosphate recovery. Acid consumption was reported to be lower when leaching phosphate from sewage sludge ash with higher iron content. Also, phosphate recovery efficiencies may be higher if a Fe-P compound like vivianite ($Fe_3(PO_4)_2\ 8H_2O$) could be harvested from sewage sludge. Developers of phosphate recovery technologies should be aware of the potential and obstacles the iron and phosphate chemistry bears.

Keywords Vivianite · Iron · Chemical phosphate removal · Struvite recovery · Sewage sludge

L. Korving (✉)
Wetsus, European Centre of Excellence for Sustainable Water Technology, Leeuwarden, The Netherlands
e-mail: Leon.Korving@Wetsus.nl

M. Van Loosdrecht
Department of Biotechnology, Delft University of Technology, Delft, The Netherlands

P. Wilfert
Wetsus, European Centre of Excellence for Sustainable Water Technology, Leeuwarden, The Netherlands

Department of Biotechnology, Delft University of Technology, Delft, The Netherlands

H. Ohtake, S. Tsuneda (eds.), *Phosphorus Recovery and Recycling*,
https://doi.org/10.1007/978-981-10-8031-9_21

21.1 Introduction: The Role of Iron in Sewage Treatment Plants

21.1.1 Today

Currently, phosphate is removed from sewage with the intention to prevent eutrophication of surface waters rather than with the objective to recover phosphate. Surface water quality regulations to protect the environment, and as a consequence stricter sewage discharge limits, have led to the introduction of advanced phosphate removal techniques. Implementation of such advanced (also known as tertiary) treatment may vary per region, country, and continent (Oleszkiewicz et al. 2015). In the European Union, sewage from 286 million population equivalent (p.e.) receives advanced treatment, whereas for 39 million p.e., tertiary treatment should still be implemented (European Commission 2016). Compliance varies strongly per country: some countries comply for 100%, while new entrants to the European Union show lower compliance rates (European Commission 2016). Only few countries start to implement regulations for phosphate recovery from sewage now (Der Schweizerische Bundesrat 2015; Deutscher Bundestag 2017). The most popular advanced phosphate removal technologies are chemical phosphate removal (CPR) using iron or aluminum salts and enhanced biological phosphorus removal (EBPR).

To meet effluent phosphate limits, EBPR is often supported by CPR. For North-West European countries, for which we could find data, CPR dominates but it is often combined with EBPR (Table 21.1). The relative use of CPR versus EBPR may

Table 21.1 Enhanced phosphate removal methods used in some North-West European countries

Country	Type of weighting	No tertiary treatment (%)	Mostly EBPR (%)	EBPR with CPR support (%)	CPR (%)	References
Germany	Population equivalents	2	6	31	61	DWA (2005)
	No. of plants	20	16	21	43	DWA (2005)
Netherlands	Sludge production	4	13	51	32	Korving (2012)
United Kingdom	Population equivalents	0.01	4.7	18.3	77.0	Cooper (2014)
	No. of plants	0.4	1.3	3.8	94.5	Cooper (2014)
Sweden	Water volume	0	0.2	3	97	SCB, 2016 Hansen, personal communication (2017)
France	No. of plants	No data	17	36	47	Paul et al. (2001)

EBPR enhanced biological phosphorus removal, *CPR* chemical phosphate removal

differ from country to country. For instance, Western Canada predominantly applies EBPR, whereas in Eastern Canada, CPR dominates (Oleszkiewicz 2014). In Japan, approximately 25% of the plants use CPR and 4% use EBPR (H. Ohtake, 2017, personal communication).

Iron salts are usually preferred because they are cheaper. For instance, 77% of German (DWA 2005) and 88% of French (Paul et al. 2001) wastewater treatment plants (WWTPs) use iron for CPR. In the United Kingdom, 88% of all WWTPs use iron salts. Here the preference for iron is not only because of lower costs but also because discharge limits are more stringent when aluminum salts are used because of concerns related to the toxicity of aluminum (Cooper 2014; Ofwat 2005). In Japan, aluminum salts are dominantly used for CPR (H. Ohtake, 2017, personal communication).

Dosing higher quantities of iron can reduce effluent phosphate levels to very low levels (Fytianos et al. 1998; Takács et al. 2006). Besides for phosphate removal, iron is often added to sludge digesters to prevent hydrogen sulfide being emitted in the biogas. Iron salts are also used to act as a coagulant to improve the effectiveness of the sludge dewatering process. It has also been suggested to add iron salts to prevent struvite formation in digesters (Mamais et al. 1994).

There are several other sources of iron in sewage and these explain why iron is present even when no iron salts are applied in the WWTPs. According to Hvitved-Jacobsen et al. (2013), iron is normally present in the raw sewage at 0.5–1.5 mg Fe/L, and dissolved inorganic phosphate (P_i) is 3–10 mg P/L, which would result in Fe/P molar ratios between 0.07 and 0.09. However, iron concentrations can be much higher because, for example, iron is added in sewage pumping stations to control odors and corrosion, and this practice may even aid the removal of phosphate in WWTPs (Gutierrez et al. 2010). Also intrusion of iron-rich groundwater can be a reason for elevated iron levels in raw sewage. Between 10% and 50% but also up to 100% of the water that is treated in WWTPs can be extraneous water with groundwater infiltration as main source (Franz 2006). In the Netherlands, iron has been found at concentrations between 1 and 10 mg/L in the influents of 19 WWTPs operated by the Waterschap Vechtstromen. The average Fe/P molar ratio was found to be 0.26 (unpublished data, Fig. 21.1). The iron concentrations in the effluents from those WWTPs were 0.1–0.3 mg/L showing that most influent iron is removed via the surplus sludge.

The different sources of iron used and, most importantly, the different methods to remove phosphate explain the wide ranges of iron concentrations that are encountered in sewage sludge (Table 21.2) and sewage sludge incineration ash (see Table 21.4, Sect. 21.3.4). Overall it seems that iron is omnipresent in sewage/WWTPs and thus that part of the phosphate is bound to iron even in plants where no iron is dosed for phosphate removal. Thus, for all phosphate recovery technologies, iron is an element that has to be dealt with. We will explain and show later how the presence of iron can also limit the recovery efficiency for struvite in EBPR plants.

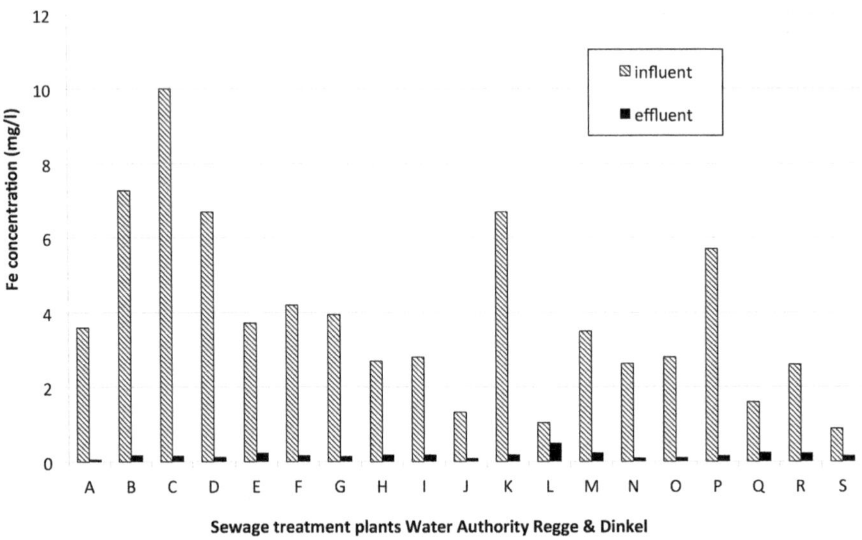

Fig. 21.1 Iron concentrations in the influent and effluent of 19 WWTPs of Waterschap Regge & Dinkel (now: Waterschap Vechtstromen) (Data from 2008 with two to five samples per WWTP)

Table 21.2 Iron concentrations that have been found in sewage sludge (in g/kg on a total solids basis)

Country	Lowest iron concentration	Highest iron concentration	Average iron concentration	Comments and reference
Germany	–	–	50	Average of 202 sludge samples (DWA 2005)
Sweden	4.4	150	49	Based on 47 sludge samples (Eriksson 2001)
Netherlands	–	–	31	Average of 28% of Dutch sewage sludge (Schipper and Korving 2009)
United States	1.6	299	–	Based on 84 biosolid samples (USEPA 2009)
United States	3.8	84	–	Based on 41 biosolid samples (Brandt et al. 2004)
New Zealand	–	–	3–25	Median values of 4 sludge types (raw, aerobic, anaerobic, pond sediment) determined in 66 samples (Ogilvie 1998)
Egypt	7	12	9.3	Based on 5 WWTPs (Lasheen and Ammar 2009)

21.1.2 Tomorrow

The presence of iron is already today very important for sewage treatment, but iron could play an even more important role in future WWTPs that act as energy and phosphate factories (Jetten et al. 1997; Wilfert et al. 2015). Iron is already widely used in energy-producing WWTPs for combined high-rate COD and phosphate removal (Böhnke 1977; Li 2005; Nowak et al. 2011). Implementing anaerobic ammonium oxidation (anammox) in the main and sidestream can further increase the energy yield in such plants. Most of the sewage treatment techniques described above are already being used or tested at the pilot scale. At the moment, the only missing process (as in other conventional treatment processes), without sludge incineration, is the economic feasible recovery of phosphate and iron from iron phosphate (FeP)-containing sewage sludge. In the following section, we will introduce how iron affects current and future phosphate recovery technologies. Along these effects, basic biological, geological, and chemical concepts will be explained that are relevant to understand the role of iron in these processes. A combination of iron and phosphate is often considered to have a negative impact when evaluating the potential for using sludge phosphate recovery (ACHS 2009; Egle et al. 2014; Morse et al. 1998; Römer 2006; Samie and Römer 2001; Schipper and Korving 2009; Schröder et al. 2010).

21.2 Chemical Phosphate Removal Using Iron Salts

21.2.1 Basic Knowledge

The chemistry of iron is very diverse (Faivre 2016; Schwertmann and Cornell 2000) and so are the possible reactions of iron compounds with phosphate during CPR (Smith et al. 2008). This has led to publications, related to the removal of phosphate using iron or on the recovery of phosphate from FePs that contain unspecific expressions such as "insoluble iron phosphate," "metal phosphates," or "iron III phosphate precipitates."

For CPR mainly ferrous, Fe(II), or ferric, Fe(III), iron is used. Ferric iron is usually soluble at low pH values ($pH<2$) only. As the pH increases dissolved ferric iron rapidly forms insoluble iron oxides, oxyhydroxides, and hydroxides (collectively termed iron oxides in the following according to Schwertmann and Cornell (2000)). Dissolved Fe(II) is usually only stable at low redox potential in the absence of oxygen; its solubility is pH dependent. In the absence of oxygen, Fe(II) is more soluble over a broader pH range than Fe(III) (Fig. 21.2).

Iron is usually added to sewage in WWTPs by dosing iron salts or iron coagulants but sometimes also via electrocoagulation (Mishima and Nakajima 2011; WEF 2011). The exact mechanisms through which P_i is initially removed after adding the Fe(II)/Fe(III) are not fully understood. The hydrolysis of Fe(III) in an aqueous

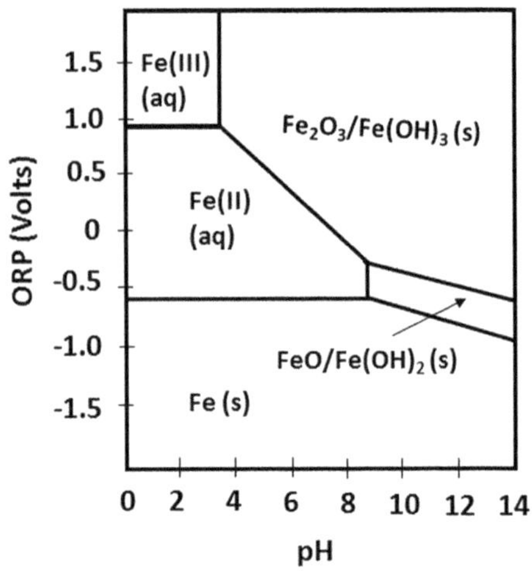

Fig. 21.2 Simplified Pourbaix diagram showing the stable iron species with different pH values under different redox conditions (Modified from Tilley 2005)

solution is usually very rapid (Wendt Von 1973). It has been suggested that the adsorption of P_i onto iron oxides is an important or even the major mechanism involved in the removal of P_i from sewage when Fe(III) salts are used (Lu et al. 2016; Mao et al. 2016; Smith et al. 2008). The situation is even more complex when Fe(II) is added because it can be partly or fully oxidized to Fe(III). Ferrous iron is usually added to aerated stages of the WWTP to allow its oxidation to Fe(III). To the best of our knowledge, it is not known to which degree the Fe(II) is oxidized during this process. The kinetics of Fe(II) oxidation strongly depend on the oxygen concentration and particularly on the pH (Stumm and Morgan 1996). Half of the Fe(II) (total Fe was about 30 mg/L) in bicarbonate-buffered water containing dissolved oxygen at 5 mg/L has been found to be oxidized to Fe(III) within 45 min at pH 7 and within 0.5 min at pH 8 (Ghassemi and Recht 1971; Stumm and Morgan 1996).

The presence of other ions (e.g., sulfate or P_i), catalysts, or dissolved organic matter can considerably influence the oxidation kinetics (Pham et al. 2004; Stumm and Morgan 1996; Theis and Singer 1974; van der Grift et al. 2016). In a full-scale WWTP, about 40% of the Fe(II) that was added was found to be rapidly oxidized to Fe(III) at a relatively high pH of 8.2 and high dissolved oxygen concentration of 4.6 mg/L (Thistleton et al. 2001). Accordingly, measurements on sewage sludge taken from WWTPs in which also Fe(II) was used to remove phosphate suggest that most of the iron in the sludge was Fe(III) (Nielsen 1996; Rasmussen and Nielsen 1996). In contrast, others found that a large fraction of the iron in surplus sewage sludge was present as Fe(II), in the form of the ferrous iron phosphate mineral vivianite, ($Fe_3(II)(PO_4)_2\ 8H_2O$) (Frossard et al. 1997; Wilfert et al. 2016).

Although often perceived differently, Fe(II) can very efficiently remove P_i. In batch tests using secondary effluents with a Fe(II):P molar ratio of 1.5 showed almost complete P_i removal (98%) at pH 8 and initial P_i concentrations of 12 mg P/L (Ghassemi and Recht 1971). Vivianite was formed in these experiments. It seems that the efficiency at which phosphate is initially removed in a WWTP by adding iron is influenced by the oxygen concentration (for ferrous salts), the concentrations of competing ions, the presence of organic matter, the pH, the alkalinity, the mixing intensity, the age of the iron oxide flocs, the type of phosphate present, and whether Fe(III) or Fe(II) salts are used (WEF 2011).

Different FePs as well as iron oxides may form when an iron-based phosphate removal process is applied. Ferrous or ferric phosphate minerals, e.g., vivianite and strengite (Fe(III)(PO_4) $2H_2O$), are examples of solid FePs that could theoretically form during CPR. However, in literature, we could not find any indications for strengite formation during sewage treatment. It is also possible that FePs with impurities form. The impurities can be organic matter inclusions or originate from the coprecipitation of iron or phosphate with other elements (calcium, potassium, magnesium, etc.). The initially formed FeP precipitates may alter during further sewage treatment, because of exposure to different oxidation reduction potentials (ORPs) and, therefore, to different microbial and chemical processes (Nielsen 1996; Nielsen et al. 2005). Thus, the iron in the iron oxides or in the minerals may be oxidized or reduced and iron oxides may age, forming a more crystalline phase over time, which might lead to an increase or decrease in the strength and capacity of phosphate binding (Lu et al. 2016; Mao et al. 2016; Smith et al. 2008).

These examples show that adsorption, mineral formation, and recrystallization may occur at different stages in and on the time scale of a WWTP. FePs are exposed to dramatic changes in ORP but also in changes of temperature over a period of about 1 month in a WWTP with an anaerobic digestion process. Little information is available on the speciation of iron and phosphate in WWTP after iron has been added. These examples illustrate that phosphate can be bound to iron in a range of ways and that the amount of phosphate bound to the iron and the strength of the bonds differ significantly for different types of FeP. To determine the type of FeP as a major phosphate phase in sewage sludge is of interest as (i) it influences the bioavailability of the phosphate when sludge is used in agriculture; (ii) it yields thermodynamic data on FeP to allow modeling, e.g., the struvite content in sewage sludge (up to now often unspecific thermodynamic data of FeP is used Flores-Alsina et al. 2016); and (iii) it affects the efficiency of phosphate release from sewage sludge upon pH changes (Sect. 21.3.3.1). Getting more insight into the types of FePs could, on the one hand, improve phosphate removal but could also lead to ideas on how to separate FePs and recover phosphate from sewage in efficient ways.

21.2.2 Vivianite As an Important Iron Phosphate Compound

Several studies showed that vivianite is present in environmental systems (reviewed by Rothe et al. 2016), but also in wastewater treatment systems, it can be a very important FeP (Azam and Finneran 2014; Frossard et al. 1997; Nriagu and Moore 1984; Singer 1972; Wilfert et al. 2016; Zhang 2012). Vivianite forms in the absence of oxygen in non-sulfidic systems and is very stable over a broad pH ranges (Nriagu 1972). It seems that vivianite is, under these conditions, more stable than most other phosphate minerals (Nriagu and Dell 1974).

Once vivianite is formed, the complete oxidation of the Fe(II) is relatively slow (Miot et al. 2009; Roldan et al. 2002). However, limited information is available on how fast vivianite actually forms and how fast it oxidizes once it is formed in sewage systems. Due to its low solubility, quick formation, and slow degradation (i.e., oxidation), it seems logically that significant amounts of phosphate are bound in vivianite in waste streams where Fe(II), phosphate, oxygen-free conditions, and relatively low concentrations of sulfide occur. Vivianite could form directly upon Fe(II) dosing or indirectly when Fe(III) is chemically/biologically reduced to Fe(II).

Iron reduction and thus vivianite can occur in activated sludge systems before the anaerobic digestion process (Frossard et al. 1997; Nielsen 1996; Wilfert et al. 2016). It is hard to predict whether vivianite dominates or not in activated sludge due to manifold treatment designs and varying redox conditions. However, we believe that during anaerobic digestion, all Fe(III) is transformed to Fe(II) and that vivianite can form. When sufficient iron is dosed, the formation of vivianite during anaerobic digestion is not hampered by the formation of insoluble iron sulfide compounds (FeS_x), induced by sulfate-reducing bacteria, since the supply of sulfate is limited in digesters (Chen et al. 2008; van den Brand et al. 2015). In digesters, optimal conditions for vivianite formation/crystallization occur.

We suggest that further research should focus on quantification and characterization (e.g., size of crystals/particles, growth rates) of vivianite in WWTPs. It has to be evaluated how pure the vivianite in the sludge is, because vivianite is notorious to contain also impurities like heavy metals (Rothe et al. 2016; Taylor et al. 2008). Similar to struvite, also vivianite can cause problems for operators by pipe clogging especially at elevated temperatures (Marx et al. 2001; Reusser 2009). Why not use current approaches to recover phosphate from sewage sludge in the form of struvite as an inspiration to recover phosphate in the form of vivianite from waste streams? It could be separated from waste streams, or its formation could be induced by, e.g., seeding. To selectively separate vivianite from waste streams, technologies based on density or magnetism could be used. This route seems attractive because vivianite could result in higher phosphate recovery efficiencies than routes based on struvite formation due to a lower solubility of vivianite (Sect. 21.3.3). Moreover, the vivianite formation obviously takes place anyway in WWTPs. For phosphate and iron recovery, the separated vivianite can be dissolved in alkaline solutions, the phosphate goes in solution, whereas the Fe(II) remains in the precipitate and forms iron oxides. Other applications are also thinkable, e.g., vivianite has been used as a slow phosphate-release fertilizer (Roldan et al. 2002), as a pigment (Čermáková et al. 2015), for anti-corrosion measures (Da Silva et al. 2004), and it also plays a role during the production of lithium iron phosphate batteries (Chen et al. 2011).

For recovery, processes relying on sludge incineration vivianite are not of relevance because it starts to decompose already at temperatures as low as 100 °C (Čermáková et al. 2015; Poffet 2007) into various iron and iron phosphate phases (Čermáková et al. 2015; Rodgers and Henderson 1986; Waerenborgh and Figueiredo 1986). Indeed, no reports on the presence of vivianite in ash have been found (Sect. 21.3.4).

21.3 Impact of Iron on Selected Phosphate Recovery Technologies

A large number of phosphate recovery technologies have been developed (Desmidt et al. 2015; Egle et al. 2014), but only few technologies have been realized at practical scale. Iron present in sewage sludge often has an impact on phosphate recovery technologies. The impact may depend on the principle of the phosphate recovery technology, and as an example, the effect on three important phosphate recovery technologies is discussed: agricultural use of sludge, recovery as struvite from sewage sludge, and recovery from sewage sludge ash.

21.3.1 Agricultural Use of Sewage Sludge

The use of sewage sludge in agriculture is a low-cost option for recycling phosphate. Sewage sludge usually needs to be hygienized before it is applied to land, and it is then often referred to as biosolids.

In 2005, 41% of all of the sewage sludge produced in the 27 EU countries was used in agriculture, 17% was composted, 19% was incinerated, 17% was landfilled, and 12% was treated in other ways (Kelessidis and Stasinakis 2012). There is currently a trend away from landfilling and toward either incineration or applying it to agricultural land in these countries (Kelessidis and Stasinakis 2012). Some countries, such as the Netherlands and Switzerland, chose to incinerate most of the sewage sludge, but others, including the United Kingdom and France, mainly rely on applying sewage sludge to agricultural land (Kelessidis and Stasinakis 2012). In the United States, 55% of all generated biosolids were applied to soils, of which 74% was used for agricultural purposes. The remaining 45% was landfilled or incinerated (Beecher et al. 2007). In Japan, sewage sludge is not applied on agricultural land (Ohtake and Okano 2015). These contrasts show that agricultural use of sewage sludge is very dependent on local policies.

The presence of iron in sewage sludge lowers the water solubility of phosphate in biosolids (Brandt et al. 2004; Krogstad et al. 2005; Miller and O'Connor 2009; O'Connor et al. 2004). This is considered to be a positive effect in some cases because it may prevent the phosphate being lost in surface runoff (Elliott and O'Connor 2007; Lu et al. 2012). However, it can also be considered a negative effect because it may limit the amount of phosphate that can directly be taken up by the plants (Kidd et al. 2007; Krogstad et al. 2005; Römer 2006; Samie and Römer 2001) although some of these studies used biosolids with high Fe/P ratios. Also, the relation between water solubility and bioavailability is debated, and other studies show that iron-bound phosphate is available to plants (Kahiluoto et al. 2015; Nanzer et al. 2014; Prochnow et al. 2008).

Regional surpluses of phosphate on agricultural lands (Macdonald et al. 2011; Schröder et al. 2011) are probably the biggest problem for practical application of biosolids. For effective use of the phosphate, the biosolids should be applied in balance with the uptake of phosphate by the crops. Manure has similar N:P ratios to sludge and is often in competition with sludge, especially in regions with animal farming. In these areas, costs related to transportation of sludge to other regions with phosphate deficit are an obstacle for sludge application.

21.3.2 *Struvite Precipitation*

21.3.2.1 Direct Crystallization

A large number of struvite precipitation techniques are available, and these techniques are often assumed to be the optimum method for recovering phosphate. Struvite precipitation is possible from sludge or water streams containing high concentrations of soluble phosphate. These are typically found when a WWTP combines EBPR and sludge digestion. Struvite can then be precipitated and recovered via a combination of pH increase and magnesium addition, either directly in the sludge or in the reject water after sewage sludge dewatering.

However, the market share taken by such techniques is limited because they can only be used in combination with EBPR (Table 21.1). It is often argued that CPR should be replaced by EBPR, but CPR is still widely used. The reasons for a dominance of CPR methods may be different in different countries. The investment costs are lower for CPR than for EBPR methods, and CPR processes are easier to control (Kang et al. 2008). The energy requirements are slightly higher for EBPR because it requires more pumping for internal recirculation, but life-cycle assessment studies have shown that the higher energy costs can be compensated for by the smaller amounts of chemicals used (Coats et al. 2011). It is more difficult to dewater sewage sludge produced by EBPR (Shimp et al. 2013), and dewatering requires more flocculants (Korving 2012).

In addition to the restricted application of EBPR, recovery efficiencies for struvite are relatively low, 10–40% of the influent phosphate load (Cornel and Schaum 2009; Egle et al. 2015; Ewert et al. 2014). For example, full-scale installations for struvite crystallization in digested sludge liquor with the AirPrex system show recovery efficiencies of 5–10% of the influent phosphate load (Ewert et al. 2014). Recovery efficiencies are slightly higher for struvite crystallization from the rejection liquid after sludge dewatering using, for example, Pearl, NuReSys, Phospaq, Phosphogreen, or Struvia technologies. For instance, Marchi et al. (2015) report 15% recovery of the total influent phosphate in a full-scale struvite recovery installation at the NuReSys plant in Leuven in Belgium. Delahaye (2017) reports 15% recovery for a WWTP in Denmark using Suez' Phosphogreen technology. Higher recovery efficiencies are possible if phosphate is stripped from the activated sludge before digestion, for instance, via the WASSTRIP process. Using this approach, phosphate recovery could be increased from 10% to 17% at the Durham Advanced WWTP in Tigard, Oregon (Cullen et al. 2013), and by Suez from 15% to

35% at the Åby WWTP in Denmark (Delahaye 2017). At the WWTP Amersfoort, the WASSTRIP approach is combined with thermal hydrolysis of the sludge before the digestion to maximize phosphate recovery via struvite precipitation. This plant intends to recover up to 40% of all the phosphate entering the plant via the influent and external sludges digested at this central digestion facility (Lycke et al. 2017).

Background concentrations of iron and aluminum can partly explain the low recovery efficiency via struvite precipitation. Iron has a higher affinity for phosphate than magnesium and is therefore often used to prevent struvite scaling in WWTPs (Baker et al. 2006; Doyle and Parsons 2002; Mamais et al. 1994; Neethling and Benisch 2004). Similarly, iron present in the sludge that can bind phosphate will make phosphate unavailable for recovery via struvite precipitation. Also aluminum salts have been used to prevent struvite formation (Baker et al. 2006).

Figure 21.1 (Sect. 21.1.1) illustrates the presence of iron in the influent of WWTPs. The majority of these plants rely on EBPR, sometimes assisted with supplemental dosage of iron salts. As discussed (Sect. 21.1.1), the iron in the influent is concentrated in the sewage sludge and is significant in relation to influent phosphate loads. The relevance of background concentrations of iron in the influent is further illustrated by a study on the EBPR plant in Derby where vivianite scaling in an acid phase digester was related to iron concentrations of 12 mg/l in the influent of this plant (Bjorn 2010). Also other WWTPs of this utility showed significant iron concentrations (3.7 ± 0.2 mg/l) in the influent.

21.3.3 Crystallization After Leaching Phosphate from the Sludge

21.3.3.1 Inducing pH Changes

Sewage and sewage sludge is usually at pH 6–8 (Tchobanoglous et al. 2013). At very high (>12) and low pH (<2), bulk phosphate should be solubilized from most FePs, i.e., irrespective if phosphate is adsorbed to iron oxides or if it is present in a mineral form. It seems logical that the amount of base or acid that needs to be invested to release all phosphate from iron also depends on the type of FeP. Usually, in literature, pH equal or above 12 are used for desorption of adsorbed P_i. However, varying levels of P_i desorption are achieved. It seems that the presence of other elements (e.g., calcium, aluminum, magnesium) can affect the net phosphate release from the sewage sludge at certain pH levels due to reprecipitation of the phosphate. In case phosphate is adsorbed to iron oxides, the pH affects P_i adsorption by affecting the charge on the adsorbent as well as the P_i ion (Stumm et al. 1992). As the pH increases beyond the PZC (point of zero charge) of the iron oxide, electrostatic repulsion leads to a decrease in P_i adsorption. Additionally, ligand exchange (by OH) is a mechanism to release adsorbed P_i (Yoon et al. 2014). If vivianite is present, a pH of around 12 should theoretically dissolve most phosphate from vivianite. At this pH, where P_i tends to desorb from adsorbents, iron oxides are formed (Nriagu 1972).

Acidification

Strong acidification will dissolve and release phosphate and Fe(III) from iron oxides and iron phosphate minerals thereby mobilizing most of the solid phosphate in sludge and ash samples. But at low pH, also heavy metals and other metals can go in solution. Several approaches have been used to leach phosphate from the sludge and thus make more of the phosphate available for recovery. The Seaborne and Stuttgart processes are examples of a strategy where the sludge is acidified to leach the phosphate from the sludge to be able to recover phosphate as struvite. The Seaborne process was operated at demonstration scale for several years at the WWTP Gifhorn, Germany (Hermanussen et al. 2012). The initial intention was to acidify the sludge to pH 2 to release phosphate and heavy metals from the sludge. After dewatering, the pH of the water phase was increased, followed by a heavy metal removal step using sulfide and a phosphate recovery step yielding struvite. During the operation of the plant, the pH of the leaching phase was first increased to 3 and later to between 5 and 5.5 to minimize acid consumption and to reduce the amount of flocculation chemicals during the dewatering phase. The Seaborne plant treated sewage sludge from the WWTP Gifhorn where iron salts were used to support EBPR. Table 21.3 shows that most of the phosphate could be released if the sludge was acidified to pH of 2 or 3. Also most of iron was solubilized.

A subsequent pH increase to recover phosphate as struvite after magnesium addition would then also lead to reprecipitation of FePs (at pH 6). The recovered struvite therefore contained significant amounts of iron (iron concentration was 63 g/kg TS (total solids), whereas the magnesium concentration was 19 g/kg). Therefore, an alternative approach was tested where the iron was first precipitated with sulfides. Experiments showed that the iron could be precipitated with sulfide at pH 5.5 with a sulfide dosage of 0.8–1.2 mol sulfide per mol iron. Overall 52–74% of all the phosphate could be recovered in this process combining acidification followed by iron removal using sulfide (Hermanussen et al. 2012).

Also the Stuttgart process acidifies digested sludge to release phosphate. This process was tested on a pilot scale at the WWTP Offenburg, Germany. This WWTP uses iron salts to remove phosphate. Acidification to pH 4–5 solubilized about 60–70% of all phosphate with solubilization efficiencies varying from 38% to 84% between batches (Antakyali et al. 2013). After a solid-liquid separation, citrate is added to complex iron and other metal salts (a molar ratio of 1:1.5) in solution, and struvite is then precipitated at pH 8.5.

Table 21.3 Release of phosphate, iron, and aluminum as a function of pH for sludge from the WWTP Gifhorn (Hermanussen et al. 2012)

	Solubilization (% of total)		
pH	Phosphate	Iron	Aluminum
4	55	40	0
3	70	70	25
2	90	95	85

The Krepro process was operated at full scale at the Helsingborg WWTP in Sweden (Karlsson 2001; Odegaard et al. 2002). In this process, dewatered iron-containing sewage sludge was acidified to a pH between 1 and 3 to leach phosphate, heavy metals, and iron into the water phase. This step was followed by a thermal hydrolysis to release organic-bound phosphate and to improve the dewaterability of the sludge. The acidified and hydrolyzed sludge was then dewatered producing a relatively dry dewatered sludge cake (ca. 35% dry matter) with low concentrations of heavy metals, iron, and phosphate. Through a pH increase a ferric phosphate precipitate containing about 75% of all phosphate from the feed material was recovered from the liquid phase. Heavy metals could subsequently be removed from the liquid phase through precipitation with sulfides or as hydroxides after further pH increase. The remaining soluble iron could be returned to the WWTP for precipitation of phosphate or could be recovered as iron oxides through further pH adjustment. Pot and field trials were performed with the recovered FeP. Although the phosphate was not water-soluble, growth yields were only slightly lower (4–5%) compared to application of mineral fertilizer.

Alkaline Treatment

Alkaline treatment of sludge is another approach to release phosphate from iron-containing sludge. An alkaline treatment has the advantage over an acidic treatment that less metals will dissolve. After solid-liquid separation, phosphate can be precipitated from the liquid phase to produce calcium phosphate as a valuable recovery product that can be directly used in the fertilizer industry. However, controversies about the solubility of FeP at alkaline pH and its efficiency to release phosphate from sewage sludge exist. Sano et al. (2012) describe phosphate recovery from FeP containing sewage sludge. The phosphate in this sludge was precipitated by slowly bringing Fe(II) into solution using electrolysis. An alkaline pH (between 13 and 14) resulted in extraction of 92% of all phosphate compared to 70% when acid was used. A more recent study of Xu et al. (2015) investigated release of phosphate from iron phosphate under alkaline conditions. The waste activated sludge was taken from a sewage treatment process with EBPR. The authors state that phosphate was liberated from the FeOOH-P by ion exchange mechanisms at pH 11, the iron content of the sludge was not reported.

Fischer et al. (2015) used dewatered FeP containing digested sludge. They used a microbial electrolysis cell to increase the pH of the sludge to 12.7. At this pH, they reported a release of 66.7% of the phosphate, but also yields up to 95% were reported. At pH higher than 12.7, a drop in phosphate release was detected. However, Maier et al. (2005) studied phosphate release from FeP-rich digested sludge at alkaline and acidic pH to produce struvite. Tests with iron-containing sludge showed that under acidic conditions at pH 2, the total phosphate release was higher than 85%, but in alkaline conditions at pH 13, only 13% of total phosphate could be released. Also Cornel et al. (2004) could only release up to 28% of the phosphate in iron-containing sludge using 1 M NaOH at unknown pH. In general, it appears that under acidic conditions more phosphate is released from iron-containing sludge compared to alkaline conditions, whereas more phosphate was released under alkaline conditions from aluminum-rich sludge (46–56% extracted at pH 13 Maier et al. 2005) and ash (27–54% extracted using 1 M NaOH Cornel et al. 2004).

21.3.3.2 Iron Reduction and Sulfide-Induced Phosphate Release

Iron-reducing bacteria are present and active in activated sludge systems (Nielsen 1996). They can reduce Fe(III) in iron oxides or iron phosphate minerals, thereby mobilizing phosphate (Heiberg et al. 2012; Patrick et al. 1973; Peretyazhko and Sposito 2005). However, it was also shown that Fe(II) compounds that are formed could, in the absence of sulfate, bind most of the released phosphate (Azam and Finneran 2014; Borch and Fendorf 2007; Cheng et al. 2015; Emerson et al. 2012). Vivianite can precipitate relatively quick, relative to SRTs (sludge retention times) in WWTPs (Azam and Finneran 2014; Roldan et al. 2002). Thus, the microbial reduction of Fe(III) in anaerobic treatment stages could lead to the initial release of phosphate from FeP. Ultimately, the reduction of Fe(III) could even act as a phosphate sink, which is suggested by the higher phosphate content (relative to iron) in vivianite (Fe:P molar ratio of 1.5) compared to ferric oxide phosphate precipitates in WWTPs which has been reported as high as 2.5 (Luedecke et al. 1989). At pH values that are encountered in WWTPs, even freshly formed precipitates showed molar Fe:P ratios greater than 1 (Fulazzaky et al. 2014; Smith et al. 2008).

Fe(III) reduction is not necessarily a reason for phosphate release from iron. However, in the presence of sulfide, iron is not able to bind phosphate any longer. Sulfide is not only capable of reducing iron, but it can also form rather insoluble FeS_x. It has been shown that this could be the main mechanism through which iron-bound phosphate is released from sediments (Caraco et al. 1989; Roden and Edmonds 1997; Smolders et al. 2006). Consequently, sulfide has been proposed to solubilize phosphate from FeP sludge. Sulfide was found to release 75% of the phosphate into solution at pH 4 from sludge collected at a water production plant (Likosova et al. 2013). Similarly, 43% of the total phosphate was found to be released from sewage sludge pre-coagulated with iron by adding sulfide (Kato et al. 2006). In another study, phosphate in sludge liquor was precipitated using iron sulfate salts. The sludge was incubated under anaerobic conditions to allow microbial sulfate reduction (Suschka et al. 2001). Subsequently, phosphate was released stoichiometrically (1.5 moles of phosphate released per mole of sulfide) over a timescale of days, without gaseous hydrogen sulfide being formed. The efficiency of the total phosphate release by sulfide was not mentioned.

When phosphate is solubilized from sewage sludge, it could also remain part of the precipitate when it is, e.g., adsorbed by other phases, precipitates with dissolved elements that occur in the sludge liquor, or reprecipitates with other metals from the sludge solids. Thus, it is possible that the mobilized phosphate cascades from one solid phosphate phase to the next, while the order of formed phases is determined by the solubility and the kinetics of precipitation. Similar observations were made for sediments when released phosphate was adsorbed to calcium, silicon, or aluminum phases (Boström and Pettersson 1982; Richardson 1985; Roden and Edmonds 1997). In other studies on sediments, a huge excess supply of sulfide was necessary to release substantial amounts of phosphate from sediments (Golterman 1995).

21.3.4 Recovery from Sewage Sludge Ash

A significant proportion of the sewage sludge produced in countries like the Netherlands, Switzerland, and Germany is incinerated in dedicated sewage sludge incinerators. These mono-incinerators produce a phosphate-rich ash from which phosphate can be recovered. Thermochemical and wet chemical technologies have been developed to recover phosphate from this ash (Cornel and Schaum 2009; Desmidt et al. 2015; Donatello and Cheeseman 2013). They are further discussed elsewhere in, e.g., Chaps. 1, 8, and 10.

Table 21.4 shows that most ashes contain a significant amount of iron in relation to the phosphate content, ranging from 0.2 to 0.7 mol Fe/mol P. This illustrates the fact that iron is omnipresent in sewage sludge as discussed in Sect. 21.1.1. The ash is a complex mixture of amorphous (40–70%) and crystalline components (Cyr

Table 21.4 Typical content for iron, phosphate, aluminum, and calcium in various ashes

References	Remark	Country	Fe (g/kg ds)	Al (g/kg ds)	Ca (g/kg ds)	P (g/kg ds)	Fe/P (mol/mol)	Al/P (mol/mol)	Ca/P (mol/mol)
Cyr et al. (2007)		France	33	67	147	65	0.28	1.2	1.8
Cyr et al. (2007)	>20 ashes	Various	64	14	106	51	0.70	0,3	1,6
Geraarts et al. (2007)	Ash from 25% of Dutch sewage sludge	Netherlands	88	55	152	84	0.58	1.0	1.4
	Low-iron ash	Netherlands	40	63	136	92	0.24	1.1	1.1
Adam et al. (2009)	7 ashes	Germany, the Netherlands	91	13	108	92	0.55	0.2	0.9
Biswas et al. (2009)	1 ash	Japan	52	60	57	88	0.33	1.1	0.5
Petzet et al. (2012)	4 ashes	Germany	66	68	127	88	0.40	1.2	1.1
Ottosen et al. (2013)	Al-rich ash	Denmark	70	67		70	0.12	1.2	
	Iron-rich ash	Denmark	98	20		98	0.33	0.4	
Magdziarz et al. (2016)	4 ashes	Poland	91	41	177	104	0.49	0.7	1.3

et al. 2007). Most researchers report whitlocktite as the most common phosphate phase in the ash (Adam et al. 2009; Cyr et al. 2007; Guedes et al. 2016; Magdziarz et al. 2016; Petzet et al. 2012). Whitlocktite is a complex phosphate-containing mineral that can contain calcium, magnesium, and iron in varying ratios in the crystal lattice. Next to this, mineral aluminum phosphate is the second most reported phosphate-bearing phase (Biswas et al. 2009; Petzet et al. 2012). Iron may also be present as hematite or even as magnetite (Adam et al. 2009; Biswas et al. 2009; Cyr et al. 2007; Guedes et al. 2016; Magdziarz et al. 2016).

Wet chemical approaches (EcoPhos, Leachphos, TetraPhos, EasyMining) leach phosphate from the ash through acidic dissolution (Egle et al. 2015). Typically phosphate starts being released from pH 3 to 4 on to reach nearly 90% release at pH 2 (Guedes et al. 2016; Ottosen et al. 2013; Petzet et al. 2012). Down to pH 2 leaching of iron from the ash is still minimal. Aluminum is released in a similar pH range as phosphate. Biswas et al. (2009) report up to 40% leaching of iron from the ash in 1 M sulfuric acid (pH not reported) at relatively high solid to liquid ratios. Both Cornel et al. (2004) and Ottosen et al. (2013) report a lower acid consumption for high iron-containing ashes compared to those with a relatively high aluminum content (ca. 30–40% less based on the report by Cornel et al. (2004)). The dissolution of aluminum is a problem for the subsequent recovery of phosphate from the solution, as the aluminum will reprecipitate back to aluminum phosphate when the pH is increased. In this approach, aluminum should therefore first be removed from the eluate to be able to get a pure calcium phosphate product. Petzet et al. (2012) proposed an alternative acid-alkaline approach. Heavy metals may leach easier from ashes with higher iron content (Ottosen et al. 2013).

Phosphate in sewage sludge ash can also be made water-soluble via dry acidulation of the ash (Langeveld and Wolde 2013; Weigand et al. 2011). Langeveld and Wolde (2013) report that ashes have a negative effect on the solubilization of phosphate when added as an additional raw material to their regular triple superphosphate production from phosphate rock. Higher iron concentrations increased the negative effect.

Significant amounts of phosphate have been recycled from ash with a low iron content using a thermal process used to produce white phosphorus (Schipper and Korving 2009). To be suitable for this process, the ash had to have a Fe/P molar ratio of less than 0.2 to minimize the amount of ferroP by-products produced (Geraarts et al. 2007; Schipper and Korving 2009). The only European company producing white phosphorus closed down in 2012, and therefore this process cannot be used anymore. In the European RecoPhos project, a new concept has been developed based on this process that can also treat high iron-containing ashes. Through a novel furnace design, the gaseous phosphorus can exit the furnace quicker, which should result in a lower formation of ferroP by-products (Rapf et al. 2012).

Furthermore, thermochemical treatment of sewage sludge ash can remove heavy metals and increase the bioavailability of the phosphate in the ash (Adam et al. 2009), and a pilot plant using this process was operated in Leoben, Austria (Hermann 2011). The phosphate in the treated ash was bioavailable in acidic and neutral soils even though the ash contained significant concentrations of iron (Nanzer et al. 2014).

21.4 Summary and Outlook

Iron is an element that is omnipresent in WWTPs. It has many useful properties that can be used during sewage management, and thus iron is very often required for modern and successful sewage treatment. Even without iron dosing in WWTPs, iron can enter WWTPs in substantial amounts due to external iron sources that are outside the scope of control of a WWTP. Iron is relatively insoluble and has a high affinity toward phosphate. Thus, the iron mainly ends up in the sewage sludge where it can bind a significant part of the total phosphate. This suggests that all phosphate recovery technologies have to deal with iron-bound phosphate. This chapter shows that the presence of iron for phosphate recovery technologies has positive, negative, and sometimes indistinct effects. For instance, recovery technologies that make use of a wet digestion to recover phosphate from ash showed lower acid consumptions for iron-rich than for aluminum-rich ash. Additionally, recovery technologies related to vivianite offer, theoretically, high efficient phosphate recovery. Vivianite-based recovery routes deserve more attention than they have had until now.

The presence of iron can reduce the efficiency of phosphate recovery technologies, e.g., when recovery of struvite is intended in EBPR plants and background iron will take away phosphate from the proposed recovery product struvite because of the higher affinity of iron for phosphate. Also some technologies for recovery of phosphate from ash show a lower performance in the presence of iron, like in the case of dry acidulation or when ash is used as raw material for white phosphate production. In some instances, the role of iron on phosphate recovery is indifferent, for example, thermochemical treatment of sewage sludge ash allows high amounts of iron and the process is independent of the iron concentration in the ash.

The role of iron is often also inconclusive. For instance, the lower solubility of phosphate bound to iron in sludge that is applied on agricultural land is sometimes considered as positive and sometimes as negative. Due to the omnipresence of iron in sewage-related waste streams and due to the inconsistent role of iron for individual phosphate recovery technologies, we suggest that the role of iron in a waste stream should be evaluated before a phosphate recovery technology is implemented. If necessary, the treatment strategies for the waste stream can be adjusted to improve phosphate recovery. In any case, developers should be aware of the potential and obstacles the iron and phosphate chemistry offers in relation to phosphate recovery.

Acknowledgments This work was performed in the TTIW-cooperation framework of Wetsus, European Centre of Excellence for Sustainable Water Technology (www.wetsus.nl). Wetsus is funded by the Dutch Ministry of Economic Affairs, the European Union Regional Development Fund, the Province of Fryslân, the City of Leeuwarden, and the EZ/Kompas program of the "Samenwerkingsverband Noord-Nederland." We thank the participants of the research theme "Phosphate Recovery" for their financial support and helpful discussions. Furthermore, we acknowledge the valuable support from Anna Jeworrek for her contribution to the section on the alkaline treatment of sewage sludge.

References

ACHS (2009) Review of the feasibility of recycling phosphates at sewage treatment plants in The UK – executive summary. Department for Environment, Food and Rural Affairs, 32 pp

Adam C, Peplinski B, Michaelis M, Kley G, Simon F-G (2009) Thermochemical treatment of sewage sludge ashes for phosphorus recovery. Waste Manag (New York, NY) 29(3):1122–1128

Antakyali D, Meyer C, Preyl V, Maier W, Steinmetz H (2013) Large-scale application of nutrient recovery from digested sludge as struvite. Water Pract Technol 8(2):256–262

Azam HM, Finneran KT (2014) Fe(III) reduction-mediated phosphate removal as vivianite (Fe3(PO4)2·8H2O) in septic system wastewater. Chemosphere 97:1–9

Baker S, Lee Y, Li W (2006) A struvite control and phosphorus removal process for centrate: full-scale testing. Proc Water Environ Fed 2006(7):5197–5208

Beecher N, Crawford C, Goldstein N, Kester G, Lono-Batura M, Dziezyk E (2007) A national biosolid regulation, quality, end use & disposal survey: final report. North East Biosolids and Residuals Association, Tamworth

Biswas BK, Inoue K, Harada H, Ohto K, Kawakita H (2009) Leaching of phosphorus from incinerated sewage sludge ash by means of acid extraction followed by adsorption on orange waste gel. J Environ Sci 21(12):1753–1760

Bjorn A (2010) Acid phase digestion at Derby STW – Context and preliminary optimisation results

Böhnke B (1977) Das Adsorptions-Belebungsverfahren. Korrespondenz Abwasser 24:33

Borch T, Fendorf S (2007) Phosphate interactions with iron (Hydr)oxides: mineralization pathways and phosphorus retention upon bioreduction. In: Adsorption of metals by Geomedia II: variables, mechanisms, and model applications, vol. 7. Developments in Earth and Environmental Sciences. Elsevier, p 321–348

Boström B, Pettersson K (1982) Different patterns of phosphorus release from lake sediments in laboratory experiments. Hydrobiologia 91–92(1):415–429

Brandt RC, Elliott HA, O'Connor GA (2004) Water-extractable phosphorus in biosolids: implications for land-based recycling. Water Environ Res 76(2):121–129

Caraco NF, Cole JJ, Likens GE (1989) Evidence for sulphate-controlled phosphate release from sediments of aquatic systems. Nature 341:316–318

Čermáková Z, Švarcová S, Hradilová J, Bezdička P, Lančok A, Vašutová V, Blažek J, Hradil D (2015) Temperature-related degradation and colour changes of historic paintings containing vivianite. Spectrochim Acta A Mol Biomol Spectrosc 140:101–110

Chen Y, Cheng JJ, Creamer KS (2008) Inhibition of anaerobic digestion process: a review. Bioresour Technol 99(10):4044–4064

Chen J, Bai J, Chen H, Graetz J (2011) In situ hydrothermal synthesis of LiFePO 4 studied by synchrotron X-ray diffraction. J Phys Chem Lett 2(15):1874–1878

Cheng X, Chen B, Cui Y, Sun D, Wang X (2015) Iron(III) reduction-induced phosphate precipitation during anaerobic digestion of waste activated sludge. Sep Purif Technol 143:6–11

Coats ER, Watkins DL, Kranenburg D (2011) A comparative environmental life-cycle analysis for removing phosphorus from wastewater: biological versus physical/chemical processes. Water Environ Res 83(8):750

Cooper J (2014) Managing phosphorus in the UK water industry to increase national resource security. PhD, Birmingham

Cornel P, Schaum C (2009) Phosphorus recovery from wastewater: needs, technologies and costs. Water Sci Technol 59(6):1069–1076

Cornel P, Jardin N, Schaum C (2004) Möglichkeiten einer Rückgewinnung von Phosphor aus Klärschlammasche: Teil 1: Ergebnisse von Laborversuchen zur Extraktion von Phosphor. GWF Wasser 145(9):627–632

Cullen N, Baur R, Schauer P (2013) Three years of operation of North America's first nutrient recovery facility. Water Sci Technol 68(4):763–768

Cyr M, Coutand M, Clastres P (2007) Technological and environmental behavior of sewage sludge ash (SSA) in cement-based materials. Cem Concr Res 37(8):1278–1289

Da Silva S, Basséguy R, Bergel A (2004) Hydrogenase-catalysed deposition of vivianite on mild steel. Electrochim Acta 49(13):2097–2103

Delahaye M (2017) Phosphorus recovery from wastewater: SUEZ's Phosphogreen technology successfully running in Denmark. IWA specialist conference on Sludge Management SludgeTech 2017

Der Schweizerische Bundesrat (2015) Verordnung über die Vermeidung und die Entsorgung von Abfällen

Desmidt E, Ghyselbrecht K, Zhang Y, Pinoy L, Van der Bruggen B, Verstraete W, Rabaey K, Meesschaert B (2015) Global phosphorus scarcity and full-scale P-recovery techniques: a review. Crit Rev Environ Sci Technol 45(4):336–384

Deutscher Bundestag (2017) Verordnung zur Neuordnung der Klärschlammverwertung

Donatello S, Cheeseman CR (2013) Recycling and recovery routes for incinerated sewage sludge ash (ISSA): a review. Waste Manag (New York, N.Y.) 33(11):2328–2340

Doyle JD, Parsons SA (2002) Struvite formation, control and recovery. Water Res 36(16):3925–3940

DWA (2005) Stand der Klarschlammbehandlung und Entsorgung in Deutschland. ISBN 3-937758-29-1, 66 p

Egle L, Rechberger H, Zessner M (2014) Endbericht Phosphorrückgewinnung aus dem Abwasser, Wien, 323 pp

Egle L, Rechberger H, Zessner M (2015) Overview and description of technologies for recovering phosphorus from municipal wastewater. Resour Conserv Recycl 105:325–346

Elliott H, O'Connor G (2007) Phosphorus management for sustainable biosolids recycling in the United States. Soil Biol Biochem 39(6):1318–1327

Emerson D, Roden E, Twining BS (2012) The microbial ferrous wheel: iron cycling in terrestrial, freshwater, and marine environments. Front Microbiol 3:383

Eriksson J (2001) Concentrations of 61 trace elements in sewage sludge, farmyard manure, mineral fertiliser, precipitation and in oil and crops. Swedish Environmental Protection Agency Stockholm, Sweden

European Commission (2016) Eighth Report on the Implementation Status and the Programmes for Implementation (as required by Article 17) of Council Directive 91/271/EEC concerning urban waste water treatment

Ewert W, Hermanussen O, Kabbe C, Mele C, Niewersch H, Paillard H, Stössel E, Wagenbach A, Steman J (2014) Sustainable sewage sludge management fostering phosphorus recovery and energy efficiency

Faivre D (2016) Iron oxides: from nature to applications. Wiley-VCH Verlag GmbH & Co. KGaA, Weinheim

Fischer F, Zufferey G, Sugnaux M, Happe M (2015) Microbial electrolysis cell accelerates phosphate remobilisation from iron phosphate contained in sewage sludge. Environ Sci: Processes Impacts 17(1):90–97

Flores-Alsina X, Solon K, Kazadi Mbamba C, Tait S, Gernaey KV, Jeppsson U, Batstone DJ (2016) Modelling phosphorus (P), sulfur (S) and iron (Fe) interactions for dynamic simulations of anaerobic digestion processes. Water Res 95:370–382

Franz T (2006) Spatial classification methods for efficient infiltration measurements and transfer of measuring results. PhD, Dresden

Frossard E, Bauer JP, Lothe F (1997) Evidence of vivianite in FeSO4-flocculated sludges. Water Res 31(10):2449–2454

Fulazzaky MA, Salim N, Abdullah NH, Yusoff A, Paul E (2014) Precipitation of iron-hydroxy-phosphate of added ferric iron from domestic wastewater by an alternating aerobic–anoxic process. Chem Eng J 253:291–297

Fytianos K, Voudrias E, Raikos N (1998) Modelling of phosphorus removal from aqueous and wastewater samples using ferric iron. Environ Pollut 101(1):123–130

Geraarts B, Koetse E, Loeffen P, Reitsma B, Gaillard A (2007) Fosfaatterugwinning uit ijzerarm slib van rioolwaterzuiveringsinrichtingen. STOWA, report 2007–31. ISBN 78.90.5773.380.2, 83 p

Ghassemi M, Recht HL (1971) Phosphate precipitation with ferrous iron. Water pollution control research series, 64 pp

Golterman HL (1995) The role of the ironhydroxide-phosphate-sulphide system in the phosphate exchange between sediments and overlying water. Hydrobiologia 297(1):43–54

Guedes P, Couto N, Ottosen LM, Kirkelund GM, Mateus E, Ribeiro AB (2016) Valorisation of ferric sewage sludge ashes: potential as a phosphorus source. Waste Manag (New York, NY) 52:193–201

Gutierrez O, Park D, Sharma KR, Yuan Z (2010) Iron salts dosage for sulfide control in sewers induces chemical phosphorus removal during wastewater treatment. Water Res 44(11):3467–3475

Hansen B. (2017) Kemira, personal communication, explanation and translation of data Statistics Sweden (SCB)

Heiberg L, Koch CBK, Charlotte J, Henning S, Hansen HBC (2012) Vivianite precipitation and phosphate sorption following iron reduction in anoxic soils. J Environ Qual 41(3):938–949

Hermann L (2011) Phosphate fertilizers from sewage sludge ash-design of an industrial manufacturing plant. Proc Water Environ Fed Nutr Recover Manag 2011(1):317–332

Hermanussen O, Müller-Schapper J, Haun E, Weichgrebe D, Rosenwinkel KH, Esemen T, Dockhorn T, Dichtl N (2012) Wissenschaftliche Begleitung der großtechnischen Anwendung Wissenschaftliche Begleitung der großtechnischen Anwendungder Seaborne-Technologie auf der Kläranlage Gifhorn: – Zusammenfassung der durchgeführten Untersuchungen und technisch-wirtschaftliche Bewertung der Verfahrenstechnik-

Hvitved-Jacobsen T, Vollertsen J, Nielsen AH (2013) Sewer processes: microbial and chemical process engineering of sewer networks, 2nd edn. CRC Press, Boca Raton

Jetten M, Horn S, van Loosdrecht MCM (1997) Towards a more sustainable municipal wastewater treatment system. Water Sci Technol 35(9):171–180

Kahiluoto H, Kuisma M, Ketoja E, Salo T, Heikkinen J (2015) Phosphorus in manure and sewage sludge more recyclable than in soluble inorganic fertiliser. In revision. Environ Sci Technol 49(4):2115–2122

Kang SJ, Olmstead K, Takacs K, Collins J (2008) Municipal nutrient removal technologies reference document, volume 1 – technical report. US Environmental Protection Agency (EPA), 268 pp. http://water.epa.gov/scitech/wastetech/upload/mnrt-volume1.pdf

Karlsson I (2001) Full scale plant recovering iron phosphate from sewage at Helsingborg, Sweden. Proc. 2nd Int.Conf. on Recovery of Phosphates from Sewage and Animal Wastes, CEEP, Holland, 12–14 March 2001

Kato F, Kitakoji H, Oshita K, Takaoka M, Takeda N, Matsumoto T (2006) Extraction efficiency of phosphate from pre-coagulated sludge with NaHS. Water Sci Technol 54(5):119

Kelessidis A, Stasinakis AS (2012) Comparative study of the methods used for treatment and final disposal of sewage sludge in European countries. Waste Manag (New York, N.Y.) 32(6):1186–1195

Kidd PS, Dominguez-Rodriguez MJ, Diaz J, Monterroso C (2007) Bioavailability and plant accumulation of heavy metals and phosphorus in agricultural soils amended by long-term application of sewage sludge. Chemosphere 66(8):1458–1467

Korving L (2012) Trends in slibontwatering. STOWA, report 2011–46. ISBN 978.90.5773.577.6, 108 p

Krogstad T, Sogn TA, Asdal Å, Sæbø A (2005) Influence of chemically and biologically stabilized sewage sludge on plant-available phosphorous in soil. Ecol Eng 25(1):51–60

Langeveld CP, Wolde KWT (2013) Phosphate recycling in mineral fertiliser production. Proceedings, 1466–1314 727. International Fertiliser Society, Leek

Lasheen MR, Ammar NS (2009) Assessment of metals speciation in sewage sludge and stabilized sludge from different wastewater treatment plants, Greater Cairo, Egypt. J Hazard Mater 164(2–3):740–749

Li J (2005) Effects of Fe(III) on floc characteristics of activated sludge. J Chem Technol Biotechnol 80(3):313–319

Likosova EM, Keller J, Rozendal RA, Poussade Y, Freguia S (2013) Understanding colloidal FeSx formation from iron phosphate precipitation sludge for optimal phosphorus recovery. J Colloid Interface Sci 403:16–21
Lu Q, He ZL, Stoffella PJ (2012) Land application of biosolids in the USA: a review. Appl Environ Soil Sci 2012:1–11
Lu J, Yang J, Xu K, Hao J, Li YY (2016) Phosphorus release from coprecipitants formed during orthophosphate removal with Fe(III) salt coagulation: effects of pH, Eh, temperature and aging time. J Environ Chem Eng 4(3):3322–3329
Luedecke C, Hermanowicz SW, Jenkins D (1989) Precipitation of ferric phosphate in activated-sludge – a chemical model and its verification. Water Sci Technol 21(4–5):325–337
Lycke D, Prasad R, Meulenkamp R, Morgenschweis CM, Steensma W (2017) Combining phosphorus recovery and ammonia removal in the Omzet.Amersfoort project. IWA specialist conference on Sludge Management SludgeTech 2017
Macdonald GK, Bennett EM, Potter PA, Ramankutty N (2011) Agronomic phosphorus imbalances across the world's croplands. Proc Natl Acad Sci U S A 108(7):3086–3091
Magdziarz A, Kosowska-Golachowska M, Kijo-Kleczkowska A, Środa K, Wolski K, Richter D, Musiał T, Filipowicz M, Dudek M, Olkuski T, Styszko K (2016) Analysis of sewage sludge ashes from air and oxy-fuel combustion in a circulating fluidized-bed. E3S Web Conf. 10, 54
Maier W, Weidelener A, Krampe J, Rott I (2005) Entwicklung eines Verfahrens zur Phosphat-Rückgewinnung aus ausgefaultem Nassschlam oder entwässertem Faulschlamm als gut pflanzenverfügbares Magnesium-Ammonium-Phosphat (MAP): Schlussbericht: Teil 1: Zusammenfassung und Wertung der Ergebnisse, 160 pp
Mamais D, Pitt PA, Cheng YW, Loiacono J, Jenkins D (1994) Determination of ferric chloride dose to control struvite precipitation in anaerobic sludge digesters. Water Environ Res 66(7):912–918
Mao Y, Yang S, Yue Q, Wang W (2016) Theoretical and experimental study of the mechanisms of phosphate removal in the system containing Fe(III)-ions. Environ Sci Pollut Res Int 23(23):24265–24276
Marchi A, Geerts S, Weemaes M, Schiettecatte W, Wim S, Vanhoof C, Christine V (2015) Full-scale phosphorus recovery from digested waste water sludge in Belgium – part I: technical achievements and challenges. Water Sci Technol 71(4):487–494
Marx JJ, Wilson TE, Schroedel RB, Winfield G, Sokhey A (2001) Vivianite nutrient removal's hidden problem? Proc Water Environ Fed 2001(8):378–388
Miller M, O'Connor GA (2009) The longer-term phytoavailability of biosolids-phosphorus. Agron J 101(4):889
Miot J, Benzerara K, Morin G, Bernard S, Beyssac O, Larquet E, Kappler A, Guyot F (2009) Transformation of vivianite by anaerobic nitrate-reducing iron-oxidizing bacteria. Geobiology 7(3):373–384
Mishima I, Nakajima J (2011) Application of iron electrolysis to full-scale activated sludge process for phosphorus removal. J Wat Environ Tech 9(4):359–369
Morse G, Brett S, Guy J, Lester J (1998) Review: phosphorus removal and recovery technologies. Sci Total Environ 212(1):69–81
Nanzer S, Oberson A, Berger L, Berset E, Hermann L, Frossard E (2014) The plant availability of phosphorus from thermo-chemically treated sewage sludge ashes as studied by 33P labeling techniques. Plant Soil 377(1–2):439–456
Neethling JB, Benisch M (2004) Struvite control through process and facility design as well as operation strategy. Water Sci Technol 49(2):191–199
Nielsen P (1996) The significance of microbial Fe(III) reduction in the activated sludge process. Water Sci Technol 34(5–6):129–136
Nielsen AH, Lens P, Vollertsen J, Hvitved-Jacobsen T (2005) Sulfide–iron interactions in domestic wastewater from a gravity sewer. Water Res 39(12):2747–2755
Nowak O, Keil S, Fimml C (2011) Examples of energy self-sufficient municipal nutrient removal plants. Water Sci Technol 64(1):1

Nriagu JO (1972) Stability of vivianite and ion-pair formation in the system fe3(PO4)2-H3PO4H3PO4-H2o. Geochim Cosmochim Acta 36(4):459–470
Nriagu JO, Dell CI (1974) Diagenetic formation of iron phosphates in recent lake sediments. Am Mineral 59:934–946
Nriagu JO, Moore PB (1984) Phosphate minerals. Springer Berlin Heidelberg, Berlin
O'Connor GA, Sarkar D, Brinton SR, Elliott HA, Martin FG (2004) Phytoavailability of biosolids phosphorus. J Environ Qual 33(2):703
Odegaard H, Paulsrud B, Karlsson I (2002) Wastewater sludge as a resource: sludge disposal strategies and corresponding treatment technologies aimed at sustainable handling of wastewater sludge. Water Sci Technol 46(10):295–303
Ofwat (2005) Water framework directive economic analysis of water industry costs
Ogilvie D (1998) National study of the composition of sewage sludge. Drainage Managers Group, a subgroup of the New Zealand Water and Wastes Association, Auckland [N.Z.]
Ohtake H (2017) CPR share in Japan
Ohtake H, Okano K (2015) Development and implementation of technologies for recycling phosphorus in secondary resources in Japan. Glob Environ Res 19:49–65
Oleszkiewicz J (2014) Options for improved nutrient removal and recovery from municipal wastewater in the Canadian context
Oleszkiewicz J, Kruk D, Devlin T, Lashkarizadeh M, Qiuyan Y (2015) Options for improved nutrient removal and recovery from municipal wastewater in the Canadian context. Canadian Water Network
Ottosen LM, Kirkelund GM, Jensen PE (2013) Extracting phosphorous from incinerated sewage sludge ash rich in iron or aluminum. Chemosphere 91(7):963–969
Patrick WH, Gotoh S, Williams BG (1973) Strengite dissolution in flooded soils and sediments. Science 179(4073):564–565
Paul E, Laval ML, Sperandio M (2001) Excess sludge production and costs due to phosphorus removal. Environ Technol 22:1363–1371
Peretyazhko T, Sposito G (2005) Iron(III) reduction and phosphorous solubilization in humid tropical forest soils. Geochim Cosmochim Acta 69(14):3643–3652
Petzet S, Peplinski B, Cornel P (2012) On wet chemical phosphorus recovery from sewage sludge ash by acidic or alkaline leaching and an optimized combination of both. Water Res 46(12):3769–3780
Pham AN, Rose AL, Feltz AJ, Waite TD (2004) The effect of dissolved natural organic matter on the rate of removal of ferrous iron in fresh waters. In: Natural organic material research: innovations and applications for drinking water. IWA Publishing, pp 213–219
Poffet MS (2007) Thermal runaway of the dried sewage sludge in the storage tanks: from molecular origins to technical measures of smouldering fire prevention. Dissertation thesis
Prochnow LI, Chien SH, Carmona G, Dillard EF, Henao J, Austin ER (2008) Plant availability of phosphorus in four superphosphate fertilizers varying in water-insoluble phosphate compounds. Soil Sci Soc Am J 72(2):462
Rapf M, Raupenstrauch H, Cimatoribus C, Kranert M (2012) A new thermo-chemical approach for the recovery of phosphorus from sewage sludge
Rasmussen H, Nielsen P (1996) Iron reduction in activated sludge measured with different extraction techniques. Water Res 30(3):551–558
Reusser SR (2009) Proceed with caution in advanced anaerobic digestion system design. Proceedings of the Water Environment Federation Session 41 through Session 50 (3065–3084)
Richardson CJ (1985) Mechanisms controlling phosphorus retention capacity in freshwater wetlands. Science 228(4706):1424–1427
Roden EE, Edmonds JW (1997) Phosphate mobilization in iron-rich anaerobic sediments: microbial Fe(III) oxide reduction versus iron-sulfide formation. Arch Hydrobiol 139(3):347–378
Rodgers KA, Henderson GS (1986) The thermochemistry of some iron phosphate minerals: vivianite, metavivianite, baraćite, ludlamite and vivianite/metavivianite admixtures. Thermochim Acta 104:1–12

Roldan R, Barron V, Torrent J (2002) Experimental alteration of vivianite to lepidocrocite in a calcareous medium. Clay Miner 37(4):709–718

Römer W (2006) Vergleichende Untersuchungen zur Pflanzenverfügbarkeit von Phosphat aus verschiedenen P-Recycling-Produkten im Keimpflanzenversuch. J Plant Nutr Soil Sci 169(6):826–832

Rothe M, Kleeberg A, Hupfer M (2016) The occurrence, identification and environmental relevance of vivianite in waterlogged soils and aquatic sediments. Earth Sci Rev 158:51–64

Samie IF, Römer W (2001) Phosphorus availability to maize plants from sewage sludge treated with Fe compounds. In: Horst WJ, Schenk MK, Bürkert A, Claassen N, Flessa H, Frommer WB, Goldbach H, Olfs H-W, Römheld V, Sattelmacher B, Schmidhalter U, Schubert S, Wirén N v, Wittenmayer L (eds) Plant nutrition. Springer Netherlands, Dordrecht, pp 846–847

Sano A, Kanomata M, Inoue H, Sugiura N, Xu K-Q, Inamori Y (2012) Extraction of raw sewage sludge containing iron phosphate for phosphorus recovery. Chemosphere 89(10):1243–1247

SCB (2016) Utsläpp till vatten och slamproduktion 2014 Kommunala reningsverk, massa- och pappersindustri samt viss övrig industri. Statistiska centralbyran

Schipper WJ, Korving L (2009) Full-scale plant test using sewage sludge ash as raw material for phosphorus production. In: Proceedings of International Conference on Nutrient Recovery, May 2009

Schröder JJ, Cordell D, Smit AL, Rosemarin A (2010) Sustainable use of phosphorus. Wageningen University and Research Centre, 140 pp

Schröder JJ, Smit AL, Cordell D, Rosemarin A (2011) Improved phosphorus use efficiency in agriculture: a key requirement for its sustainable use. Chemosphere 84(6):822–831

Schwertmann U, Cornell RM (2000) Iron oxides in the laboratory: preparation and characterization, 2nd completely rev. and extended edn. Wiley-VCH, Weinheim

Shimp GF, Barnard JL, Bott CB (2013) Seeking to understand and address the impacts of biological phosphorus removal on biosolids dewatering. Proc Water Environ Fed 2013(9):5668–5685

Singer PC (1972) Anaerobic control of phosphate by ferrous iron: anaerobic control of phosphate by ferrous iron. J Water Pollut Control Fed 44(4):663

Smith S, Takacs I, Murthy S, Daigger GT, Szabo A (2008) Phosphate complexation model and its implications for chemical phosphorus removal. Water Environ Res 80(5):428–438

Smolders AJP, Lamers LPM, Lucassen ECHET, Van Der Velde G, Roelofs JGM (2006) Internal eutrophication: how it works and what to do about it – a review. Chem Ecol 22(2):93–111

Stumm W, Morgan JJ (1996) Aquatic chemistry: chemical equilibria and rates in natural waters, Environmental science and technology, 3rd edn. Wiley, New York

Stumm W, Sigg L, Sulzberger B (1992) Chemistry of the solid-water interface: processes at the mineral-water and particle-water in natural systems, A Wiley-Interscience publication. Wiley, New York

Suschka J, Machnicka A, Poplawski S (2001) Phosphate recovery from iron phosphate sludge. Environ Technol 22:1295–1301

Takács I, Murthy S, Smith S, McGrath M (2006) Chemical phosphorus removal to extremely low levels: experience of two plants in the Washington, DC area. Water Sci Technol 53(12):21

Taylor KG, Hudson-Edwards KA, Bennett AJ, Vishnyakov V (2008) Early diagenetic vivianite [Fe3(PO4)2·8H2O] in a contaminated freshwater sediment and insights into zinc uptake: a μ-EXAFS, μ-XANES and Raman study. Appl Geochem 23(6):1623–1633

Tchobanoglous G, Burton FL, Stensel HD (2013) Wastewater engineering: treatment and reuse, 5th edn. McGraw-Hill Higher Education; McGraw-Hill [distributor], New York

Theis TL, Singer PC (1974) Complexation of iron(II) by organic matter and its effect on iron(II) oxygenation. Environ Sci Technol 8(6):569–573

Thistleton J, Clark T, Pearce P, Parsons SA (2001) Mechanisms of chemical phosphorus removal. Process Saf Environ Prot 79(6):339–344

Tilley (2005) Supplementary material to part 3: reactions and transformations. In: Understanding solids. Wiley, pp 531–542

USEPA (2009) Targeted national sewage sludge survey sampling and analysis technical report (January)

van den Brand TPH, Roest K, Chen GH, Brdjanovic D, van Loosdrecht MCM (2015) Occurrence and activity of sulphate reducing bacteria in aerobic activated sludge systems. World J Microbiol Biotechnol 31(3):507–516

van der Grift B, Behrends T, Osté LA, Schot PP, Wassen MJ, Griffioen J (2016) Fe hydroxyphosphate precipitation and Fe(II) oxidation kinetics upon aeration of Fe(II) and phosphate-containing synthetic and natural solutions. Geochim Cosmochim Acta 186:71–90

Waerenborgh JC, Figueiredo MO (1986) X-ray powder diffraction and 57 Fe Mössbauer spectroscopy study of the thermal breakdown of vivianite, Fe3(PO4)2x8H2O. Hyperfine Interact 29(1):1101–1104

WEF (2011) Nutrient removal, WEF manual of practice no. 34. McGraw-Hill; WEF Press, New York

Weigand H, Bertau M, Bohndick F, Bruckert A (2011) RECOPHOS: recophos: full scale recovery of phosphate from sewage sludge ash. Sardinia 2011, Thirteenth International Waste Management and Landfill Symposium

Wendt Von H (1973) Die Kinetik typischer Hydrolysereaktionen von mehrwertigen Kationen. Chimia 27:575–588

Wilfert P, Suresh Kumar P, Korving L, Witkamp GJ, van Loosdrecht MCM (2015) The relevance of phosphorus and iron chemistry to the recovery of phosphorus from wastewater: a review. Environ Sci Technol 104:449

Wilfert P, Mandalidis A, Dugulan I, Goubitz K, Korving L, Temmink H, Witkamp GJ, van Loosdrecht M (2016) Vivianite as an important iron phosphate precipitate in sewage treatment plants. Water Res 104:449

Xu Y, Hu H, Liu J, Luo J, Qian G, Wang A (2015) pH dependent phosphorus release from waste activated sludge: contributions of phosphorus speciation. Chem Eng J 267:260–265

Yoon SY, Lee CG, Park JA, Kim JH, Kim SB, Lee SH, Choi JW (2014) Kinetic, equilibrium and thermodynamic studies for phosphate adsorption to magnetic iron oxide nanoparticles. Chem Eng J 236:341–347

Zhang X (2012) Factors influencing iron reduction–induced phosphorus precipitation. Environ Eng Sci 29(6):511–519

Part IV
Steelmaking Slag

Chapter 22
Phosphorus Separation and Recovery from Steelmaking Slag

Takahiro Miki

Abstract Iron (Fe) is a metal element which is most abundantly produced in the world. Steel is an alloy of iron and other elements and plays a crucial role in the development of sustainable society. In the iron manufacturing process, the content of phosphorus (P) in the raw materials is relatively low but is concentrated into molten iron and then removed nearly completely into steelmaking slag. Hence, steelmaking can be viewed as a P enrichment process and generates slag which has a potential to serve as a secondary resource of P. The fundamental system of steelmaking slag is CaO-FeO-SiO_2 which is formed inside the primary 2CaO · SiO_2 region, indicating that this phase precipitates at the early stage of cooling slag. P is unevenly distributed in steelmaking slag and generally concentrated in the 2CaO · SiO_2 phase as a 2CaO · SiO_2-3CaO · P_2O_5 solid solution. This evidence reveals the potential to separate P from steelmaking slag. In this chapter, various technologies for P separation and recovery from steelmaking slag are described. They include magnetic separation, capillary action, dissolution, and carbothermic reduction. Since each method has its own merits and demerits, the best choice of single or combination of the technology options is critical to effective P recovery and separation from steelmaking slag.

Keywords Steelmaking slag · Magnetic separation · Capillary action · Elution · Dissolution · Carbothermic reduction

22.1 Introduction

Iron (Fe) is the most abundantly produced metal in the world. For instance, more than 1600 million tons (Mt) of Fe were produced worldwide in 2014. This is more than 30 times larger than aluminum (Al) of which the world production is the second largest after Fe. It is also well recognized that steel, which is an alloy of iron and

T. Miki (✉)
Tohoku University, Sendai, Japan
e-mail: miki@material.tohoku.ac.jp

H. Ohtake, S. Tsuneda (eds.), *Phosphorus Recovery and Recycling*,
https://doi.org/10.1007/978-981-10-8031-9_22

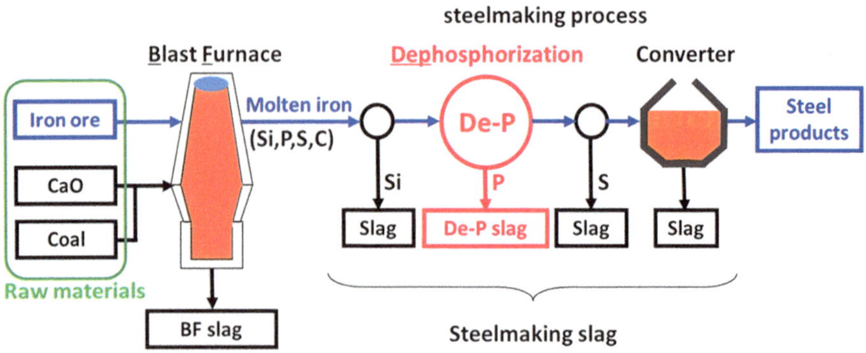

Fig. 22.1 The flow of elements during a steelmaking process in an integrated steel mills

other elements, plays a crucial role in the development of sustainable society. Various elements have a positive effect on steel properties so that they have been used to produce high value-added Fe-based alloys. However, phosphorus (P) is known to give a negative effect on the plasticity, ductility, and impact strength of steel. Therefore, P has been considered as an unwanted impurity in the steel industry.

Steel is produced by reducing iron ore and also by remelting and refining Fe scrap. Figure 22.1 shows the flow of major elements during a steelmaking process in an integrated steel mill. Iron ore, lime (CaO), and coal are fed into the blast furnace where molten iron and BF(blast furnace) slag are generated. Molten iron discharged from the blast furnace contains impurities such as Si, P, S, and C. These elements are then removed in the steelmaking process. Since the amount of Fe produced is huge, the total mass of impurities is also very large. On a yearly basis, the amount of P removed into steelmaking slag (dephosphorization slag in Fig. 22.1) is almost 2.5 times more than that of P imported as phosphate rock in Japan. Although the P content in the raw materials is relatively low, it is concentrated into molten iron and then is removed into steelmaking slag. Hence, steelmaking process can be viewed as a P enrichment process and generates slag which can serve as a secondary resource of P. This chapter describes various technologies for P separation and recovery from steelmaking slag.

22.2 Steelmaking Slag

Steelmaking slag is a by-product from the processes where molten iron and steel scrap are converted to steel with excellent toughness and workability (Nippon Slag Association 2017a). Steelmaking slag consists of converter slag that is generated by a steelmaking converter (see Fig. 22.1) and electric arc furnace (EAF) slag that is generated by the process for recycling steel scrap as a raw material. Further, EAF slag consists of oxidizing slag from oxidation refining and reducing slag from reduction refining. Approximately 110 kg of converter slag is generated for each ton

Table 22.1 The typical composition of converter slag (BOF) and EAF slag (mass%) (Nippon Slag Association 2017b)

	CaO	SiO_2	T-Fe	MgO	Al_2O_3	S	P_2O_5	MnO
BOF	45.8	11.0	17.4	6.5	1.9	0.06	1.7	5.3
EAF	22.8	12.1	29.5	4.8	6.8	0.2	0.3	7.9

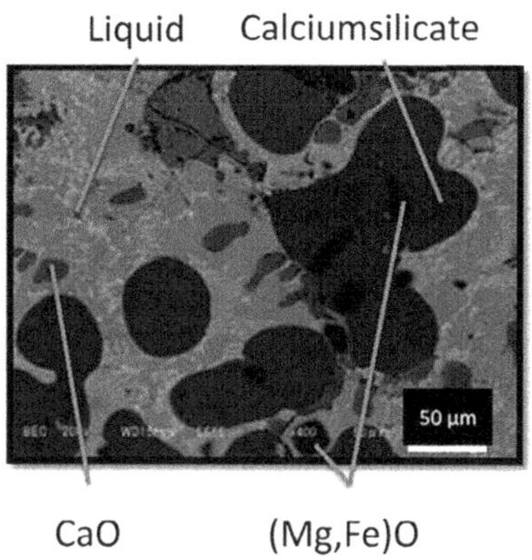

Fig. 22.2 SEM image of laboratory-prepared steelmaking slag (Miki and Kaneko 2015)

of converter steel produced. On the other hand, approximately 70 and 40 kg of EAF oxidizing slag and reducing slag, respectively, are generated for each ton of EAF steel produced.

The typical composition of steelmaking slag is shown in Table 22.1 (Nippon Slag Association 2017b). P is present in both converter slag (BOF) and EAF oxidizing slag. The fundamental system of steelmaking slag is CaO-FeO-SiO_2 which resides inside the primary 2CaO · SiO_2 region, suggesting that this phase precipitates at the early stage of cooling slag. P is typically concentrated in the 2CaO · SiO_2 phase as a 2CaO · SiO_2-3CaO · P_2O_5 solid solution.

Figure 22.2 shows a SEM image of simulated steelmaking slag which was prepared in laboratory and quenched after holding at 1623 K (Miki and Kaneko 2015). The SEM image shows that calcium silicate (2CaO · SiO_2) exists like an island in the liquid matrix phase. Since P is unevenly distributed in the steelmaking slag, there is a chance to effectively separate P from other elements. Magnetization curves for the liquid matrix phase, 2CaO · SiO_2 (C_2S), and 3CaO · P_2O_5 (C_3P) are shown in Fig. 22.3 (Yokoyama et al. 2007). Obviously, the magnetic properties of the matrix phase, 2CaO · SiO_2, and 3CaO · P_2O_5 are different from each other, suggesting the possibility to separate them using a magnetic force.

Importantly, the 2CaO · SiO_2 phase has characteristics which are potentially useful for separating it from the matrix phase. Namely, P is enriched in the 2CaO · SiO_2 phase which shows little interaction with a magnetic force. By contrast, the matrix phase, which is rich in FeO, has a distinctive interaction with magnetic force. In addition, the 2CaO · SiO_2 phase shows a relatively high solubility in water and has a small density compared to the matrix phase.

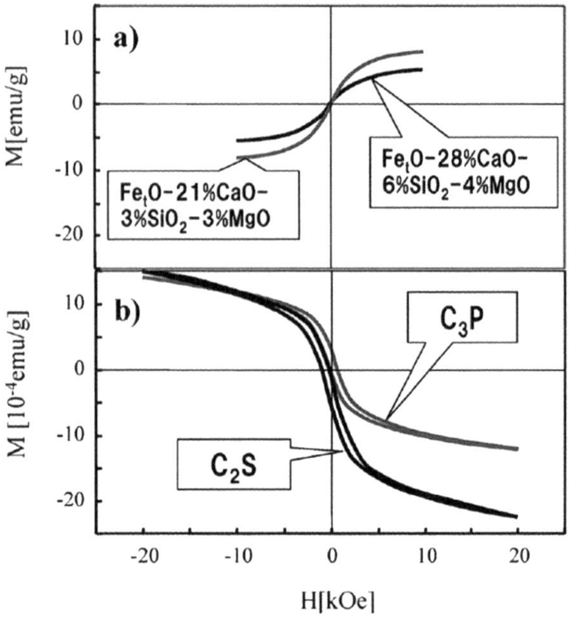

Fig. 22.3 The magnetization curves for the liquid matrix phase, $2CaO \cdot SiO_2$, and $3CaO \cdot P_2O_5$ (Yokoyama et al. 2007)

22.3 P Separation and Recovery

Several technologies have been available to separate and recover P from steelmaking slag. They involve magnetic separation, capillary action, dissolution, and carbothermic reduction.

22.3.1 Magnetic Separation (Yokoyama et al. 2009)

The P-rich $2CaO \cdot SiO_2$ phase is separated from the matrix phase using a magnetic force. After pulverization of steelmaking slag, powders are suspended into water. Magnetic separation can be conducted using an equipment as shown in Fig. 22.4. The matrix phase is preferentially magnetized and separated from the P-rich $2CaO \cdot SiO_2$ phase.

The effect of magnetic field strength and slag particle size on P recovery is shown in Fig. 22.5. Obviously, decreasing the particle size of slag increases the P recovery ratio. On the other hand, increasing the strength of magnetic field reduces the P recovery ratio, since the P-rich $2CaO \cdot SiO_2$ phase is gradually magnetized as the strength of magnetic field increases.

The effects of magnetic field strength on the ratio of unmagnetized and magnetized slag particles are shown in Fig. 22.6. Increasing the strength of magnetic field significantly decreases the recovery ratio of P-rich particles. However, the

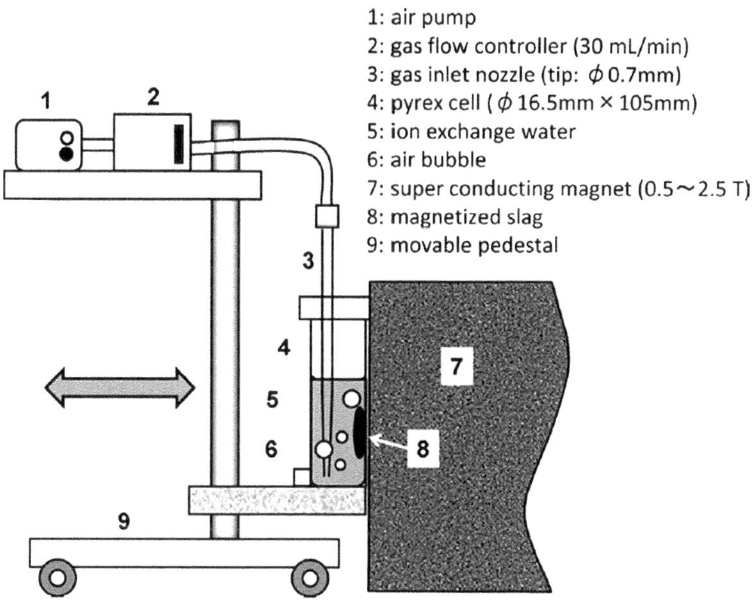

Fig. 22.4 An equipment for magnetic separation of steelmaking slag (Yokoyama et al. 2009)

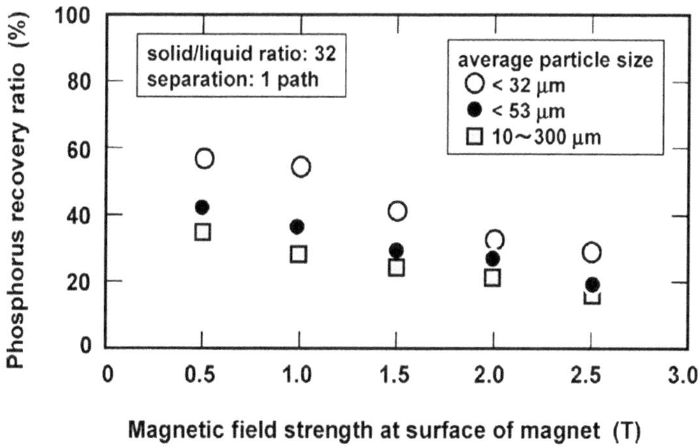

Fig. 22.5 Effects of magnetic field strength and slag particle size on P recovery (Yokoyama et al. 2009)

selectivity of P-rich particles increases with increasing the strength of magnetic field, because FeO-rich particles are removed by the magnetic force more effectively. Consequently, a trade-off exists between the P recovery ratio and the product selectivity for the magnetic separation technology. This technology can recover P from steelmaking slag at room temperature without the need to heat the aqueous suspension of slag particles.

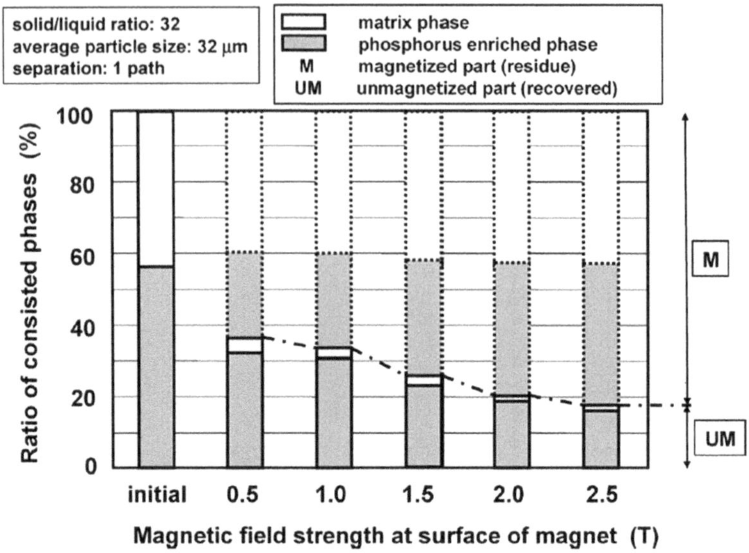

Fig. 22.6 Effects of magnetic field strength on the ratio of unmagnetized and magnetized slag particles (Yokoyama et al. 2009)

22.3.2 Capillary Action (Miki and Kaneko 2015)

Capillary action is the ability of a liquid to flow into a narrow space without any assistance of external force. The liquid matrix phase of steelmaking slag can be absorbed into a sintered CaO sphere due to the capillary action. Figure 22.7 shows the absorption of liquid matrix phase by a CaO sphere which was placed in a slag sample melted at 1773 K. After the liquid matrix phase is absorbed by the CaO sphere, approximately 90% of P can be recovered from the remaining solid $2CaO \cdot SiO_2$.

Figure 22.8 schematically shows the on-site separation between liquid and solid phases in steelmaking slag by capillary action. If the capillary action can be applied to the melting slag, no additional heat energy is required for the P enrichment.

22.3.3 Dissolution into Aqueous Solution (Numata et al. 2014)

Based on the high water solubility of P, it can be leached from the $2CaO \cdot SiO_2$ phase of steelmaking slag using acid. Figure 22.9 shows the SEM image of slag particles after P leaching. It can be seen that the $2CaO \cdot SiO_2$ phase is partially dissolved in water, making small holes on the surface of slag particles after P leaching.

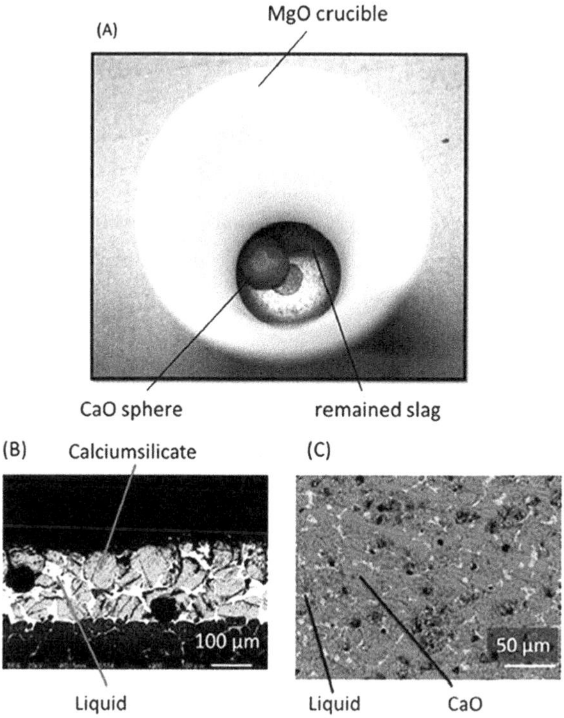

Fig. 22.7 The absorption of the liquid matrix phase of a slag sample by a CaO sphere. (**a**) An experimental apparatus used for the absorption of the liquid matrix phase by a CaO sphere due to capillary action. SEM images of the slag (**b**) and the CaO sphere (**c**) were taken after the absorption of the liquid matrix phase by the CaO sphere due to capillary action (Miki and Kaneko 2015)

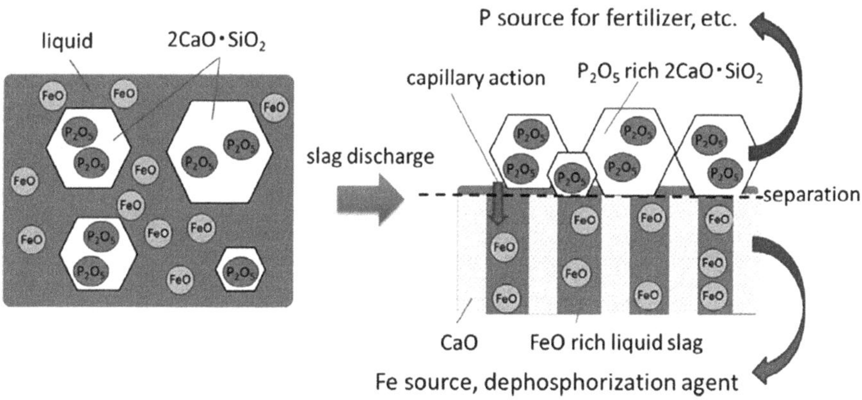

Fig. 22.8 The schematic diagram showing the on-site separation of liquid and solid phases in steelmaking slag due to capillary action (Miki and Kaneko 2015)

Figure 22.10 shows the time-course data on the release of Ca, Si, P, and Fe from slag in water at pH 7 (a) and 3 (b). Approximately 12% and 25% of slag were dissolved by 2 h at pH 7 and 3, respectively. This technology enables the recovery of P from steelmaking slag at room temperature.

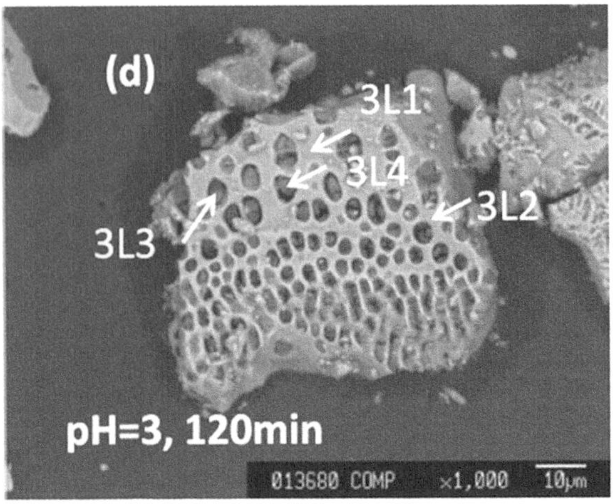

Fig. 22.9 SEM image of slag particle after P leaching with acid (Numata et al. 2014)

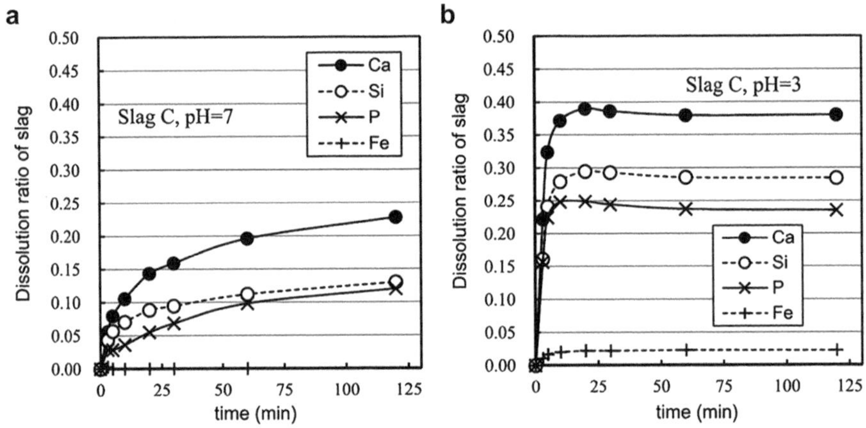

Fig. 22.10 Release of various elements from slag into aqueous solution at pH 7 (**a**) and 3 (**b**) (Numata et al. 2014)

22.3.4 Carbothermic Reduction

P oxides (P_2O_5) in slag can be reduced to elemental P (P_2 or P_4) by carbon at high temperatures, while iron oxides are reduced to metallic Fe. The carbothermic reduction of steelmaking slag has been examined using microwave heating (Morita et al. 2002). Up to 80% of P can be recovered into the Fe-P-C alloy. However, approximately 20% of P is estimated to be evaporated as gas by the microwave heating. The carbothermic reduction technology has a potential to produce P gas directly or to enrich P into Fe-C-P alloy. Further research is needed for optimizing the carbothermal reduction technology for P recovery from steelmaking slag.

22.4 Conclusions

Steelmaking slag is promising as a secondary P resource. Various technologies are potentially applicable to P separation and recovery from the P-rich $2CaO \cdot SiO_2$ phase of steelmaking slag. They involve magnetic separation, capillary action, dissolution, and carbothermic reduction. Since each method has its own merits and demerits, the best choice of single or combination of technology options is critical to effective P recovery and separation from steelmaking slag.

References

Miki T, Kaneko S (2015) Separation of FeO and P_2O_5 from steelmaking slag utilizing capillary action. ISIJ Int 55:142–148
Morita K, Guo M, Oka N, Sano N (2002) Resurrection of the iron and P resource in steel-making slag. J Mater Cycl Waste Manag 4:93–101
Nippon Slag Association (2017a) http://www.slg.jp/e/slag/kind.html. Accessed 30 Oct 2017
Nippon Slag Association (2017b) http://www.slg.jp/e/slag/character.html. Accessed 30 Oct 2017
Numata M, Maruoka N, Kim SJ, Kitamura S (2014) Fundamental experiment to extract phosphorous selectively from steelmaking slag by leaching. ISIJ Int 54:1983–1990
Yokoyama K, Kubo H, Nagasaka T (2007) Separation and recovery of P from steelmaking slags with the aid of a strong magnetic field. ISIJ Int 47:1541–1548
Yokoyama K, Kubo H, Nagasaka T (2009) Magnetic separation of P enriched phase from multiphase dephosphorization slag. Tetsu-to-Hagane 95:300–305

Chapter 23
Extraction of Phosphorus from Dephosphorization Slag

Takaiku Yamamoto and Masashi Nakamoto

Abstract In Japan, steelmaking slag is one of the most important secondary phosphorus (P) resources. The annual amount of P removed into steelmaking slag (c. 114 kt P/a) is approximately 2.8 times more than that of P imported as phosphate (P_i) rock to Japan. Increasing attention has been paid to P recovery from dephosphorization slag for mitigating the Japan's dependency on P import. This chapter describes the smelting reduction technology for extracting P from dephosphorization slag into iron. The potential of this technology has been examined using a pilot-scale test furnace having the capacity to treat dephosphorizing slag at 2 t/day. P could be extracted from dephosphorization slag, forming Fe-P alloy with the P content of approximately 10 wt%.

Keywords Dephosphorization slag · Smelting reduction process · P extraction · Fe-P alloy

23.1 Introduction

Phosphorus (P) is one of essential elements for life and a major component of fertilizer. P is also used for the manufacture of a wide variety of industrial products, including automobiles, electronics, medicines, fuel cells, and processed foods (Scholz et al. 2014). Today, P is mostly obtained from mined phosphate (P_i) rock. However, natural reserves of P_i rock are concentrated in only a limited number of countries such as Morocco, Iraq, and China. Japan has no significant reserve of P_i rock and is most entirely dependent on imported P. To mitigate the Japan's dependency of imported P, it is critical to develop technologies for recovering P from untapped secondary P resources.

T. Yamamoto (✉)
Kyoto University, Kyoto, Japan

M. Nakamoto
Osaka-University, Siuta, Osaka, Japan
e-mail: nakamoto@mat.eng.osaka-u.ac.jp

H. Ohtake, S. Tsuneda (eds.), *Phosphorus Recovery and Recycling*,
https://doi.org/10.1007/978-981-10-8031-9_23

In Japan, steelmaking is a big industry with a production output of 110 million tons of crude steel per year. In the steelmaking process, P resides in the raw materials such as iron ore, coal, and lime at concentrations as low as 0.1 wt%. Since P can cause a detrimental effect on the quality of steel produced, it must be removed nearly completely into steelmaking slag. The annual amount of P removed into steelmaking slag is estimated to be 114 kt/a (10^3 tons per year), which is about 2.8 times larger than that of P imported as P_i rock to Japan (Matsubae et al. 2014). Hence, increasing attention has been paid to P recovery from P-rich slag, called dephosphorization slag, for mitigating the Japan's dependency of P import. This chapter describes the smelting reduction technology for extracting P from dephosphorization slag into iron (Fe). The potential of this technology has been examined using a pilot-scale test furnace having the capacity to treat dephosphorizing slag at 2 t/day. Using this technology, P could be effectively extracted from dephosphorization slag into iron, forming Fe-P alloy with the P content up to approximately 10 wt%.

23.2 P Extraction from Steelmaking Slag

23.2.1 High-P Iron Alloy Manufacturing

Figure 23.1 shows the schematic diagram of a steelmaking process, focusing on the generation of dephosphorization slag and the process for the manufacture of P-rich iron alloy. In the steelmaking process, P resides in iron ore at concentrations of

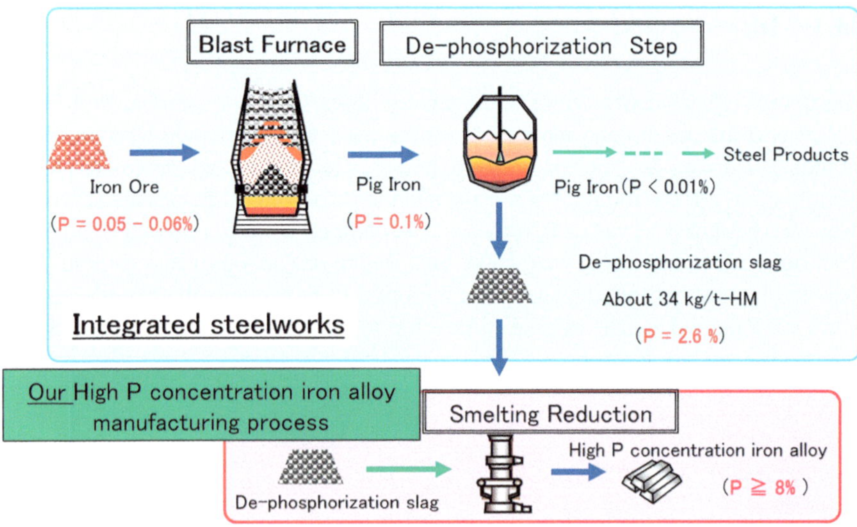

Fig. 23.1 Schematic diagram of steelmaking process focusing on the generation of dephosphorizing slag and the process for the P-rich iron alloy manufacturing

0.05–0.06 wt%. P_i is reduced to elemental P and captured by pig iron in the blast furnace. The concentration of P in the pig iron is typically about 0.1 wt%. P is then oxidized to P_2O_5 and removed from pig iron into dephosphorization slag having the P content of 2.0–3.5 wt% and the basicity (CaO/SiO_2) of 2.0–3.0. Typically, about 30–40 kg of dephosphorization slag is generated from each ton of pig iron produced. Then, the smelting reduction of dephosphorization slag allows to form Fe-P alloy containing P of 8 wt% or more. Iron can serve as a dissolving medium for P enrichment at this stage.

23.2.2 Metallurgical Experiments

Dephosphorization slag, which was obtained from a full-scale steelmaking plant, was pulverized into a fine powder of 1 mm or less. Ductile metal Fe was mechanically crushed into small pieces of 1 mm or less. The typical composition of dephosphorization slag used is shown in Table 23.1. The major components of dephosphorization slag are CaO (44.1 wt%) and SiO_2 (23.4 wt %). The total Fe content was 12.1 wt%, of which FeO and metal Fe contents accounted for about 60 and 40%, respectively. The content of P_2O_5 was 4.2 wt% (1.8 wt% as P).

As shown in Fig. 23.2, the melting temperature of dephosphorization slag is approximately 2000 °C (Inoue et al. 2012). To lower the melting temperature, SiO_2 and Al_2O_3 were added to dephosphorization slag to form SiO_2-CaO-Al_2O_3 system having SiO_2, CaO, and Al_2O_3 of 33.3, 45.6, and 21.0 wt%, respectively. This mixture was expected to melt at about 1400 °C (Inoue et al. 2012). A graphite crucible (28 mm ϕ) was filled with 30 g of the mixture and held for 1, 3, 5, or 17 h at 1450 °C under Ar flow of 100 ml/min. The graphite crucible could serve not only as a container but also as a reducing agent for P_2O_5 and FeO. The mixture was then cooled down using water in the crucible. For chemical analyses, the direct combustion infrared absorption method was used for determining carbon (C). Other elements were determined by the ICP emission spectroscopy.

Figure 23.3a shows the time dependence of total-Fe and FeO concentrations in slag after the start of smelting reduction. Both total-Fe and FeO concentrations in the slag decreased soon after the start of experiment, indicating that FeO could be readily reduced to metal Fe and removed from the slag. However, the reduction of FeO ceased around 17 h. Figure 23.3b shows the changes of P_2O_5 and MnO concentrations in slag. The content of P_2O_5 sharply decreased soon after the start of experiment. However, the rate of P_2O_5 reduction became low at 1 h. The MnO concentration in the slag was decreased relatively slowly compared with FeO and P_2O_5, suggesting that MnO is more difficult to reduce than FeO and P_2O_5. Actually, in the

Table 23.1 The typical composition of dephosphorization slag used in the present study

CaO	SiO_2	Al_2O_3	MgO	T-Fe	FeO	M-Fe	MnO	TiO_2	P_2O_5	S
44.1	23.4	1.1	1.9	12.1	6.1	4.1	5.9	1.4	3.1	<0.01 (wt%)

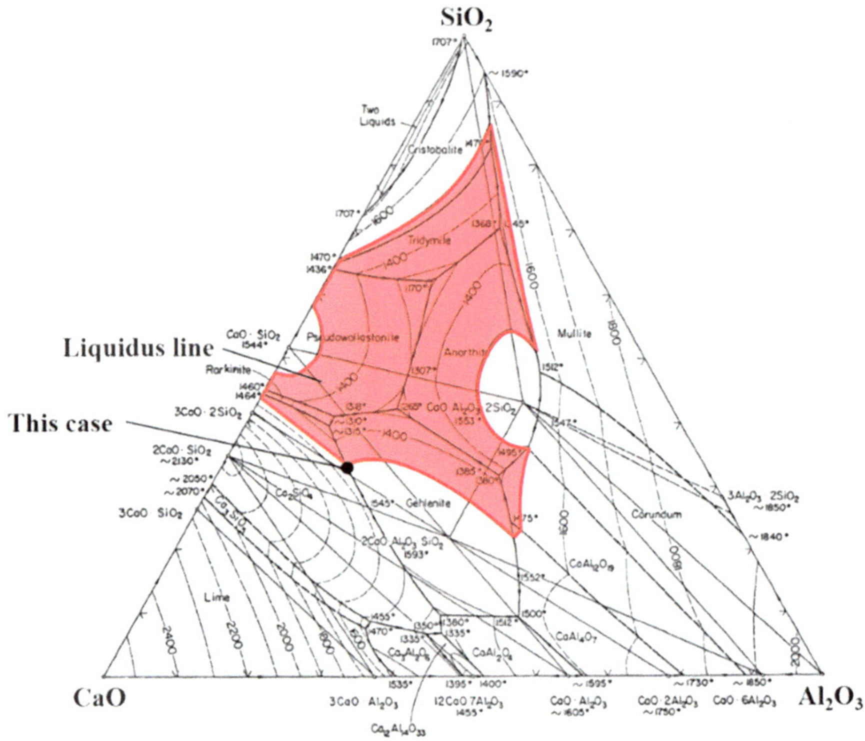

Fig. 23.2 SiO_2-CaO-Al_2O_3 state diagram from dephosphorization slag

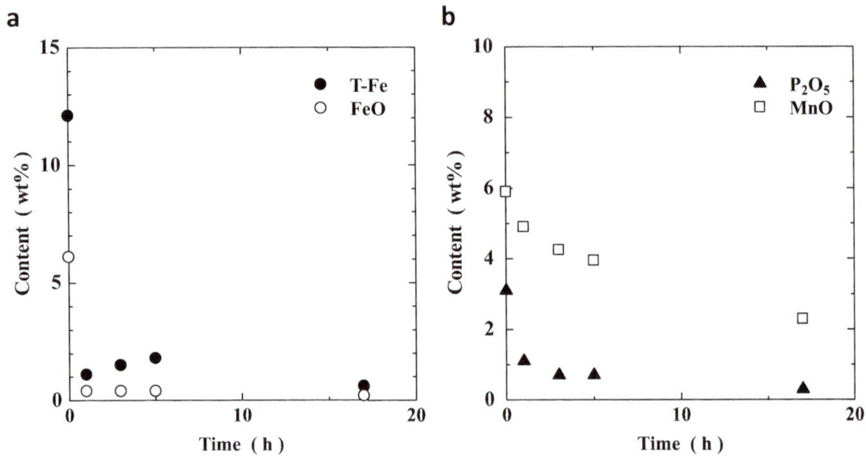

Fig. 23.3 Time courses of changes in total Fe and FeO (**a**) and P_2O_5 and MnO contents in slag (**b**)

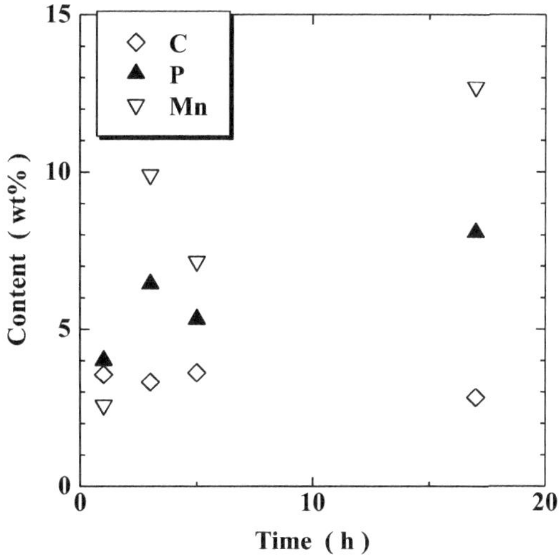

Fig. 23.4 Change of C, P, and Mn contents in the metal during smelting reduction treatment

Ellingham diagram, the equilibrium of $2MnO = 2Mn + O_2$ is located below those of $2FeO = 2Fe + O_2$ or $2/5P_2O_5 = 4/5P + O_2$.

Figure 23.4 shows the concentration changes of C, P, and Mn in metal. The C, P, and Mn concentrations increased to about 3.5, 4.0, and 3.8 wt% at 1 h. The concentrations of P and Mn increased up to about 8 and 13 wt%, respectively, at 17 h. As a result, the smelting reduction treatment could generate Fe alloy having a P concentration of about 8 wt% at 17 h.

23.3 Pilot-Scale Furnace Test

23.3.1 Pilot-Scale Furnace and Test Conditions

The demonstration experiments were carried out using a pilot-scale melting test furnace having the capacity to treat dephosphorization slag at 2 t/day (Fig. 23.5).

In the smelting reduction treatment, coke was used as a reducing agent. The bed coke was filled up to the secondary tuyere level and preheated with oxygen gas blasting from the primary and secondary tuyeres. After preheating, dephosphorization slag and additional coke were fed to the furnace. Oxygen was blasted into the furnace from both tuyere levels, thereby smelting dephosphorization slag and reducing Fe, Mn, and P to metal. The tests were conducted by varying conditions as shown in Fig. 23.6.

In experiments 3–1 and 3–2, 10% of silica stone was added as a melt aid. About 600 and 400 kg of coke were used in experiments 3–1 and 3–2, respectively. In experiments 4–1 and 4–2, coke was filled up to the upper part of the secondary tuyere to lengthen the reaction time. In addition, limestone was added to the slag in

Fig. 23.5 Pilot-scale module with a test furnace (**a**), the lower part of the test furnace (**b**), the inside view of the operating furnace (**c**), and the smooth tapping of molten iron with a high P content (**d**)

experiment 4–2 in order to lower the slag viscosity. In experiment 5–1, dephosphorization slag and coke were filled up to the secondary tuyere level to lengthen the reaction time.

23.3.2 Results of Pilot-Scale Tests

Figure 23.7a shows the relationship between the P and C concentrations in the metal. Obviously, there was a negative correlation between the P and C concentrations. This is consistent with the prediction from the liquid phase line of the Fe-C-P

<table>
<tr><td>Case</td><td>3 - 1</td><td>3 - 2</td></tr>
<tr><td>Material</td><td>Mixed Charge of Dephosphorization Slag 1000kg, Coke 600kg and Silica sand 100kg</td><td>Mixed Charge of Dephosphorization Slag 1000kg,Coke 400kg and Silica sand 100kg</td></tr>
<tr><td>Height of Material</td><td>Intermediate Height of Primary and Secondary Tuyere</td><td>Intermediate Height of Primary and Secondary Tuyere</td></tr>
<tr><td>Time</td><td>2hr</td><td>2hr</td></tr>
<tr><td>Schematic</td><td>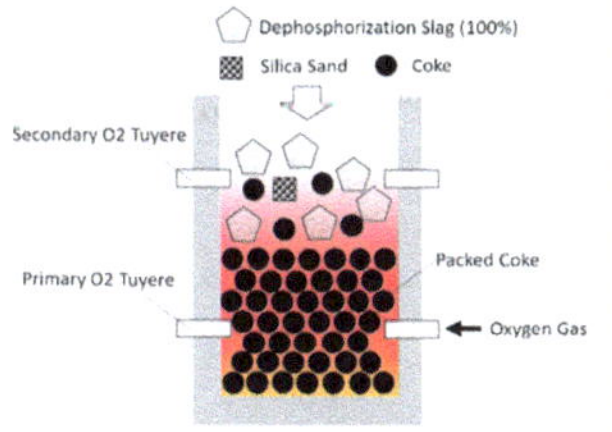
</td><td>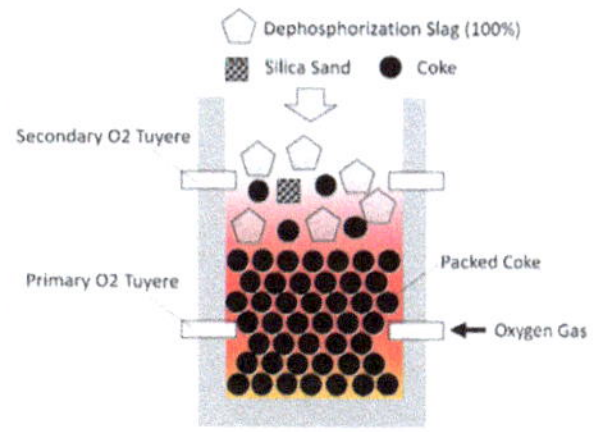
</td></tr>
</table>

<table>
<tr><td>Case</td><td>4 - 1</td><td>4 - 2</td></tr>
<tr><td>Material</td><td>Mixed Charge of Dephosphorization Slag 1000kg and Coke 600kg</td><td>Mixed Charge of Dephosphorization Slag 1000kg, Coke 400kg and Lime Stone 100kg</td></tr>
<tr><td>Height of Material</td><td>Upper of Secondary Tuyere</td><td>Upper of Secondary Tuyere</td></tr>
<tr><td>Time</td><td>2hr</td><td>2hr</td></tr>
<tr><td>Schematic</td><td>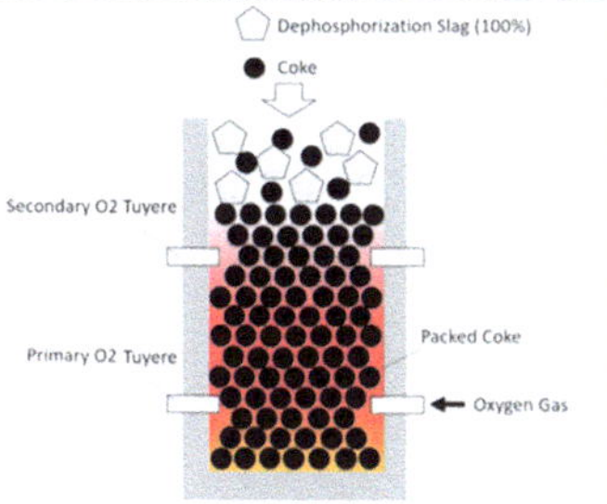
</td><td>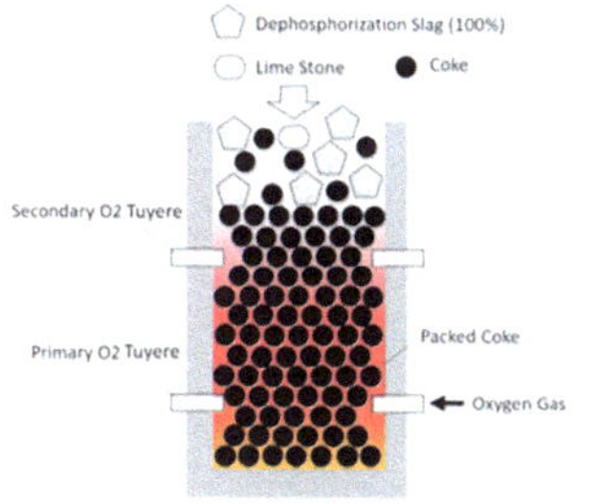
</td></tr>
</table>

<table>
<tr><td>Case</td><td>5 - 1</td></tr>
<tr><td>Material</td><td>Layered by Layer Charge of Dephosphorization Slag 600kgand Coke 300kg</td></tr>
<tr><td>Height of Material</td><td>Upper of Secondary Tuyere</td></tr>
<tr><td>Time</td><td>2hr</td></tr>
<tr><td>Schematic</td><td>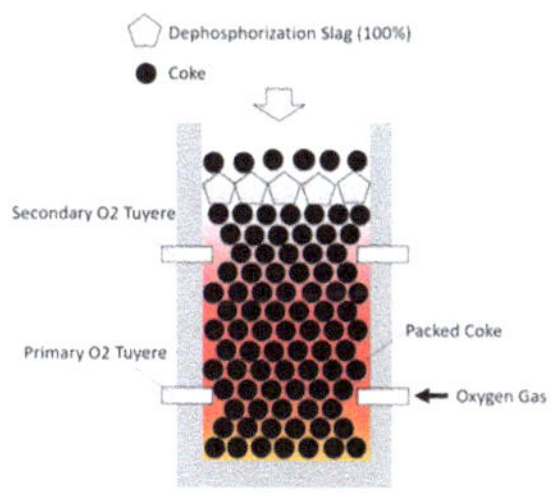
</td></tr>
</table>

Fig. 23.6 Five different conditions of smelting reduction experiments using the pilot-scale furnace

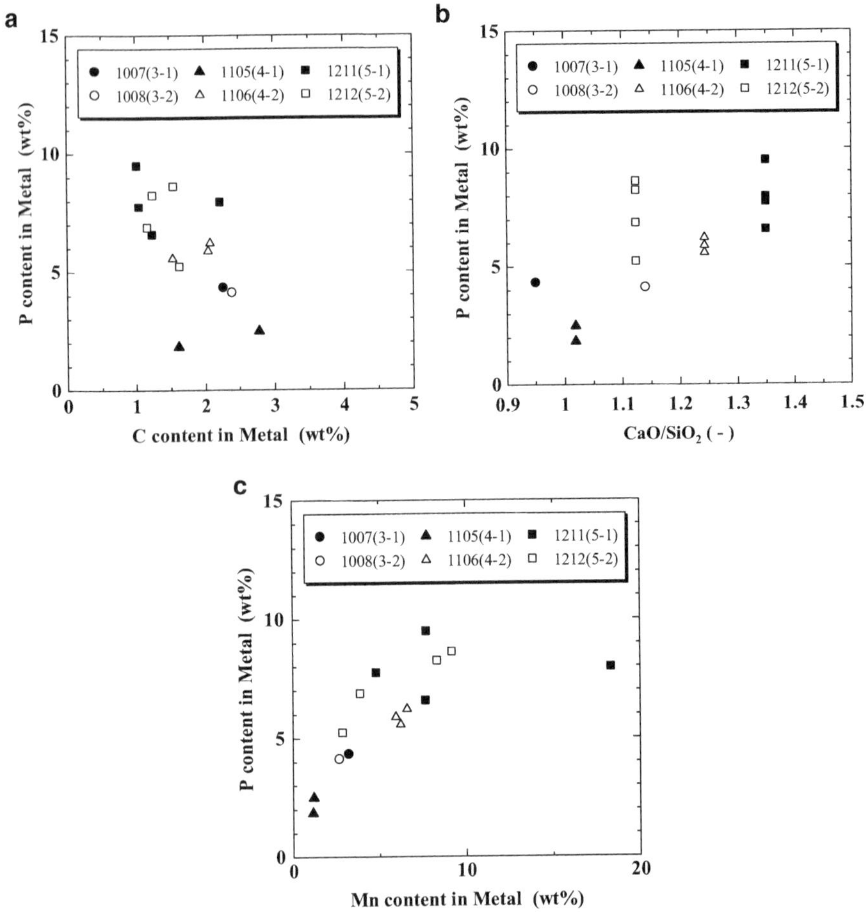

Fig. 23.7 (**a**) The relationship between the C and P contents in metal, (**b**) the relationship between the P content in metal and the CaO/SiO_2 ratio in slag, and (**c**) the relationship between Mn and P contents in metal in the pilot-scale experiments

system at 1515 °C (Atlas 1995). The maximum P concentration in the metal was about 9.5 wt%. Figure 23.7b shows the relationship between the P concentration in the metal and the basicity (CaO/SiO_2) of the slag. The concentration of P in the metal was likely to increase with increasing the ratio of CaO/SiO_2, thereby enhancing the fluidity of slag and thus the rate of smelting reduction. Figure 23.7c shows the relationship between the P and Mn concentrations in metal. Clearly, there was correlation between the P and Mn concentrations in the metal. The maximum concentration of Mn was about 18 wt%. The concentration of P_2O_5 in the slag was decreased to as low as 0.4 wt%.

23.4 Conclusions

The smelting reduction technology was applied to the extraction of P from dephosphorization slag into iron. The performance for P extraction was examined using a pilot-scale test furnace having the capacity to treat dephosphorizing slag at 2 t/day. P could be extracted from dephosphorization slag into iron, forming Fe-P alloy with the P content of approximately 10 wt%. The next step is to separate Mn from iron-P alloy.

References

Inoue R, Nakamoto M, Pahlevani F, Yamamoto T, Kitamura S, Ariyama T (2012) Proceedings of 5th international congress on science and technology of steelmaking, Dresden, Germany. CD-ROM

Matsubae K, Kubo Y, Yamasue H, Nagasaka T (2014) J Environ Conserv Eng 43:79

Scholz RW, Roy AH, Brand FS, Hellums DT, Ulrich AE (2014) Sustainable phosphorus management

Slag Atlas (1995) 2nd ed, p 105. Verlag Stahleisen GmbH, Düsseldorf

Chapter 24
Phosphorus in Steelmaking Processes

E. Yamasue and K. Matsubae

Abstract This chapter examines the role of iron ore phosphorus in the development of steelmaking. The prosperity of steelmaking nations strongly depended on how they strategically dealt with iron ores with various phosphorus contents. At the beginning of steelmaking, the ore's phosphorus content was immaterial. However, as the quality of steel improved, the effect of phosphorus became obvious. In Europe, iron ores with higher and lower phosphorus content were minable, prompting the development of the Thomas and Bessemer steelmaking processes, respectively. At the dawn of the Industrial Revolution, the main steel producer was the UK using the Bessemer process, but then it shifted to Germany adopting the Thomas process, followed by the USA, which used both open-hearth furnaces for scrap steels and the Thomas process until the mid-twentieth century. In these transitions, Germany found that slag, a steelmaking by-product with higher phosphorus content, could be used as phosphate fertilizer. Japan invented a phosphorus-oriented steelmaking process charging phosphate ore in addition to iron ore; in this process, slag with higher phosphorus was used as fertilizer. After World War II, Japan developed a highly efficient iron ore transpiration system using bulk carrier, and the introduction of the so-called Linz-Donawitz converter process with oxygen gas strongly promoted Japan to the steelmaking mainstream up to the present; most of their products contain extremely low phosphorus. Thus, innovative changes in steelmaking can be explained from the standpoint of the iron ore phosphorus content. From these summaries, future strategies for steel industries are discussed.

Keywords Iron ore · Thomas converter · Bessemer converter · Steel industry

E. Yamasue (✉)
Ritsumeikan University, Shiga, Japan
e-mail: yamasue@fc.ritsumei.ac.jp

K. Matsubae
Department of Environmental Studies for Advanced Society,
Graduate School of Environmental Studies, Tohoku University, Sendai, Japan

H. Ohtake, S. Tsuneda (eds.), *Phosphorus Recovery and Recycling*,
https://doi.org/10.1007/978-981-10-8031-9_24

24.1 Introduction

Although it seems that phosphorus (P) as a resource is not bound up with steel industries, in fact both are closely related. Because phosphorus content generally influences the mechanical properties of steel materials (mostly negatively), the issues still remain regarding the removal of phosphorus during the smelting and refining processes in steel production. Thus, one aspect of steel industry development can be regarded as the development of phosphorus removal (dephosphorization) processes. On the other hand, the removed phosphorus is distributed into steelmaking by-products called slags after the dephosphorization process. This slag would be one of the potential secondary resources of phosphorus in nations with mass steel production such as Japan, whereas sewage sludge is a common and important secondary P resource in other nations. That is, phosphorus has two aspects: it is unfavorable in terms of steel materials but a favorable element in terms of resource issues. Thus, it is of great interest to review the role of phosphorus in the history of the steel industry.

In this chapter, the development and innovation of steel industries involving phosphorus from past to present are summarized (Matsubae et al. 2016). Then, various iron ores used and their phosphorus content are briefly reviewed, followed by the introduction of countermeasures to remove phosphorus. At the early phase of steelmaking history, no attention was paid to phosphorus, but the development of finer steels could not ignore the existence of phosphorus. On the other hand, there are interesting examples of active utilization of phosphorus in the steel industries in a positive sense.

24.2 Phosphorus in Iron Ore

Here, the outline and characteristics of iron ore deposits are discussed. A reasonable definition of iron ore would be "ore that enables economical production of iron and steel reflecting the steelmaking technology in each era" (Inazumi 2009). Even if an iron ore has excellent quality and properties, it is merely a stone unless iron and steel can be obtained affordably. Iron reserves are also controlled by markets and technologies (Shimomura 1995). From this perspective, we should realize that an iron ore in one era is different from that in another era.

24.2.1 Sedimentary Deposit

According to the 1955 United Nations statistical survey (United Nations 1955), reserves of iron ore in Europe alone totaled 6413 billion metric tons (t) of metallic iron. Disregarding grade and scale, iron ore deposits can be found not just in Europe,

Table 24.1 The scale of the iron ore deposit and the ore grade

Iron ore deposit		Typical deposit	Ore grade(%) Finest >66	High 65–63	High middle 63–58	Middle ~50~	Low 35–25	Major iron ore	Impurities
Sedimentary deposit	Banded iron formation (BIF)	Hamersley	Middle	Large			Super large	Hematite	
		Carajas						Magnetite	
		Bailadila							
	Channel iron deposits (CID)	Yandi	–	Large	Large			Hematite	
		Robe River						Magnetite	
	Minette deposit	Lorraine-Clinton				Middle		Siderite	**P**, S
								Goethite	
	Carboniferous deposit	England coal mine				Middle		Siderite	S, **P**
Magmatic/hydrothermal deposit	Orthomagmatic deposit	Bushveld				Small		Hematite	Ti, V, Cr
		Panzhihua							
	Skarn deposit	Magnitoya			Middle			Hematite	S, Cu
	Iron oxide copper-gold (IOCG) deposit	Kiruna				Large	Small	Magnetite	**P**, Cu,
								Hematite	REE
	Hydrothermal deposit	Bilbao			Small			Magnetite	Mn
								Hematite	
Laterite deposit		Cuba				Small		Hematite	Cr, Ni
Iron sand deposit		New Zealand				Small		Hematite	Ti
Scale	Small, <0.1 giga ton; middle, 0.1–1 giga ton; large, 1–10 giga ton; super large, >10 giga ton								

but all over the world, and carry the distinctive feature that small-scale steel manufacturers in any region can aim for local production and consumption. Although there are incomparably large deposits with other metallic elements, extremely high-graded ores, almost the same as the theoretical grade of pure iron oxide, are mined.

Table 24.1 shows the major types of iron ore deposits around the world and their respective scales (Kassner and Ozier 1950; Shimamura 1953; Parak 1985). The largest iron ore reserve is the Carajás iron deposit, containing 18 billion t of iron ore with an average iron grade greater than 67%. The excellent deposit formation properties of iron set it apart from other metals and can be attributed to the specific geochemical nature of iron as an element. There is also a remarkable difference in solubility in water between bivalent and trivalent iron ions. With each dissolution, diffusion, and precipitation of iron in the Earth's crust, various sedimentary deposits were formed in the seabed, rivers, lakes, and marshes dating back to ancient times. Among these are large-scale deposits, known as banded iron formations (BIFs), where the grade of iron in alternating bands of iron oxide and silica averages 35%. These accumulated in the Earth's seabed from approximately 2.7 to 1.8 billion years ago during the Precambrian era and are estimated to have reached a scale of 40 trillion t of iron ore.

Some BIFs on the seabed upheaved and became dry land; others remain buried beneath the ocean floor. These deposits have undergone hydrothermal processes induced by rainwater and igneous rock. The leaching of silica and other impurities has enriched the iron, leading to high-grade iron ore deposits. The steelmaking industry in the modern world primarily utilizes this type of ore, with 370 billion t of reserves having been proved, including 170 billion t of economically feasible ore (Japan Oil Gas and Metals National Corporation…). There are, however, regional differences in the nature of formation. The formation of European BIF-type deposits is limited to strata dating back to approximately 1.9 billion years ago during the Precambrian era. These reserves of BIF deposits in Europe are small, at 34 million t, and, even when expressed as a percentage by deposit type within Europe, stand at a low 5% (Besta et al. 2011). The natural distribution of these resources, with fewer high-purity BIF-type deposits in Europe than in other areas of the world, coupled with the fact that these deposits were limited to Northern Europe, influenced the manner in which the European steelmaking industry developed.

Many of the other sedimentary deposits in Europe are deposits that have accumulated since hundreds of millions of years ago in geological time. Of these, the so-called Minette deposits are perhaps the most well-known, formed when bivalent iron in river water was deposited as trivalent iron along the coasts. Most iron ores are composed of iron oxide minerals including hematite, goethite, limonite, and magnetite. Hematite is mainly Fe_2O_3 and is 70% of the weight of iron, goethite is FeOOH and is 63% of the weight of iron, limonite is a mixture of hydrated iron oxides with up to 60% of the weight of iron, and magnetite is Fe_3O_4 and is 72% of the weight of iron. The phosphorus content is high in limonite ore mixed with hematite. A representative example is the deposit in Lothringen in the Alsace-Lorraine region of France near the German border, with an average iron content in the ore of 31% and a total iron content of 6.7 billion t. The United Kingdom (UK) had the

Fig. 24.1 A picture of bog iron ore mined in Finland (by authors)

Cleveland mines, which rivaled those of France, and Germany had the Salzgitter and various other iron deposits. In terms of amount of iron, European reserves total around 4 billion t, and from a global perspective, it is a region with considerable production.

Although smaller in scale than Minette deposits, there also are the sedimentary-type iron ore deposits of carbonaceous siderite formed in the Carboniferous coal beds in Europe. Although they can be found all over continental Europe, they are especially common in the UK (reserves of over 1 billion t). This type of ore helped relieve the supply limits of conventional bog iron resources during the first half of the Industrial Revolution and supported the expansion in production.

Bog iron ore deposits are small-scale limonite deposits formed in lakes and marshes, where portions become solid layers. Figure 24.1 shows an example of common bog iron ore mined in Finland. Many exist all over Europe and were the primary raw materials for steelmaking during the Middle Ages. Iron in the exposed areas of massifs and in the gravel surrounding deposits is dissolved by reductive surface water and flows into lakes and marshes where it is deposited as an oxide containing phosphorus. It is said to have continued in its small-scale form even to the present day. Although the ore can be easily mined and reduced, the phosphorus content is not always low. In Sweden, where lakes and marshes in forested areas are particularly plentiful, the deposits are also abundant. Although there is a range of variance in phosphorus levels, low-phosphorus iron ore deposits are numerous compared to those of other regions as shown in Table 24.1.

24.2.2 Igneous Deposits

Iron ore deposits from magmatic igneous activity were also formed throughout Europe but are limited to regions with past igneous activity. Igneous deposits can be classified as follows: magmatic deposits, formed when iron in magma segregates and is deposited in the deep strata; contact metasomatic deposits, formed when magma rises and comes into contact with limestone and other materials; and hydrothermal deposits and pneumatolytic deposits, formed when magma rises even further into the Earth's crust and the iron in hydrothermal fluids is deposited. With the exception of magmatic deposits, igneous deposits are usually smaller in scale than the submarine sedimentary BIF and Minette deposits.

The grade of ore from igneous deposits tends to be inferior in purity. However, although few in number, certain deposits produce high-grade ore with good purity. In Europe, magmatic deposits can be found in the Kiruna district and central regions of Sweden, and although high in phosphorus, they are high in iron ore content and are also large in scale with reserves of 1.5 billion t. Besides these deposits, a wide variety of medium-scale contact metasomatic deposits can be found in central Sweden (reserves 3.8 million t) (Shimamura 1953). The Dannemora mine is a contact metasomatic deposit in limestone famous for its low phosphorus content. In this region, because the phosphorus content differs so dramatically from one deposit to the next, and there are bog iron deposits as well, it is believed that the mode of deposition changed over the ages.

Compared to other areas in Northern Europe, hydrothermal deposits are formed in geologically young regions. They are the Bilbao-type deposits found by low-phosphorus hematite or siderite as a result of magmatic hot water enacting its effect in the limestone parent rock. Reserves are recorded at 400 million t (United Nations 1955) and have been Europe's high-grade iron source since ancient times (Beck 1895). Well-known examples of similar deposits are the Cumberland in the UK, the Bilbao in Spain, and the Erzberg mine in Austria. In the UK, hematite is formed in limestone, whereas in continental Europe, siderite ore bodies are formed when the crust is weathered by hematite-limonite; thus, although the deposit type is the same, the formation processes are different. In addition to Bilbao-type deposits, there are small-scale deposits of ore veins primarily formed from hot water in a number of countries, most notably in Germany. The siderite deposits in Siegerland (reserves of 40 million t) are well-known, are high in manganese, and were formed by the weathering of limonite near the Earth's surface.

24.3 Steelmaking Technology Development

24.3.1 Early Steelmaking Without Considering P in Iron Ore

How did civilizations utilize the iron ores described in the previous section? In the ironmaking by means of bloomery, such as the Stuck furnace and Renn furnace in the Middle Ages (Agricola 2013), iron was locally produced and consumed. Minable ores were widely and equally used for ironmaking. Although even a minute amount of phosphorous deteriorates steel properties, less attention was paid to phosphorus content in the ore used at that time, until Sven Rinman clarified its effect on steel properties in the eighteenth century in Sweden. According to modern metallurgy, phosphorus does not solute into solid iron but into molten slag under equilibrium between the solid iron and the molten slag at lower temperatures (Hino 2009). Because ancient bloomery was operated under lower temperatures and higher oxygen potential and slag was removed during the smithing (forging) process, the effect of phosphorus would not be very high, and products could be applied to daily use, unless iron ore with extraordinarily high phosphorus content was used. Nevertheless, it seems that some differences in strength and brittleness by mining deposits were empirically recognized, although it would be multiple effects not only by phosphorus but also by carbon, sulfur, and other elements. However, long experience finally determined the superiority of iron ore with lower phosphorus. In particular, Swedish iron ore was highly rated.

In the era of the blast furnace, where operational conditions shifted to higher temperatures and lower oxygen potential, reduced phosphorus dissolved into the molten iron (hot metal). At that time, a puddle process was employed for steel production. In the puddle process, molten (pig) iron was "puddled" to promote a reaction of carbon in the molten iron with oxygen in the air, resulting in the production of semi-molten wrought iron. During the puddling, phosphorus was converted to iron phosphide, resulting in dissolution into puddle slag. Because the wrought iron was then subjected to a slag-off process, the amount of phosphorus (0.4%) in molten pig iron was decreased to 0.03% in wrought iron (Inazumi 2009).

As described previously, the effects of phosphorus were not revealed. Nevertheless, steel with lower phosphorus content was empirically known to exhibit better properties. Despite this knowledge, the center of ironmaking in the UK shifted from mining districts producing iron ore with lower phosphorus to those producing iron ore with poor grade (higher phosphorus) but with a large scale from the seventeenth to the nineteenth centuries. This implies that they faced a lack of iron ore due to the drastic increase in demand from the Industrial Revolution and therefore had to use lower-grade ore.

24.3.2 Early Steelmaking Considering P in Iron Ore

In the middle of the nineteenth century, the Bessemer converter, which used acidic refractories, was invented in Britain. It represented a radical change in the method of steelmaking. The key principle behind the converter is the removal of impurities from the iron by oxidation with air being blown through the molten iron. No further energy is needed because of the exothermic reaction between the carbon in molten pig iron and the oxygen in the air.

In the Bessemer converter period, the smelting temperature became higher, and even phosphates (P_2O_5) in acidic refractories were reduced into phosphorus, leading to dissolution into molten steel. That is, in principle, the phosphorus is scarcely removed by the Bessemer converter using acidic refractories. When molten iron with higher phosphorus by the puddling method was supplied to the Bessemer converter, the resulting steel was of no use. Swedish Dannemora ore with lower phosphorus contributed to overcome such difficulties and simultaneously proved the effect of phosphorus in the molten process. However, because reserves of lower-phosphorus ore in Europe were quite limited, a novel process, which could treat molten iron (hot metal) with higher phosphorus, was required. After a string of failures (Beck 1895), the Thomas converter was finally invented in 1878.

The bottom-blown design of the Thomas converter was almost the same as that of the Bessemer converter, but it employed a basic refractory, resulting in the efficient removal of the phosphorus. This development led to further increases in the amount of steel being produced. It should be noted that the drastic changes in the amount of production and in the methods of steelmaking existed from country to country, taking advantage of the invention of the Thomas converter.

The UK continued using the Bessemer converter, although iron ore with higher phosphorus such as Minette ore was domestically minable. This is because the UK practically monopolized the global steel market with high-grade Bessemer steel at that time, and conservative forces that were afraid of the use of higher-phosphorus ore were reluctant to adopt the Thomas converter (Beck 1895). They kept up the production of Bessemer steel, importing lower-phosphorus ore from Spain and Sweden, and relatively slowed the increase in steel production, even after 1880. However, the UK was not able to keep pace with Germany, which rather quickly adopted the Thomas converter, leading to a decrease in the UK's market share. Notably, the use of domestic (and therefore inexpensive) iron ore was economically attractive for the UK too, but their attempts to use higher-phosphorus ore caused unstable operation, because of the difference in ore composition other than phosphorus. Namely, because iron ore from the UK contained higher sulfur and lower lime (Einecke 1907), lime addition and high-temperature operation were required, which was difficult to control with the technological level of the day. As a result, the UK lost price competitiveness and shifted to open-hearth furnace technology with basic refractory in twentieth century, which is suitable for molten iron with intermediate phosphorus content.

On the other hand, Germany made significant progress by introducing the Thomas converter with Minette ore. At that time, Germany was experiencing a high level of economic growth and imported a large amount of steel from the UK. Thus, they were very hungry for steel production using domestic iron ore. It was lucky for Germany to be able to use untapped Swedish and Minette ores with higher phosphorus in the Alsace-Lorraine region, the latter of which was ceded from France as a result of the Franco-Prussian War. This indicates that the invention of the Thomas converter contributed to the rapid progress of German steelmaking. It should be noted that Germany was also blessed with domestic coal mining. Another factor that helped Germany gain ground on the UK at the end of the nineteenth century was that the converter slag, with its high phosphorus content, was used as fertilizer. This reduced the county's dependence on imported fertilizer and led to significant gains in agricultural productivity (Beck 1895; Trinder 1992). The Bessemer converter was still used, aiming at differentiation to produce high-grade steels; the combination of the Thomas converter and an open-hearth furnace successfully established a technology to produce high-grade steels. It is of great interest that the phosphorus content in the ore was one of the factors that influenced the production structure of steelmaking.

Later, a different competitor appeared outside the European countries. That is, the USA made rapid advances for different reasons from those of Germany. In the middle of the nineteenth century, a large-scale iron ore deposit with lower phosphorus and higher iron content was discovered around the Great Lakes. The Bessemer converter was exceptionally effective, and the US surpassed the UK. Around the end of the nineteenth century, the reserves of deposit satisfying the Bessemer converter ($P < 0.08\%$) had been pushed to the limit. To use iron ore in which the phosphorus content was higher for the Bessemer converter and lower for the Thomas converter, the Bessemer converter was gradually replaced with an open-hearth furnace with basic refractory that could also treat scrap steels. In this manner, the US came to produce half of steel in the world, supplanting the UK at the beginning of the twentieth century. Then, the mainstream of steel production in the middle of the twentieth century used both an open-hearth furnace and the Thomas converter with basic refractories; in other words, the basic (steelmaking) process or the Thomas process was followed by modern steelmaking.

24.3.3 Modern Low Phosphorus-Oriented Steelmaking

In the period of postwar recovery after World War II, the Linz-Donawitz (LD) converter process with forced oxygen gas was developed in Austria to address their iron source circumstance. They had less scrap steel and sufficient domestic (pig) iron. The terms Linz and Donawitz are the names of Austrian cities. Unlike the Bessemer converter process with forced air, the production rate of the LD converter was remarkably high, about five times as fast as an open-hearth furnace, and the saved energy by the nitrogen-less process could allow molten iron with lower silicon and/or phosphorus; a basic process could accept molten iron with relatively higher phosphorus, and the product with lower nitrogen was cost-competitive with that produced by a basic open-hearth furnace. Because the Japanese iron source circumstance was similar to that of Austria at that time, the LD converter was suitable for Japan.

When Japan introduced the LD converter process, it could use BIF-hosted iron ore with phosphorus less than 0.5%, a value that was comparable to the standard for the Bessemer converter ($P < 0.045\%$). The appearance of bulk carriers such as the very large ore carrier (VLOC), Capesize, and Panamax overcame the cost issue due to long-haul transportation, and then Japan, a resource-poor country, could produce fine steels. The content of phosphorus in the molten iron produced in Japan was actually changed from 0.3% to 0.8% in the 1950s to more or less 0.1% in the 1960s (Inazumi 2009). A large amount of molten iron with higher quality strongly promoted the LD converter process, and it was highly competitive with the Thomas converter and open-hearth furnace with basic refractories in the Western countries. On the basis of its stable operation, Japan constructed a continuous casting system for the first time in the world, and the refining process to purify steel was being improved. The Western countries followed the Japanese LD converter process, and the use of iron ore with higher phosphorus finally receded.

24.4 Application of Lower-P Steel to High-Performance Pipeline

The reason for dephosphorization of steel is that elemental phosphorus tends to segregate in steels, leading to a decrease in toughness and weldability. Thus, lower phosphorus content is generally preferable for many steels. Then, how low is the required phosphorus concentration? In this section, some concrete values are introduced by referring to examples of pipeline steels.

The global energy supply system uses a large amount of pipelines made of steel. Pipelines are especially indispensable for the extraction and transportation of natural gas, which is the primary energy source in the world. The properties of the pipeline determine the safety and cost of the entire transportation system, and therefore the required quality level of pipelines is very high. Various characteristics are required for pipelines. In particular, corrosive gas resistivity and arrest property against propagating shear fracture are two important properties. The former strongly relates to the occurrence of hydrogen-induced cracking (HIC). To prevent HIC, the dispersion of inclusions and segregation of components should be well controlled. From this point of view, desulfurization is effective under a pH >5, but dephosphorization is required when pH = 3. Lower phosphorus further contributes to an increase in steel strength by raising the permissible concentration of manganese (Takeuchi et al. 2004).

In a high-pressure pipeline, once a crack is initiated, the crack may propagate several kilometers with a speed of 100–400 m/s despite the ductile fracture. This does not occur in the case of liquid transportation and is called a propagating shear fracture. To arrest (prevent) the propagating shear fracture, the control of sulfur and phosphorus concentration in the steel is of great importance. A pipeline with X65 grade, which is one of the grades regulated by the American Petroleum Institute (The American Petroleum Institute 2007), containing $C = 0.08$, Mn = 1.38, $P = 0.021$, and $S = 0.003$ mass% (Hara et al. 2012), shows a lower ability to arrest the propagating shear fracture. Only when the contents of phosphorus and sulfur satisfy $P = 0.006$ mass% and $S = 0.005$ mass%, corresponding to X100 grade (Makino and Takeuchi 2010), can good performance be achieved to arrest the propagating shear fracture. Although a concentration of 0.021 mass% generally sounds like a sufficiently low value, such a value does not always satisfy required performance.

24.5 Effective Use of Phosphorus in the Steel Industry

Steelmakers attempt to distribute phosphorus to slag as much as possible, resulting in the production of a large amount of slag with higher phosphorus content. How do (or did) they utilize the slag? In this section, some examples of phosphorus stocks and flows in the slag in Japan, which is a mass producer of steel, are introduced.

24.5.1 Dissipation and Use of Phosphorus in the Steelmaking Slag

Stocks and flows of phosphorus have been well summarized in the case of Japan (Kubo et al. 2008; Matsubae-Yokoyama et al. 2009). According to Matsubae and colleagues (2011), in 2005, the amount of phosphorus imported to Japan as phosphate ore (101 kt (kilo tons) P) was almost the same as the phosphorus in the steelmaking slags generated in Japan (104 kt P). In 2016, the latter is about 2.5 times larger than the former (see Chap. 1). Although the amount of sewage treatment sludge (55 kt P), which also includes phosphorous, is large enough to be considered similar to that of other countries, the slag seems to have an advantage because its generation is localized around steelmaking plants. It would be a characteristic and advantage for nations equipped with large-scale steelmaking industries.

However, the most common use of the slag is as a raw material for the cement industry or as road construction material. The intrinsic value of phosphorus as a fertilizer ingredient is lost when the slags are used as such materials. This indicates "dissipation" or "infinite dilution" of phosphorus. It should be noted, however, that many researchers from academia as well as from steel industries energetically engage to discover effective utilizations of steelmaking slag as a phosphorus resource. The use of phosphorus in slag can be classified into two approaches: the use of slag as a phosphorus donor without direct extraction of phosphorus and the extraction of phosphorus using chemical means. Detailed examples of the latter have been mentioned in previous sections.

As an example of the former, some processed slags from blast furnaces are commercialized as fertilizers, which supply not only phosphorus but also calcium, magnesium, manganese, boron, silicon, and other elements. More concretely, they are commonly used in rice farming, but in this case, they mainly supply silicon and alkalis rather than phosphorus. It is reported that they contribute to the soil stability at higher pH levels to prevent blights such as a clubroot, because converter (basic oxygen furnace) slags contain moderate alkalis (Murakami et al. 2004). However, this use is still under investigation, and the use of phosphorus in the slags is not the main purpose. In Japan, molten iron is generally pretreated to thoroughly remove phosphorus, silicon, and other elements, resulting in the generation of large amount of slag with higher phosphorus content. Reflecting the recent rise in phosphate ore price, slag containing more than 3% phosphorus can now be commercialized as a phosphate fertilizer by the revision of the fertilizer standard in March 2005, which opened the way for the use of slag (Yao 2005). As another example, the use of the converter slag as a breeding ground (reef) for fish was proposed (JFE Website…). Dissolved iron from the slag contributes to the increase in phytoplankton, leading to an increase in zooplankton, and finally bringing about the recovery of fishing ground.

24.5.2 *Effective Use of Phosphorus in the Steel Industry*

An interesting steelmaking process considering phosphorus production was operated in past Japan (1935–1957) in the former NKK corporation (currently JFE Steel after combination), which was called the "Japan-style Thomas process" (Iron and Steel Institute of Japan Steel Science and Technological History Committee 1982). This process modified the Thomas process using iron ores with higher phosphorus so that Japan could produce Thomas phosphate (a slag containing higher phosphorus content made from the Thomas converter), even using iron ores with lower phosphorus by adding phosphate ores. This was a kind of interindustrial process, pursuing both steel and fertilizer production.

At that time, it was difficult for Japan to obtain iron ores including higher phosphorus. The use of the Thomas converter was not realistic, and therefore an open-hearth furnace was employed to process stably supplied scrap steel from the US. The Bessemer converter using iron ores with lower phosphorus and sulfur was additionally used to produce fine steels at that time. In 1923, however, a novel process charging both iron ores and phosphate ore to the Thomas converter had been proposed in Japan (Imaizumi 1927, 1940). This process could simultaneously produce molten (pig) iron with higher phosphorus and the Thomas phosphate, namely, phosphate fertilizer.

As a phosphate ore to be charged into the Japan-style Thomas process, guano from oceanic islands containing 21.5–23.5% P_2O_5 as well as domestic iron ore with higher phosphorus and slag from open-hearth furnaces were used. Around 1942–1943, 350–380 kt of pig iron containing 2% phosphorus were produced. It seemed that there was no remarkable difference in the operability compared with an open-hearth furnace, and the higher fluidity due to the higher phosphorus content was rather an advantage.

The Japan-style Thomas process had the following advantages (Imaizumi 1940; Kikuchi and Iwafuji 1940; Horie 1939a, b): first, the Thomas phosphate with fertilizer response was replaceable with phosphate ore to produce phosphate fertilizers. Then, the sulfuric acid to produce phosphate fertilizers could be reduced. From the viewpoint of steelmaking, higher productivity was expected; namely, the refining process itself ended in 15 min regardless of the charged molten iron amounts, followed by 15 min of incidental processes. Thirty minutes of steelmaking contributes to a reduction in equipment costs. Further, limitations of iron ore were relaxed, and the combination with the open-hearth furnace brought about diversity in the steelmaking process.

The quality of the steel produced by the Japan-style Thomas process is an object of interest. It seemed that decarburization and dephosphorization were easy, but the phosphorus content in the steel was actually higher to produce the Thomas phosphate with higher phosphorus. However, the nitrogen content in the steel was four to five times higher than that produced by open-hearth furnaces because of the bottom-blown air. Higher nitrogen content causes brittleness, and the demand for finer steel revealed this issue. At that time, an oxygen converter was invented, and the usefulness of the converter drove the Japan-style Thomas converter out of Japan's steel industry, resulting in the operation of the LD converter.

24.5.3 Phosphorus As an Alloying Element in Steel Materials

Ordinary steels contain phosphorus less than 0.03–0.04 mass %. Generally, phosphorus is known as an unfavorable element because it reduces the steel's toughness, but some positive effects are also evident. Good examples of active utilization of phosphorus as an alloying element for steel materials include solid solution-strengthened sheet steel with $P \geq 0.05$ mass % and precipitation-strengthened austenitic stainless steel with $P = 0.2$–0.4 mass %. In terms of the strength-elongation relationship, the solid solution strengthening by phosphorus shows a better trade-off between the decrease in rupture elongation and the increase in tensile strength than precipitation strengthening and work hardening. Further, the performance of solid solution hardening by phosphorus for a steel is outstandingly better than that by other elements such Si, Mn, and Cr (Katoh et al. 1984; Matsufuji et al. 1983; Ushioda et al. 1994).

Phosphorus is well-known as an alloying element for steel to improve corrosion resistance and weatherability. The "COR-TEN A" by Nippon Steel & Sumitomo Metal containing 0.07–0.15% phosphorus and 0.25–0.5% Cu is a good example (Nippon Steel & Sumitoo Metal. COR-TEN...). Nakayama reported that phosphorus-alloyed or phosphorus-and-copper-alloyed hot rolled steel for vehicles shows slightly lower or similar weatherability performance compared with galvanized sheet steel (Nakayama et al. 1990). As other examples, phosphorus improves the machinability of steels. Onishi reported that the addition of phosphorus into even sulfur-based free-cutting steel resulted in better performance (Onishi et al. 1955). In addition, the effects of phosphorus on creep deformation are of interest. Although phosphorus induces boundary segregation in the case of ferritic heat-resisting steel, leading to embrittlement during creep deformation and tempering, it contributes to the refinement of carbides in the case of austenitic steel, resulting in the improvement creep rupture strength and ductility (Tanaka et al. 1970; Matsuo et al. 1973; Kegg et al. 1974).

24.6 Discussion

24.6.1 Available Iron Ore in the Future

Because of recent mass production in China, a shortage trend of iron ore with lower phosphorus is seen, which causes an increase in the ore price. This could be regarded as a sign of innovation needed to use a new generation of iron ore. Iron ore that remains unused because it is difficult to use is still abundant. Few reserve statistics for such unused iron ores are available because they are not regarded as iron ore. A part of the BIF-type deposits is a candidate as an abundant next-generation iron ore. The BIF-type deposit consists of iron-rich and iron-poor lithofacies, the latter of which contains lower iron with lower phosphorus. Although some BIF-type deposits contain highly graded iron ore, they are presently unused because of their higher phosphorus content. Although iron oxide copper-gold (IOCG) ore deposits are massive despite their igneous nature, they are still unused because they include nonferrous metals and phosphorus. Presently, the depletion of phosphorus resources is an urgent issue on a mass global scale, so the effective utilization of iron ore with higher phosphorus content can be important and challenging subjects in near future.

24.6.2 Phosphorus Price and Role in the Steel Industry

Because the phosphorus content in steelmaking slags is generally low and the current phosphorus price is relatively low, the value of slag as a phosphorus resource is low. Therefore, of course, steel materials are the main products in the current steel industry. Will our society enter a new epoch when the phosphorus price increases in the future? Given that the steelmaking process is not only the concentrating process of iron but also that of phosphorus, the steelmaking process can be regarded as rather a phosphorus making process in an imaginary extreme case. The steelmaking slag by-produced from the dephosphorization process of molten pig iron generally contains approximately 5% P_2O_5 (Kubo et al. 2010), which is the highest value in the whole process. Nevertheless, this value is regarded as far from satisfactory from an industrial perspective, compared with natural phosphate ore containing more or less 30% P_2O_5. However, that is a misunderstanding. Even in the case of natural mines, the grade of crude ore is not always high. For example, because the amount of tailings to obtain a unit (kg) of refined ore in Jordan is 7.7 (kg) (Kobayashi and Sago 2001), the overall grade of phosphate ore in the mine is equivalent to 3.5% P_2O_5 at most, which is lower than that in the dephosphorization slag described previously. Although the other impurities in slag are different from those in phosphate ore, the slag is attractive enough to be a secondary resource of phosphorus, if the phosphorus price increases drastically (Yamasue et al. 2013; Yamasue 2015). The fact that there are no naturally occurring radioactive materials (NORMs) would further promote the value of slag.

24.6.3 Phosphorus-Oriented Steelmaking

Among the possible upcoming issues in the steel industry is the deterioration of iron ore. In other words, we must consider the use of iron ore with higher phosphorus content, which the mainstream of modern steelmaking hesitates to use. In the current steelmaking process, especially in Japan, a two-step dephosphorization process is performed to produce fine steels with extremely low phosphorus content. In the former dephosphorization process, the phosphorus content in the slag is relatively high but is still insufficient as a secondary phosphorus resource. In the latter process, a large amount of slag with rather lower phosphorus content is generated because the distribution coefficient is constant. To improve this situation, a kind of flexible pig iron production with higher phosphorus content should be taken into account as was seen in Japan, followed by thorough dephosphorization.

References

Agricola GA (2013. (originally 1556) De Re Metallica. Dover Publications, New York

Beck L (1895) Die Geschichte des Eisens in technischer und kulturgeschichtlicher Beziehung. F. Vieweg und sohn

Besta FF et al (2011) Petrology and geochemistry of the banded iron formation (BIF) of Wadi Karim and Um Anab, Eastern Desert, Egypt: implications for the origin of Neoproterozoic BIF. Precambrian Res 187:15

Einecke G (1907) Der Eisenerzbergbau und der Eisenhüttenbetrieb an der Lahn, Dill und in den benachbarten Revieren: eine Darstellung ihrer wirtschaftlichen Entwicklung und gegenwärtigen Lage (translated in Japanese in 1953 by T. Shimamura). The Japan Iron and Steel Federation

Hara T, Shinohara Y, Doi N (2012) The conditions of occurrence of separation for line pipe. In: 22nd international offshore and polar engineering conference. Rhodes, Greece

Hino M (2009) Nishiyama Kinen Kouza. NMS-ISIJ 199:1–18

Horie M (1939a) On the manufacturing process of the Thomas-phosphate manure. Jpn J Soil Sci Plant Nutr 13(1):21–36

Horie M (1939b) On the manufacturing process of the Thomas-phosphate manure. Jpn J Soil Sci Plant Nutr 13(2):93–102

Imaizumi K (1927) Rational reform of our steel industry, and its relation to the policy of fertilizer. Tetsu to Hagane 13(9):701–711

Imaizumi K (1940) On the start of basic Thomas steelmaking process (in Japanese) (translated by author). Tetsu to Hagane 26(3):183–191

Inazumi T (2009) Kinzoku 79(7): 648–654

Iron and Steel Institute of Japan Steel Science and Technological History Committee (1982) History of a oxygen steelmaking way in Japan. Iron and Steel Institute of Japan

Japan Oil Gas and Metals National Corporation. Material flow on mineral resources. Available from: https://www.nssmc.com/product/catalog_download/pdf/A006.pdf. (in Japanese)

JFE Website. Available from: http://www.kobelco.co.jp/pr/1183662.html

Kassner JL, Ozier MA (1950) Determination of phosphorus in iron ore. Anal Chem 22:194–195

Katoh H et al (1984) In: Proceedings of symposium on metallurgy of continuously annealed cold-rolled sheet steel. AIME, Detroit

Kegg GR, Silcock JM, West DRF (1974) The effect of phosphorus additions and cooling rate on the precipitation of M23C6 in austenite. Metal Sci 8(1):337–343

Kikuchi K, Iwafuji M (1940) Comparative study of the quality of Thomas Steel and open hearth furnace steel. Tetsu-to-Hagane 26(12):865–875

Kobayashi H, Sago R (2001) A study on life cycle assessment of energy consumption and CO2 emissions in the manufacturing and transportation processes of nitrogen and phosphate fertilizers. Jpn J Farm Work Res 36(3):141–151

Kubo H et al (2008) Application for material flow accounting to phosphorus in Japan with a focus on its accumulation. J Environ Eng Manag 18(1):47–53

Kubo H, Matsubae-Yokoyama K, Nagasaka T (2010) Magnetic separation of phosphorus enriched phase from multiphase dephosphorization slag. ISIJ Int 50(1):59–64
Makino H, Takeuchi I (2010) Full scale burst test of high strength line pipe-part 2: fracture arrestability of gas transmission pipeline by X100 Line Pipe. In: International seminar on application of high strength line pipe 2010. Xi'an, China
Matsubae K et al (2011) Virtual phosphorus ore requirement of Japanese economy. Chemosphere 84(6):767–772
Matsubae K et al (2016) Innovations in steelmaking technology and hidden phosphorus flows. Sci Total Environ 542:1162–1168
Matsubae-Yokoyama K et al (2009) A material flow analysis of phosphorus in Japan. J Ind Ecol 13(5):687–705
Matsufuji K et al (1983) Effect of phosphorus on the deep drawability of cold rolled steel sheets. Tetsu-to-Hagane 69(10):1303–1311
Matsuo T, Shinoda T, Tanaka R (1973) Effect of nitrogen, boron and phosphorous on high temperature strength of 18Cr-10Ni and 18Cr-10Ni·Mo austenitic steels bearing small amounts of titanium and niobium. Tetsu-to-Hagane 59(7):907–918
Murakami K et al (2004) J Soil Sci Plant Nutr 75:53
Nakayama T et al (1990) Effect of alloying elements on perforation corrosion resistance of hot rolled steel for automotive bodies. Tetsu-to-Hagane 76(8):1333–1340
Nippon Steel & Sumitoo Metal. COR-TEN. Available from: https://www.nssmc.com/product/catalog_download/pdf/A006.pdf. Accessed on 1 Aug 2017
Onishi T, Hiura T, Masuda T (1955) Some properties of free-cutting steel, ~low-carbon resulphurized steel and rephosphorized steel~. Tetsu-to-Hagane 41(9):973–974
Parak T (1985) Phosphorus in different types of ore, sulfides in the iron deposits, and the type and origin of ores at Kiruna. Econ Geol 80:646–665
Shimamura T (1953) World iron ore resource. Iron and Steel Institute, Japan
Shimomura Y (1995) Topics of the technological history of iron and steel in the twentieth century. The Japan Iron and Steel Federation, Tokyo
Takeuchi I et al (2004) Development of high strength line pipe for sour service and full ring evaluation in sour environment. In: 23rd international conference on offshore mechanics and arctic engineering
Tanaka R, Shinoda T, Tobe Y (1970) The effect of Mo, N, and C on high temperature properties of 25%Cr-28%Ni heat resisting steels. Tetsu-to-Hagane 56(8):1014–1033
The American Petroleum Institute (2007) Specification for line pipe ANSI/API specification 5L, 44th edition
Trinder B (1992) The Darbys of Coalbrookdale, 2nd edn. Phillimore, Chichester
United Nations (1955) Survey of world iron ore resources. United Nations, New York
Ushioda K et al (1994) Proceeding of international forum on IF steels. ISIJ, Tokyo
Yamasue E et al (2013) Using total material requirement to evaluate the potential for recyclability of phosphorous in steelmaking dephosphorization slag. J Ind Ecol 17(5):722–730
Yamasue E, Matsubae K, Ishihara KN (2015) Weight of land use for phosphorus fertilizer production in Japan in terms of total material requirement. Glob Environ Res 19:97–104
Yao Y (2005) Shinhiryo "Rinsan-Keisan-Hiryo" ni tsuite. Kikan Hiryo 102:128

Part V
Animal Manure

Chapter 25
Energy-Effective Carbonization Technology

Kozo Ucda

Abstract Pig farms in Japan tend to concentrate in the countryside owing primarily to livestock feed transport costs. Nearly all livestock waste generated in these areas is converted into composts or dry fertilizing materials according to Japanese agricultural regulation, but because the demand for composts and dry materials of livestock waste arises only in spring and autumn, a substantial need exists for developing new technologies to effectively use surplus pig slurry. This chapter describes the development of a sustainable system for recycling pig slurry into organic fertilizer through carbonization. The Energy-Free Carbonizing for Resource Recovery (EFCaR) system can self-generate enough heat for carbonization and produce biochar that is useful as a fertilizing material for organic fertilizer like steamed bone meal. The biochar generated from pig slurry shows the water solubility higher than those of phosphate rock and pig slurry burning ash (cinder). The EFCaR system is a viable solution to address pig slurry surplus in concentrated swine feeding areas.

Keywords Livestock · Pig manure · Pig slurry · Biochar · Pyrolysis · Carbonization

25.1 Introduction

The Japanese livestock industry relies on the import of grains from abroad. In particular, pig feed is nearly all imported. There are several ports of discharge for grains, and pig farms are concentrated in these areas to save transport costs. Pig slurry contains high levels of phosphorus (P), potassium, and nitrogen which are potentially recyclable for growing crops. Conventionally, pig slurry has been mostly composted and distributed as an organic fertilizer to neighboring crop producers. This system has been an efficient, environmentally friendly option for valorizing pig manure. In recent years, however, with an upswing in large-scale pig farming induced by the hope of gaining larger profits, the volume of compost exceeds the

K. Ueda (✉)
Hitachi Zosen Corporation, Environment Business Headquarters, Osaka, Japan
e-mail: ueda_k@hitachizosen.co.jp

H. Ohtake, S. Tsuneda (eds.), *Phosphorus Recovery and Recycling*,
https://doi.org/10.1007/978-981-10-8031-9_25

capacity of nearby crop fields, requiring pig farmers to cover considerable costs of transport and ship their compost to more remote areas. In addition, overfertilization of crop fields has led to the pollution of groundwater. To solve these problems, the Japanese government introduced the Act on the Appropriate Treatment and Promotion of Utilization of Livestock Manure in 1999, obligating farmers to adequately manage, treat, store, and recycle pig slurry. At present, more than 99.8% of pig farmers practice some treatment options such as composting, aerobic fermentation, and drying (Katiku haisestubutu no kanri to riyou no gennjo to taisaku nitsuite 2015), but the volume of dried and composted manure is still increasing in concentrated areas, making it difficult for farmers to efficiently manage and use pig manure.

Meanwhile, Japan relies most entirely on the import of phosphate (P_i) rock and chemical fertilizer. Rises in food demand stemming from global population growth, however, have driven up the CIF value of phosphate fertilizer to about twice the level of 15 years ago (Sinyou joukasouodei karano rin kaishu rikatsuyouno suisin ni mukete 2013). Thus securing the supply of phosphorus resources with a reasonable price is very important for the agricultural sector in Japan. Although intensive research has been conducted to develop the technology for recovering P_i from sewage sludge and steelmaking slag, they are not economically feasible options for pig farmers. The Energy-Free Carbonizing for Resource Recovery (EFCaR) system has been developed to solve both the environmental problems caused by livestock manure and the challenge of resource recycling for stable food supply.

25.2 Livestock Waste Management in Japan

25.2.1 Conventional Waste Management System

The livestock industry in Japan flourishes in several areas, the largest being the southern region of the island of Kyushu. The number of pigs bred in Kagoshima Prefecture alone is more than 1.3 million, while that in Miyazaki Prefecture is over 0.8 million (Dote 2010). The two prefectures combined generate an estimated 5 million tons of pig slurry each year. Nearly all of the slurry is converted into compost by aerobic fermentation, but the demand for compost in the areas is smaller than the supply and is limited to spring and autumn. As a result, pig farmers are pressed to spend a great deal of money either to transport the compost to more remote areas or to build large storage tanks. Based on the nitrogen content, the two prefectures combined supply more than 250 kg of livestock manure per hectare of arable field (Wada 2013), meaning it is difficult to use any more compost in the areas. Two methods of aerobic fermentation are typically employed to treat pig slurry. One is pile up aerobic fermentation (Fig. 25.1) and the other vertical aerobic fermentation (Fig. 25.2). The pile up method uses a machine with rotary paddles that move back and forth on the pile for air mixing. Although this entails relatively low costs and low energy consumption, it requires a large yard and long treatment time extending to about

Fig. 25.1 Pile up aerobic fermentation

Fig. 25.2 Vertical aerobic fermentation

2 months for compost production. The moisture content of the slurry ranges from 90% to 95%, while that of the compost is less than 40%. The vertical method uses a smaller cylindrical tank with a stir bar arranged from bottom to top. The stir bar rotates slowly to mix the pig slurry. The slurry is fed from the top of the tank and air from the bottom. This requires only 2 weeks to produce compost but consumes a large amount of electric power.

25.2.2 New Approach by Hitachi Zosen

Hitachi Zosen analyzed samples of compost produced from pig slurry at different pig farms ($N = 38$) (Table 25.1) and, based on the results, calculated the P_2O_5 content of compost produced throughout Miyazaki Prefecture. The volume of pig slurry generated in Miyazaki amounts to 640,000 tons per year. Presuming that 143,000 tons (about 22.4%) of this is dry matter and that the average P_2O_5 content of dry matter is 3.5%, the quantity of P_2O_5 generated in Miyazaki is calculated to be 5000 tons per year. Japan imports from countries such as China, Jordan, South Africa, Morocco, and Vietnam about 240,000 tons of phosphate rock annually (Trade Statistics of Japan 2016). The quantity of P_2O_5 in imported phosphate rock is estimated to be about 72,000 tons (assuming the P_2O_5 content as 30 wt%), indicating that Miyazaki produces P_2O_5 in the quantity equivalent to about 6.9% of the imported P_2O_5. Similarly, Kagoshima generates annually 1 million tons of pig slurry, of which 8000 tons is P_2O_5. Together, the two prefectures produce P_2O_5 in the quantity equivalent to about 18% of imported P_2O_5.

Hitachi Zosen has interviewed several pig farmers in Miyazaki and Kagoshima about the practical issues related to surplus pig slurry. Most common answers are as follows:

1. Transport costs are higher than the price of compost.
2. Demand for compost arises only in spring and autumn so the storage is needed during summer and winter.
3. Long-term stockpiling is difficult because composts absorb moisture and emit an unpleasant odor.

To solve these problems, the volume of the compost and the content of volatile matter must be reduced. An effective option may be thermal treatment. Then the study analyzed the value of thermal treated materials as fertilizer by comparing the total, citrate-soluble, and water-soluble P_2O_5 contents with those of imported phosphate rock, burned ash, and char produced from pig slurry (Table 25.2).

Table 25.1 The main components of the compost generated from pig slurry in Miyazaki Prefecture

	DM	Ash	Organic matter	C	N	H	O	P_2O_5	K_2O
	%								
Miyazaki ($n = 38$)	22.4	13.8	86.2	44.8	3.6	6.4	31.4	3.5	2.0

Table 25.2 Comparison of phosphate rock, char, and burned ash from pig slurry

Sample	① Total P_2O_5 (%)	② Ammonium citrate soluble P_2O_5(%)	Water soluble P_2O_5 (%)	②/① × 100 (%)
Phosphate rock	31.6	2.1	<0.01	6.7
Biochar	10.0	6.9	0.49	68.6
Cinder (ash)	22.9	13.3	0.19	58.1

Table 25.3 Major phosphate salts detected from X-ray diffraction analysis of biochar and burned ash from pig slurry

Sample	Major phosphate salts
Biochar	$KMgPO_4$, $Mg_3(PO_4)_2$, $AlPO_4$, $FePO_4$, $Ca_9MgNa(PO4)_7$ etc.
Cinder (ash)	$Ca_9MgK(PO_4)_7$, $Ca_9MgNa(PO_4)_7$, $Ca_{18}Na_3Fe(PO_4)_{14}$, $Ca_{18}Mg_2H_2(PO_4)_7$, $KMgPO_4$ etc.

Phosphate rock contained the highest concentration of total P_2O_5 but a very low concentration of citrate- and water-soluble P_2O_5. Cinder ash had a higher citrate-soluble P_2O_5 but a lower P_i solubility (citrate- and water-soluble P_2O_5/total P_2O_5) than biochar. Unlike biochar, burned ash contained no organic carbon and thus a higher content of P_2O_5. However, biochar had the highest P_i solubility. The P_2O_5 of burned ash presumably binds with calcium. X-ray diffraction analysis showed that the elements binding with phosphate changed from magnesium to calcium during burning process (Table 25.3). Based on these results, Hitachi Zosen has decided to convert pig slurry into biochar for the use as a fertilizing material.

25.3 Hitz Carbonization System

25.3.1 Energy-Effective Carbonization

In a common carbonization process, the burning reaction consumes oxygen, thereby increasing the temperature in the retort. However, this process generates ash (cinder) along with char in the retort. Since ash has the low solubility of P_2O_5 as shown in Table 25.2, its generation decreases the bioavailability of P_i in biochar. To avoid the occurrence of burning reaction in the retort, air supply to the retort must be shut off. The retort, therefore, needs to be heated from the outside. In general, the gas generated from the pyrolysis of organics has enough calories for heating the retort. Thus, the indirect heating kiln method that uses pyrolysis gas as an energy source for heating the retort from the outside was employed.

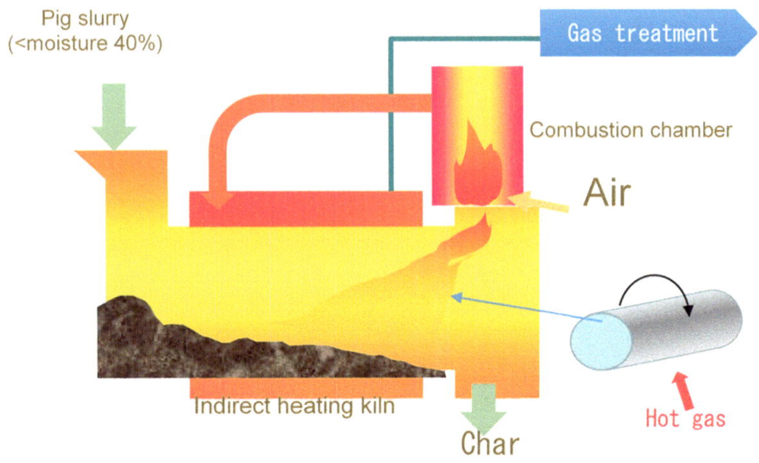

Fig. 25.3 Indirect heating kiln process using the EFCaR system

A pilot-scale test was conducted to study the energy balance of the indirect heating kiln system. The kiln used in the test was 4.5 m long and 0.58 m in diameter and had no insulation at inside and outside. It had a rotating in the outside of retort whose inner wall was covered with fire-resistant material. Both edges of the kiln were fitted with a housing, and the rotating and stationary components were sealed to shut out air. The rotation speed was set at 1 rpm, and the retention time of pig slurry in the kiln was about 1 h. The rotary kiln was heated indirectly from the outside of retort to maintain the temperature at 400–450 °C. The pig slurry was continuously fed to the kiln, converted to biochar in the kiln, and then withdrawn to an outlet housing with a jacket screw feeder. The biochar was cooled down from 450 to 15 °C (Fig. 25.3).

Pyrolysis gas from the kiln is sent to a combustion chamber mounted on the outlet housing. Air was supplied to this combustion chamber to raise the temperature up to at least 800 °C. The flue gas from the combustion chamber was mixed with air for cooling the gas below 700 °C. This gas was used to heat the rotary kiln from the outside. As a result, the gas temperature was cooled down to about 300–350 °C. The gas was sprayed with water in a gas cooler to cool the temperature down to 170 °C. The remaining dust was collected and neutralized with slaked lime in a bag filter. As mentioned above, the EFCaR system can use organics as a heat-energy source to generate biochar. To that end, the moisture content of pig slurry needs to be lower than 40%. About 2% of the total energy consumption of the EFCaR needs to be supplied from a pilot burner.

Fig. 25.4 EFCaR system pilot plant

25.3.2 *On-Site Experiments*

To sell pig slurry char to fertilizer companies, several points must be clarified:

1. Stable supply throughout a year
2. Quality meeting the regulation standards
3. Quantity sufficient for the demand

With the financial support from the Ministry of Agriculture, Forestry, and Fisheries, Hitachi Zosen implemented a pilot-scale plant (Fig. 25.4) at a compost center operated by Kobayashi City in Miyazaki Prefecture to conduct demonstration tests.

Kobayashi City, located in southern Miyazaki, has a total land area of 56,000 hectares, of which 12% is farmland, 73% is forest, and 15% is dedicated to residential quarters. The annual production value of the agriculture cluster amounts to 30 billion yen, of which 70% comes from the livestock industry. The compost center in Kobayashi collects 2300 tons of pig slurry, 2400 tons of cow slurry, 1300 tons of chicken slurry, and 2000 tons of wastewater sludge annually. They are composted by aerobic fermentation. The annual production of livestock manure is about 510,000 tons, of which 130,000 tons is pig slurry. In this area, almost all of pig slurry is composted using aerobic fermentation. Since the volume exceeds the demand in the farmland, pig farmers need to cover the costs of transporting the compost to remote areas. Moreover, since the demand for compost arises only in spring and autumn, farmers need to pay additional costs for storage during the off-season. The demonstration tests of EFCaR system were conducted for 1 month each season from autumn 2014 to summer 2015. The demonstration tests used pig slurry (solidified layer) with a moisture content of 90%. The pig slurry collected from pig farms was dried by aerobic fermentation in a vertical reactor (Fig. 25.2) to a moisture content of 40%. The reactor was equipped with an air blow system that continuously supplied air from the bottom of the tank to the reactor, thereby increasing the temperature to 60 °C and reducing the moisture content by evaporation. The retention time of the slurry in the reactor was about 2 weeks.

25.4 Fertilizer Evaluation Test

25.4.1 Seasonal Variation in Pig Slurry Quality

Figure 25.5 shows the seasonal change of the P_2O_5 content in pig slurry char. The P_2O_5 content in pig slurry char showed only a small fluctuation between seasons. The test results showed:

1. The P_2O_5 content of char averaged 9.5% ($n = 36$). The coefficient of variation was with 6.3%. This enables production of biochar with stable ingredients.
2. Of the heavy metals regulated for fertilizer, cadmium was not detected (<1.0 mg/kg) and arsenic was negligibly low (1.2 mg/kg) compared to the Japanese standard for fertilizer.
3. Varying temperatures between 400 and 450 °C did not effect on the quality of char, indicating that 400 °C is enough to obtain stable P_2O_5 levels.

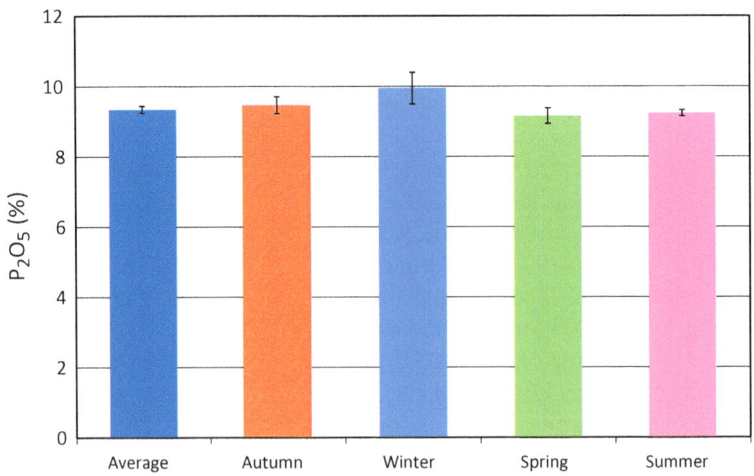

Fig. 25.5 Seasonal changes in the P_2O_5 level of pig slurry char

25.4.2 Field Test

In cooperation with Osaka Prefectural Research Institute of Environment, Agriculture, and Fisheries, Hitachi Zosen has conducted field tests to study the fertilizer efficacy of pig slurry. The results showed that the char is a viable alternative to organic fertilizer as a source of phosphorous and that the continuous use causes neither a decline in the quality of product nor a deficiency or excess of inorganic nutrients. Moreover, the char could serve as a potential source of potassium in organic fertilizer. More than 90% of phosphorus in the char was citric acid-soluble. Since the levels of toxic heavy metals are below those of the Japan's fertilizer regulation, the char can also be used as a fertilizing material without chemical treatment. Germination tests in a Neubauer pot showed that the char was a viable alternative to steamed bone meal as a source of phosphorous for Japanese mustard spinach. Growth tests were also conducted using a Wagner pot, confirming the previous results that the continuous use of the char causes neither a decline in the quality of product nor a deficiency or excess of inorganic nutrients (Table 25.4).

Furthermore, fertilizer manufacturers have also conducted evaluation tests on the char for functions such as fertilizer effect, water absorption, and portability. The results were as follows:

1. Cadmium was not detected (<1.0 mg/kg), and arsenic concentrations were extremely low (1.2 mg/kg) compared to the company standards for fertilizing materials.
2. Pig slurry contains copper and zinc because pig feed contains these minerals for body weight gain. Average (n = 17) levels in the char were 377 mg Cu/kg and 1098 mg Zn/kg. Although they were not exceptionally low, the levels were acceptable under the Japanese regulations even when organic fertilizer containing 5% char is continuously applied to soil.
3. Citric-acid solubility of more than 90% is a substantial advantage for fertilizer manufacturers, as the char requires no chemical treatment, thereby saving chemical costs.

Table 25.4 Germination test with char, SBM, and chemical fertilizer

	Weight (dry g)	Number of leaves	Max. length of leaf (cm)	Color of leaf
Control without phosphate	0.07	0.6	2.1	46.4
Biochar (4 kg)	0.85	4.9	11.3	45.4
SBM[a] (4 kg)	0.75	4.9	11.0	47.0
Chem. fertilizer (4 kg)	1.03	4.9	11.7	43.7
Biochar (8 kg)	1.22	5.5	12.8	51.7
SBM[a] (8 kg)	1.12	5.4	12.2	52.2
Chem. fertilizer (8 kg)	1.20	5.1	12.6	43.6

[a]*SBM* steamed bone meal

4. The char passed physical property tests as a fertilizing material such as workability and hardness.
5. Average P_2O_5 concentration was 9.5%, which is equivalent to the level of fish meal. The potassium content was about 5.8%, equivalent to that of plant burning ash. Hence, the char is suitable as a fertilizing material for organic fertilizer.

25.5 Conclusions

Hitachi Zosen has surveyed the circumstances surrounding livestock waste treatment in concentrated areas of the livestock industry in Japan and also conducted tests for resource recycling by carbonization employing the EFCaR system. The char has been proved to have a higher solubility than burning ash and phosphate rock. Pilot plant tests have verified the high energy efficiency of the EFCaR system and that the system requires no external fuel when treating materials with a moisture content of less than 40%. The char produced from pig slurry have maintained a stable level of phosphate concentration with a seasonal fluctuation of only 10%. Based on these results, it is concluded that the EFCaR system is a viable solution for resource recycling and waste management in the livestock sector.

References

Dote Y (2010) Report of Ministry of Agriculture, Forestry and Fisheries. https://www.miyazaki-u.ac.jp/education/files/0713_2.pdf

Katiku haisestubutu no kanri to riyou no gennjo to taisaku nitsuite (2015) Ministry of Agriculture, Forestry and Fisheries. https://www.env.go.jp/council/09water/y0917-03/mat03.pdf

Sinyou joukasouodei karano rin kaishu rikatsuyouno suisin ni mukete (2013) Ministry of the Environment, Government of Japan. https://www.env.go.jp/recycle/misc/jokaso/pamphlet.pdf

Trade Statistics of Japan (2016) Ministry of Finance. http://www.customs.go.jp/toukei/srch/index.htm

Wada T (2013) Report of Livestock Industry's Environmental Improvement Organization. 48:1–8

Chapter 26
Hydrothermal Process for Extracting Phosphate from Animal Manure

Huanhuan Li, Xueli Zhao, Tao Zhang, and Andrea Kruse

Abstract To mitigate the risk of the secured supply of phosphate rock, increasing attention has been paid to phosphorus (P) recovery from untapped secondary resources. Animal manure is one of the most important secondary resources for P in light of the large quantity and high P content. This chapter describes the hydrothermal technology to efficiently extract P from animal manure. Laboratory experiments showed that the hydrothermal treatment of pig manure at 180 °C under the oxygen partial pressure of 1 MPa could increase the P extractability by approximately 150% compared to the untreated control. Under this condition, the concentration of P in the eluate reached up to 114 mg/L. The kinetics of extracting P from pig manure by the hydrothermal technology could be given by the first-order reaction equation.

Keywords Hydrothermal treatment · Animal manure · P extraction kinetics

26.1 Introduction

Phosphorus (P) is an irreplaceable element for all types of living organisms and is widely used in agriculture as fertilizers and feed additives (Kauwenbergh et al. 2013). Today, P is mostly obtained from mined phosphate (P_i) rock which is a non-renewable resource on a human time scale, since its formation requires millions of years and extremely complex metallogenic conditions (Fang et al. 2016). Hence,

H. Li · T. Zhang (✉)
Beijing Key Laboratory of Farmland Soil Pollution Prevention and Remediation,
Key Laboratory of Plant-Soil Interactions of Ministry of Education,
College of Resources and Environmental Sciences, China Agricultural University,
Beijing, China
e-mail: taozhang@cau.edu.cn

X. Zhao · A. Kruse
Institute for Agricultural Engineering, University of Hohenheim, Stuttgart, Germany

H. Ohtake, S. Tsuneda (eds.), *Phosphorus Recovery and Recycling*,
https://doi.org/10.1007/978-981-10-8031-9_26

Table 26.1 The number of livestock and the estimated amount of manure produced in 2013 (FAO 2015; Monreal and Schnitzer 2012)

	Cattle		Pigs		Chicken		Sheep and goats	
Country	Number $\times 10^3$	Manure $\times 10^6$ Mg	Number $\times 10^3$	Manure $\times 10^6$ Mg	Number $\times 10^3$	Manure $\times 10^6$ Mg	Number $\times 10^3$	Manure $\times 10^6$ Mg
Argentina	51,095	474.00	2440	2.04	107	9.14	18,950	8.06
Australia	29,291	382.57	2098	2.92	99	8.41	79,098	50.24
Brazil	211,764	1964.53	36,744	31.58	1249	107.05	26,070	11.03
Canada	12,215	159.58	12,879	18.18	167	14.23	922	0.57
China	113,645	1000.51	482,398	414.1	4835	414.52	357,871	151.55
Germany	12,587	168.7	27,690	42.43	161	13.76	2024	1.25
India	189,000	958.61	10,130	8.71	709	60.72	197,800	83.77
Mexico	32,402	300.59	16,202	13.89	524	44.97	17,162	7.32
Pakistan	38,299	194.23	–	–	413	35.44	93,613	39.65
Russia	19,930	246.69	18,816	28.87	448	38.4	24,180	10.22
UK	9844	131.94	4885	7.54	152	13.05	32,954	20.92
USA	89,300	1166.4	64,775	91.35	1917	164.33	8146	5.17
Sum	809,372	7148.36	679,057	661.61	10,780	924.02	858,789	389.74

– data not available

Table 26.2 The typical content of P in manure from various livestock animals (Zhongqi et al. 2016; Dou et al. 2000; Sharpley and Moyer 2000; He and Honeycutt 2001; Ajiboye et al. 2004; He et al. 2004a, b; Mcdowell and Stewart 2005; Pagliari and Laboski 2012; Pagliari and Laboski 2013; Li et al. 2014)

Livestock	Number samples	Pt (g kg^{-1})	Pi (%)	Po (%)	P_R (%)
Dairy	37	2.8–18.3	28–96	8–61	0–24
Beef	11	2.5–14.2	48–91	8–44	1–32
Swine	9	3.9–48.7	51–92	8–37	0–28
Turkey	7	10.1–28.2	49–73	13–50	0–10
Poultry	34	8.6–30.4	32–84	14–68	0–24
Sheep	3	7.2–10.7	48–81	11–23	8–13
Goat	2	10.3–13.1	36–44	18–33	9
Horse	2	5.4–12.4	45–49	11–25	5–11
Deer	1	7.4	78	12	10

Pt total P, *Pi* inorganic P, *Po* organic P, *PR* residual P

closing the loop of the anthropogenic P cycle is critical to sustaining the ever-increasing global population. Meanwhile, the current use of P is inefficient and leaky, particularly in the agricultural sector, thereby causing pollution problems known as eutrophication problems in natural bodies of water. Therefore, from an environmental perspective, it is also critical to efficiently remove and recycle P from waste streams such as sewage sludge, animal manure, and industrial wastes (Zhongqi et al. 2016).

Animal manure is one of the most important secondary P resources in light of the large quantity and high P content (see Tables 26.1 and 26.2). Obviously, the

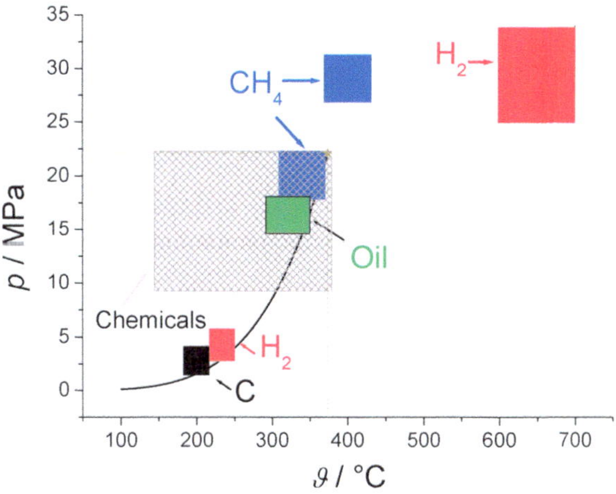

Fig. 26.1 Hydrothermal conversion of biomass (Reproduced with permission from Kruse 2008)

recycling of P from animal manure has the potential to make a great contribution to the realization of sustainable agriculture. For instance, Szogi and Vanotti (FAO 2015) have estimated that up to 15% of P fertilizer being used in the USA could be substituted by P recovered from poultry litter. However, the direct use of animal manure as fertilizer is often hampered by factors such as imbalanced manure distribution, potential hygiene problem, and its relatively high moisture content. Hence, various technologies have been developed to recover P_i from animal manure in a more condensed form, thereby making it easy to handle, store, and use (Monreal and Schnitzer 2012; Azuara et al. 2013; Szogi and Vanotti 2009; Lehmann and Joseph 2015). The proper choice of these technology options could lead to the efficient use of P_i in waste streams, while minimizing the risk of pollution.

26.2 Hydrothermal Technology

The hydrothermal technology is capable of producing a carbon-rich char, called hydrochar, with a low moisture and reduced odor from organic wastes (Pagliari et al. 2010; Wnetrzak et al. 2013; Tsai and Liu 2015). The main advantage of hydrothermal technology is to directly convert wet biomass into a product having the high carbon and low moisture contents. This makes the transportation of bulky biomass easy and cost-effective (Dael et al. 2014). The hydrochar can be used as an energy carrier, as soil amendment, or as a precursor to functional carbon materials. Thus, the hydrothermal technology has been developed for the valorization of waste streams. In general, hydrothermal conversion is conducted under an elevated temperature and autogenetic pressure to accomplish the decomposition and polymerization of organic matters in biomass. This leads to the generation of hydrochar, bio-crude oil, and various gases such as CO_2, H_2, CO, and CH_4. The properties and types of the final products can be tailored by adjusting the reaction parameters (Fig. 26.1).

Table 26.3 Recent examples on the application of hydrothermal technology in animal manure

Authors	Reaction conditions	Targets	Feedstock
Ghanim et al. (2017)	250 °C, 2 h	The effects of different pH on the composition of various substances in hydrochar were investigated	Chicken manure and straw mat
Catalkopru et al. (2016)	225 °C, 10–30 min	Effect of waste liquid circulation on the properties of hydrochar	Grape residue, orange peel, chicken manure mat
Liu et al. (2017)	180–300 °C, 1.5 h	The characteristics of pig manure hydrochar under different HTC temperature were investigated	Pig manure
Ekpo et al. (2016)	170–250 °C	The effects of different acid and alkali treatments on the release of nitrogen and phosphorus in swine manure were studied during HTC	Pig manure
Toufiq Reza et al. (2016)	180–280 °C, 5–30 min	The distribution of C and N in solid-liquid at different temperatures and times	Cow manure
Dai et al. (2015)	200 °C	Changes of P in the process of HTC	Cow manure
Theegala and Midgett (2012)	250–350 °C	Biomass oil and energy accounting of dairy manure during HTL	Cow manure
Cao et al. (2016)	500–620 °C	Effect of different temperature, catalyst, and concentration on gas components after chicken manure HTG	Chicken manure

Previously, Medoll (1985) has reported the unique function of sub- and supercritical water in the hydrothermal transformation of biomass. At a pressure of 22 MPa and a temperature of 374.3 °C, water reaches its critical point where the liquid-gas boundary vanishes and only one phase exists. As temperature reaches the critical point, the characteristics of water change drastically and significantly. Since the dielectric constant of water declines, the property of water as a solvent remarkably changes. As a consequence, the solubility of organic matters in water increases dramatically. As such, water shows both liquid and gas properties at the critical point, enabling its use as an environment-friendly solvent with characteristics similar to those of organic solvents.

According to the level of reaction severity and the types of target products, the hydrothermal technology can be divided into three categories: (1) hydrothermal carbonization (HTC), (2) hydrothermal liquefaction (HTL), and (3) hydrothermal gasification (HTG). HTL and HTG are mainly used for recovering combustible gases and bio-crude oil, respectively, while HTC is targeted on producing solid hydrochars. Besides producing energy and carbon materials, it has been found that the hydrothermal technology can retain most of the plant-available nutrients in its products in the form of inorganic salts. Therefore, many researchers have started to investigate the potential to recover nutrients from nutrient-rich biomass using the hydrothermal technology. Table 26.3 summarizes the recent literature on the application of hydrothermal techniques relevant to the nutrient recovery from animal manure. Recently, the hydrothermal carbonization has attracted increasing attention on the separation of P_i from biomass. The reason is that it requires a relative low temperature for carbonization and makes the process relative facile compared to dry carbonization processes.

26.3 P Extraction from Animal Manure

In a hydrothermal process, the extractability of P from animal manure is influenced by various parameters including reaction temperature, pressure, and pH as well as the choice of catalysts.

26.3.1 Temperature

We have investigated the release of P_i from animal manure at an oxygen partial pressure of 1.0 MPa with varying temperature from 60 to 180 °C (Fig. 26.2). Obviously, the release of P_i increased with increasing the reaction temperature.

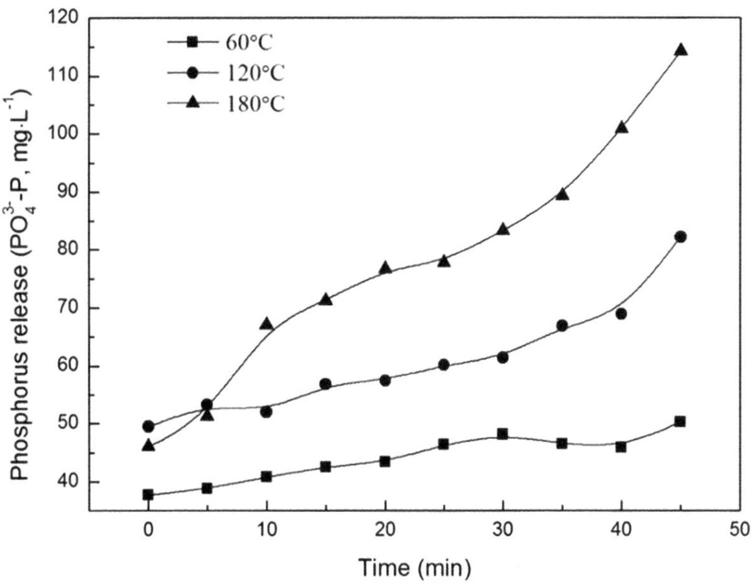

Fig. 26.2 P_i release from animal manure at different temperatures

This was consistent with the results reported by Ekpo et al. (2016). They have recovered 94% of P by the hydrothermal hydrolysis of swine manure at 170 °C in the presence of H_2SO_4. It has also been found that a large proportion of P in animal manure could be retained in the solid product of hydrothermal treatment. Dai et al. (2015) have found that the P content in the solid fraction of cow manure increased by about 20% after HTC at 200 °C, while the water-extractable and the Mehlich-3 (leaching agent)-extractable P reduced by 80% and 50%, respectively. Heilmann et al. (2014) have carried out the HTC of poultry, swine, and cattle manure at 200, 225, and 250 °C. They have reported that more than 90% of P in animal manure could be retained in the hydrochars produced by HTC at temperature of 250 °C. Most of the P was found to form Ca, Mg, Al, or Fe salts. Ekpo et al. (2016) have also reported that the highest rate of P recovery (i.e., 94% of total P in biomass) from animal manure could be realized by HTC using H_2SO_4 as catalyst at 170 °C.

26.3.2 pH

The release of P_i from hydrochar is strongly influenced by pH. Heilmann et al. (2014) have recovered P_i from hydrochar by treating it with an acid. In HTC, P_i is accumulated on the surface of hydrochar and can be leached at acidic pH. After separating hydrochar from the acidic solution, P_i can be precipitated from the solution by adding sodium hydroxide to adjust the pH to 9 or higher. The product is mainly calcium P_i, which can be further separated from the basic solution by filtration. Dai et al. (2015) have observed the pattern of P_i release at 200 °C and showed that pH decreased concomitantly with the P_i release from hydrochar. Ekpo et al. (2016) have also studied the influence of pH on releasing N and P_i from the liquor generated from the hydrothermal hydrolysis of animal manure at 120 and 170 °C. The results have showed that the recovery rate of P was high under acidic conditions. At 170 °C, the highest concentration of P (2200 mg/L) was observed after treating with H_2SO_4. This showed that 94% of P in the animal manure was extracted into the acidic solution. The lowest concentration of P (240 mg/L) was observed when using deionized water in place of H_2SO_4.

It can be concluded that HTC can be viewed as an efficient immobilization technology for P in animal manure. To recycle P, animal manure can be treated by a cascaded process consisting of HTC, acid treatment for P_i release from hydrochar, and P_i precipitation at alkaline pH. Comparing to traditional P fertilizer, the recovered P product is more suitable for the application to soil, because of its relatively slow rate of P_i release (Angst and Sohi 2013).

26.3.3 Pressure

In the airtight space, the pressure changes with changing the temperature. A higher partial pressure of oxygen can enhance the mass transfer across the gas-liquid interface, keeping oxygen molecules in the liquid phase for a longer period of time. A higher pressure can also accelerate the release of P_i into the liquid phase during hydrothermal treatment. We have showed that the total amount of P_i dissolved in the liquid phase at 1.5 MPa was significantly higher than that at 1 MPa treatment (Fig. 26.3).

26.3.4 The Kinetics of P_i Release

To indicate the temperature dependence of the hydrothermal reactions, the Arrhenius equation (Eftaxias et al. 2014; Tembhekar et al. 2015) has been frequently used in various studies:

$$\mathrm{d}C / \mathrm{d}t = k0\left(\exp-\frac{Ea}{RT}\right)C^m \tag{26.1}$$

where *Ea*, reaction activation energy (kJ/mol); *T*, thermodynamic temperature (K); *C*, organic matter concentration (mol/L); *m*, reaction order; *t* – reaction time (min); and *R*, gas constant (8.314 J/(mol, K)). The kinetics rate constant *k* is given by

$$k = k0\left(\exp-\frac{Ea}{RT}\right) \tag{26.2}$$

Using the kinetics rate constant *k*, the above empirical model can be simplified to

$$-\frac{\mathrm{d}[C]}{\mathrm{d}t} = k[C]^m \tag{26.3}$$

When $m = 1$ and $t = 0$, $[C] = [C]_0$, the above equation can be integrated to obtain a first-order dynamic equation. When $m = 2$ and $t = 0$, $[C] = [C]_0$, the above equations can be obtained by integrating the two-order kinetic equations:

$$\frac{1}{[C_0]} - \frac{1}{[C]} = -kt \tag{26.4}$$

The kinetics formulas were fitted to the experimental data on the P_i release from pig manure (Fig. 26.4). Table 26.4 shows the parameters gained by fitting and the corresponding reaction kinetics. As shown in the table, the first-order and the second-order kinetic approach fits well to the experimental data with R^2 higher than

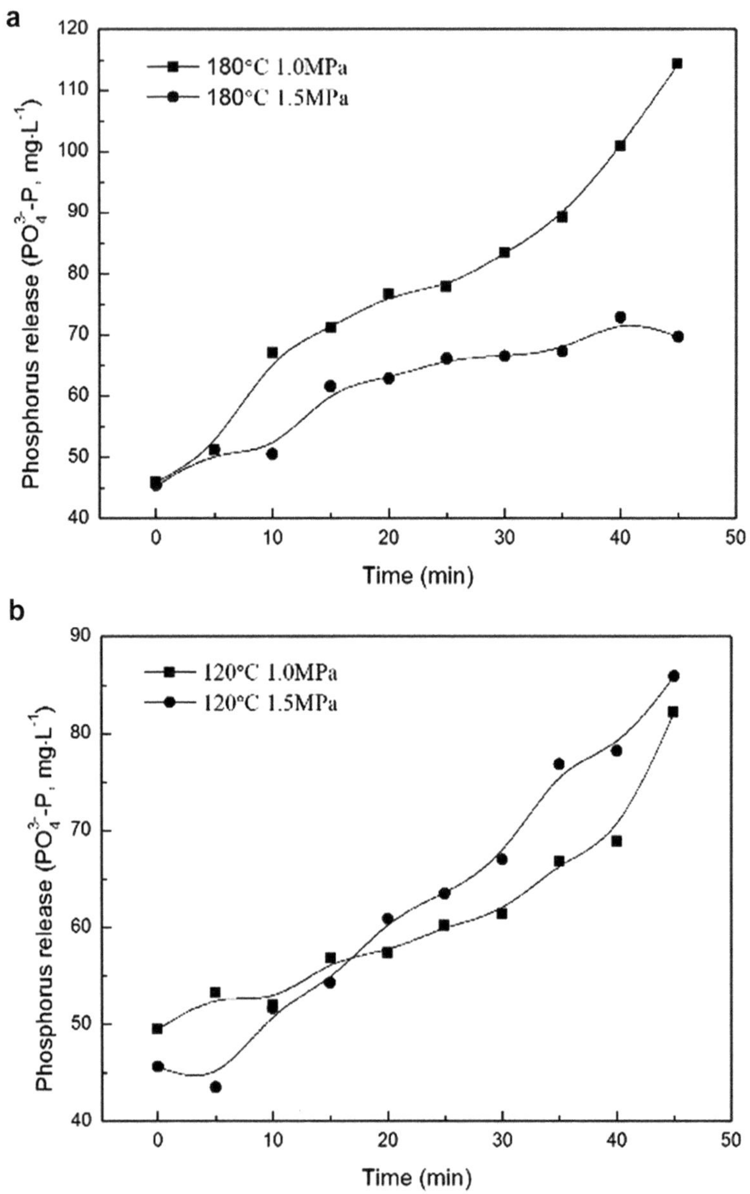

Fig. 26.3 P_i release from pig manure at different oxygen pressures by hydrothermal treatment at 180 (**a**) and 120 °C (**b**)

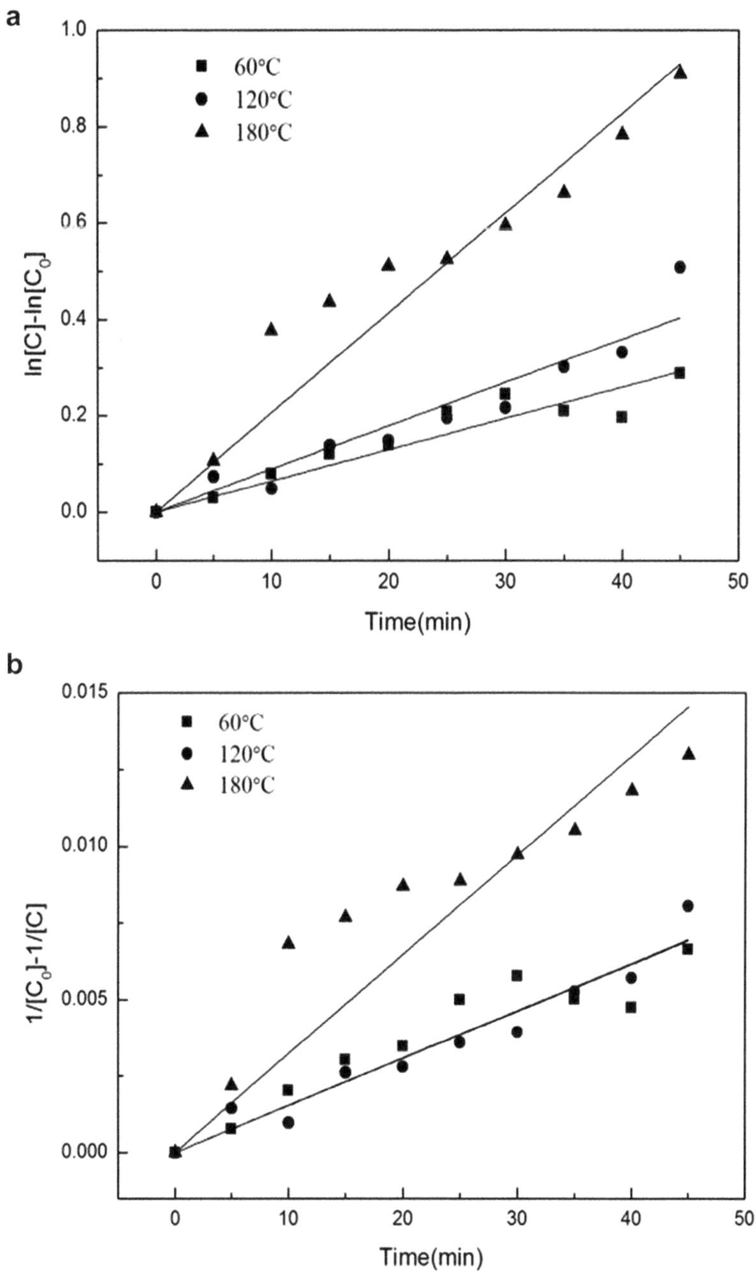

Fig. 26.4 Kinetics equation fitting for P_i release from swine manure by wet air oxidation treatment. (**a**) First-order reaction kinetics; (**b**) second-order reaction kinetics)

Table 26.4 Fitting parameters of phosphorus release from swine manure by hydrothermal carbonization treatment

Reaction order	T (°C)	$-k$ (L·mol^{-1}·min^{-1})	R^2	Reaction equation
The first-order reaction ($m = 1$)	60	0.00649	0.96495	$\frac{d[C]}{dt} = 0.00649[C]$
	120	0.00898	0.96484	$\frac{d[C]}{dt} = 0.00898[C]$
	180	0.02069	0.97824	$\frac{d[C]}{dt} = 0.02069[C]$
	60	1.54486E-4	0.96365	$\frac{d[C]}{dt} = (1.54486E-4)[C]^2$
The second-order reaction ($m = 2$)	120	1.53648E-4	0.98077	$\frac{d[C]}{dt} = (1.53648E-4)[C]^2$
	180	3.23629E-4	0.95557	$\frac{d[C]}{dt} = (3.23629E-4)[C]^2$

95%. At 120 °C, the R^2 of the first-order reaction equation was lower than that of the second-order reaction. In conclusion, the first-order kinetics approach fits best. Therefore, the first-order reaction kinetics equation should be used for the P_i release from pig manure. This means that the P_i releasing rate is proportional to the concentration of P_i in pig manure.

26.4 Outlook

The hydrothermal technology is becoming increasingly attractive in recycling nutrients from animal manure. However, further study is needed to optimize the process for recovering P_i and to address emerging issues relevant to the application at a full scale. Among the unsolved issues is how to prevent the contamination of hazardous substances such as heavy metals and persistent organics from animal manure. The following challenges need to be addressed:

1. What is the fate of the potentially hazardous substances in the hydrothermal conversion of animal manure? Although certain substances such as antibiotics and hormones can be degraded under the severe conditions of hydrothermal treatment, the risk of heavy metal contamination is relatively high. In HTC, a large proportion of heavy metals in biomass can be adsorbed by the matrix of hydrochar. A care should be given to the possibility prior to the application of hydrochar. To minimize the risk of contamination, appropriate reaction conditions need to be explored for HTC.

2. New detection technologies (e.g., synchrotron radiation) should be explored to understand the mechanism underlying the P_i release from animal manure. New technologies should also be tested and applied to realize the optimum control of HTC process and to boost the quality control of end products.
3. The economic benefits brought by the technology application need to be maximized by reducing the cost for implementation and operation. In particular, the high-pressure equipment is often expensive and requires much money for implementation.
4. The hydrothermal treatment requires a relatively high energy input in full-scale operation. To reduce the cost of operation, the opportunity of heat recovery should be fully exploited.
5. A priority should be given to the development of catalysts which enables to lower the reaction temperature and shorten the reaction time.

The research on recycling P_i from livestock and poultry manure using the hydrothermal technology has been attracting increasing attention. However, the full-scale application has yet been limited. The collaboration between academia and industry is a key to achieve innovation in the development of hydrothermal technology.

Acknowledgments The work was supported by a grant from the National Key Technology Research and Development Program of China (No. 2016YFD0501404-6, 2017YFD0800202), the National Natural Science Foundation of China (No. 31401944), the China Agricultural University Education Foundation "Da Bei Nong Group Education Foundation" (No. 1031-2415005), and the Beijing Municipal Natural Science Foundation (No. 6144026).

References

Ajiboye B, Akinremi OO, Racz GJ (2004) Laboratory characterization of phosphorus in fresh and oven-dried organic amendments. J Environ Qual 33(3):1062–1069

Angst TE, Sohi SP (2013) Establishing release dynamics for plant nutrients from biochar. Glob Change Biol Bioenergy 5(2):221–226

Azuara M, Kersten SRA, Kootstra AMJ (2013) Recycling phosphorus by fast pyrolysis of pig manure: concentration and extraction of phosphorus combined with formation of value-added pyrolysis products. Biomass Bioenergy 49(2):171–180

Cao W, Cao C, Guo L, Jin H, Dargusch M, Bernhardt D et al (2016) Hydrogen production from supercritical water gasification of chicken manure. Int J Hydrog Energy 41(48):22722–22731

Catalkopru AK, Kantarli IC, Yanik J (2016) Effects of spent liquor recirculation in hydrothermal carbonization. Bioresour Technol 226:89

Dael MV, Márquez N, Reumerman P, Pelkmans L, Kuppens T, Passel SV (2014) Development and techno-economic evaluation of a biorefinery based on biomass (waste) streams – case study in the netherlands. Biofuels Bioprod Biorefin 8(5):635–644

Dai L, Tan F, Bo W, He M, Wang W, Tang X et al (2015) Immobilization of phosphorus in cow manure during hydrothermal carbonization. J Environ Manag 157:49–53

Dou Z, Toth JD, Galligan DT, Ramberg CFJ, Ferguson JD (2000) Laboratory procedures for characterizing manure phosphorus. J Environ Qual 29(2):508–514

Eftaxias A, Font J, Fortuny A, Giralt J, Fabregat A, Stüber F (2014) Kinetic modelling of catalytic wet air oxidation of phenol by simulated annealing. Int J Chem React Eng 33(1):175–190

Ekpo U, Ross AB, Camargo-Valero MA, Fletcher LA (2016) Influence of ph on hydrothermal treatment of swine manure: impact on extraction of nitrogen and phosphorus in process water. Bioresour Technol 214:637–644

Fang C, Tao Z, Jiang R, Ohtake H (2016) Phosphate enhance recovery from wastewater by mechanism analysis and optimization of struvite settleability in fluidized bed reactor. Sci Rep 6:32215

FAO (2015) FAOSTAT. Available online at http://www.fao.org/statistics/en/. Verified on 20 Sept 2016

Ghanim BM, Kwapinski W, Leahy JJ (2017) Hydrothermal carbonisation of poultry litter: effects of initial ph on yields and chemical properties of hydrochars. Bioresour Technol 238:78–85

He Z, Honeycutt CW (2001) Enzymatic characterization of organic phosphorus in animal manure. J Environ Qual 30(5):1685–1692

He Z, Griffin TS, Honeycutt CW (2004a) Phosphorus distribution in dairy manures. J Environ Qual 33(4):1528–1534

He Z, Griffin TS, Honeycutt CW (2004b) Enzymatic hydrolysis of organic phosphorus in swine manure and soil. J Environ Qual 33(1):367–372

Heilmann SM, Molde JS, Timler JG, Wood BM, Mikula AL, Vozhdayev GV et al (2014) Phosphorus reclamation through hydrothermal carbonization of animal manures. Environ Sci Technol 48(17):10323–10329

Kauwenbergh SJV, Stewart M, Mikkelsen R (2013) World reserves of phosphate rock. a dynamic and unfolding story. Better Crops Plant Food 97:18–20

Kruse A (2008) Supercritical water gasification. Biofuels Bioprod Biorefin 2(5):415–437

Lehmann J, Joseph S (2015) Biochar for environmental management. Sci Technol Earthscan 25(1):15801–15811(11)

Li G, Li H, Leffelaar PA, Shen J, Zhang F (2014) Characterization of phosphorus in animal manures collected from three (dairy, swine, and broiler) farms in China. PLoS One 9(7):e102698

Liu Y, Yao S, Wang Y, Lu H, Brar SK, Yang S (2017) Bio- and hydrochars from rice straw and pig manure: inter-comparison. Bioresour Technol 235:332–337

Mcdowell RW, Stewart I (2005) Phosphorus in fresh and dry dung of grazing dairy cattle, deer, and sheep: sequential fraction and phosphorus-31 nuclear magnetic resonance analyses. J Environ Qual 34(2):598–607

Modell M (1985) Gasification and liquefaction of forest products in supercritical water. Fundamentals of thermochemical biomass conversion. Springer, Netherlands

Monreal CM, Schnitzer MI (2012) Bio-oil production from poultry litter: potential energy, environmental and chemical opportunities. Applied research of animal manure. Challenges and opportunities beyond the adverse environmental concerns

Pagliari PH, Laboski CAM (2012) Investigation of the inorganic and organic phosphorus forms in animal manure. J Environ Qual 41(3):901–910

Pagliari PH, Laboski CAM (2013) Dairy manure treatment effects on manure phosphorus fractionation and changes in soil test phosphorus. Biol Fertil Soils 49(8):987–999

Pagliari P, Rosen C, Strock J, Russelle M (2010) Phosphorus availability and early corn growth response in soil amended with turkey manure ash. Commun Soil Sci Plant Anal 41(11):1369–1382

Sharpley A, Moyer B (2000) Phosphorus forms in manure and compost and their release during simulated rainfall. J Environ Qual 29(6):1462–1469

Szogi AA, Vanotti MB (2009) Prospects for phosphorus recovery from poultry litter. Bioresour Technol 100(22):5461–5465

Tembhekar PD, Padoley KV, Mudliar SL, Mudliar SN (2015) Kinetics of wet air oxidation pretreatment and biodegradability enhancement of a complex industrial wastewater. J Environ Chem Eng 3(1):339–348

Theegala CS, Midgett JS (2012) Hydrothermal liquefaction of separated dairy manure for production of bio-oils with simultaneous waste treatment. Bioresour Technol 107(2):456–463

Toufiq Reza M, Freitas A, Yang X, Hiibel S, Lin H, Coronella CJ (2016) Hydrothermal carbonization (htc) of cow manure: carbon and nitrogen distributions in htc products. Environ Prog Sustain Energy 35(4):1002–1011

Tsai WT, Liu SC (2015) Thermochemical characterization of cattle manure relevant to its energy conversion and environmental implications. Biomass Convers Biorefinery 6(1):1–7
Wnetrzak R, Kwapinski W, Peters K, Sommer SG, Jensen LS, Leahy JJ (2013) The influence of the pig manure separation system on the energy production potentials. Bioresour Technol 136(136C):502
Zhongqi HE, Pagliari PH, Waldrip HM (2016) Applied and environmental chemistry of animal manure: a review. Pedosphere 26(6):779–816

Chapter 27
Recovery of Calcium Phosphate from Composted Chicken Manure and Industrial Waste

S. Sugiyama

Abstract Bench-scale experiments were carried out to examine the potential of recovering calcium phosphates from composted chicken manure. Ca^{2+} and phosphate ions were eluted from composted chicken manure using HNO_3 that released less K^+ from composted chicken manure compared to HCl and H_2SO_4. After filtration, the eluate pH was increased to 6.0 using aqueous ammonia to precipitate calcium phosphates, forming calcium hydroxyapatite which is known as a principal component of phosphate rock. The purity of recovered calcium phosphates was more than 95 mol%, suggesting that composted chicken manure could potentially be used as a secondary source of calcium phosphates for industrial applications. Using this technology, calcium phosphates could also be recovered from industrial wastes such as dephosphorization slag, chemical industry sludge, and spent fluorescent phosphorus powder (called bag powder).

Keywords Calcium phosphates · Composted chicken manure · Dissolution-precipitation technique · Industrial wastes

27.1 Introduction

Phosphorus (P) is one of essential elements for all living organisms and widely used in fertilizer and feed additives. P is also used in a wide range of high-value-added industrial products including human body-friendly biomaterials, food additives, surface treatment agents, fuel cells, fire retardants, fire distinguishers, catalysts, boiler compounds, and detergents. Today, P is mostly obtained from mined phosphate (P_i) rock which contains calcium phosphate (CaP) as a major component. Various technologies have been developed for recovering P from secondary

S. Sugiyama (✉)
Department of Applied Chemistry, Graduate School of Science and Technology, Tokushima University, Tokushima, Japan
e-mail: sugiyama@tokushima-u.ac.jp

H. Ohtake, S. Tsuneda (eds.), *Phosphorus Recovery and Recycling*,
https://doi.org/10.1007/978-981-10-8031-9_27

resources such as sewage sludge, incinerated sludge ash, and animal manure. However, most of them have been developed for recycling P as a fertilizing material for agricultural purposes. To our knowledge, relatively little attention has been paid to recover P from secondary resources for industrial applications. CaP is an important starting material for the manufacture of a wide range of industrial products, including advanced biomaterials for medical, pharmaceutical, and dentifrice applications. Since high-grade P_i rock is becoming deleted by ongoing mining, it seems necessary to explore and examine the potential of alternative CaP sources for industrial applications.

This chapter describes a simple, potentially cost-effective technique (designated as the dissolution-precipitation technique) to recover CaP from composted chicken manure (Sugiyama et al. 2016). Ca^{2+} and P_i were eluted from composted chicken manure with nitric acid (HNO_3). Then CaP was precipitated as calcium hydroxyapatite (CaHAp or $Ca_{10}(PO_4)_6(OH)_2$) by increasing the eluate pH to 6 or greater with the addition of aqueous ammonia. This technology was also applied to the recovery of CaP from dephosphorization slag (Sugiyama et al. 2014), chemical industry sludge (Sugiyama et al. 2012), and spent fluorescent P powder (called bag powder) (Sugiyama et al. 2015).

27.2 Composted Chicken Manure

Chicken manure is a suitable candidate for the regeneration of CaP since the main components are P, Ca, K, and N (Sugiyama et al. 2016; Dikinya and Mufwanzala 2010). Chicken manure has been widely used as a fertilizing material for agricultural purposes. It is often composted to break down organic matter, aiming to suppress plant diseases and to prevent offensive odors. During composting, P_i is gradually released from organic P compounds in chicken manure (Nakasaki 2014). Table 27.1 shows the typical composition of composted chicken manure.

In the present study, the dissolution-precipitation technique was used to recover CaP from composted chicken manure (Fig. 27.1). Composted chicken manure was dried for 2 days at 80 °C and milled into small particles of less than 0.85 mm in diameter. Then the composted chicken manure was dissolved with acid solution to elute Ca^{2+} and P_i at 25 °C with a constant stirring (130 rpm) (Fig. 27.1). Strong acids, including HNO_3, HCl, and H_2SO_4, were examined at a concentration of 0.1 mol/L for their ability to elute Ca^{2+} and P_i from composted chicken manure. All the acids could readily elute P_i from composted chicken manure. However, H_2SO_4 was less effective in eluting Ca^{2+} compared to HNO_3 and HCl, because it formed insoluble $CaSO_4$ (Fig. 27.2). In addition, HNO_3 eluted less K^+ compared to HCl and H_2SO_4. Accordingly, HNO_3 was chosen to elute Ca^{2+} and P_i for further study.

No significant release of silicates was observed with the addition of HNO_3 at concentrations lower than 1.0 mol/L (data not shown). Iron cations (Fe^{2+} and Fe^{3+}) were not released by HNO_3 at concentrations below 0.1 mol/L. While the elution of K^+, Mg^{2+}, Mn^{2+}, and Zn^+ was not significantly influenced by the concentration of

Table 27.1 Typical composition of composted chicken manure (Sugiyama et al. 2016)

Content [mol%]									
P	Ca	K	Mg	Fe	Mn	Zn	Al	Si	Others
11.1	50.9	17.5	3.0	1.4	0.3	0.3	1.3	3.4	10.8

Fig. 27.1 CaP recovery from composted chicken manure using the dissolution-precipitation technique

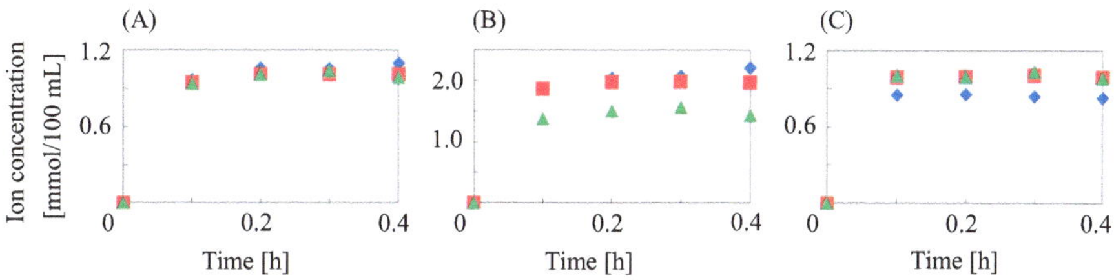

Fig. 27.2 The elution of P_i (**a**), Ca^{2+} (**b**), and K^+ (**c**) from composted chicken manure with 0.1 mol/L HNO_3 (◆), HCl (■), or H_2SO_4 (▲)

Fig. 27.3 Changes of the concentration of PO_4^{3-} (◆), Ca^{2+} (■), K^+ (▲), and Mg^{2+} (×) in the pH range of 1.3–10.0

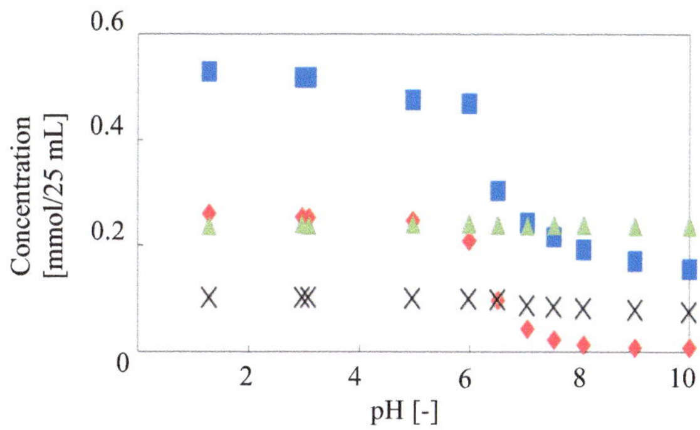

HNO_3, the elution of Ca^{2+} and P_i was somewhat increased with increasing the concentration of HNO_3 up to 0.1 mol/L. Therefore, HNO_3 was used at the concentration of 0.1 mol/L for further study. After 0.1 h of acid dissolution, the reaction mixture was filtered using a glass filter to remove solid residues. Then, the filtrate pH was adjusted to various values of 1.3–10.0 using aqueous ammonia (NH_3 (aq)). Figure 27.3 shows the change of P_i, Ca^{2+}, K^+, and Mg^{2+} concentrations in the pH

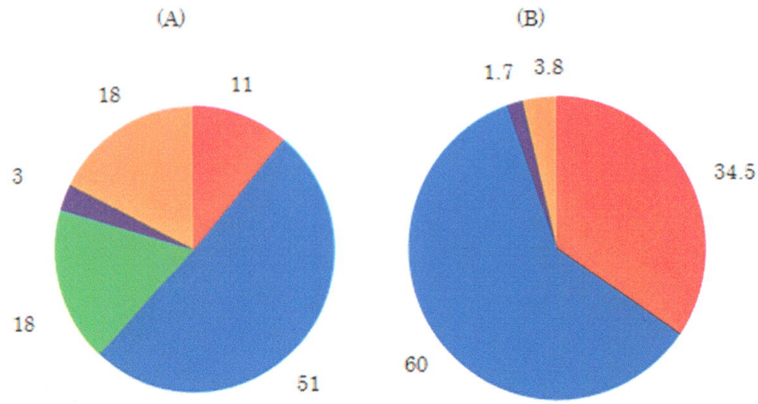

Fig. 27.4 The typical composition [mol%] of P (red), Ca (blue), K (green), Mg (violet), and others (orange) in composted chicken manure (**a**) and recovered product from composted chicken manure at pH 7.5 (**b**)

range from 1.3 to 10.0. When the pH reached around 6.0, P_i and Ca^{2+} concentrations were significantly decreased due to the formation of CaP precipitates. Nearly all P_i was precipitated at pH 10.0. By contrast, the concentrations of K^+ and Mg^{2+} remained essentially constant in the pH range of 1.3–10, indicating that neither K^+ nor Mg^{2+} precipitated in this range of pH.

The composition of recovered product at pH 7.5 was analyzed using X-ray fluorescence (XRF) and compared with that of composted chicken manure (Fig. 27.4). The CaP content of recovered products was 97 mol% on average. The content of Mg was less than 1.0 mol%. No significant amount of K was detected with the recovered product. The P:Ca molar ratio of the recovered product was estimated to be 0.57. This value was similar to those of $CaHPO_4$ (0.5), $Ca_{10}(PO_4)_6(OH)_2$ (0.6), and $Ca_3(PO_4)_2$ (0.67) which are known to be major components of P_i rock.

X-ray diffraction (XRD) analysis showed broad signals corresponding to $Ca_{10}(PO_4)_6(OH)_2$ in the recovered product (Fig. 27.5). When the recovered product was calcined at 800 °C for 5 h, sharp signals corresponding to $Ca_3(PO_4)_2$ were observed. It has been reported that $Ca_{10}(PO_4)_6(OH)_2$ is readily converted to $Ca_3(PO_4)_2$ by calcination (Sugiyama et al. 2016). These results indicated that composted chicken manure could potentially be used as a secondary source of CaP.

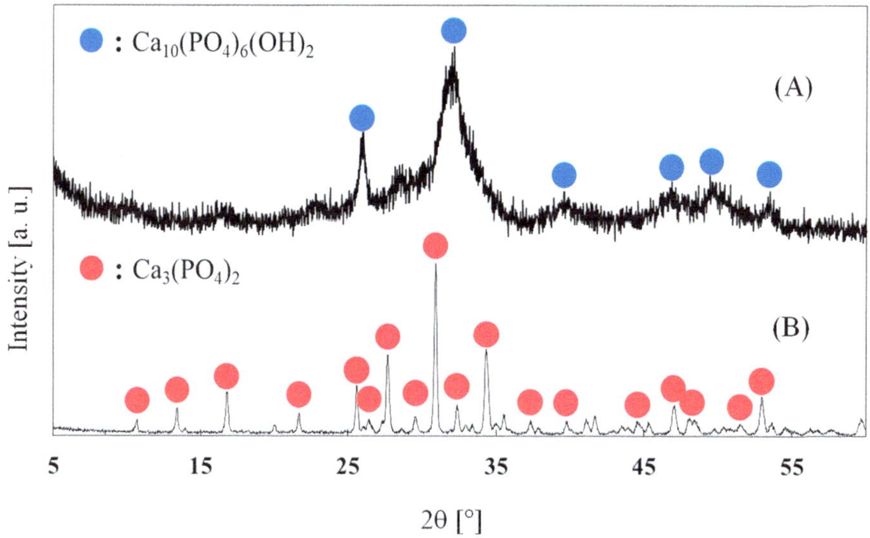

Fig. 27.5 XRD patterns of the recovered CaP product before (**a**) and after (**b**) calcination at 800 °C for 5 h

27.3 Industrial Wastes

27.3.1 Dephosphorization Slag

In the steel industry, P is present in raw materials such as iron ore, coal, and lime at concentrations as low as 0.1 mass%. Since even small amounts of P can cause a detrimental effect on high-grade steel quality, it must be removed nearly completely to steelmaking slag. The annual amount of P removed into steelmaking slag is estimated to be approximately 100 kt/a, which is two times more than that of P imported as P_i rock to Japan in 2016. Although steelmaking slag has occasionally been considered as a secondary P resource, little attention has been paid to the recovery of P from steelmaking slag for industrial applications (Kubo et al. 2010). Table 27.2 shows the typical composition of a hot metal pretreatment slag (called dephosphorization slag) which was supplied from a steel company in Japan.

Dephosphorization slag was dissolved with 1.0 mol/L HNO_3 (Fig. 27.6). Approximately 80% of slag was dissolved with HNO_3 for 30 min. After removing solid residues by filtration, the filtrate was further treated with 1.0 mol/L HNO_3 for 4 h. This allowed to effectively elute Ca^{2+} and P_i from slag. However, the filtrate also contained approximately 15 mol% Fe and 5 mol% Mn (Fig. 27.7a). To remove iron, 1.0 g of CaHAp was added to the filtrate at 25 °C and pH 1.5–1.8. This was intended to precipitate Fe ions in the form of insoluble Fe-CaHAp (Moriguchi et al. 2008).

Table 27.2 The typical composition of dephosphorization slag (Sugiyama et al. 2014)

Content [mmol/g-slag]						
P	Al	Si	Ca	Mn	Fe	Mg
0.407	0.248	4.73	7.71	0.958	2.52	0.476

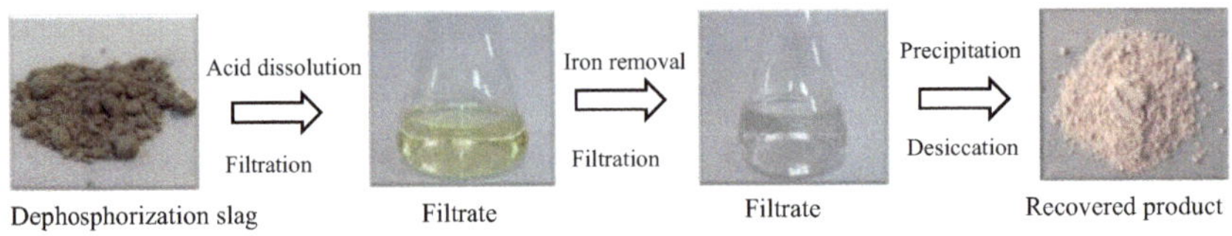

Fig. 27.6 CaP recovery from dephosphorization slag

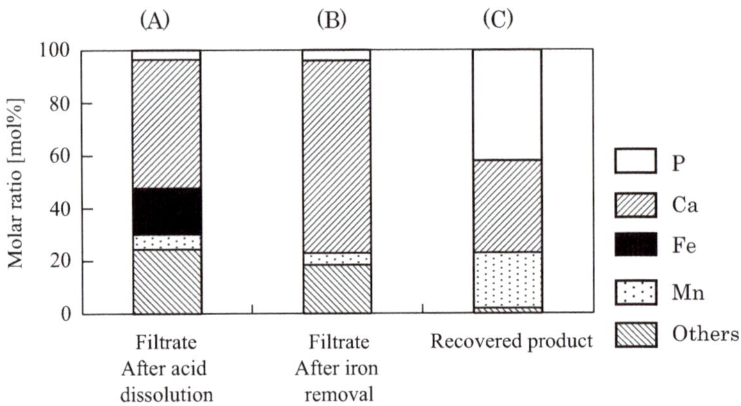

Fig. 27.7 The molar ratio of P, Ca, Fe, Mn, and others in the filtrate after dissolving slag in HNO_3 (**a**), the filtrate after removing iron (**b**), and the recovered product (**c**)

As expected, Fe ions were effectively removed from the filtrate, forming insoluble Fe-CaHAp (Fig. 27.7b). After Fe-CaHAp precipitation followed by filtration, the filtrate pH was increased to 6.0 using aqueous ammonia. This readily formed pale pink-colored precipitates as shown in Fig. 27.6. Meanwhile, CaHAp could be regenerated by treating Fe-CaHAp with an aqueous $Ca(NO_3)_2$ solution (Sugiyama et al. 2009). XRF analysis showed that the recovered product still contained considerable amounts of Mn (Fig. 27.7c), suggesting the formation of ammonium manganese phosphate ($MnNH_4PO_4$). Although various attempts were made to remove Mn, it was not possible to obtain Mn-free products. To use dephosphorization slag as a secondary source of CaP, it seems likely to need additional pretreatment to remove Mn prior to the acid dissolution treatment.

Table 27.3 The typical inorganic components of a chemical industry sludge (Sugiyama et al. 2012)

Content [mmol/g-sludge]						
P	Al	Si	Ca	Fe	Mg	K
0.481	0.053	0.033	0.117	0.466	0.044	0.059

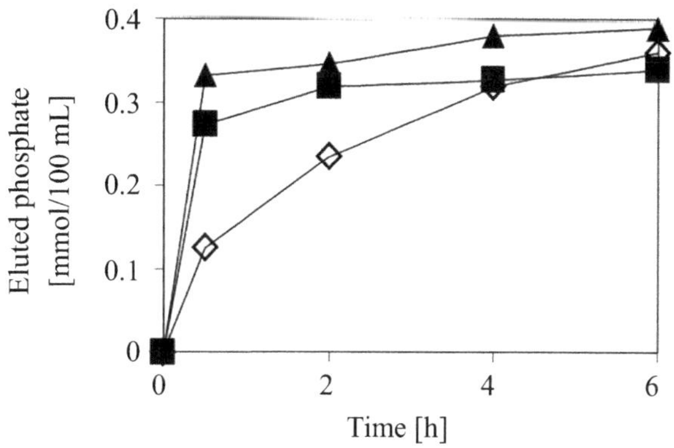

Fig. 27.8 The elution of P_i from untreated chemical sludge (diamond), pulverized sludge (square), and calcined sludge (triangle) with 1.0 mol/L HNO_3 at 25 °C

27.3.2 Chemical Industry Sludge

It seems interesting to examine the potential of recycling P-rich industrial wastes as a secondary source of CaP. Table 27.3 shows the inorganic components of a solid waste from a chemical company. In addition to the inorganic components, the sludge contained approximately 85% of its dry weight as organics. It contained relatively high P_i which was used to neutralize alkaline waste streams from a chemical manufacturing process.

To examine the potential of the P-rich chemical industry sludge as a secondary source of CaP, dry sludge samples were pulverized, dried for 0.5 h and calcined at 550 °C for 6 h to remove organics. CaP was then recovered from the calcined sludge using the dissociation-precipitation technique. P_i was eluted from calcined sludge at 25 °C with 1.0 mol/L HNO_3 (Fig. 27.8). Obviously, the rate of P_i elution was increased by pulverization and also by calcination. As is the case of dephosphorizing slag, Fe ions could be effectively removed from the eluate by the addition of CaHAp (data not shown). XRF analysis showed that the calcined sludge contained Ca, P, and Fe at 7, 25, and 49 mol%, respectively, while the recovered product contained CaP at 96 mol% on average (Fig. 27.9). XRD analysis of the recovered product showed broad signals corresponding to $Ca_{10}(PO_4)_6(OH)_2$ (Fig. 27.10). Finally, $Ca_3(PO_4)_2$ could be obtained by the calcination of recovered products at 800 °C for 5 h

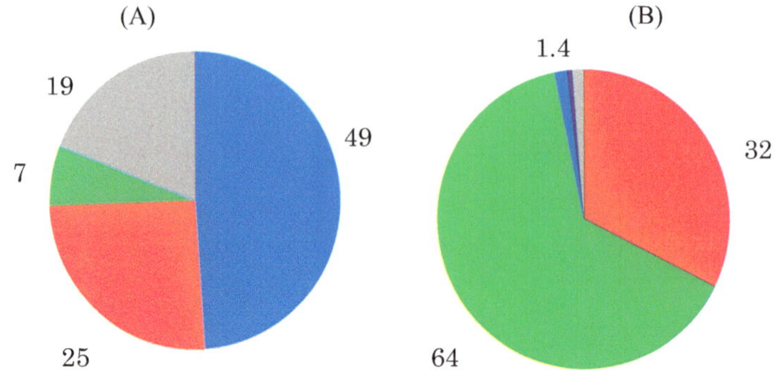

Fig. 27.9 The composition [mol%] of calcined chemical sludge (**a**) and recovered product (**b**). P (red), Ca (green), Fe (blue), Mn (violet), and others (gray)

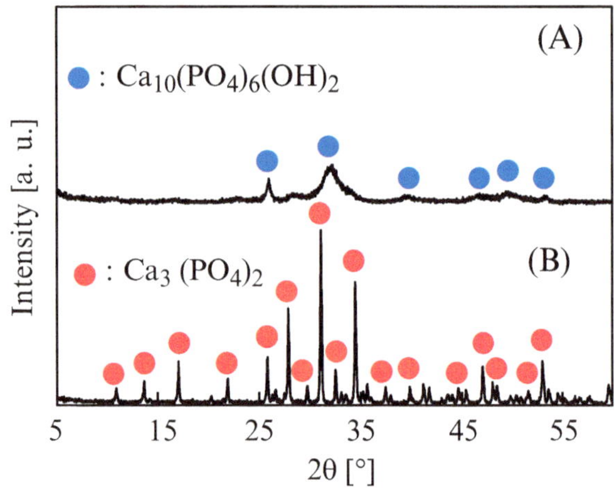

Fig. 27.10 XRD patterns of the CaP product recovered from the chemical sludge before (**a**) and after (**b**) calcination at 800 °C for 5 h

27.3.3 *Bag Powder*

Bag powder is a P-rich waste which is collected from gas streams by a bag filter in the recycling process for used fluorescence tube. According to the Japan Lighting Manufacturers Association, approximately 0.7–6 g of bag powder can be collected from each fluorescence tube (Sugiyama et al. 2015). A total of 200 t of bag powder is generated annually in the recycling industry of Japan. Table 27.4 shows the typical composition of bag powder. P and Ca account for approximately 60 mol% of the bag powder. Bag powder also contains silicon (Si), yttrium (Y), strontium (Sr), and lanthanides such as lanthanum (La) and cerium (Ce). Mercury is not detected with

Table 27.4 Typical components of bag powder (Sugiyama et al. 2015)

Content [mol%]							
P	Ca	Y	Si	Sr	La	Ce	Others
23	34	17	3.0	4.5	3.8	1.7	13

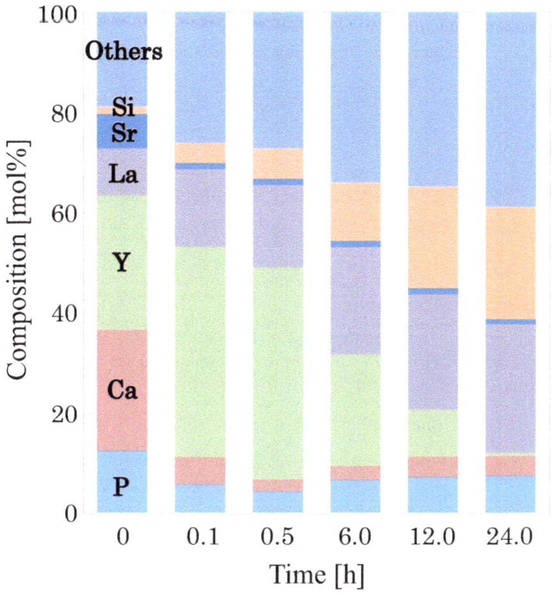

Fig. 27.11 The changes of the composition of bag powder after being subjected to 1.0 mol/L HNO_3

the bag powder. Bag powder has been considered as a secondary source of Y, Sr, La, and Ce. The recycling of these elements may help reduce the cost of CaP recovery from bag powder.

Bag powder samples were dissolved in 1.0 mol/L HNO_3 at 25 °C for 6 h (Fig. 27.11). Ca, P, and Y were readily released from bag powder, while other elements remained undissolved. After 4 h of acid treatment, yttrium could be removed as P_i salts by adjusting the eluate pH to 4.0 where no significant CaP precipitation occurred. After removing yttrium P_i salts by filtration, the filtrate pH was increased to 8.0 to precipitate CaP. XRF analysis showed that the recovered product contained more than 90 mol% CaP (Fig. 27.12). XRD analysis showed peak signals corresponding to $Ca_{10}(PO_4)_6(OH)_2$ in the recovered product (Fig. 27.13). $Ca_3(PO_4)_2$ was obtained by the calcination of recovered products at 800 °C for 5 h.

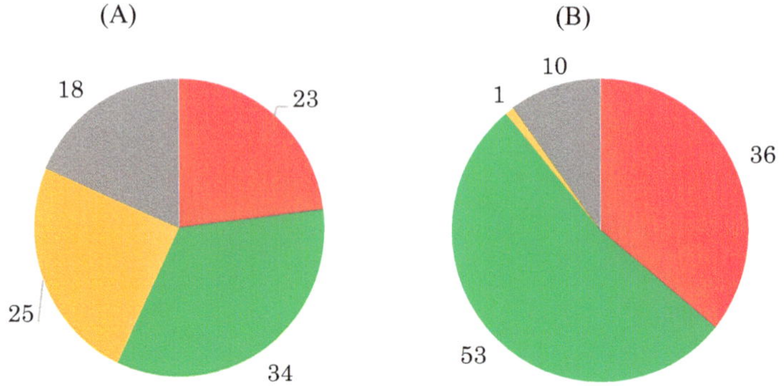

Fig. 27.12 The typical composition [mol%] of bag powder (**a**) and recovered product (**b**). P (red), Ca (green), rare earth (yellow), and others (gray)

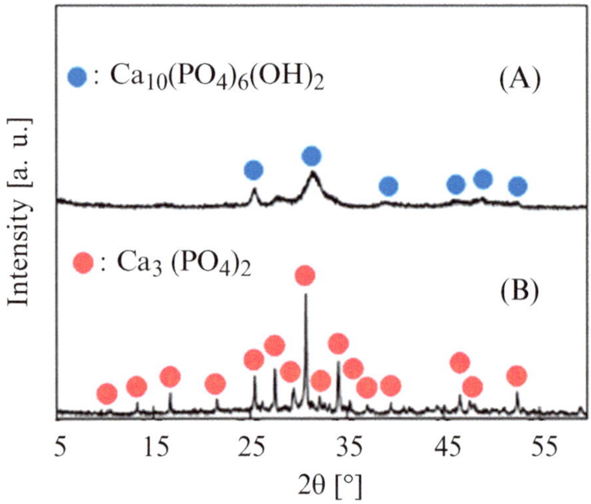

Fig. 27.13 XRD patterns of recovered CaP product before (**a**) and after (**b**) calcination at 800 °C for 5 h

27.4 Conclusions

Using the dissolution-precipitation technique, CaP could be recovered from composted chicken manure, dephosphorization slag, industrial sludge, and bag powder. Except dephosphorization slag, the purity of recovered CaP was 95 mol% or higher, suggesting that they could potentially be used as a substitute for P_i rock to prepare CaP for industrial applications. Further study is needed to purify recovered CaP to the level required for industrial applications.

References

Dikinya O, Mufwanzala N (2010) Chicken manure-enhanced soil fertility and productivity: effects of application rates. J Soil Sci Environ Manag 1:46–54

Kubo H, Matsubae-Yokoyama K, Nagasaka T (2010) Magnetic separation of phosphorus enriched phase from multiphase dephosphorization slag. ISIJ Int 50:59–64

Moriguchi T, Nakagawa S, Kaji F (2008) Reaction of Ca-deficient hydroxyapatite with heavy metal ions along with metal substitution. Phosphorus Res Bull 22:54–60

Nakasaki K (2014) Phosphorus recovery from chicken droppings treated with advanced compositing techniques. Agric Hortic 89:524–530

Sugiyama S, Shimizu Y, Manabe T, Nakagawa K, Sotowa K-I (2009) Preparation of hydroxyapatite film and its application in the removal and regeneration of aqueous cations. J Colloid Interface Sci 332:439–443

Sugiyama S, Hayashi T, Shinomiya I, Nakagawa K, Sotowa K-I (2012) Recovery and enrichment of aqueous phosphate from the slag released by a chemical factory. Phosphorus Res Bulletin 27:23–27

Sugiyama S, Shinomiya I, Kitora R, Nakagawa K, Katoh M (2014) Recovery and enrichment of phosphorus from the nitric acid extract of dephosphorization slag. J Chem Eng Jap 47(6):483–487

Sugiyama S, Kinoshita H, Shinomiya I, Kitora R, Nakagawa K, Katoh M, Masumoto K (2015) Recovery of the phosphorus from the nitric acid extract of powder collected in a bag filter during the recycling of used fluorescence tubes. J Chem Eng Jap 48(2):99–103

Sugiyama S, Kitora R, Kinoshita H, Nakagawa K, Katoh M, Nakasaka K (2016) Recovery of calcium phosphates from composted chicken manure. J Chem Eng Jap 49(2):224–228

Chapter 28
Outotec Manure, Slurry, and Sludge Processing Technology

Ludwig Hermann and Tanja Schaaf

Abstract Relevant waste flows like sewage sludge, farmyard manure, digestion residues, and humid residues from food and feed production are known for food safety issues and for environmental and waste management problems. If used as a resource for crop nutrients and soil fertility, distribution is the main issue: urbanization and intensive livestock farming produce mass flows requiring extended cropland typically not available in the densely populated regions of our planet. Thermal conversion is an acknowledged option for concentration and recycling of mineral residues including phosphates, but the typical moisture content of >70 wt% makes it difficult to yield relevant surplus energy flows. This challenge is approached by increasing the efficiency of drying and replacing combustion by gasification, in essence by making effective use of the hydrogen (H_2) molecules of water in the process chain. Outotec's technology approach aims at keeping H_2 molecules in the loop and eventually using them in the form of a hydrogen-rich gas in a variety of energy and biochemical applications. The approach is intrinsically circular, and the related processes – closed-loop steam drying and steam gasification – are well known but have not been applied to the waste flows and in the configuration as outlined in this book. If successfully implemented, waste flows in the order of 1–1.5 billion cubic meters in the EU28 may be recycled to a relevant building block of a future hydrogen economy with a vast array of applications in the energy and biochemistry sector.

Keywords Sewage sludge · Farmyard manure · Digestion residues · Slurries · Drying · Gasification · Product gas · Hydrogen · Nutrient recovery · Phosphate recycling · Energy conversion

---

L. Hermann (✉) · T. Schaaf
Outotec GmbH & Co. KG, Oberursel, Germany
e-mail: l.hermann@proman.pro

H. Ohtake, S. Tsuneda (eds.), *Phosphorus Recovery and Recycling*,
https://doi.org/10.1007/978-981-10-8031-9_28

28.1 Introduction

High mass flows of nutrient-rich, wet, organic residues disposed of on cropland frequently cause nutrient losses, eutrophication, and eventually dead zones in inland and coastal waters. Phosphate losses are identified as single most important cause for most of the EU28 member states missing their water quality targets. But not only the EU, the USA and China also have a massive water quality problem.

Phosphorus is likewise a vital component of every cell in all living organisms. No other element can replace phosphorus as a primary nutrient essential for plant growth. Consequently, we must avoid all phosphate losses, including those to cement and construction materials as a result of current sewage sludge management practices. Nutrient losses may be effectively prevented by thermally processing nutrient-rich residues to mineral fertilizers with higher nutrient concentrations for easy transport and handling. Efficient processing may save and yield up to 5% of the annual European energy supply and about 50% of the annual European phosphate and nitrogen consumption.

In one of its relevant R&D efforts in recent years, Outotec has developed and built a closed-loop steam dryer aiming at highly efficient drying of "wet" biomass such as sewage sludge, farmyard manure, effluents from anaerobic digesters, and many other organic waste materials suitable for conversion to gas, heat, electricity, and as feedstock for biorefineries. Apart from the abovementioned European target materials, test work will be extended to wet "straw-type" biomass like sugarcane residues (bagasse), rice straw, and humid industrial waste flows. The dryer is highly appropriate for efficiently drying feedstock aimed at the production of H_2-rich gas for combustion or other purposes, facilitating energy and nutrient recovery.

28.2 Background

28.2.1 Societal, Economic, and Political Background

Intensive livestock farming with limited cropland for using nutrient-rich farmyard manure for crop production has become the most prominent source for excessive nutrient loads to soils, eutrophication, algae blooms, and eventually dead zones in aquatic bodies. Huge mass flows – over 1 billion wet tonnes in the EU28 – and low concentrations of tradable substances do not allow much of handling and transport, causing the widespread practice of "dumping" the waste on cropland in the vicinity of livestock farms, a practice euphemistically called recycling.

Limiting the livestock density per hectare of available cropland has been the regulative response in many countries, but – as in any other sector of our market economy – production and productivity needs growing to provide a decent level of income for farmers, particularly due to farmers not having the market power to determine the prices of food and feed (Potocnik 2017). Moving animal farms from high density areas to other regions sounds like a meaningful response but is facing other barriers: animal rearing has been growing organically accompanied by an appropriate infrastructure (meat processing, food production, veterinary clinics and doctors, etc.) and know-how that cannot be simply transplanted to other regions (Thornton 2010).

The currently most promising solution is farmers setting up cooperatives or entrepreneurs launching agricultural service companies and jointly implementing technical facilities to reap some added value from agricultural by-products and produce fertilizing products with nutrient concentrations that allow transporting them to regions with a nutrient deficit (Hermann 2014). Researchers propose processes that are partly successfully applied to sewage sludge and include anaerobic digestion, ammonia stripping, and recycling to ammonium sulfate or ammonium nitrate, struvite crystallization, sieving, dissolved air flotation, solid/liquid separation (dewatering), drying, and – occasionally – thermal conversion (Ma et al. 2013; Hermann 2011).

In response to European and national initiatives for a circular economy and more sustainable nutrient management, Switzerland and Germany have banned thermal valorization of nutrient-rich waste if phosphorus is not recovered prior or post combustion. Direct use of sewage sludge has been prohibited in Switzerland in 2006 and quite restricted in Germany in 2017 by the fertilizing and fertilizer regulations and related legislation (UMWELT-MATERIALIEN Nr 2004; Bundesministerium für Umwelt, Naturschutz und Reaktorsicherheit im Einvernehmen mit dem Bundesministerium für Verbraucherschutz 2003). Replacement of fossil fuels, reduction of CO_2 emissions, and relevant gate fees have motivated utilities and cement industries to use dry sewage sludge and processed animal by-products (meat and bone meal) as fuels. The practice as such is meaningful, but the associated loss of nutrients is unsustainable and therefore addressed by the abovementioned regulatory framework.

In Switzerland, Germany, Austria, and other countries, up to 50% of sewage sludge and most of category 1 animal by-products are combusted in power or waste-to-energy plants and cement kilns (Bundesministerium für Umwelt, Naturschutz und Reaktorsicherheit im Einvernehmen mit dem Bundesministerium für Verbraucherschutz 2003; Wiechmann et al. 2013). These resources would have to be replaced by a considerable number of wastewater treatment plant-based recovery and recycling facilities and/or new mono-incinerators – combusting exclusively nutrient-rich sludge and animal by-products – if P cycles have to be closed. Finding locations and getting permits for a large number of new processing plants may be a challenge.

Leaving the waste flows and transports relatively unchanged and continuing to use the existing waste-to-energy, power, and cement plants as licensed sludge handling locations may increase the societal acceptance of the transition toward a circular nutrient economy. This hypothesis has motivated Outotec to invest in an alternative, thermal processing chain for sewage sludge, farmyard manure, and the organic fraction of municipal solid waste.

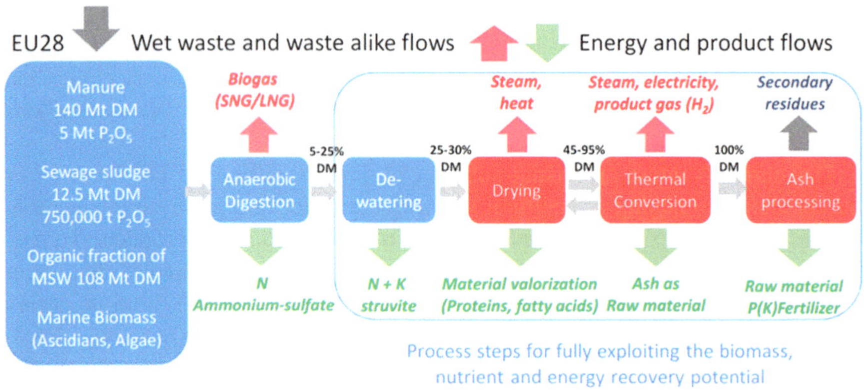

Fig. 28.1 Nutrient-rich waste flows and energy and nutrient recovery value chain (Source Outotec)

28.2.2 Nutrient-Rich Biomass as a Fuel

The feedstock for combined energy and nutrient recovery in the EU28 comprises farmyard manure (140 million tonnes (Mt) dry matter (DM)), sewage sludge (12.5 Mt DM), the organic fraction of municipal solid waste (108 Mt DM), and animal by-products from meat processing (1.5 Mt DM, category 1) (Hermann 2011; Schoumans et al. 2015). With the gradual implementation of integrated multi-trophic aquaculture (IMTA), i.e., the use of excessive nutrients from aquaculture for deliberate cultivation of by-products (seaweed, algae, ascidians), marine by-products may become an additional resource (Kelly and Dworjanyn 2008).

Figure 28.1 exhibits the energy and nutrient recovery value chain that may be implemented to various degrees by water boards, utilities, and industries with regard to municipal and industrial waste flows as well as livestock cooperatives or service enterprises with regard to agricultural residues in regions with high livestock density (Hermann 2011). It does not show processed animal by-products due to their low humidity and outright suitability for thermal combustion or gasification.

The recommended value chain includes anaerobic digestion, dewatering, drying, thermal conversion (combustion or gasification), and ash processing. Outotec's innovative thermal processing portfolio covers drying, thermal conversion, and ash processing (shown in red in Fig. 28.1) (Beckmann 2015).

In response to the challenge of (i) making efficient use of the energy potential of the materials listed above as nutrient-rich feedstock, (ii) facilitating effective phosphate recovery from nutrient-rich residues, and (iii) creating a win-win situation whereby waste-to-energy (WTE), power, and cement industries could continue or even expand the use of biomass-based fuels as well as wastewater treatment plants could continue to supply dry sludge to the cement plants, Outotec proposes a field-tested and reliable fluidized bed gasifier as a converter of nutrient-rich biomass and sewage sludge to a gaseous energy carrier suitable for direct combustion in WTE plants, power plants, and all kinds of industrial processing plants interested in

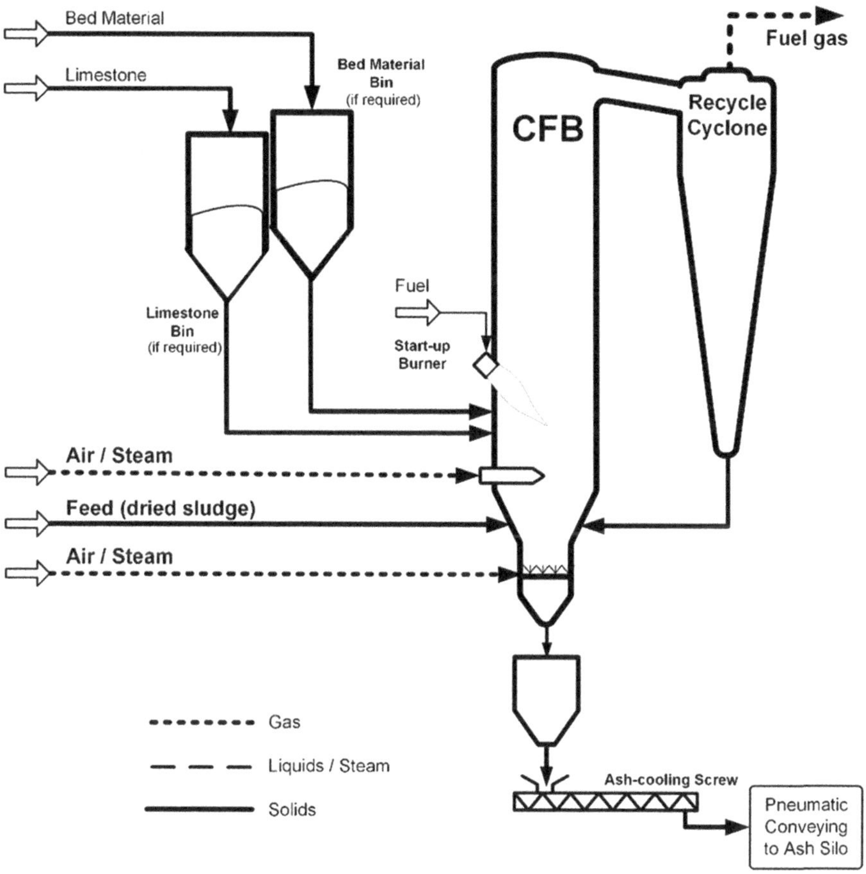

Fig. 28.2 Scheme of Outotec's gasification system (Source: Outotec)

replacing fossil by biomass-based fuels (Fig. 28.2). Alternatively to the – foolproof – direct combustion of the hydrogen-rich product gas, its use as feed to microbes for the production of chemical building blocks will be considered in upcoming research and test programs.

28.2.3 *Outotec's Drying and Gasification Technologies*

The closed-loop steam dryer concept was developed in the 1990s and was installed in two large combined heat-and-power (CHP) and pellet plants in northern Sweden. After a series of start-up problems had been solved, the dryers operated reliably and with an unprecedented efficiency – consuming less than 300 kWh per tonne of water evaporated or 70% less than state-of-the-art dryers. Departing from the idea that this dryer could be a perfect fit for preparing wet waste biomass feedstock including sludges and slurries for gasification, Outotec engineered, built, and commissioned a 700 kg/h pilot dryer in Skellefteå, Sweden. Test work has started in Qu2/2016 and should be terminated in Qu4/2017 due to a number of modifications that had to be developed and installed. The main challenge was to design a feeding nozzle avoiding clogging and scaling of sewage sludge (the most critical feedstock) in the lower sections of the drying pipe. In the early test runs, discharging the fluffy or fibrous dry material was also difficult, but this problem was mastered within a few weeks. The team tested drying of different waste materials, mainly sewage sludge and farmyard slurries from Sweden and Finland that had undergone various pretreatment processes, including anaerobic digestion, hydrolyzation, and chemical modification.

Outotec's gasification roots go back to Lurgi's development of fluidized bed gasification for a variety of fuels – different coals, lignite, pet coke, plastic waste, biomass, municipal waste, and sewage sludge – between 1977 and 1995 in its 700 mm CFB test plant in Frankfurt. Biomass gasification in a fluidized bed gasifier is performed in a gasification medium (air, steam, or oxygen) and a hot bed of solid materials (for instance, sand or olivine) to which the dry biomass is fed. The materials are intensively mixed in the upstreaming medium leading to high heat exchange rates and a uniform and easily controllable temperature of 750–950 °C in the reactor.

Lurgi has supplied three industrial gasification plants between 1985 and 1993 – a 27 MW gasifier for pulp and paper slurries in Pölz (Austria), a 100 MW gasifier for refuse-derived fuels (RDF) in Rüdersdorf (CEMEX cement plant close to Berlin), and an 84 MW RDF gasifier in Geertruidenberg (the Netherlands). All of these plants are operational and continue to generate energy, albeit not always processing the biomass or waste the facilities were originally designed for. In 2016, seven so-called advanced-staged RDF gasification units have been sold to investors in the UK – they will be commissioned in Qu4/2017 and Qu1-Qu2/2018. More contracts of this type are currently negotiated.

An alternative, dual-reactor fluidized bed gasification process originally developed by the Vienna University of Technology has been successfully implemented in three Austrian and one German biomass gasification plants since 2008 (Kern et al. 2013). The dual-reactor gasification system has been continuously developed by Bioenergy2020+ and Umeå University toward steam reforming gasification of phosphorus-rich materials (Kuba et al. 2016; Skoglund et al. 2014; Skoglund 2014). The largest facility of this type (planned final capacity 100 MW) is currently operating at 20 MW in Gothenburg, Sweden, producing pipeline grade CNG (compressed natural gas) from wooden biomass as a vehicle fuel (Thorbjörnsson 2017). Gas upgrading to CNG, however, remains a challenge.

This type of challenge is irrelevant in the present concept. Since the main purpose being the separation of the energy carrier and the nutrient (phosphate) carrier, the product gas will be directly combusted in furnaces, reactors, or processing plants – in all cases being exposed to high temperatures where the tars will be totally combusted. Solid particles containing the phosphorus and secondary nutrients are separated and collected as (hot) ash for further treatment by the AshDec or other processes for nutrient recovery and recycling.

In conclusion, fluidized bed biomass gasification today can be considered as mature process and – in contrast to incineration – producing synthetic gas as an energy carrier that can be used as fuel in different combustion and high temperature processes or as a feedstock for the production of chemical building blocks.

28.3 Process Description

28.3.1 Closed-Loop Steam Dryer

Wet material (sewage sludge or manure) with a moisture content of approximately 75–80% enters a raw material silo. The silo ensures a steady flow of material into the feeding pump. The feeding pump distributes the right amount of sludge into the pressurized dryer loop (Fig. 28.3). The sludge enters the dryer loop in a vertical pipe where superheated steam at a pressure of approximately 3.5 bar (g) flows upward. When the sludge enters the steam flow, it is accelerated and entrained by the steam along the drying section.

Before the sludge is added, the steam in the loop is superheated in the heat exchanger and therefore carries energy that evaporates the water in the sludge. Evaporation of the water takes energy from the superheated steam, and therefore

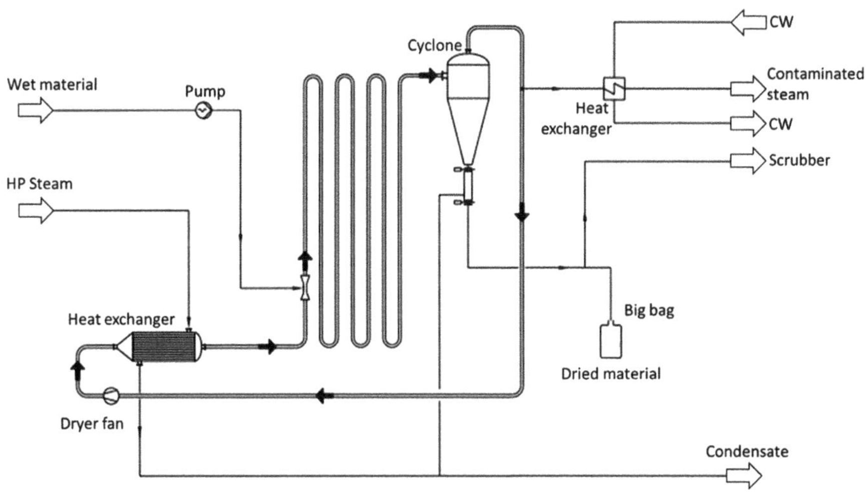

Fig. 28.3 Scheme of the CLS dryer used before gasification – superheated steam at 3.5 bar and 200 °C is applied to evaporate water from the feedstock. The dirty steam at 3,5 bar and 160 °C is purged, and the physical conditions in the loop are kept constant through the application of a saturated steam at 24 bar and 222 °C (Source: Outotec)

the steam temperature decreases. At the end of the drying section, the steam enters a cyclone that separates solids (dried sludge) from steam. Solids are collected at the bottom of the hopper and discharged through the outlet valve. When the warm dried sludge is discharged to atmospheric pressure, some of the remaining water will transfer to gas phase; this steam is separated from the dried sludge in a flash chamber. Dried sludge is discharged through a cell feeder to a bin.

The steam leaving the cyclone now contains the evaporated water from the sludge and therefore is considered as contaminated. Contaminated steam at an amount that is equivalent to the evaporated water must therefore be discharged to a condenser via a valve that controls the pressure in the loop. In a full-scale dryer, the steam will enter the condenser at a pressure of approximately 3.5 bar (g). The saturation temperature at this pressure is about 160 °C. The transfer from liquid-to-gas or gas-to-liquid requires or gives a relatively large amount of energy. This energy can be transferred from the steam to another media, e.g., clean water or steam that can be used for other purposes, for instance, as a medium for steam gasification. In simple terms, this means that the energy used to dry the sludge can be reused after the condenser. In the pilot plant, the steam will be condensed and cooled by cooling water (CW) and let back to the wastewater treatment plant.

The rest of the steam is recirculated from the cyclone to the heat exchanger via the fan. The fan distributes the steam along the loop and compensates for the pressure drop. The heat exchanger is heated by saturated steam at 24 bar (g) that is condensed and discharged to a condensate tank. The condensate thereafter is released to a tank at atmospheric pressure where some of the water evaporates and the rest is fed back to the steam boiler feed water tank.

28.3.2 Dual Fluidized Bed Steam Gasifier

In many applications for gasification of solid feedstock, including those supplied by Outotec as described above, air is used as gasification agent, driving the process autothermally. The drawback of using air is its high nitrogen content diluting the product gas and lowering its heating value to about 3–6 MJ/Nm^{3}db (dry basis). When using steam or CO_2 as gasification agent, the product gas is free of nitrogen, and the calorific value of the gas is higher: for steam, product gas heating values of 10–18 MJ/Nm^{3}db are achieved. In addition, the reactivity of steam is about four times higher than that of CO_2. With steam as a gasification agent, the process becomes allothermal, so the heat for the endothermic gasification reactions has to be provided externally. In fluidized beds, gasification typically takes place at temperatures between 800 and 900 °C (Kern et al. 2013).

Dual fluidized bed (DFB) gasification solves the problem of providing the energy externally by combusting char (gasification residues) and gas cleaning residues (if containing carbon) in a separate combustion reactor where circulating bed material is heated and carried to the gasification reactor, transferring the heat to the solid material that is gasified. Apart from its function as a heat carrier, the bed material is

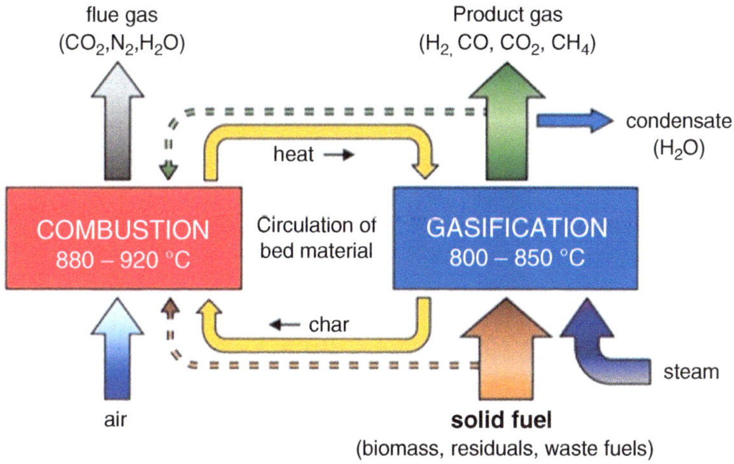

Fig. 28.4 Scheme of dual fluidized bed (DFB) gasifier (Source: Vienna University of Technology)

Table 28.1 Experimental results (Schweitzer et al. 2014)

Process	–	Steam gasification	
Fuel		Wood pellets	Sewage sludge
$T_{gasifier}$	°C	810	810
S/C ratio	mol_{H2O}/mol_C	1.6	1.5
Gas yield	$m^3_{STP,\ db}/kg_{fuel,\ daf}$	1.1	1.04
H_2	Vol.-$\%_{db}$	45.3	51.2
CO	Vol.-$\%_{db}$	17.9	12.2
CH_4	Vol.-$\%_{db}$	10.5	9.2
C_xH_y	Vol.-$\%_{db}$	1.5	3.7
CO_2	Vol.-$\%_{db}$	24.8	23.7
LHV	$MJ/m^3_{STP,db}$	12.85	13.49

STP standard temperature and pressure, *db* dry basis, *daf* dry ash-free, *LHV* lower heating value

selected for its catalytic activity regarding the water-gas-shift reaction and decomposition of biomass tars. Olivine, a natural magnesium iron silicate, is frequently used as bed material because of its catalytic activity, resistance to abrasion and satisfactory behavior in terms of agglomeration (Kuba et al. 2016). The process principle is shown in Fig. 28.4.

In industrial gasifiers the water introduced for conversion of the feedstock is not converted completely. The steam flow is usually higher than stoichiometrically required to provide sufficient flow volume for fluidization. In addition, many reactions take place apart from the gas-water shift reaction: pyrolysis, gasification, reforming, and cracking as well as recombination reactions for tar compounds that lead to the actual product gas composition (Kern et al. 2013). Table 28.1 shows the fuel gas composition after DFB gasification of wood pellets and sewage sludge (Schweitzer et al. 2014).

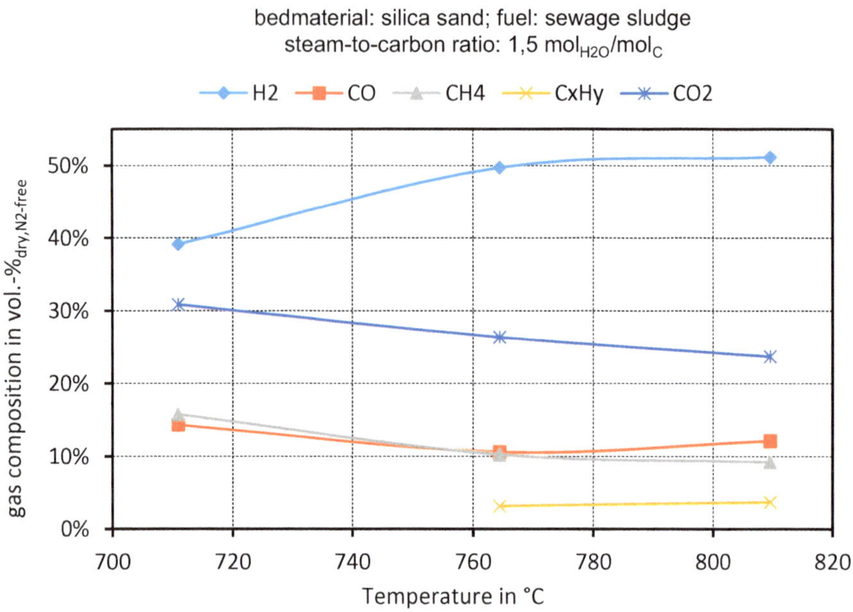

Fig. 28.5 Influence of temperature on the gas composition (Schweitzer et al. 2014)

At a temperature of 810 °C and a S/C ratio of 1.5 mol H_2O /mol C, the synthetic gas of sewage sludge gasification had a slightly higher H_2 concentration (51 vs 45 vol-% dry, N_2-free) and a lower CO concentration (12 vs 18 vol-% dry, N_2-free) than syngas of similar experiments using wood pellets. Figure 28.5 shows the influence of temperature on gas composition during steam gasification of sewage sludge.

Absorption-enhanced reforming (AER) gasification has yielded up to 72 vol-% H_2 in the synthetic gas, albeit with a lower total gas yield due to the temperature limits (700 °C) for AER gasification (Schweitzer et al. 2014). However, more test work is needed to find optimal gasification conditions for sewage sludge and agricultural and industry slurries.

All figures refer to dry, ash-free substrate. Due to the high (30–46 wt%) inorganic, ash-forming mass fractions of sewage sludge and the high water content before drying, the gas yield of the feedstock as received is of course significantly lower compared to wood pellets. This is, however, compensated by gate fees that are usually paid for processing the materials and by the relevant phosphate, potassium, and trace element concentrations in the ash.

Most of the test work was performed within the framework of the EIT InnoEnergy project "DeBugger" in cooperation of Outotec Sweden, Outotec Germany, and the Institute of Combustion and Power Plant Technology at the University of Stuttgart (Hermann et al. 2017). Sewage sludge and agricultural and industry slurries proved being suitable substrates for the production of H_2-rich synthetic gas and a raw material for fertilizer or elementary phosphorus production.

28.4 Industrial Implementation Concept

The goal of using P-rich biomass and sludge as fuel in WTE, power, and thermal processing plants while symbiotically recovering phosphorus from solid residues can be achieved by separating biomass conversion to a gaseous energy carrier (syngas) from combustion and biomass ash from combustion residues of mixed, low nutrient fuels. In the proposed concept, combustion is performed in the WTE, power, or high temperature processing plant, whereas conversion is performed upstream in the fluidized bed gasifier. As only organic compounds are converted to gas, phosphates and other inorganic mass fractions remain as solid gasification residues, in our concept as an ash with 1% carbon. As an example of the proposed concept, its application in a cement plant is shown in Fig. 28.6.

Depending on the gas and residue quality required, a single or double fluidized bed system may be installed. Gasification may be performed in air or steam – the latter offering an excellent opportunity for efficient use of the secondary steam of the closed-loop steam dryer. The evident benefit of the system is converting a solid, low energy density waste to a gaseous, higher density energy carrier which may be used (combusted) wherever appropriate. In addition, gasification facilitates separation of pollutants from the feed. Volatile components may be evaporated and transferred to the user plant (e.g., the waste-to-energy plant), whereas the nutrient-rich ash remains purified.

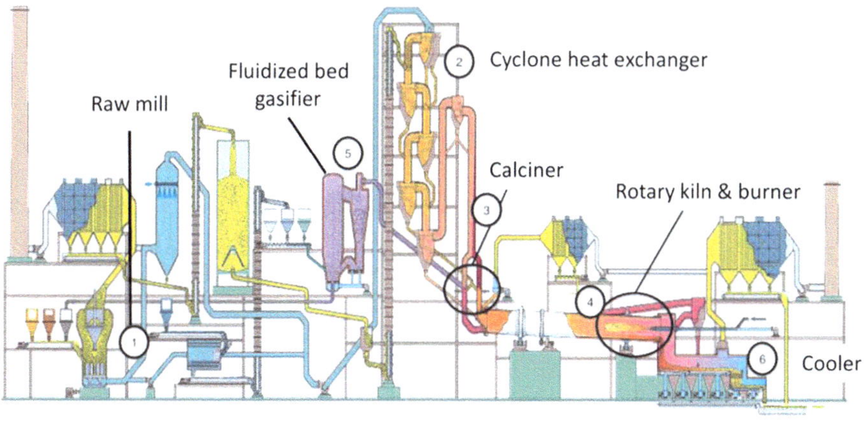

Fig. 28.6 Installation of a fluidized bed gasifier in a cement plant (Source Outotec)

28.5 P Recycling

In contrast to single-stage gasification, DFB gasification produces a carbon-free ash as a solid residue. The ash is a raw material for fertilizer production. Depending on the chemical characteristics of the biomass feedstock, the ash may be processed by wet chemical processes including those in conventional fertilizer manufacturing plants or by the thermal AshDec® process. Typical ash characteristics of different phosphate-rich substrates are shown in Table 28.2

Residues presenting characteristics similar to rock phosphate such as ash from animal by-products and – within certain limits – poultry manure are perfectly suitable for conventional wet chemical processing and could be supplied as raw material to fertilizer industries. Alternatively they could be processed by recently developed wet chemical methods such as the EcoPhos, Tetraphos, or Phos4Life technologies or thermochemical processes such as the optimized AshDec process (Joint EU Research Project P-REX 2012).

The AshDec process is comprehensively explained in Chap. 15 of this book. It has been designed to remove most of the (toxic) heavy metal fractions like As, Cd, and Pb and to improve the plant availability of the ash-borne phosphates that are typically hardly available to crops.

Table 28.2 Chemical composition of ash from different biomass wastes and sewage sludge in comparison to rock phosphate

Substance	Kouribga phosphate rock	Animal by-products (D)[a])	Poultry manure (NL)[b])	Pig manure 1 (NL)	Pig manure 2 (NL)	Sewage sludge (NL)
P_2O_5	32.97	32.50	22.71	23.60	22.00	21.30
CaO	51.34	44.10	37.19	17.90	16.80	15.70
SiO_2	2.35	1.80	3.19	15.70	10.90	21.60
Al_2O_3	0.40	1.10	0.79	1.70	1.00	10.80
Fe_2O_3	0.20	1.00	1.05	2.10	11.90	16.30
MgO	0.30	3.00	6.67	13.10	11.00	2.90
Na_2O	0.80	7.90	3.59	1.60	1.80	1.00
K_2O	0.10	2.90	17.17	6.80	7.70	1.00
SO_3%	1.70	4.30	6.71	6.10	11.70	5.00
Cd mg/kg P_2O_5	51.60	0.92	21.09	10.17	12.72	13.76
Pb mg/kg P_2O_5	9.10	26.77	475.56	33.89	36.36	943.67
Zn mg/kg P_2O_5	700.00	1415.38	12091.59	31355.93	25000.00	10239.43

[a]Kley G. (2004)
[b]http://www.biodat.eu

A recent agricultural research project (Ylivainio et al. 2017) has surprisingly shown a dramatic increase of the P solubility in gasification residues from sewage sludge after treatment by the AshDec method if the solubility is tested by the diffuse gradient in thin film (DGT) method. If this is confirmed by future test work, the combination of gasification and AshDec may produce excellent fertilizer products that virtually outperform most conventional fertilizing products currently available.

28.6 Conclusions and Outlook

Outotec has developed an alternative process chain for recovering energy and nutrient from sewage sludge and industrial and agricultural slurries including a drying and thermal gasification step. The closed-loop steam dryer promises unprecedented energy efficiency by converting the evaporated water into a secondary steam that can be used for different purposes, preferably as a medium for the thermal gasification proposed as a second step of the process chain. Thermal gasification was intended for enabling energy and nutrient recovery from nutrient-rich feedstock if the energy conversion is performed in so-called co-incineration facilities, mainly power, cement, or waste incineration plants. Gasification separates the energy carrier from the solid feedstock material and thus keeps the nutrient carrier apart from the – eventually gaseous – energy carrier. Directly combusting the synthetic gas avoids the typical challenges associated with gas purification and has achieved industrial maturity. The process chain has been developed and tested in a joint research project funded by EIT InnoEnergy and could be implemented as soon as the target feedstock will have been analyzed and tested.

Direct combustion of the product gas is the simple and fast application of the new process chain but limited to convert biomass to steam and electricity. Gasification, however, offers many more options that are subject to subsequent development projects. The underlying objective is to be at the forefront of the development of "next generation" sludge and slurry conversion technologies with the potential to produce products of higher commercial value justifying the significant investments required for large conversion plants. Outotec's first goal is to produce a gas with a high hydrogen concentration. H_2-rich gas has proven to be an appropriate substrate for feeding microbes from which a large variety of fuels and chemical building blocks can be produced. Outotec aims at using its expertise in fluidized bed conversion technologies to provide processes and equipment for producing hydrogen from renewable resources at competitive costs. Close cooperation with global players in the microbiology and chemistry sectors will foster the role of municipal, industrial, and agricultural waste flows with high water content as feedstock in a circular economy.

References

Beckmann K (2015) New process to gasify sludge and slurry may turn farmers into energy producers. Energypost:14–10

Bundesministerium für Umwelt, Naturschutz und Reaktorsicherheit im Einvernehmen mit dem Bundesministerium für Verbraucherschutz (2003) Ernährung und Landwirtschaft, Verordnung über das Inverkehrbringen von Düngemitteln, Bodenhilfsstoffen, Kultursubstraten und Pflanzenhilfsmitteln (Düngemittelverordnung – DüMV), Bundesrecht, Bundesrepublik Deutschland, Bonn, 26.11

Hermann L (2011) How energy from livestock manure can reduce eutrophication, Ecoregion Perspectives "Sustainable agriculture in the Baltic Sea Region in times of peak phosphorous and global change". pp 58–62

Hermann L (2014) A review of innovations in mineral fertilizer production. 16th World Fertilizer Congress of CIEC, Rio de Janeiro

Hermann L, Schaaf T Johansson R (2017) EIT InnoEnergy. KIC InnoEnergy SE, 28 07. [Online]. Available: http://www.innoenergy.com/innovationproject/our-innovation-projects/debugger/. Accessed 28 July 2017

Joint EU Research Project P-REX (2012–2015) P-REX. 05 04 2016. [Online]. Available www.p-rex.eu

Kelly MS, Dworjanyn S (2008) The potential of marine biomass for anaerobic biogas production. Scottish Association for Marine Science, Oban

Kern S, Pfeifer C, Hofbauer H (2013) Gasification of wood in a dual fluidized bed gasifier: influence of fuel feeding on process performance. Chem Eng Sci 90:284–298. https://doi.org/10.1016/j.ces.2012.12.044

Kuba M, He H, Kirnbauer F, Skoglund N, Boström D, Öhman M, Hofbauer H (2016a) Mechanism of layer formation on olivine bed particles in industrial-scale dual fluid bed Gasi fi cation of wood. Energy Fuel 30:7410–7418. https://doi.org/10.1021/acs.energyfuels.6b01522

Ma J, Kennedy N, Yorgey G, Frear C (2013) Review of emerging nutrient recovery technologies for farm-based anaerobic digesters and other renewable energy systems. Washington State University, Seattle

Potocnik J (2017) Circular change. Circular Change Platform, 31 07. [Online]. Available http://www.circularchange.com/janez-potocnik-introductory-speaker-world-circular-economy-forum-helsinki/. Accessed 31 July 2017

Schoumans OF, Bouraoui F, Kabbe C, Oenema O, van Dijk KC (2015) Phosphorus management in Europe in a changing world. Ambio 44(Suppl. 2):180–192. https://doi.org/10.1007/s13280-014-0613-9. Springer

Schweitzer D, Stack V, Gredinger A, Lorenz M, Dieter H, Scheffknecht G (2014) Dual fluidized bed steam gasification of dried sewage sludge, Fifth International Symposium on Energy from Biomass and Waste, Venice

Skoglund N (2014) Ash chemistry and fuel design focusing on combustion of phosphorus-rich biomass (Thesis), Umea: Umea University, Thermochemical Energy Conversion Laboratory

Skoglund N, Grimm A, Öhman M, Boström D (2014) Combustion of biosolids in a bubbling fluidized bed, part 1: main ash-forming elements and ash distribution with a focus on phosphorus. Energy Fuel 28:1183–1190. https://doi.org/10.1021/ef402320q

Thorbjörnsson H (2017) "Gobigas," Göteborg Energi, 28 07. [Online]. Available https://gobigas.goteborgenergi.se/English_version/Start. Accessed 28 July 2017

Thornton PK (2010) Livestock production: recent trends, future prospects. Philos Trans B 365(1554):2853–2867. https://doi.org/10.1098/rstb2010.0134. The Royal Society Publishing

UMWELT-MATERIALIEN Nr (2004) 181 – Abfall, "Klärschlammentsorgung in der Schweiz – Mengen- und Kapazitätserhebung," BUWAL Bundesamt für Umwelt. Wald und Landschaft, Bern

Wiechmann B, Dienemann C, Kabbe C, Brandt S, Vogel I, Roskosch A (2013) Sewage sludge management in Germany. Umweltbundesamt (UBA), Dessau-Roßlau

Ylivainio K, Albihn A, Bloem E, Elving J, Hermann L, Lehmann L, Schaaf T, Schick J, Turtola E (2017) The final publishable summary report. BONUS PROMISE PROJECT 1.4.2014–31.3.2017, Helsinki

Chapter 29
Bone Char As a Novel Phosphorus Fertilizer

Peter Leinweber, Philipp Hagemann, Lutz Kebelmann, Katharina Kebelmann, and Mohsen Morshedizad

Abstract Bone char is the product of a thermochemical conversion of defatted bones. This chapter summarizes the state of the art in the technical pyrolysis process, resulting physicochemical properties and other characteristics of bone chars and possible applications. Special emphasis is put on the solubility of P compounds, which in general characterize bone chars as potentially slow-release P fertilizers. The P release into soil can be improved by an "internal activation" through adsorption of reduced S compounds. Other agronomically relevant properties originate from the porosity that promotes water retention and the habitat function for soil microorganisms. Bone char effects on crop yields are summarized, giving the impression that field crops with long vegetation period and intensive rooting systems benefit most from this material. In conclusion, the carbonization by pyrolysis and formulation of bone char-based products is a reasonable approach in the recycling of P-rich bones from slaughterhouses. However, longer-term agronomic trials are required to fully evaluate the fertilizer potential of bone chars.

Keywords Biochar · Crop yield · Fertilization · Plant nutrition · Soil

P. Leinweber (✉) · M. Morshedizad
Soil Science, University of Rostock, Rostock, Germany
e-mail: peter.leinweber@uni-rostock.de

P. Hagemann
Schwerin, Germany

L. Kebelmann
Berlin, Germany

K. Kebelmann
Hamburg, Germany

H. Ohtake, S. Tsuneda (eds.), *Phosphorus Recovery and Recycling*,
https://doi.org/10.1007/978-981-10-8031-9_29

29.1 Introduction

Biochars, i.e. chars obtained from biomass heating in the absence of oxygen, have many beneficial effects on soil properties (Qian et al. 2015, Lehmann et al. 2011). The most fertile soils worldwide such as chernozems (Schmidt et al. 1999), various anthrosols (Acksel et al. 2017) and the famous terra preta in Amazonia (Glaser et al. 2001) contain substantial amounts of char-derived carbon (C). Since biochars normally are not particularly rich in plant nutrient elements, chars obtained from the pyrolysis of bone materials may have a greater value for soil fertility because bones are rich in phosphorus (P), calcium (Ca) and magnesium (Mg) – all major elements for plant nutrition. Traditionally, bone char has been used as "black bone" for paints and as filter materials in sugar purification. Since actually about 339 × 10^6 kg P are lost from the food sector in EU 27 countries (van Dijk et al. 2016) and the utilization of meat and bone meal in animal feeding is not allowed anymore, attempts have been made to find alternative applications for bones and other slaughterhouse waste in the EU (Vassilev et al. 2013). This chapter summarizes the state of the art in the production and agricultural application of bone chars. Charred materials originating from meat and bone meal are not considered, because they may be not relevant for agriculture due to environmental toxicity.

29.2 Manufacturing of Bone Char by Technical Pyrolysis

Pyrolysis is a thermochemical degradation reaction, which forms molecules smaller than the starting material through decomposition caused by temperature increase within the feedstock under inert atmosphere. Products obtained by pyrolysis are vapours, partly condensable to liquids and solid carbonaceous residues called biochar (Babu 2008).

Different types of pyrolysis processes exist, varying in operation conditions, including reaction temperature, heating rates and residence times of vapours and feedstock leading to pyrolysis products with different characteristics. Slow pyrolysis at low reaction temperatures (200–400 °C) and long residence times for vapours and feedstock, lasting from hours to days (charcoal kiln), produces charcoal from wood. Fast pyrolysis was developed to maximize the yields of pyrolysis liquids from wood and other lignocellulosic materials that may replace fossil fuels. For this purpose feedstock with a fine particle size is rapidly heated up to 500 °C and kept at a very short residence time (approx. 1 s). Consequently, lower amounts of gases and char are obtained (Babu 2008; Panwar et al. 2012). Intermediate pyrolysis operates between fast and slow pyrolysis, applying reaction temperatures of 350–500 °C, hot vapour residence times of 2–4 s and feedstock residence times of 0.5–25 min at moderate heating rates of 1–1000 °C/s. This affects the composition of the liquid phase as well as the characteristics of the biochar. Hence, slower reaction rates lead to a gentler decomposition of the molecules, subsequently prevent formation of

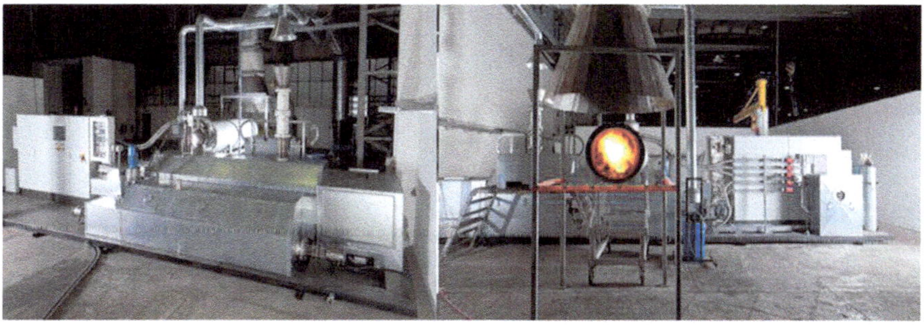

Fig. 29.1 EREKA bioreactor installed in Schwerin, Germany (Hagemann and Kebelmann 2017)

high molecular tars and produce dry chars that fix alkali metals and ash contained in the feedstock. Notably the pyrolytic processing of ash-rich material like bones is possible, if process temperatures lower than ash fusion temperatures are applied. Another advantage is the possible application of feedstock with larger particle sizes due to sufficient heating given through moderate heating rates and longer residence times. The product distribution obtained by this process is 40–60% pyrolysis liquids, 20–30% noncondensable vapours and 20–35% biochar, depending on the feedstock (Hornung et al. 2011).

Bone materials with 1–2 mm particle size were pyrolysed in batch processes in muffle furnaces (Patel et al. 2015) or Carbolite tube ovens (Reidsma et al. 2016; Rojas-Mayorga et al. 2015), applying reaction temperatures between 220 and 750 °C and residence times of 45 min up to 2 h. Furthermore, bone chars have been produced by intermediate pyrolysis at the laboratory scale and the impact of pyrolysis conditions including temperature and residence time on the yield and quality of bone char investigated (Hornung et al. 2011; Patel et al. 2015; Reidsma et al. 2016; Rojas-Mayorga et al. 2015; Zwetsloot et al. 2015).

More recently, upscaling to the industrial level has been demonstrated (Hagemann and Kebelmann 2017). The industrial-scale EREKA bioreactor (Fig. 29.1) has a capacity to thermochemically processing 500 kg/h feedstock. This pyrolysis unit consists of two coaxial screws that convey the material through the reactor. This unique set-up ensures a self-cleaning of the decomposition area and prevents material blockings. An exceptional feature of the EREKA bioreactor is the heating system. A heat carrier flowing through the inner core of the two screws delivers the main heat substantial for an efficient decomposing of the feedstock (Warchol and Kebelmann 2012). In addition, an electric outer heating prevents heat loss through the reactor body. This combination leads to a low energy input into the decomposition process. The reactor set-up includes an automatic feeding system consisting of two pinch valves, which open and close alternately to prevent gas exchange between the reactor and atmosphere and gradually feed the process with material. Produced vapours and gaseous products are lead to a burner and fired instantly. The char is discharged at the end of the reactor applying a discharge screw.

Table 29.1 Physicochemical properties of bone chars as compiled from various publications

References	Specific surface area (m^2/g)	Porosity (cm^3/g)	Total Ca (g/kg)	Total P	Total C	Hydroxyapatite (%)
Cheung et al. (2001)	100	0.22	–	–	90–110	70–76
Flores-Cano et al. (2016)	87	0.24	239	157	–	89
Iriarte-Velasco et al. (2016)	76	0.26	242	115	380	–
Medellin-Castillo et al. (2007)	104	0.3	–	–	110	76
Medellin-Castillo et al. (2014)	104	0.3	–	150	63	85
Morshedizad et al. (2016)	–	–	185	149	112	–
Patel et al. (2015)	58–114	0.29–0.32	–	–	–	–
Siebers and Leinweber (2013)	42	–	280	152	130	–
Robinson et al. (2018)	–	–	238	130	104	56
Rojas-Mayorga et al. (2015)	69	0.23	394	206	–	–
Warren et al. (2009)	82	–	–	134	125	–
Zwetsloot et al. (2016) (Pyrolysis at 350 °C)	–	–	271	127	180	–
Zwetsloot et al. (2015) (Pyrolysis at 550 °C)	–	–	307	140	–	–
Zwetsloot et al. (2016) (Pyrolysis at 750 °C)	–	–	337	153	82	–

The size of the applied bone fragments was approximately 10–15 mm in width and in length, and continuous processing with a feed rate of 40–50 kg/h was carried out under industrial conditions. The pyrolysis reaction temperature ranged between 350 and 500 °C; feedstock residence time between 10 and 20 min and moderate heating rates were applied. From the total amount of applied feedstock, 65% of dry bone char has been obtained, whereas the other 35% were incondensable gases and pyrolysis liquids. This bone char has been characterized for its elemental composition and P speciation (Robinson et al. 2018) and is currently tested in agronomic application trials.

29.3 Characteristics of Bone Chars

The characteristics of bone char vary in wide ranges, depending on characteristics of source materials and pyrolysis conditions. The light yellow colour of bone turns to brown and black during the pyrolysis at low to high temperatures (300–900 °C) (Reidsma et al. 2016), but Rojas-Mayorga et al. (2015) reported colour changes from black to grey and white at 800 °C and 900 °C, respectively. These colour changes originate from thermal degradation of organic matter of the bone matrix (Patel et al. 2015; Rojas-Mayorga et al. 2015). Protein and collagen are transformed to cyclic organic molecules with increasing temperature (Reidsma et al. 2016). Because of the absence of volatile material, pyrolysis-field ionization mass spectrometry (Py-FIMS) (Leinweber et al. 2009) of bone chars yielded only very weak mass spectra.

Bone chars are characterized by a mesoporous structure with a specific surface area between 42 and 114 m^2/g (Table 29.1). The porosity and crystal structure of bone chars depend on the mineral composition (Rothwell et al. 1980), animal species and age (Wu et al. 2003) and pyrolysis conditions (Larsen et al. 1994). Increase in mineralization of bones and mineral maturity during growth of the animal can improve the bone crystallinity (Novotny et al. 2012; Wu et al. 2003). Moreover, it has been reported that poorly crystalline fresh bones (Aue et al. 1984; Rogers and Daniels 2002; Mkukuma et al. 2004; Rajendran et al. 2013) experience microstructural changes during the heating. Patel et al. (2015) found a pronounced alteration in crystal size and porosity of bone char with increasing pyrolysis temperature and residence time. In that study the porous structures were interconnected with each other, creating large pore diameters with rising temperature from 400 to 600 °C while the specific surface area was reduced. Similarly, it has been shown that the interlocked porous network remained after removal of organics from heated bones (Etok et al. 2007; Ooi et al. 2007).

According to Figueiredo et al. (2010), elevated temperatures can result in higher crystallinity degrees, larger crystallite sizes and a less porous structure. The gas atmosphere around animal bones during pyrolysis is an important parameter affecting the properties of produced bone char. Rojas-Mayorga et al. (2015) showed that a CO_2 atmosphere accelerated the crystallinity of bone char and rearranged the molecular structure with higher density as compared with bone chars obtained by pyrolysis under N_2 atmosphere. Pyrolysis temperature had the most significant effect on bone char characteristics, and at higher temperature, the porous structure was condensed and the bone char bulk density increased (Figueiredo et al. 2010; Rojas-Mayorga et al. 2015). Patel et al. (2015) obtained the maximum porosity of 0.32 cm^3/g at 400 °C pyrolysis temperature and 3-h residence time.

The major chemical elements of bone chars are calcium (Ca), phosphorus (P) and carbon (C) (Table 29.1). The proportions of Ca and Mg compounds result in alkaline pH of bone chars in the range of about 7–10. Dissimilar to most biochars, bone char has only about 10 w/w % of C (Cheung et al. 2002; Fuller et al. 2003; Wilson et al. 2003; Medellin-Castillo et al. 2007). Based on ash-free dry matter, the C content and the C/H and C/N ratios increased with increasing pyrolysis temperature, indicating the progressive formation of cyclic C compounds (Rojas-Mayorga et al. 2015).

According to the Ca and P contents, hydroxyapatite is the predominating mineral constituent of bone char as determined by X-ray diffraction (Medellin-Castillo et al. 2014; Rojas-Mayorga et al. 2015) and X-ray near edge fine structure analysis (Zwetsloot et al. 2015, 2016; Robinson et al. 2018). The elemental contents and ratios of bone char can be influenced by the temperature and residence time of the pyrolysis. Zwetsloot et al. (2015) reported a total P enrichment with increasing pyrolysis temperature. Patel et al. (2015) obtained a mean Ca/P ratio of 2.2 at pyrolysis temperatures of 400–600 °C and 1 h residence time which increased to 2.6 at 2-h residence time. Other elements such as N, Mg and S occur in low concentrations, and bone chars are almost free of soil contaminants like Cd and U (Siebers and Leinweber 2013). These properties stimulated research into the potential use of bone chars as soil amendments and alternative P fertilizers.

29.4 Extractability and Availability of P Compounds in Soil

The lower crystalline structure and higher carbonate concentration of bone apatite result in greater solubility in comparison to the mineral hydroxyapatite (Wopenka and Pasteris 2005; Pan and Darvell 2010). Nevertheless, the extractability of P from bone char is low in water (3–20% of total P) but larger in formic acid (92–95% of total P), with a general tendency of a lower solubility in water but a larger solubility in formic acid with increasing pyrolysis temperature between 220 and 750 °C (Zwetsloot et al. 2015). P concentrations after 1–4 h in buffered ammonium acetate solutions ranged from 230 to 320 mg/L (pH 4) and 50–100 mg/L (pH 6). In mixtures of bone char with soil, the strict pH dependency of the P solubility has been confirmed several times (Warren et al. 2009; Siebers and Leinweber 2013). Comparing various extraction methods common in agronomy and soil protection, Morshedizad et al. (2016) obtained the order H_2O-P < NH_4NO_3-P < $NaHCO_3$-P for soils amended with bone char, but in general not more than 3.7% of the total added P was extracted after 34 d of incubation. In response to the bone char solubility as a pH-dependent process, Morshedizad et al. (2016) reported that surface modification of bone char by reduced sulphur (S) compounds ("internal activation") promoted the P release from fertilized soils over a wide pH range. This surface-modified bone char contains more $Ca(H_2PO_4)_2$ and less hydroxyapatite than the original bone char (Morshedizad et al. 2018). This modified mineral composition and reductions in pH values (4.9 vs. 7.8) along with the bacterial oxidation of S compounds seem to promote the P dissolution from these bone chars. Consistent with these findings, Zwetsloot et al. (2016) reported that the P availability in P-fixing soils was stimulated by root-mycorrhizae-bone char interactions, which may be explained by acidification in the soil rhizosphere. Warren et al. (2009) showed in an incubation experiment that the P sorption capacity of soil had an important influence on bone char dissolution. This was supported by Siebers and Leinweber (2013) who revealed a positive correlation between P release from bone char and the P sorption capacity of studied soils. Since bone chars are essentially slow-release fertilizers, highly dependent on P sorption capacity of soil, any discharge of the P-saturated soil solution can lead to further P release via bone char dissolution. Morshedizad and Leinweber (2017) recently reported that non-equilibrium conditions through repeated leaching steps mobilized more P in comparison to a continuous

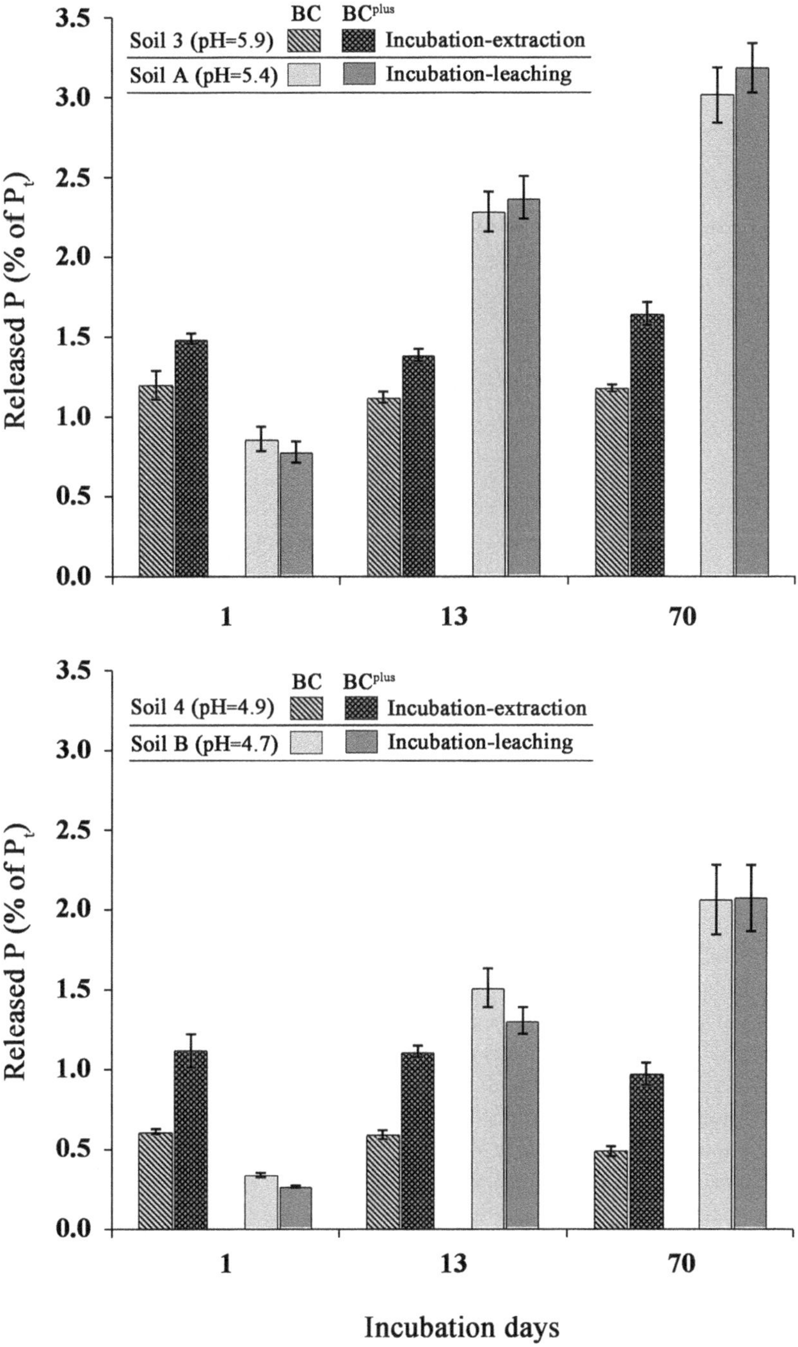

Fig. 29.2 Proportions of released P (as percentage of total P (P_t) added) from bone char-amended soils obtained by incubation-extraction in a previous study (soils *3* and *4* in Morshedizad et al. 2016) and by incubation leaching in the present study (soils *A* and *B*) (Morshedizad and Leinweber 2017) (With permission from John Wiley and Sons)

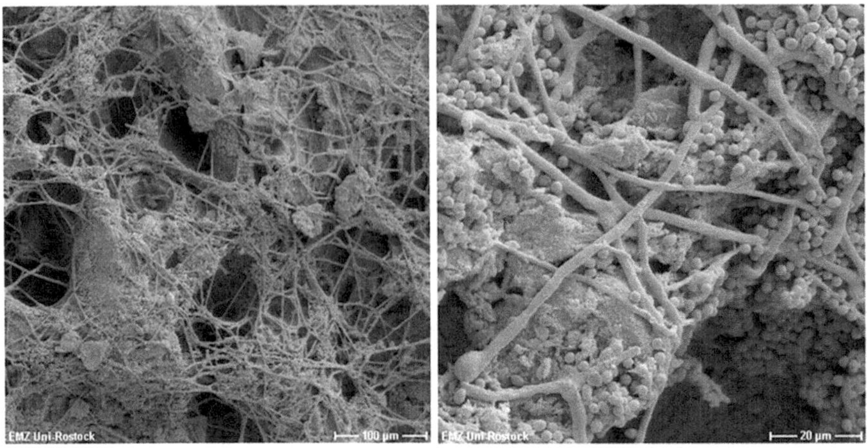

Fig. 29.3 Scanning electron microscopic pictures of bone char particles that have been inoculated with *Trichoderma* spp. and are colonized by this fungus

incubation-extraction experiment that established equilibrium conditions in the soil solution (Fig. 29.2).

29.5 Other Bone Char Effects on Soil

Bone char addition to soil improves the water retention capacity. In a simple experiment with addition of 1, 5 and 10 w/w% of bone char to a sandy soil, we found a substantial increase of the water holding capacity at the permanent wilting point of plants ($-10^{4.2}$ hPa pressure) to 1.4, 3.2 and 5.5 w/w%, respectively (control without bone char 1.3 w/w%) (Leinweber 2017). This indicates that bone char improves the physical soil conditions by increasing its porosity and water retention. This may be particularly beneficial if crops at sandy soils suffer from dry conditions. The porous structure of bone chars also supports the soil habitat function. For instance, beneficial fungi have successfully been cultivated and propagated at bone char particles (Fig. 29.3). These fungi had no P source other than the bone char, confirming the ability to acquire P from the char particles and utilize it in their metabolism.

This habitat function has stimulated research into functional inoculants for biocontrol. Postma et al. (2013) successfully tested strains of the genera *Pseudomonas* to suppress *Pythium* and *Fusarium* species which cause damping off and crown and root rot in tomato. They concluded that bone char is a valuable carrier for delivery of biocontrol bacteria into soil or substrate and combines biocontrol with recycling a phosphorus-rich waste product.

The retention of contaminants by bone char is described in many publications. The material successfully has been tested for the removal of fluoride from water (Medellin-Castillo 2007), and the adsorption increased with decreasing pH as explained by electrostatic interactions (Medellin-Castillo 2014). Since the point of zero charge of bone char is at around pH 7.7, it can be also used for the removal of cationic contaminants from water. Experiments with Cr(III) (Flores-Cano et al. 2016); Cu and Zn (Wilson et al. 2003); Cd, Cu and Zn (Cheung et al. 2001 2002); and U (Fuller et al. 2003) demonstrated the adsorption capacity of bone chars and its potential use in water purification. Thus, not surprisingly, bone char has also been tested for the retention of contaminants from soil. For instance, Chen et al. (2006) showed that Pb uptake by Chinese cabbage was significantly reduced in the presence of 1.6 w/w% of bone char in soil. Siebers and Leinweber (2013) showed a reduced Cd mobility in bone char-amended soils and explained this by the increased pH. Subsequently, Siebers et al. (2014) reported from pot experiments that even small bone char applications tended to decrease the Cd uptake of arable crops, although this reduction was mostly insignificant. This Cd immobilization capacity of bone chars added to moderately contaminated soils was confirmed by Morshedizad et al. (2016). Furthermore, the most recent research showed that the 0.5–1 mm particle size of bone chars was optimal for releasing a maximum of P along with relatively small Cd concentrations into soil solution and this was valid for different types of bone chars (Morshedizad and Leinweber 2017). As this particle size can be spread with common agricultural machinery, the way is open to test bone chars at the farm scale in practical agriculture.

29.6 Crop Yield Effects and Agronomic Efficiency

Not many publications are available about crop yield effects of bone chars, and almost all are from short-term pot experiments. Siebers et al. (2012) reported yield increases for potato, no effects on wheat and yield decreases for onions compared to 0-P controls. In a study by Siebers et al. (2014), the yield of wheat heads responded positive to bone char application, but lettuce and potato tubers yielded indifferent results for two soils with moderate and high P levels. Little et al. (2015) investigated the effect of bone char on the growth of various weeds and arable crops in comparison to composted poultry manure in biological farming systems. They reported that bone char did not adequately mimic the P supply from composted poultry manure. Zwetsloot et al. (2016) reported positive effects of bone char in combination with arbuscular mycorrhizae (AM) inoculation on maize yield in P-fixing soil. Their inoculated maize showed similar P accumulation when fertilized with either bone char (750 °C pyrolysis temperature) or triple superphosphate (TSP). Actually, the author's group and collaborators obtained slightly increased grass yields due to bone char application in a current pot experiment with a low P, sandy soil (K. Panten; Julius-Kühn Institute Braunschweig, Germany, unpublished results). In summary, it is evident that bone char sometimes but not always has a positive fertilization effect.

The agronomic effectiveness of bone chars, relative to the commercially available and widely used standard P fertilizer TSP from the above cited studies, is compiled in Fig. 29.4. Negative values mean that the bone char treatments yielded less

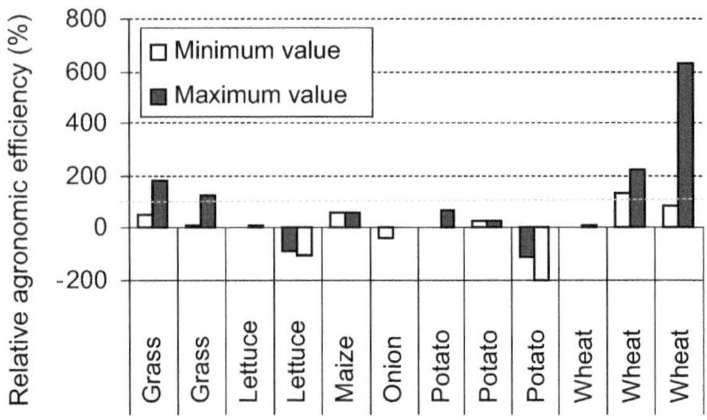

Fig. 29.4 Ranges of relative agronomic efficiency of bone chars as compiled from various publications and unpublished data of the authors. Only the yield data for consumable plant parts were considered

than the 0-P-controls, indicating yield depressions. Values from 0% to 100% indicate positive effects of bone char to the crop yields which are less pronounced than in the TSP treatments, and values >100% result from bone char treatments performing better than TSP. It seems that the relative agronomic effectiveness is better for grasses and grain crops that have a relatively long vegetation period and intensive root system enabling the exploitation of slowly soluble P sources. Vegetables and potato seemingly are rather inhibited in growth and yield by bone chars and perform better under the supply of easily soluble P from TSP. This different behaviour of the plant is plausible considering the mineralogy, surface charge and P release characteristics of bone chars, but longer-term studies, e.g. of crop rotations, are required to assess the agronomic value of bone chars.

29.7 Conclusions

Bone char can be ascribed as a plant nutrient-element-rich biochar that is produced by pyrolysis of defatted bone materials. Thus, all beneficial effects that are established for soil amendments with biochars can be expected for bone chars as well plus a fertilizer value. Technically, the intermediate pyrolysis at reaction temperatures between 220 and 750 °C, hot vapour residence times of 2–4 s and feedstock residence times up to 25 min has been proven well suited for the production of bone chars. Industrial upscaling to a capacity of 500 kg feedstock h^{-1}, pyrolysis reaction between 350 and 500 °C and feedstock residence time between 10 and 20 min has been successfully demonstrated. Generally, the pyrolysis conditions affect the product quality and important properties such as mineralogy, crystallinity and the solubility of P compounds. Therefore, from an agricultural point of view, more research and technology development is required to optimize the char production conditions in order to obtain relatively low crystalline P compounds that are moderately soluble which seems to be favoured by low pyrolysis temperatures.

Bone chars are generally characterized by a relatively large specific surface area in the same order of magnitude as soil clay minerals. Therefore, if applied to soil, they likely contribute to ad- and desorption processes of ions, organic molecules and colloidal substances and consequently enhance soil fertility. This can be also attributed to their porosity, which enhances the water retention and supports the habitat function for soil microorganisms. The latter became obvious when using bone chars as carrier for beneficial microorganisms that exert biocontrol functions against harmful fungi which cause economically relevant plant diseases in modern agriculture. The chemical composition and speciation of P compounds, mainly hydroxyapatites of low solubility, resulted in a slow release of phosphates into soil solution. However, the liming effect from Ca and Mg ions reduced the mobility, plant availability and uptake of contaminants such as Cd and therefore its transfer into the nutrient chain. Although the few agronomic fertilization experiments sometimes showed yield reductions and positive effects on crop yield only under certain circumstances, these and the relative agronomic efficiency sometimes above the commercial highly soluble P fertilizer are promising and call for the intensification of research in order to disclose the best application scenarios. In brief, bone chars can be ascribed as slow-release, multi-element fertilizers or soil amendments with liming and Cd immobilization by-effects.

Finally, the research on the properties and agronomic applications of bone chars still is in its infancy. Therefore, these conclusions based on publications from the past 20 years have a preliminary character, and new evidence and more detailed insights can be expected for the future because the P recycling now is at the political agenda.

Acknowledgements Parts of this work have been performed within the InnoSoilPhos project (http://www.innosoilphos.de/default.aspx), funded by the German Federal Ministry of Education and Research (BMBF) in the frame of the BonaRes-program (No. 031A558). Mohsen Morshedizad acknowledges a PhD grant from the state of Mecklenburg-Western Pomerania, Germany. This research was conducted within the scope of the Leibniz ScienceCampus Phosphorus Research Rostock.

References

Acksel A, Kappenberg A, Kühn P, Leinweber P (2017) Human activity formed deep, dark topsoils around the Baltic Sea. Geoderma Reg 10:93–101

Aue WP, Roufosse AH, Glimcher MJ, Griffin RG (1984) Solid-state phosphorus-31 nuclear magnetic resonance studies of synthetic solid phases of calcium phosphate: potential models of bone mineral. Biochemistry 23:6110–6114

Babu BV (2008) Biomass pyrolysis: a state-of-the-art review. Biofuels Bioprod Biorefin 2(5):393–414

Chen SB, Zhu YG, Ma YB, McKay G (2006) Effect of bone char application on Pb bioavailability in a Pb-contaminated soil. Environ Pollut 139(3):433–439

Cheung CW, Chan CK, Porter JF, McKay G (2001) Combined diffusion model for the sorption of cadmium, copper, and zinc ions onto bone char. Environ Sci Technol 35(7):1511–1522

Cheung CW, Porter JF, McKay G (2002) Removal of Cu(II) and Zn(II) ions by sorption onto bone char using batch agitation. Langmuir 18(3):650–656

van Dijk KC, Lesschen JP, Oenema O (2016) Phosphorus flows and balances of the European Union Member States. Sci Total Environ 542:1078–1093

Etok SE, Valsami-Jones E, Wess TJ, Hiller JC, Maxwell CA, Rogers KD, Manning DAC, White ML, Lopez-Capel E, Collins MJ, Buckley M, Penkman KEH, Woodgate SL (2007) Structural and chemical changes of thermally treated bone apatite. J Mater Sci 42:9807–9816

Figueiredo M, Fernando A, Martins G, Freitas J, Judas F, Figueiredo H (2010) Effect of the calcination temperature on the composition and microstructure of hydroxyapatite derived from human and animal bone. Ceram Int 36(8):2383–2393

Flores-Cano JV, Leyva-Ramos R, Carrasco-Marin F, Aragón-Piña A, Salazar-Rabago JJ, Leyva-Ramos S (2016) Adsorption mechanism of Chromium (III) from water solution on bone char: effect of operating conditions. Adsorption 22(3):297–308

Fuller CC, Bargar JR, Davis JA (2003) Molecular-scale characterization of uranium sorption by bone apatite materials for a permeable reactive barrier demonstration. Environ Sci Technol 37(20):4642–4649

Glaser B, Haumaier L, Guggenberger G, Zech W (2001) The 'Terra Preta' phenomenon: a model for sustainable agriculture in the humid tropics. Naturwissenschaften 88:37–41

Hagemann P, Kebelmann L (2017) Klärschlammpyrolyse mit dem EREKA Bio-Reaktor. In: Sauermann U, Klätte M (Eds) Thermopolyphos Steinbeis Edition Stuttgart, p 49–57

Hornung A, Apfelbacher A, Sagi S (2011) Intermediate pyrolysis: a sustainable biomass-to-energy concept – biothermal valorisation of biomass (BtVB) process. J Sci Ind Res 70:664–667

Iriarte-Velasco U, Sierra I, Zudaire L, Ayastuy JL (2016) Preparation of a porous biochar from the acid activation of pork bones. Food Bioprod Process 98:341–353

Larsen MJ, Pearce EIF, Ravnholt G (1994) The effectiveness of bone char in the defluoridation of water in relation to its crystallinity, carbon content and dissolution pattern. Archs Oral Bid 39(9):807–816

Lehmann J, Rillig MC, Thies J, Masiello CA, Hockaday WC, Crowley D (2011) Biochar effects on soil biota – a review. Soil Biol Biochem 3(9):1812–1836

Leinweber P (2017) Pyrolyse von Schlachtknochen – ein attraktiver Ansatz im Phosphorrecycling. In: U. Sauermann, M. Klätte (Eds) Thermopolyphos Steinbeis Edition Stuttgart, p 59–65

Leinweber P, Jandl G, Eckhardt K-U, Schlichting A, Hofmann D, Schulten H-R (2009) Analytical pyrolysis and soft-ionization mass spectrometry. In: Senesi N, Xing B, Huang PM (eds) Biophysico-chemical processes involving natural nonliving organic matter in environmental systems. Wiley, New York, pp 539–588

Little NG, Mohler CL, Ketterings QM, DiTommaso A (2015) Effects of organic nutrient amendments on weed and crop growth. Weed Sci 63(3):710–722

Medellín-Castillo NA, Leyva-Ramos R, Ocampo-Pérez R, García de la Cruz RF, Aragón-Piña A, Martínez-Rosales JM, Guerrero-Coronado RM, Fuentes-Rubio L (2007) Adsorption of fluoride from water solution on bone char. Ind Eng Chem Res 46:9205–9212

Medellin-Castillo NA, Leyva-Ramos R, Padilla-Ortega E, Perez RO, Flores-Cano JV, Berber-Mendoza MS (2014) Adsorption capacity of bone char for removing fluoride from water solution. Role of hydroxyapatite content, adsorption mechanism and competing anions. J Ind Eng Chem 20(6):4014–4021

Mkukuma LD, Skakle JMS, Gibson IR, Imrie CT, Aspden RM, Hukins DWL (2004) Effect of the proportion of organic material in bone on thermal decomposition of bone mineral: an investigation of a variety of bones from different species using thermogravimetric analysis coupled to mass spectrometry, high-temperature X-ray diffraction, and fourier transform infrared spectroscopy. Calcif Tissue Int 75(4):321–328

Morshedizad M, Leinweber P (2017) Mobilization and leaching of phosphorus and cadmium in soils amended with different bone chars. Clean – Soil Air Water 45:1600635. https://doi.org/10.1002/clen.201600635

Morshedizad M, Zimmer D, Leinweber P (2016) Effect of bone chars on phosphorus-cadmium-interactions evaluated by three extraction procedures. J Plant Nutr Soil Sci 179:388–398

Morshedizad M, Klysubun W, Panten K, Leinweber P (2018) Alteration of bone chars and amended soils as revealed by sequential fractionation and XANES spectroscopy. SOIL 4:23–35. https://doi.org/10.5194/soil-4-23-2018

Novotny EH, Auccaise R, Velloso MHR, Corrêa JC, Higarashi MM, Abreu VMN, Rocha JD, Kwapinski W (2012) Characterization of phosphate structures in biochar from swine bones. Pesq Agropec Bras 47:672–676

Ooi CY, Hamdi M, Ramesh S (2007) Properties of hydroxyapatite produced by annealing of bovine bone. Ceram Int 33:1171–1177

Pan H, Darvell BW (2010) Effect of carbonate on hydroxyapatite solubility. Cryst Growth Des 10:845–850

Panwar NL, Kothari R, Tyagi VV (2012) Thermo chemical conversion of biomass – eco friendly energy routes. Renew Sust Energ Rev 16(4):1801–1816

Patel S, Han J, Qiu W, Gao W (2015) Synthesis and characterisation of mesoporous bone char obtained by pyrolysis of animal bones, for environmental application. J Environ Chem Eng 3(4):2368–2377

Postma J, Clematis F, Nijhuis EH, Someus E (2013) Efficacy of four phosphate-mobilizing bacteria applied with an animal bone charcoal formulation in controlling *Pythium aphanidermatum* and *Fusariumoxysporum* f. sp. *radicis lycopersici* in tomato. Biol Control 67(2):284–291

Qian K, Kumar A, Zhang H, Bellmer D, Huhnke R (2015) Recent advances in utilization of biochar. Renew Sust Energ Rev 42:1055–1064

Rajendran J, Gialanella S, Aswath PB (2013) XANES analysis of dried and calcined bones. Mater Sci Eng C 33(7):3968–3979

Reidsma FH, van Hoesel A, van Os BJH, Megens L, Braadbaart F (2016) Charred bone: physical and chemical changes during laboratory simulated heating under reducing conditions and its relevance for the study of fire use in archaeology. J Archaeol Sci Rep 10:282–292

Robinson S, Baumann K, Kebelmann L, Hagemann P, Hu Y, Leinweber P (2018) Phosphorus transformations in plant- and bio-waste feedstocks induced by pyrolysis: implications for fertilizer replacement potential. Ambio 47(Suppl. 1):73–82. https://doi.org/10.1007/s13280-017-0990-y

Rogers KD, Daniels P (2002) An X-ray diffraction study of the effects of heat treatment on bone mineral microstructure. Biomaterials 23(12):2577–2585

Rojas-Mayorga CK, Silvestre-Albero J, Aguayo-Villarreal IA, Mendoza-Castillo DI, Bonilla-Petriciolet A (2015) A new synthesis route for bone chars using CO_2 atmosphereand their application as fluoride adsorbents. Micropor Mesopor Mater 209:38–44

Rothwell WPJ, Waugh S, Yesinowski JP (1980) High-resolution variable-temperature phosphorus-31 NMR of solid calcium phosphates. J Am Chem Soc 102(8):2637–2643

Schmidt MWI, Skjemstad JO, Gehrt E, Kögel-Knabner I (1999) Charred organic carbon in German chernozemic soils. Eur J Soil Sci 50:351–365

Siebers N, Leinweber P (2013) Bone char – a clean and renewable fertilizer with cadmium immobilizing capacity. J Environ Qual 42:405–411

Siebers N, Godlinski F, Leinweber P (2012) Utilization of bone char phosphorus for potato, wheat, and onion production. Landbauforschung – vTI Agric Forest Res 62:59–64

Siebers N, Godlinski F, Leinweber P (2014) Bone char as phosphorus fertilizer involved in Cd immobilization in lettuce, wheat, and potato cropping. J Plant Nutr Soil Sci 177:75–83

Vassilev N, Martos E, Mendes G, Martos V, Vassileva M (2013) Biochar of animal origin: a sustainable solution to the global problem of high-grade rock phosphate scarcity? J Sci Food Agric 93:1799–1804

Warchol G, Kebelmann L (2012) Screw and method for producing same. Patent WO 2012/126574

Warren GP, Robinson JS, Someus E (2009) Dissolution of phosphorus from animal bone char in 12 soils. Nutr Cycl Agroecosyst 84:167–178

Wilson JA, Pulford ID, Thomas S (2003) Sorption of Cu and Zn by bone charcoal. Environ Geochem Health 25:51–56

Wopenka B, Pasteris JD (2005) A mineralogical perspective on the apatite in bone. Mater Sci Eng C 25:131–143

Wu Y, Ackerman JL, Strawich ES, Rey C, Kim H-M, Glimcher MJ (2003) Phosphate ions in bone: identification of a calcium–organic phosphate complex by ^{31}P solid-state NMR spectroscopy at early stages of mineralization. Calcif Tissue Int 72(5):610–626

Zwetsloot MJ, Lehmann J, Solomon D (2015) Recycling slaughterhouse waste into fertilizer: how do pyrolysis temperature and biomass additions affect phosphorus availability and chemistry? J Sci Food Agric 95:281–288

Zwetsloot MJ, Lehmann J, Bauerle T, Vanek S, Hestrin R, Nigussie A (2016) Phosphorus availability from bone char in a P-fixing soil influenced by root-mycorrhizae-biochar interactions. Plant Soil 408(1):95–105

Part VI
Solid Adsorbents

Chapter 30
Phosphate Recovery Using Amorphous Calcium Silicate Hydrates

K. Okano, H. Ohtake, M. Kunisada, H. Takano, and M. Toda

Abstract Amorphous calcium silicate hydrates (A-CSHs) can be chemically synthesized using abundantly available, inexpensive materials such as siliceous shale and calcium hydroxide. A-CSHs can serve as a bifunctional adsorption-aggregation agent for phosphate (P_i) recovery from aqueous solution. A-CSHs can also be prepared by soaking recyclable calcium silicates such as concrete sludge and steelmaking slag in a dilute hydrochloric acid solution. Since A-CSHs show high P_i removability, settleability, and filterability, they have the potential to offer a simple, cost-effective option to the recovery of P_i from P_i-rich waste streams. On-site experiments using a mobile, pilot-scale plant have showed that A-CSHs can recover approximately 80% P_i from a P_i-rich sidestream in a wastewater treatment plant. This chapter describes a simple process for P_i recovery from aqueous solution using A-CSHs as a bifunctional adsorption-aggregation agent.

Keywords Amorphous calcium silicate hydrates · Bifunctional adsorption-aggregation agent · Mobile pilot-scale plant · By-product phosphate fertilizer

K. Okano (✉)
Department of Biotechnology, Graduate School of Engineering, Osaka University, Suita, Osaka, Japan
e-mail: okano@bio.eng.osaka-u.ac.jp

H. Ohtake
Phosphorus Atlas Research Institute, Waseda University, Shinjuku-ku, Tokyo, Japan

M. Kunisada
Mikuni Pharmaceutical Industrial Co., Ltd., Toyonaka, Osaka, Japan

H. Takano
Research & Development Center, Taiheiyo Cement Co., Sakura, Chiba, Japan

M. Toda
Research & Development Laboratory, Onoda Chemical Industry Co., Ltd., Tokyo, Japan

H. Ohtake, S. Tsuneda (eds.), *Phosphorus Recovery and Recycling*,
https://doi.org/10.1007/978-981-10-8031-9_30

30.1 Introduction

Calcium silicate hydrates (CSHs) are the main product of the hydration of Portland cement and serve as the main binding phase in concrete (Andersen et al. 2003). Their general chemical formula is given as $(CaO)_x(SiO_2)_y(H_2O)_z$. CSHs form many metastable and amorphous disordered structures from which stable and highly crystalline CSHs are generated by heating (Houston et al. 2009). Crystalline CSHs have a wide range of structures with various Ca/Si molar ratios (Shaw et al. 2000). Crystalline CSHs with a low Ca/Si ratio of 1.0–1.2, such as tobermorite ($Ca_5Si_6O_{16}(OH)_2 \cdot 4H_2O$) (Berg et al. 2005) and xonotlite ($Ca_6Si_6O_{17}(OH)_2$) (Chen et al. 2009), have been used for phosphate (P_i) recovery from wastewater. Amorphous-type calcium silicate hydrates (A-CSHs) could be chemically synthesized using abundantly available, inexpensive materials such as siliceous shale and $Ca(OH)_2$ (Okano et al. 2013). A-CSHs, which have a Ca/Si molar ratio of about 1.0, can serve not only as a P_i adsorbent but also as an aggregation agent in aqueous solution (Okano et al. 2015). P_i removed by A-CSHs exhibits better settleability, filterability, and dewaterability than P_i precipitated with conventional $CaCl_2$ and $Ca(OH)_2$. Moreover, unlike $CaCl_2$ and $Ca(OH)_2$, no significant carbonate inhibition is observed with P_i removal by A-CSHs.

Importantly, A-CSHs can also be prepared by soaking recyclable calcium silicates such as concrete sludge (CS) and steelmaking slag in a dilute hydrochloric acid solution. CS is abundantly available at construction sites where more concrete materials are generated than required (Iizuka et al. 2012). Conventionally, CS has been disposed of as a landfill material after solid-liquid separation followed by neutralization with a strong acid (Tsunashima et al. 2012). However, since this is costly and wasteful, it is desirable to develop an alternative technology option that can valorize unwanted CS for the cement and construction industry. On the other hand, steelmaking slag is the most abundantly available by-product which contains calcium silicates as a major component in the steel industry (Matsubae et al. 2015). Since the amount of steelmaking slag produced by a steelmaking process is very large, it is a matter of great concern for the steel industry to effectively recycle this by-product.

This chapter describes the chemical structure of A-CSHs, the possible mechanism for P_i recovery from aqueous solution, and the P_i recoverability from P_i-rich sidestream at a wastewater treatment plant.

30.2 The Structure of A-CSHs

Okano et al. (2013) have synthesized A-CSHs from a natural siliceous shale, M-rite, and $Ca(OH)_2$. M-rite and $Ca(OH)_2$ are abundantly available, inexpensive materials in Japan. Soluble silicates were extracted by treating M-rite with 0.5% (w/v) NaOH at 60 °C for 10 min. After the alkaline treatment, the solid fraction was removed

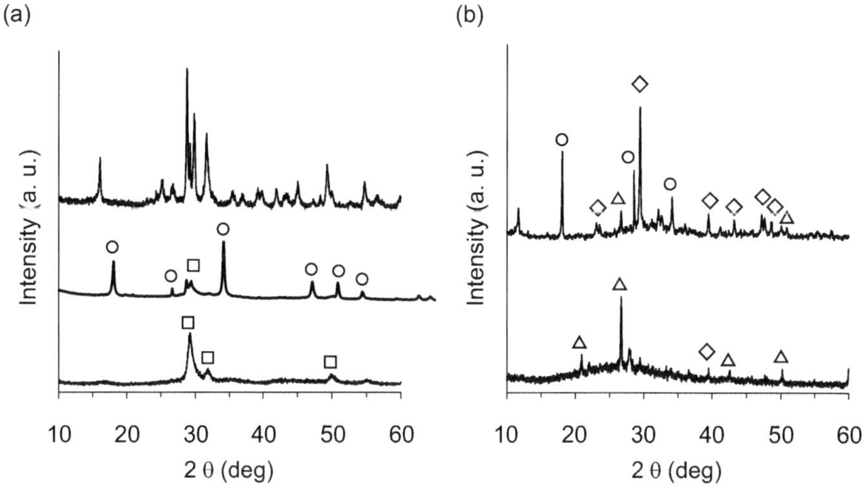

Fig. 30.1 Powder X-ray diffraction patterns. (**a**) Tobermorite (top) and chemically synthesized A-CSHs having Ca/Si molar ratio of 3.5 (middle) and 1.0 (bottom). (**b**) CS before (top) and after (bottom) acid treatment. Symbols: $Ca(OH)_2$ (circles), semicrystalline CSHs (squares), quartz (triangles), $CaCO_3$ (diamonds) (Okano et al. 2013, 2016b)

from the reaction mixture by sedimentation. A-CSHs slurry could be obtained by adding $Ca(OH)_2$ to the supernatant and incubating the mixture for 5–6 h at 60 °C with continuous stirring. A-CSHs having a wide range of Ca/Si molar ratio could be obtained by varying the amount of $Ca(OH)_2$ added.

The salient feature of A-CSHs is the low crystallinity compared with that of crystalline CSHs. Figure 30.1a shows the result of powder X-ray diffraction (P-XRD) analysis of tobermorite (crystalline-type CSHs) and the chemically synthesized A-CSHs. Although tobermorite exhibited several sharp peaks corresponding to its lattice planes, A-CSHs having a Ca/Si molar ratio of 3.5 (A-CSHs 3.5) showed no peaks corresponding to tobermorite. They exhibited several peaks characteristic of $Ca(OH)_2$. Small peaks corresponding to semicrystalline CSHs were observed at approximately 0.304 nm (corresponding to 29.4° 2θ). When A-CSHs were prepared at Ca/Si molar ratio of 1.0 (A-CSHs 1.0), no peaks corresponding to $Ca(OH)_2$ were detected. Instead, peaks corresponding to semicrystalline CSHs became more visible at 0.304, 0.280, and 0.182 nm (corresponding to 29.4, 32.0, and 50.1° 2θ, respectively).

The main components of CS particles were SiO_2 (19.3 mass%) and CaO (41.0 mass%) (data not shown), indicating that the Ca/Si molar ratio was 2.28. The CS particles also contained significant amounts of Al_2O_3 and Fe_2O_3 (6.3 and 2.3 mass%, respectively). However, no significant amounts of heavy metals such as As, Cd, Pb, Ni, and Cr were detected with CS particles. P-XRD analysis showed that CS particles exhibited several peaks corresponding to quartz, $Ca(OH)_2$, and $CaCO_3$ (Fig. 30.1b). None of the other peaks were clearly assigned because of the complexity and low crystallinity of hydrated cement (Iizuka et al. 2012). When CS was treated

Table 30.1 The chemical composition of vacuum-dried A-CSHs (Okano et al. 2015)

The content of main components (mass%)						
Na_2O	MgO	Al_2O_3	SiO_2	CaO	Others[a]	Ca/Si molar ratio
8.5	0.2	0.1	42.0	39.4	9.8	1.0

[a]Others are mainly hydrated water

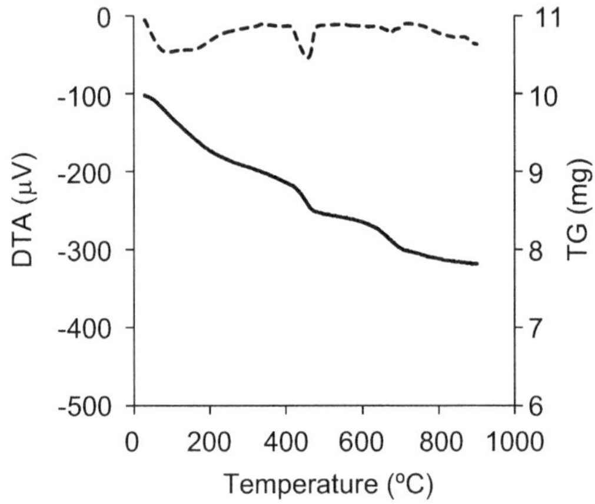

Fig. 30.2 Simultaneous thermogravimetry (TG) and differential thermal analysis (DTA) of synthesized A-CSHs. Solid and dashed lines indicate TG and DTA curves, respectively

with 1 M HCl solution, it showed several peaks corresponding to quartz. However, no significant peak was detected for $Ca(OH)_2$ and $CaCO_3$. Acid-treated CS showed a broad peak at 15–40° 2θ, suggesting that the amorphous structure of the CS particles became more prevalent after the acid treatment.

The composition of chemically synthesized A-CSHs was determined by X-ray fluorescence spectrometry (Table 30.1). The synthesized A-CSHs consisted mainly of SiO_2 (42.0 mass%) and CaO (39.4 mass%). The Ca/Si molar ratio of A-CSHs was approximately 1.0. Thermogravimetry and differential thermal analysis (TG-DTA) showed that A-CSHs had no distinct weight loss in the temperature range of 400–500 °C corresponding to $Ca(OH)_2$ dehydration (Fig. 30.2). In addition, no endothermic peak was detected at this temperature range in the DTA curves of A-CSHs. This indicated that the synthesized A-CSHs contained no detectable amounts of free $Ca(OH)_2$. It was also observed that no significant TG-DTA peak corresponding to $CaCO_3$ decarboxylation was detected at approximately 730–770 °C.

^{29}Si MAS-NMR analysis of wet A-CSHs showed three distinct peaks located at −80.2, −83.6, and −86.4 ppm (Fig. 30.3a). The standard Q^n nomenclature has been used to describe silicate tetrahedron sites (Cong and Kirkpatrick 1993). In this nomenclature, Q stands for a given silicate tetrahedron, and the superscript (n) refers to the number of associated bridging oxygens. Bridging oxygens bind to other tetrahedra, while non-bridging ones bind to a low-charge cation such as Ca^{2+} or H^+. Namely, Q^0 sites are present as monomers. Q^1 sites can be present either in pairs of

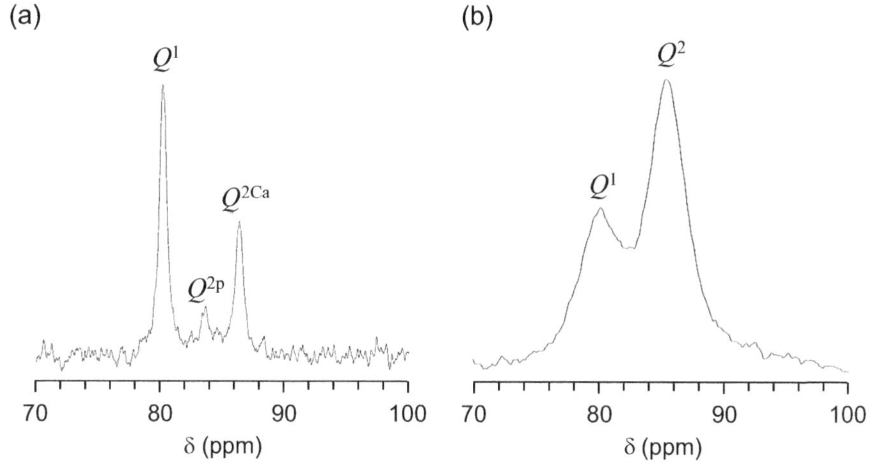

Fig. 30.3 ^{29}Si MAS-NMR spectra of wet (**a**) and vacuum-dried (**b**) A-CSHs. The Q^n notation of Si tetrahedron is shown above each peak (Okano et al. 2015)

linked silicate tetrahedra (dimers) or at the ends of silicate chains. Q^2 sites are present in the middle of silicate chains and are further divided into three types by ligand ion: Q^{2p} (H^+ coordination), Q^{2i} (H^+ and Ca^{2+} coordination), and Q^{2Ca} (Ca^{2+} coordination) (Aono et al. 2007). Q^3 and Q^4 sites are tetrahedra forming silicate networks. Peaks located at −80.2, −83.6, and −86.4 ppm corresponded to Q^1, Q^{2p}, and Q^{2Ca} sites, respectively. No significant signals corresponding to Q^0, Q^{2i}, Q^3, or Q^4 were detected with wet A-CSHs. The peak intensity ratio of Q^1:Q^{2p}:Q^{2Ca} was 4.6:1:2.4. In case that Q^3 and Q^4 are negligible compared with Q^1 and Q^2 (viz., linear polymers), the average chain length of silicate polymers N can be estimated by Eq. (30.1) (Aono et al. (2007):

$$N = 2 \times \Sigma Q^n / Q^1, \tag{30.1}$$

where $\sum Q^n = Q^1 + Q^{2i} + Q^{2p} + Q^{2Ca}$. Based on the peak intensity ratio of Q^1:Q^{2p}:Q^{2Ca}, the average chain length of silicate polymers in the wet A-CSHs was estimated to be approximately 3.5.

On the other hand, ^{29}Si MAS-NMR analysis of dried A-CSHs revealed only two distinct peaks located at −81.1 and −86.5 ppm, corresponding to Q^1 and Q^2, respectively (Fig. 30.3b). It was difficult to resolve Q^{2i}, Q^{2p}, and Q^{2Ca} peaks in the NMR spectra. The peak intensity of Q^2 was distinct, compared with the signals of wet A-CSHs. Although the Q^1 and Q^2 peaks overlapped with each other, the peak intensity ratio of Q^1:Q^2 was estimated as 1:2.4 by curve fitting. Consequently, the average chain length of silicate polymers in dried A-CSHs was estimated as 6.9 which was approximately two times longer than that of wet A-CSHs. This suggested that two silicate polymers of wet A-CSHs bound together when A-CSHs were dried with vacuum.

^{29}Si MAS-NMR analysis also showed that approximately 70% of Q^2 tetrahedra coordinated to Ca^{2+} (Q^{2Ca}) in wet A-CSHs. If only Q^{2Ca} tetrahedra coordinated to Ca^{2+}, the Ca/Si molar ratio of A-CSHs could be estimated as

$$\mathrm{Ca\,/\,Si\,molar\ ratio} = Q^{2\mathrm{Ca}} / \sum Q^{n} \tag{30.2}$$

However, this gave the Ca/Si molar ratio of approximately 0.3, which was significantly smaller than that determined by XRF spectrometry (viz., 1.0) (see Table 30.1). This suggested that Q^1 tetrahedra also coordinated to Ca^{2+}, although it was difficult to resolve Q^1 and Q^{1Ca} (Ca coordination) by ^{29}Si MAS-NMR spectroscopy (Fig. 30.3). To satisfy the Ca/Si molar ratio of 1.0, approximately 60% of oxygen atoms in Q^1 tetrahedra must bind to Ca^{2+}. Since one of four oxygen atoms in Q^1 linked to Q^2 tetrahedra, the remaining 15% of oxygen atoms in Q^1 tetrahedra was likely to form OH groups. Taken together, the molecular formula of A-CSHs may be given as $(Ca_{3.5}(SiO_{3.3})_{3.5}H_2)_n$ (n indicates the degree of polymerization). A possible structure of the synthesized A-CSHs is shown in Fig. 30.4.

30.3 Possible Mechanism for P_i Recovery

A-CSHs release Ca^{2+} and soluble silicates, when they are dispersed in aqueous solutions. Chen et al. (2004) have reported that the dissociation of amorphous CSHs occurred according to equilibrium reactions involving Ca^{2+}, soluble silicates, and CSHs in water. Since the first dissociation constant pK_{a1} of silicic acid (H_4SiO_4) is approximately 9.7, a large portion of the silicate polymers is likely protonated after being released from A-CSHs in aqueous solution at pH below 9.0. The overall reaction may be given as

$$\begin{aligned}&\left(\mathrm{Ca}_{3.5}\left(\mathrm{SiO}_{3.3}\right)_{3.5}\mathrm{H}_2\right)_n + 10.4m\mathrm{H_2O} \rightleftarrows (\mathrm{Ca}_{(3.5-1.7m/(n-m))}\left(\mathrm{SiO}_{3.3}\right)_{3.5}\\&\mathrm{H}\left(_{(2+0.6m/(n-m))}\right)_{(n-m)} + m\left(\mathrm{SiO}_{3.3}\mathrm{H}_{2.6}\right)_{3.5} + 5.2m\mathrm{Ca}^{2+} + 10.4m\mathrm{OH}^-\end{aligned} \tag{30.3}$$

where m indicates the number of silicate polymers released from A-CSHs. The stoichiometric coefficient for Ca^{2+} (namely, 5.2) was determined from the experimental data as described previously (Okano et al. 2015).

P_i may be removed by either the chemical precipitation with free Ca^{2+} or the formation of Ca-P_i-silicates ion aggregates. Guan et al. (2013) have reported that crystalline CSHs such as tobermorite and xonotlite could remove P_i by chemical precipitation with released Ca^{2+}. If this is the prevailing mechanism of P_i removal by A-CSHs, the following equilibrium reactions must be considered:

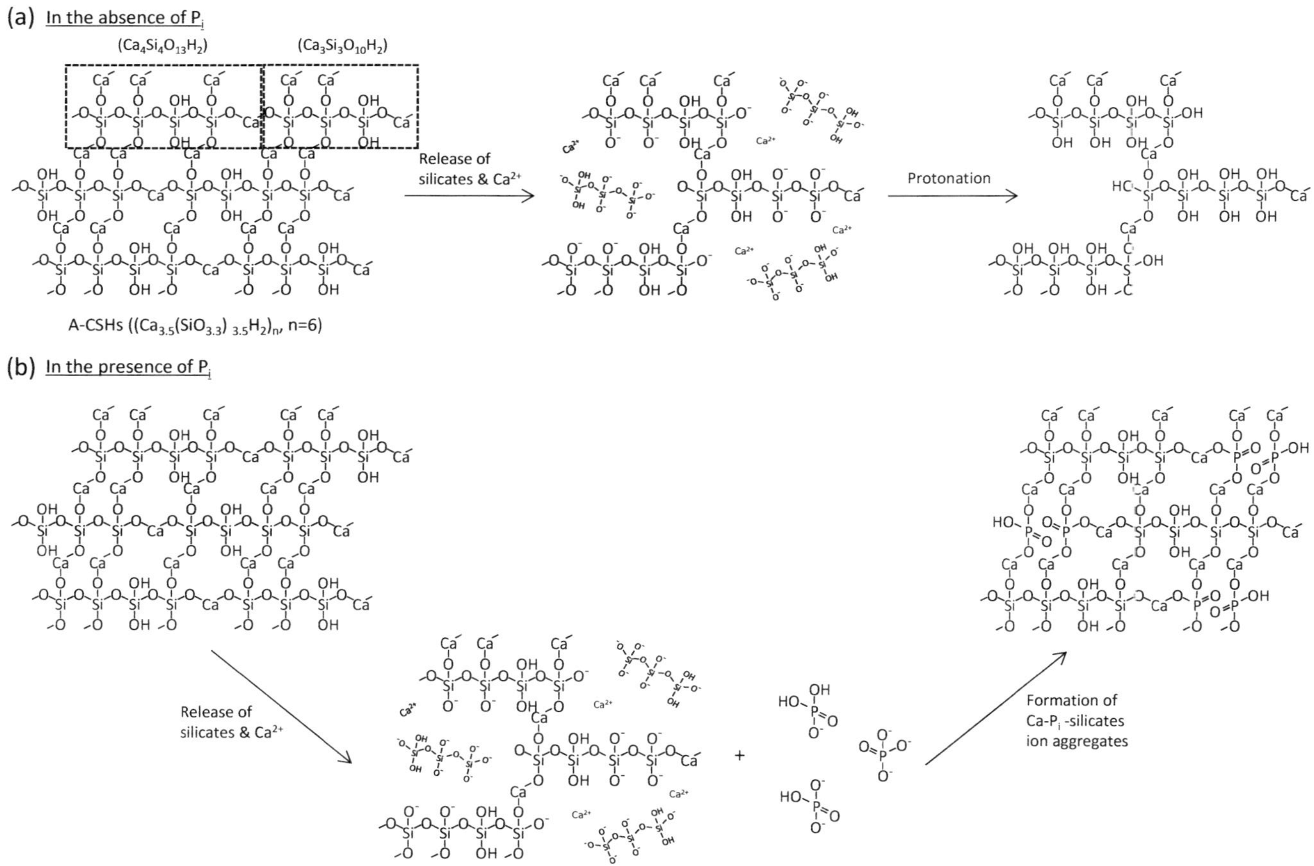

Fig. 30.4 A possible chemical structure of synthesized A-CSHs and its dissociation model in aqueous solutions in the absence of **(a)** and the presence of P_i **(b)** (Okano et al. 2015)

$$H_3PO_4 \rightleftarrows H^+ + H_2PO_4^- \quad K_{a1} = [H^+][H_2PO_4^-]/[H_3PO_4] = 1.1 \times 10^{-2} \tag{30.4}$$

$$H_2PO_4^- \rightleftarrows H^+ + HPO_4^{2-} \quad K_{a2} = [H^+][HPO_4^{2-}]/[H_2PO_4^-] = 7.5 \times 10^{-8} \tag{30.5}$$

$$HPO_4^{2-} \rightleftarrows H^+ + PO_4^{3-} \quad K_{a3} = [H^+][PO_4^{3-}]/[HPO_4^{2-}] = 4.8 \times 10^{-13} \tag{30.6}$$

$$Ca^{2+} + 2H_2PO_4^- \rightleftarrows Ca(H_2PO_4)_2\,(aq) \tag{30.7}$$

$$Ca^{2+} + HPO_4^{2-} \rightleftarrows CaHPO_4\,(aq) \tag{30.8}$$

$$3Ca^{2+} + 2PO_4^{3-} \rightleftarrows Ca_3(PO_4)_2\,(aq) \tag{30.9}$$

$$Ca(H_2PO_4)_2\,(aq) \rightleftarrows Ca(H_2PO_4)_2\,(s) \quad K_{sp1} = [Ca^{2+}][H_2PO_4^-]^2 = 2.1 \times 10^{-3} \tag{30.10}$$

$$CaHPO_4\,(aq) \rightleftarrows CaHPO_4\,(s) \quad K_{sp2} = [Ca^{2+}][HPO_4^{2-}] = 1.3 \times 10^{-7} \tag{30.11}$$

$$Ca_3(PO_4)_2\,(aq) \rightleftarrows Ca_3(PO_4)_2\,(s) \quad K_{sp3} = [Ca^{2+}]^3[PO_4^{3-}]^2 = 1.2 \times 10^{-29} \tag{30.12}$$

where (aq) and (s) denote aqueous and solid forms, respectively. K_a and K_{sp} are the acid dissociation constant and the solubility product, respectively (the activity coefficients of ionic species are not considered).

$CaHPO_4$ (s) and $Ca_3(PO_4)_2$ (s) are thermodynamically metastable and undergo transformation into more stable hydroxyapatite (HAP) (Johnsson and Nancollas 1992). The acid dissociation constants indicate that the dominant phosphoric ion species are $H_2PO_4^-$ and HPO_4^{2-} in the pH range of 7–9. However, the formation of $Ca(H_2PO_4)_2$ is negligible because of its large solubility in water ($K_{sp1} = 2.1 \times 10^{-3}$). Conversely, $Ca_3(PO_4)_2$, which is the most widely found in amorphous Ca-P_i precipitates in alkaline media (pH > 9), is likely formed due to its low solubility ($K_{sp3} = 1.2 \times 10^{-29}$), even though the abundance ratio of PO_4^{3-} is extremely low (<0.05%) in the pH range of 7–9). The formation of $CaHPO_4$ (s) and $Ca_3(PO_4)_2$ (s) removes free Ca^{2+} from the aqueous solution (reactions (30.8–30.12)), thereby driving the dissociation of A-CSHs according to reaction (30.3). Hence, if the prevailing mechanism for P_i removal is the precipitation with free Ca^{2+}, more Ca^{2+} must be released from A-CSHs in the presence of P_i than in the absence of P_i. However, experimental evidence showed that the total amount of released Ca^{2+} (Ca^{2+} consumed by the Ca-P_i formation plus the remaining free Ca^{2+}) was significantly smaller in aqueous solutions containing P_i than in P_i-free aqueous solutions (Okano et al. 2015). Hence, P_i precipitation with free Ca^{2+} could be ruled out as the prevailing mechanism for P_i removal by A-CSHs.

The initial steps of nucleation of Ca-P_i in aqueous solution have been investigated using quantum/classical molecular mechanics simulation (Zahn 2004). The molecular mechanics simulation has revealed that a $[Ca^{2+}\text{-}(HPO_4)^{2-}\text{-}Ca^{2+}]^{2+}$ aggregate can be formed at the initial stage of nucleation of Ca-P_i in aqueous solution in the range $pK_{a2} < pH < pK_{a3}$. Since A-CSHs can acquire negative electrical charges after releasing Ca^{2+} and silicates, it seems possible that the $[Ca^{2+}\text{-}(HPO_4)^{2-}\text{-}Ca^{2+}]^{2+}$ ion triple ionically binds to the negatively charged A-CSHs, forming Ca-P_i-silicates ion aggregates (Fig. 30.4). This hypothesis could reasonably explain the high settleability of P_i removed by A-CSHs.

30.4 P_i Recovery by A-CSHs

On-site P_i recovery tests were carried out at a full-scale wastewater treatment plant (WWTP) using chemically synthesized A-CSH and acid-treated CS. The mobile plant was used to recover P_i from anaerobic sludge digestion liquor (the reject water from a membrane-type solid-liquid separator for digested sludge) at the WWTP (Fig. 30.5a).

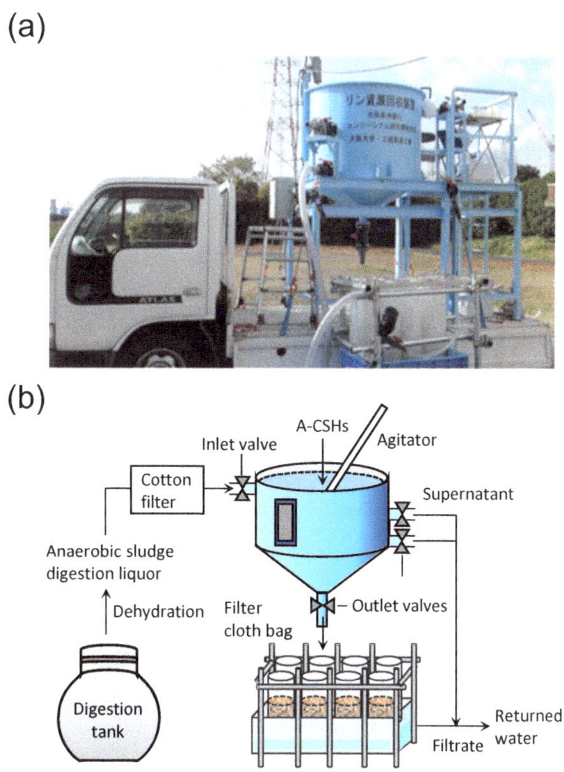

Fig. 30.5 Mobile plant (**a**) and the on-site process to recover P_i from anaerobic sludge digestion liquor in a WWTP (**b**) (Okano et al. 2016a; Ohtake et al. 2018)

A 1000-L reactor of 1.3-m height and 1.2-m diameter was made of 1-cm-thick reinforced plastic and set up on a 1.5-tonne motor truck using a steel frame (Fig. 30.5a). The lower part of the reactor, which had the shape of an inverted circular cone, was designed to collect P_i-rich solids by free sedimentation. The reactor was equipped with an agitator having two stirring blades for mechanical mixing. One inlet valve, which was fitted on the top of the reactor, was connected with an accordion hose for inlet water. Another end of the accordion hose was connected to the outlet of a membrane-type solid-liquid separator for digested sludge in a full-scale WWTP. The sludge digestion liquor was fed into the reactor through the accordion hose by gravity (Fig. 30.5b).

Two outlet valves were fitted on the side of the reactor to drain the supernatant after the free sedimentation of P_i-rich solids. P_i-rich solids were withdrawn from the bottom of the reactor through an outlet valve. All valves were manually operated when needed. P_i-rich solids were filtered using a self-made filter system on which no mechanical pressure was applied. To prepare the filter system, a 25-cm-diameter filter cloth tube was tied at one end with a flexible plastic band and hooked on a steel pipe frame. The filter cloth tube was made of high tenacity yarns of synthetic fibers, having a thickness of 0.36 mm. Since the filter cloth tube is widely used for civil engineering work, it is tough, easy to handle, and available at a cheap price. A 250-L plastic container was placed under the steel pipe frame to collect the filtrate. This filtration system was set up on the ground beside the motor truck before being used.

After measuring the initial P_i concentration, either chemically synthesized A-CSHs or acid-treated CS were added to the reactor at a Ca/P molar ratio of 2.0. The Ca/P ratio is a critical factor to be considered for P_i recovery. The P_i removal efficiency is increased with increasing the Ca/P ratio. However, the P_i content of the recovered product decreased as the Ca/P molar ratio increase. Therefore, the Ca/P ratio must be optimized. Both chemically synthesized A-CSHs and acid-treated CS could remove approximately 80% P_i from the reject water after 10 min (Fig. 30.6). Then, P_i removal gradually increased to 82% by 60 min. The pH of the reaction

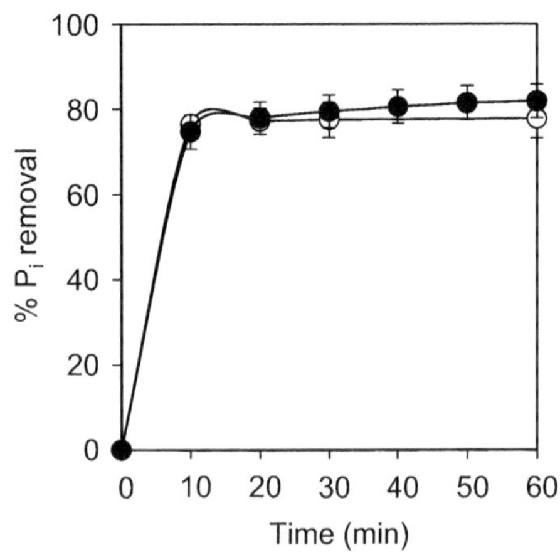

Fig. 30.6 Time-course data on the % P_i removal by chemically synthesized A-CSHs (open circles) and acid-treated CS (closed circles) in on-site P_i recovery experiments. Data points represent the means ± standard deviations for three independent experiments (Ohtake et al. 2018)

mixture increased concomitantly with P_i removal in the first 10 min and reached 8.6–9.0 after 60 min (data not shown).

After 20 min of P_i removal, P_i was recovered by 30 min of free sedimentation followed by 90 min of filtration using filter cloth bags. When about 830 L of the supernatant was drained from the reactor, sediments were observed at the reactor bottom. They were easily poured into filter cloth bags. Then, the filtration was carried out without mechanical pressure being applied. The filtrate in the plastic container was clear, suggesting that P_i-rich solids were effectively recovered by the filter cloth bags. After 90 min of filtration, the filter cake was removed from the bags for chemical analyses. Mass balance calculation showed that 72–85% P_i was recovered from the reject water, while 13–23% P_i was lost in the drainage water from the reactor. The loss of P_i due to filtration was only 2–5%. There was no significant difference in the P_i recovery efficiency between chemically synthesized A-CSHs and acid-treated CS. Despite the high total CO_2 of the anaerobic sludge digestion liquor (1280 mg L^{-1}), no carbonate inhibition was observed with P_i removal in the on-site experiments.

The recovered P_i product contained 83–87% moisture before drying for chemical analyses. The ignition loss of the product ranged from 19% to 23%. After drying at 105 °C for 24 h, the product contained 19–24% of the dry weight as P_2O_5. The P_2O_5 content was nearly identical to the citrate-soluble P_2O_5 (C-P_2O_5) content (18–24%). On average, total potassium (K_2O), total nitrogen (T-N), and CaO contents were 0.1, 0.5, and 38%, respectively, on a dry weight basis. Cd was never detected in the recovered products. The contents of As, Pb, Ni, and Cr were less than 4.0, 4.0, 2.0, and 12.0 mg kg^{-1}, respectively. These values were much lower than their regulatory levels (840, 100, 300, and 500 mg kg^{-1}, respectively) for fertilizer in Japan. Therefore, P_i recovered by A-CSHs can be directly used as P_i fertilizer.

The efficacy of the recovered P_i as a by-product P_i fertilizer was examined in plant cultivation tests (Okano et al. 2016a). The efficacy of the recovered P_i products as a P_i fertilizer was compared with those of two commercial P_i fertilizers, Gifu-no-daichi (JA Gifu, Japan) and calcium superphosphate (Kureha Co., Tokyo, Japan). When the recovered P_i product was applied to andosol at a dose of 100 mg C-P_2O_5 per pot, the germination rate of Komatsuna (*Brassica rapa* L. var. *perviridis*) seeds was higher than 90% after 5 days (data not shown). There was no significant difference in the germination rate between the recovered P_i product and the commercial P_i fertilizers (Gifu-no-daichi and superphosphate). Increasing the dose of the recovered P_i product had no significant effect on the germination rate. On the other hand, the leaf length and live weight were 8.1 cm and 9.1 g per pot, respectively, at 22 days of growth testing, when the recovered P_i product was applied at a dose of 100 mg C-P_2O_5 per pot. The growth of Komatsuna was enhanced by increasing the dose of the recovered P_i product. No abnormal plant growth was observed with Komatsuna at the dose of 400 mg C-P_2O_5 per pot (Fig. 30.7).

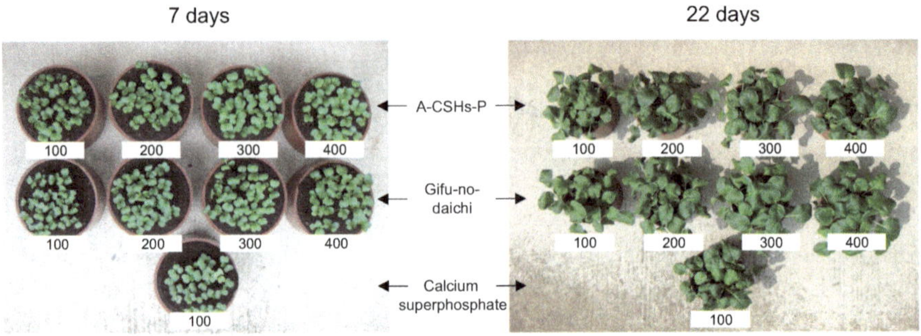

Fig. 30.7 Photographs of the leaf vegetable Komatsuna (*Brassica rapa* L. var. *perviridis*) after 7 days (left) and 22 days (right) of cultivation. Either P_i recovered by A-CSHs or Gifu-no-daichi was applied to the pots at doses of 100, 200, 300, and 400 mg of C-P_2O_5 per pot. Calcium superphosphate was applied at a dose of 100 mg of C-P_2O_5 per pot (Okano et al. 2016a)

30.5 Conclusions

A-CSHs can be synthesized using abundantly available, inexpensive materials such as siliceous shale and calcium hydroxide. Unlike crystalline CSHs, A-CSHs can serve as a bifunctional adsorption-aggregation agents. The bifunctional agent can be also prepared simply by soaking recyclable CSHs such as steelmaking slag and concrete sludge in hydrochloric acid solution. Because of their high P_i removability, settleability, and filterability, a simple technology for P_i recovery was developed. On-site experiments using a mobile plant showed that approximately 80% P_i could be recovered from a reject water from the membrane-type solid-liquid separator for anaerobically digested sewage sludge by a process consisting of short time of mixing, settling, and filtration. The recovered products could be directly used as a by-product P_i fertilizer. Therefore, A-CSHs have great potential as novel, beneficial materials for P recovery and recycling.

References

Andersen MD, Jakobsen HJ, Skibsted J (2003) Incorporation of aluminum in the calcium silicate hydrate (C–S–H) of hydrated Portland cements: a high-field ^{27}Al and ^{29}Si MAS NMR investigation. Inorg Chem 42:2280–2287

Aono Y, Matsushita F, Shibata S, Hama Y (2007) Nano-structural changes of C-S-H in hardened cement paste during drying at 50°C. J Adv Concr Technol 5:313–323

Berg U, Donnert D, Ehbrecht A, Bumiller W, Kusche I, Weidler PG, Nüesch R (2005) "Active filtration" for the elimination and recovery of phosphorus from waste water. Colloid Surf A 265:141–148

Chen JJ, Thomas JJ, Taylor HFW, Jennings HM (2004) Solubility and structure of calcium silicate hydrate. Cem Concr Res 34:1499–1519

Chen X, Kong H, Wu D, Wang X, Lin Y (2009) Phosphate removal and recovery through crystallization of hydroxyapatite using xonotlite as seed crystal. J Environ Sci 21:575–580

Cong XD, Kirkpatrick RJ (1993) ^{29}Si MAS NMR spectroscopic investigation of alkali silica reaction products gels. Cement Concr Res 23:811–823

Guan W, Ji F, Chen Q, Yan P, Pei L (2013) Synthesis and enhanced phosphate recovery property of porous calcium silicate hydrate using polyethylene glycol as pore-generation agent. Materials 6:2846–2861

Houston JR, Maxwell RS, Carroll SA (2009) Transformation of meta-stable calcium silicate hydrates to tobermorite: reaction kinetics and molecular structure from XRD and NMR spectroscopy. Geochem Trans 10:1

Iizuka A, Sasaki T, Hongo T, Honma M, Hayakawa Y, Yamasaki A, Yanagisawa Y (2012) Phosphorus adsorbent derived from concrete sludge (PAdeCS) and its phosphorus recovery performance. Ind Eng Chem Res 51:11266–11273

Johnsson MSA, Nancollas GH (1992) The role of brushite and octacalcium phosphate in apatite formation. Crit Rev Oral Biol M 3:61–82

Matsubae K, Webeck E, Nansai K, Nakajima K, Tanaka M, Nagasaka T (2015) Hidden phosphorus flows related with non-agriculture industrial activities: a focus on steelmaking and metal surface treatment. Resour Conser Recy 105:360–367

Ohtake H, Okano K, Kunisada M, Takano H, Toda M (2018) Simple technology for recycling phosphate from wastewater to farmland in rural areas. Ambio 47(S1):83–92

Okano K, Uemoto M, Kagami J, Miura K, Aketo T, Toda M, Honda K, Ohtake H (2013) Novel technique for phosphorus recovery from aqueous solutions using amorphous calcium silicate hydrates (A-CSHs). Water Res 47:2251–2259

Okano K, Miyamaru S, Kitao A, Takano H, Aketo T, Toda M, Honda K, Ohtake H (2015) Amorphous calcium silicate hydrates and their possible mechanism for recovering phosphate from wastewater. Sep Purif Technol 144:63–69

Okano K, Miyamaru S, Yamamoto Y, Kunisada M, Takano H, Toda M, Honda K, Ohtake H (2016a) A mobile pilot-scale plant for in situ demonstration of phosphorus recovery from wastewater using amorphous calcium silicate hydrates. Sep Purif Technol 170:116–121

Okano K, Yamamoto Y, Takano H, Aketo T, Honda K, Ohtake H (2016b) A simple technology for phosphorus recovery using acid-treated concrete sludge. Sep Purif Technol 165:173–178

Shaw S, Clark SM, Henderson CMB (2000) Hydrothermal formation of the calcium silicate hydrates, tobermorite ($Ca_5Si_6O_{16}(OH)_2{\cdot}4H_2O$) and xonotlite ($Ca_6Si_6O_{17}(OH)_2$): an in situ synchrotron study. Chem Geol 167:129–140

Tsunashima Y, Iizuka A, Akimoto J, Hongo T, Yamasaki A (2012) Preparation of sorbents containing ettringite phase from concrete sludge and their performance in removing borate and fluoride ions from waste water. Chem Eng J 200–202:338–343

Zahn D (2004) Mechanisms of calcium phosphate ion association in aqueous solution. Z Anorg Allg Chem 630:1507–1511

Chapter 31
High-Performance Phosphorus Adsorbent Based on Concrete Sludge

A. Iizuka, H. Yoshida, and Y. Hayakawa

Abstract This chapter describes the high-performance phosphorus adsorbent, PAdeCS®, which is produced from concrete sludge. The concrete sludge is a slurried, strong alkali waste that is generated in a concrete pole and pile manufacturing plant in Japan. PAdeCS® is a powdery granulate that can be used as a phosphorus-recovery agent for dissolved phosphate. The precipitation of hydroxyapatite is the dominant mechanism for phosphate removal from aqueous solution by PAdeCS®. Because of the high calcium content of PAdeCS®, the recovered phosphorus product can be used as a slow-release phosphorus fertilizer. The phosphorus-removal performance of PAdeCS® was compared with that of conventional phosphorus-removal agents such as polyferric sulfate, aluminum polychloride, and aluminum sulfate. PAdeCS® showed the highest phosphorus-removal performance among them. PAdeCS®, which is produced from concrete sludge, can be used widely as an inexpensive phosphorus-recovery agent.

Keywords Concrete sludge · Phosphorus adsorbent · PAdeCS® · Hydroxyapatite · Ettringite · Fertilizer

31.1 Introduction

Currently, the recovery of phosphorus (P) from secondary P resources is less cost-effective than the extraction of P from phosphate (P_i) rock. This hampers the implementation of phosphorus recovery from secondary P resources (Molinos-Senante et al. 2011). A promising option for cost reduction is the use of low-cost materials for phosphorus capture (Boyer et al. 2011). Various waste materials have been examined for their potential to serve as an inexpensive phosphorus-removal

A. Iizuka (✉)
Tohoku University, Sendai, Japan
e-mail: atsushi.iizuka.e4@tohoku.ac.jp

H. Yoshida · Y. Hayakawa
Nippon Concrete Industries Co., Ltd., Tokyo, Japan

H. Ohtake, S. Tsuneda (eds.), *Phosphorus Recovery and Recycling*,
https://doi.org/10.1007/978-981-10-8031-9_31

agent. Cementitious waste materials are promising candidates for phosphorus adsorbents, because they are rich in alkali calcium compounds which are required for phosphorus precipitation as calcium hydroxyapatite (Mohara et al. 2009, 2011; Iizuka et al. 2012a, b, 2017; Sasaki et al. 2014). In this chapter, the high-performance phosphorus adsorbent, PAdeCS® (pronounced "Padex"), which is produced from concrete sludge, is introduced. The concrete sludge is abundantly available from a concrete pole and pile manufacturing plant in Japan.

31.2 Concrete Sludge

Concrete sludge is a slurried, strong alkali waste that is generated from fresh concrete usage. In definition, it includes excess fresh concrete, wastewater contaminated with fresh concrete, and substandard fresh concrete that cannot meet required construction standards. Concrete sludge contains coarse and fine aggregates, hydrating cement, and water. The liquid phase of concrete sludge is a saturated calcium hydroxide solution with pH of approximately 12. Currently, several percent of fresh concrete is inevitably disposed of as concrete sludge. Since concrete is one of the most abundantly used materials, large quantities of concrete sludge are generated worldwide. Typical sources of concrete sludge are ready-mixed concrete plants and concrete product manufacturing plants.

Concrete sludge waste management is difficult and costly because it is strongly alkaline and readily solidified. Typically, concrete sludge is disposed of as waste or is reused in low-grade construction materials, e.g., road beds or backfill materials after solidification. Coarse and fine aggregates in concrete sludge are often recycled as low-grade aggregates. After solid/liquid separation, the alkali liquor is discharged with/without neutralization with strong acid such as sulfuric acid and hydrochloric acid. Alternatively, it is recycled for the concrete product manufacturing plant or used as washing water for relevant facilities. The cost of concrete sludge waste management is expensive and increases the construction cost of concrete buildings and products.

31.3 High-Performance Phosphorus Adsorbent (PAdeCS®)

PAdeCS® is a powdery granulate that can be used as a phosphorus-recovery agent for dissolved phosphate in solution (Fig. 31.1). The PAdeCS® is produced from concrete sludge generated from a concrete pole and pile manufacturing plant in Japan. The detailed information on PAdeCS® has been described elsewhere (Sasaki et al. 2014; Iizuka et al. 2017).

Fig. 31.1 Appearance of PAdeCS® (SP-00)

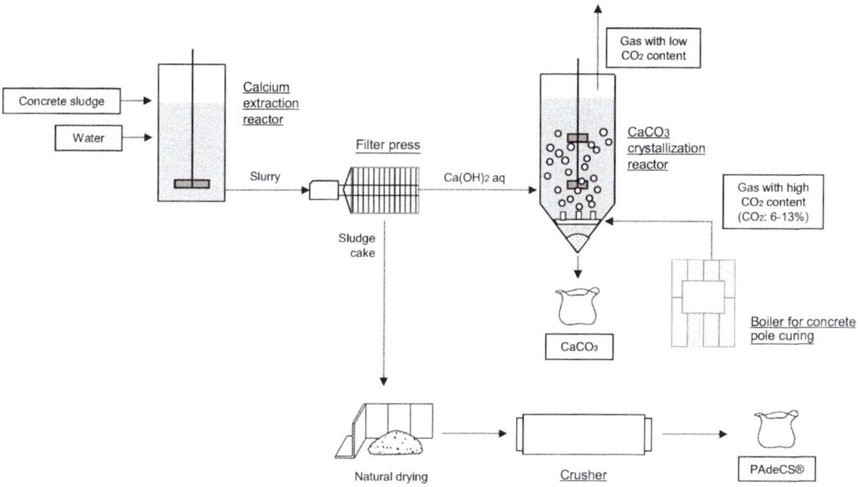

Fig. 31.2 Flow diagram of pilot-scale plant (Iizuka et al. 2017)

31.3.1 Production

The phosphorus adsorbent, PAdeCS®, can be produced by the process consisting of dilution, agitation, solid/liquid separation, drying, and particle size adjustment. PAdeCS® can be produced from a variety of concrete sludges having different water contents, because the production process starts with suspending concrete sludge in water. Figure 31.2 shows the production process in the Kawashima second factory of the NC East Japan Concrete Industry Co., Ltd. This factory manufactures concrete pole and pile by centrifugal molding. Sludge with less coarse and fine

Table 31.1 Typical chemical composition of concrete sludge (Iizuka et al. 2012b)

Component	CaO	SiO_2	Al_2O_3	Fe_2O_3	Water[a]
Content [wt%]	17.2	3.2	0.6	1.9	77.1

[a]Water is present in the concrete sludge as either free or hydrated water

aggregates can be obtained by this process, because large aggregates remain in pole and pile products during centrifugal molding. The obtained sludge, which consists mostly of diluted hydrating cement, is suspended in water and stirred for a predetermined period of time to facilitate the hydration of cement and the release of alkali calcium into the solution. Table 31.1 shows the typical chemical composition of the concrete sludge that is generated in the factory (Iizuka et al. 2012a). The dilution ratio and stirring time in the Ca extraction reactor are important parameters that determine the physical properties of PAdeCS® produced. After solid/liquid separation, the sludge cake is dried and crushed by sieving to obtain the phosphorus adsorbent, PAdeCS®. The filtrate, a saturated $Ca(OH)_2$ solution, is sent to the $CaCO_3$ crystallization reactor. Fine $CaCO_3$ powder is produced by feeding the exhaust CO_2 gas into the reactor. The exhaust gas is obtained from boilers that supplies steam to cure the concrete product. The CO_2 concentration in the exhaust gas ranges from 6% to 13%. The feeding of CO_2 gas and the subsequent $CaCO_3$ precipitation neutralize the alkali solution, and the neutralized water can be recycled for the PAdeCS® production.

The physical strength of PAdeCS® can be controlled by adjusting the dilution ratio and the hydration time in the Ca extraction reactor. The particle size distribution of PAdeCS® can be controlled by crushing, sieving, and granulation. As mentioned above, concrete sludge is readily available from ready-mixed concrete plants. Since large amounts of concrete sludge (c. several million t/a) are generated in Japan, it is possible to supply sufficient amounts of PAdeCS® on request. Concrete sludge is available throughout Japan, and thus PAdeCS® can be supplied practically everywhere. PAdeCS® can be produced from concrete sludge by the abovementioned simple procedure under ambient pressure and temperature without any chemical addition. The phosphorus adsorbent PAdeCS®, therefore, can be supplied cheaply and stably.

31.3.2 Characteristics

PAdeCS® consists mainly of cement hydrates (e.g., calcium hydroxide, calcium silicate hydrates, and ettringite). Table 31.2 shows the typical elemental composition of the produced PAdeCS®. Calcium was the most abundant element in PAdeCS® (28.9 wt%). It also contained silicon, aluminum, sulfur, and iron. These elements were derived from ordinary Portland cement that is used for concrete production.

Table 31.2 Typical chemical composition of PAdeCS® (Iizuka et al. 2017)

Element	Content [wt%]	Element	Content [wt%]
Ca	28.9	As	0.0018
Si	8.9	Cd	0.0005
Al	2.4	Hg	<0.0003
Fe	1.6	Ni	0.002
Mg	0.61	Cr	0.005
S	2.3	Pb	0.006
Na	0.20	Others[a]	54.6
K	0.23		
Ti	0.12	Total	100
P	0.094		
Mn	0.056		
Zn	0.028		

[a]Main elements are oxygen and hydrogen that are present in oxides and water

The heavy metal contents were small and less than 0.01 wt%. Because concrete sludge is a by-product that is generated from the construction and manufacturing process, its composition and origin are obvious. The Brunauer–Emmett–Teller specific surface area of the produced PAdeCS® powder varied from 11 to 27 m^2/g depending on the production condition.

PAdeCS® can be used as an agent for recovering dissolved phosphate from aqueous solution. PAdeCS® can be used in the same way as conventional agents such as polyferric sulfate, aluminum polychloride, and aluminum sulfate. The chemical reaction of phosphorus removal by PAdeCS® may be expressed as follows:

$$10Ca^{2+} + 6PO_4^{3-} + 2OH^- \rightarrow Ca_{10}\left(PO_4\right)_6\left(OH\right)_2 \downarrow$$

The precipitation of hydroxyapatite can occur at the surface of PAdeCS® particles or in aqueous solution depending on the precipitation conditions. In addition, anion exchange by ettringite ($3CaO{\cdot}Al_2O_3{\cdot}3CaSO_4{\cdot}32H_2O$) that is present in PAdeCS® is likely a mechanism for phosphate removal from aqueous solution. Because of the high calcium content of PAdeCS®, the recovered product can be used as a slow-release phosphorus fertilizer.

PAdeCS® can also be used as an agent for environmental remediation. For example, PAdeCS® can increase pH in aqueous solution by releasing hydroxyl ions mainly from $Ca(OH)_2$. PAdeCS® can also serve as a calcium donor in aqueous solution, leading to calcium-salt precipitation. Moreover, various anions, including AsO_4^{3-}, F^-, and $B(OH)_4^-$, can be removed by ion exchangeable ettringite. Thus, PAdeCS® can be applied to the neutralization of acidic wastewater, the removal of toxic substances from polluted water, and the immobilization of pollutants in contaminated soils.

31.4 Phosphorus Recovery from Wastewater

The phosphorus-removal performance of PAdeCS® (SP-00) was compared with that of conventional phosphorus-removal agents, i.e., polyferric sulfate, aluminum polychloride, and aluminum sulfate. PAdeCS® (SP-00) is a product type with a particle size of 15–40 μm (volume-based diameter), a specific surface area of 4000–9000 cm^2/g (Blaine), a moisture content of about 10 wt%, and a bulk density of approximately 0.6–0.7 kg/L. Each phosphorus-removal agent was added to 1.0 L of a phosphate solution of 16.5 mg-P/L at a concentration of 0.03 wt %. The reaction time was set to 30 min.

PAdeCS® (SP-00) showed the highest phosphorus-removal performance among the agents examined (Fig. 31.3). The rate of phosphorus removal by PAdeCS® reached 92.3%, and the final phosphorus concentration in the solution was 1.3 mg-P/L. Figure 31.4 compares the quantities of phosphorus removed by each agent. PAdeCS® (SP-00) showed the highest phosphorus removal amount of 50.8 mg-P/g-PAdeCS® which is equivalent to 116 mg-P_2O_5/g-PAdeCS®. The phosphorus content of the recovered product increased with increasing the phosphorus concentration in aqueous solution. More information on the phosphorus-removal performance of PAdeCS® is available in the literature (Sasaki et al. 2014).

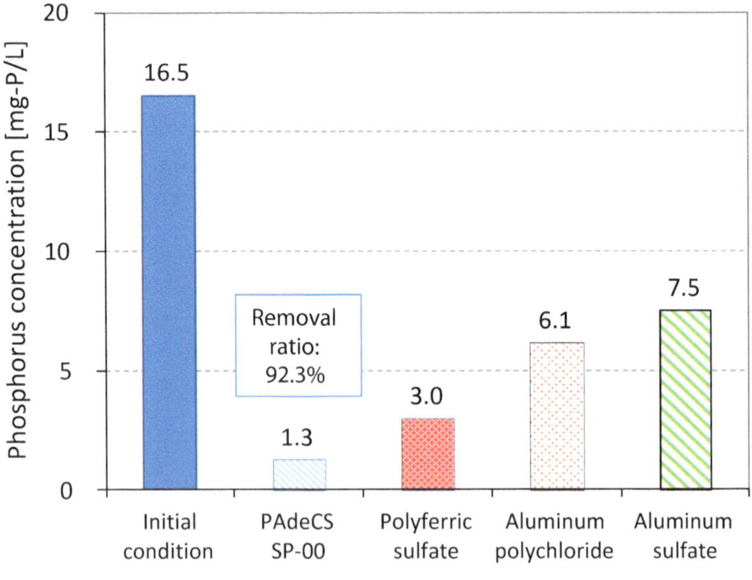

Fig. 31.3 Phosphorus removal from aqueous solution by various agents. The concentration of PO_4-P in the reaction mixture was measured after 30 min of phosphorus removal by each agent

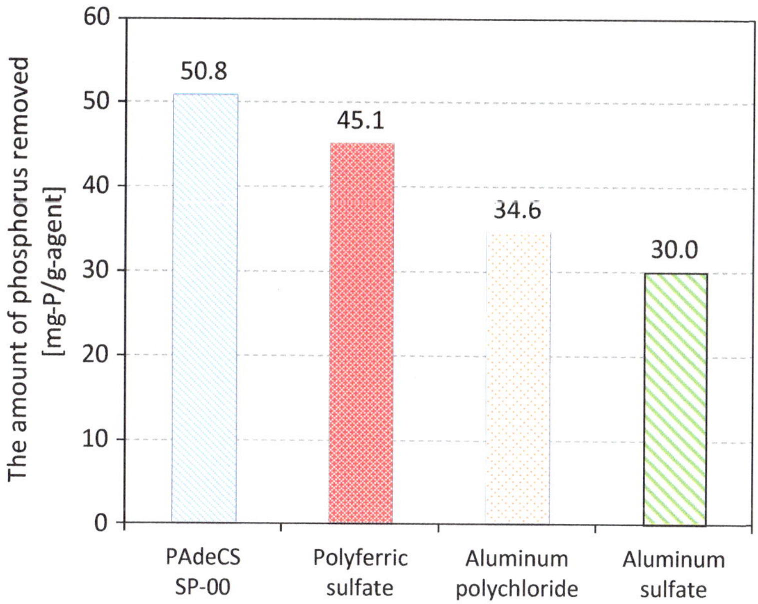

Fig. 31.4 Comparison of quantities of phosphorus removed by each agent

PAdeCS® has been applied to phosphorus removal in an open circulating water system of a public park and phosphorus removal from a food-processing wastewater in Japan. The recovered phosphorus products from the food-processing plant are used as a fertilizing material.

31.5 Conclusions

Cementitious waste materials are a promising candidate for the development of an inexpensive phosphorus adsorbent, because they contain alkali calcium compounds that are effective in phosphorus recovery. In this chapter, the high-performance phosphorus adsorbent, PAdeCS®, which is produced from concrete sludge, was introduced. PAdeCS® is a powdery granulate that can be used as a phosphorus-recovery agent for dissolved phosphate. Hydroxyapatite precipitation is the dominant mechanism for phosphate removal from aqueous solution by PAdeCS®. Because of the high calcium content of PAdeCS®, the recovered phosphorus products can be used as a slow-release phosphorus fertilizer. PAdeCS® showed the high phosphorus-removal performance compared to conventional phosphorus-removal agents. PAdeCS® can be used widely as an inexpensive phosphorus-recovery agent.

References

Boyer TH, Persaud A, Banerjee P, Palomino P (2011) Comparison of low-cost and engineered materials for phosphorus removal from organic-rich surface water. Water Res 45:4803–4814

Iizuka A, Sakai Y, Yamasaki A, Honma M, Hayakawa Y, Yanagisawa Y (2012a) Bench-scale operation of a concrete sludge recycling plant. Ind Eng Chem Res 51:6099–6104

Iizuka A, Sasaki T, Hongo T, Honma M, Hayakawa Y, Yamasaki A, Yanagisawa Y (2012b) Phosphorus adsorbent derived from concrete sludge (PAdeCS) and its phosphorus recovery performance. Ind Eng Chem Res 51:11266–11273

Iizuka A, Sasaki T, Honma M, Yoshida H, Hayakawa Y, Yanagisawa Y, Yamasaki A (2017) Pilot-scale operation of a concrete sludge recycling plant and simultaneous production of calcium carbonate. Chem Eng Commun 204:79–85

Mohara G, Iizuka A, Nagasawa H, Yamasaki A, Kumagai K, Yanagisawa Y (2009) Phosphorus recycling process from wastewater treatment plant utilizing concrete waste. Kagaku Kogaku Ronbunshu 35:12–19 (in Japanese)

Mohara G, Iizuka A, Nagasawa H, Kumagai K, Yamasaki A, Yanagisawa Y (2011) Phosphorus recovery from wastewater treatment plants by using waste concrete. J Chem Eng Jpn 44:48–55

Molinos-Senante M, Hernández-Sancho F, Sala-Garrido R, Garrido-Baserba M (2011) Economic feasibility study for phosphorus recovery processes. J Hum Environ 40:408–416

Sasaki T, Iizuka A, Honma M, Yoshida H, Hayakawa Y, Yanagisawa Y, Yamasaki A (2014) Phosphorus recovery from waste water by a continuous flow type reactor with phosphorus adsorbent derived from concrete sludge (PAdeCS). Kagaku Kogaku Ronbunshu 40:443–448 (in Japanese)

Chapter 32
Valorisation of Nutrients in Wastewaters Using Reactive Inorganic Sorbents

M. Hermassi, D. Guaya, O. Gibert, C. Valderrama, and J. L. Cortina

Abstract The recovery and removal of nutrients from the main anthropogenic flows (e.g. urban/industrial wastewater and animal manure) could be crucial to maintain the ecosystems and to secure a renewable source of nutrients. Technology options for recovering phosphorus (P) from wastewater have often been limited by the need to treat large volume of diluted waste streams. The solid-liquid technology, which uses an adsorbent in conventional sorption-regeneration processes, is one of the most realistic solutions for P recovery from wastewater. In particular, the approaches based on a low-cost inorganic adsorbent have attracted increasing attention for recovering P from wastewater treatment plants. This chapter describes the use of coal combustion fly ash and its derivative zeolites as low-cost inorganic adsorbents for P recovery from wastewater. Laboratory experiments showed that the presence of aluminium and iron oxides as well as calcium (Ca)-based minerals was critical to promote the complexation with phosphate (P_i) on the solid structure, thereby forming brushite ($CaHPO_4{\cdot}2H_2O$) deposits. Bioavailability assays have demonstrated that the P_i-loaded inorganic sorbents could serve as a slow-release P_i fertiliser in both basic and acidic soils. Furthermore, the use of synthetic zeolite mixtures has demonstrated their ability to simultaneously remove P_i and ammonium from wastewater. The recovered products could be used not only as mineral fertiliser but also as soil amendment, since the zeolite-based particles have a structure capable of improving the water retention capacity of soil. The potential

M. Hermassi · C. Valderrama
Chemical Engineering Department, UPC-Barcelona TECH, Barcelona, Spain

Barcelona Research Center for Multiscale Science and Engineering, Barcelona, Spain

D. Guaya
Chemical Engineering Department, UPC-Barcelona TECH, Barcelona, Spain

Departament of Chemistry, Universidad Técnica Particular de Loja, Loja, Ecuador

O. Gibert · J. L. Cortina (✉)
Chemical Engineering Department, UPC-Barcelona TECH, Barcelona, Spain

Barcelona Research Center for Multiscale Science and Engineering, Barcelona, Spain

Water Technology Center CETaqua, Cornellà de Llobregat, Spain
e-mail: jose.luis.cortina@upc.edu

H. Ohtake, S. Tsuneda (eds.), *Phosphorus Recovery and Recycling*,
https://doi.org/10.1007/978-981-10-8031-9_32

implementation of the zeolite-based sorbents as P_i recovery technology has also been evaluated using a hybrid process consisting of P_i sorption and membrane ultrafiltration.

Keywords Zeolite · Inorganic sorbents · Slow-release Pi fertiliser · Valorisation · Wastewater

32.1 Introduction

During the last decades, the presence of anthropogenic reactive nitrogen (N) in the environment has strongly influenced the biogeochemical nitrogen cycle, resulting in major environmental concerns. Similarly, the biogeochemical phosphorus (P) cycle is currently in a very critical state as demonstrated by the fact that phosphate rock has been recently listed as a critical raw material in Europe. P is an important element in industry and agriculture and is frequently present in domestic, industrial, and agricultural waste streams. In the last decades, P_i has been considered an environmental concern because of its role in the eutrophication of water bodies (Ma and Zhu 2006). Currently, it is also becoming increasingly an economic concern because its natural deposits are diminishing due to the accelerated mining to support the ever-growing world population. Environmental impacts of the urban and industrial sectors are attracting increasing attention due to the production of huge volumes of wastewater with high concentrations of organics and minerals such as N, P, and potassium (K) at more diluted concentrations (Lu et al. 2009; Ledda et al. 2013; Petersen et al. 2007). The recovery of nutrients from the main anthropogenic flows, i.e. animal manure, urban/industrial wastewater, and agricultural run-offs, is crucial not only to reduce the environmental impact of eutrophication but also to secure a renewable source of nutrients (Kumar and Pal 2013; Taddeo et al. 2016).

Chemical or biological processes currently being practised for Pi removal from wastewater are no longer deemed adequate in meeting new discharge limits. For example, in the United Kingdom, P_i-discharge limits have been approaching to 0.1 mgP/L under the Water Framework Directive (WFD), making conventional chemical/biological processes insufficient. Meanwhile, P_i removal and recovery through sorption has proven to be a cost-effective technique. The main advantages of using sorption are that less sludge is produced, compared to chemical precipitation; and it offers a reversible and reusable option for P_i recovery. This development has led to the exploration of a wide range of potential sorbents for P_i recovery. In this respect, the use of polymeric ion exchangers impregnated with hydrated metal oxides (HMO) is likely a suitable option, since their stable mechanical and hydraulic properties could be synergistically leveraged with the high selectivity of HMO towards oxyanions as P_i (Zhu and Jyo 2005; Awual et al. 2011). Among a variety of hydrated metal oxides, hydrated ferric oxide (HFO) is an innocuous, inexpensive, and readily available substance. It is also chemically stable over a wide pH range and can provide freshly precipitated amorphous particles having a diameter of 10–100 nm (Demarco et al. 2003).

To maintain the economic feasibility in a full-scale process, the sorbents should (1) be abundantly available and low-cost, (2) display high selectivity, (3) be able to recover Pi in a suitable form for recycling, and (4) work efficiently in sorption-desorption cycles. To date, however, such challenges have not yet been fully addressed. There is no commercial application on the recovery of Pi from wastewater using polymeric sorbents. The use of natural or synthetic materials has been an area of increasing interest as it is a potentially low-cost solution and has the potential to recover Pi that can be directly applied as a fertiliser. Natural sorbent materials, which are locally and abundantly available, include environmental friendly by-products such as biochar (De Gisi et al. 2016), fly ash (He and Xu 2012), and slag (Barca et al. 2012; Safa Jaouadi 2014). These materials can be used as agricultural amendments, because they contain both P and other useful elements (e.g. organic carbon and minerals such as calcium and potassium) after being used in wastewater treatment processes (Scholz et al. 2013).

The reuse and valorisation of recovered products would be readily possible if low-cost sorbents can be enriched by valuable elements (N, P, K) at levels required for its use as a secondary resource in the fertiliser industry or for direct application to farmland, forestry, and pasture (Evelyn Desmidt et al. 2014; Boyer et al. 2011). In the former case, the choice of sorbents needs to be made in consideration of the compatibility with the fertiliser-manufacturing process. For the latter case, in addition to the material cost and high sorption capacities, it is critical to avoid the contamination of toxic substances during the sorption stage.

There is a potential in the use of wastes generated in the energy, mining, metallurgical, and chemical industries as low-cost sorbents. However, they are often contaminated with toxic metals (e.g. Cd, Pb, and Hg) and other hazardous substances (e.g. As, Se) (Hermassi et al. 2017a). Therefore, it is critical to assure the safety of using these recyclable materials by minimising the contamination of toxic substances. Meanwhile, levels of organic micro-contaminants associated with personal care products and drugs, both legislated and not legislated, are continuously increasing, and their potential transfer from wastewater to the natural environment is becoming one of the main limitations and concerns for the use of urban sewage sludge in agriculture.

In the case of Europe, due to the limited capacity for land applications, incineration of sewage sludge is the main management route, providing the opportunity for P recovery from ash. By contrast, in the south of Europe and United Kingdom, the land application is the main route for sludge disposal. The use of inorganic sorbents having the limited affinity for organic molecules provides a reliable solution to reduce the accumulation of organic micro-pollutants on sorbents and then limited their transfer to the environment. In the design and selection of a suitable inorganic sorbent as a P carrier, one of major challenges to be addressed is bioavailability. Pi is preferentially present in the form of ionic species $HPO_4^{2-}/H_2PO_4^-$ at pH 6–8. In a solid-liquid process, P_i can be transferred from the aqueous phase to the solid phase by two main mechanisms: (1) the formation of weak complexes with functional groups or (2) the formation of P_i-containing minerals. Both reactions need to be favoured in the P_i-sorption stage, but after then, they should be weak enough to moderately release P_i in soil, thereby guaranteeing the bioavailability of P_i for plant.

This chapter describes the use of powder inorganic sorbents generated from combustion fly ash and its derivative zeolites to selectively recover P_i from waste-water streams. Special attention has been paid to the use of sorbents modified with metal hydroxides (Al, Fe, Mn) and Ca-based minerals to promote the above-mentioned mechanisms. The use of mixtures of synthetic apatites (hydroxyapatite (Hap, $Ca_{10}(PO_4)_6(OH)_2$)) and natural zeolites as solid media for growing plants has been postulated (Liu and Lal 2014). However, because of the low P_i bioavailability of crystalline Hap in soil, efforts have been directed towards the preparation of relatively soluble Ca-phosphate minerals as brushite ($CaHPO_4{\cdot}2H2O$) by capturing Pi on the surface of Ca-containing minerals. These materials include Ca silicates (Liu and Ding 2002) and Ca-Al-layered double hydroxides (Watanabe et al. 2010). The modification of sorbents with metal hydroxides (Al, Fe, Mn) and Ca-based minerals could promote the formation of more bioavailable brushite rather than Hap.

In this chapter, the development of sorbents capable of simultaneously recovering N, K, and P_i from wastewater will be also described. Since N and K are typically present as cationic forms (i.e. NH_4^+ and K^+), the sorbents need to have sorption properties for both anion (P_i) and cations (NH_4^+ and K^+). To this end, two solutions have been evaluated: (1) the modification of zeolites with Ca^{2+} and hydrated metal oxides and (2) the use of mixtures of different inorganic sorbents. The inorganic sorbents could be used in the form of small-size powders for recovering nutrients from wastewater. The use of powdered sorbents in wastewater treatment processes requires solid/liquid separation using technologies such as hydro-cyclone and hybrid sorption-filtration. In the latter case, the continuous progress in pressure-driven membrane processes has facilitated the integrated use of powdered activated carbon with ultrafiltration (UF) or nanofiltration (NF) membranes (Wang et al. 2016). In this chapter, initial studies on the application of a hybrid process consisting of P_i sorption and membrane ultrafiltration will be also presented.

32.2 Inorganic Reactive Sorbents for P_i Recovery

32.2.1 Fly Ash from Carbon Combustion Plants

Increasing energy demands worldwide have led to the increased use of coal and, thus, the production of large quantities of fly ash (FA) as a waste product (Hui and Chao 2006). Recycling coal fly ash could be a good alternative disposal option, providing significant economic and environmental benefits. The global average FA utilisation ratio is estimated to be nearly 25% (Bhattacharjee and Kandpal 2002). FA is mostly an alkaline mineral and has its surface negatively charged at high pH; thus, it could be used to remove metal ions from solutions by precipitation (Penill et al. 2006) or by sorption (Wu et al. 2006). Coal FA has attracted substantial

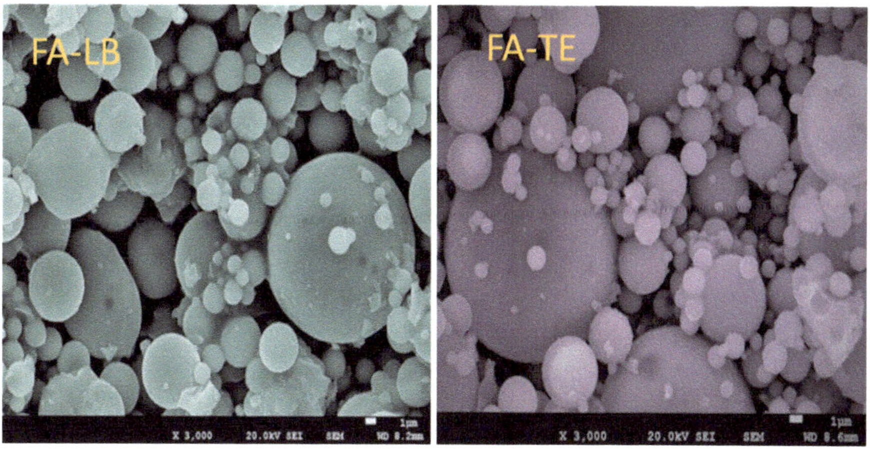

Fig. 32.1 SEM analysis of coal combustion fly ash samples from Teruel and Los Barrios coal power plants: FA-TE; FA-LB (Hermassi et al. 2017a)

attention as a potential material for P_i removal because it is easily available and cost-effective (Tsitouridou and Georgiou 1988). The presence of aluminium, iron, calcium, and magnesium oxides imbues FA with suitable properties for Pi removal by complexation or precipitation of Ca/Mg-P_i (Grubb et al. 2000). Although it is generally accepted that P_i removal by FA involves sorption and/or precipitation mechanisms, the interaction between P_i and Ca^{2+} remains partially unsolved (Lu et al. 2009). Two different types of FA from two coal power stations with different CaO(s) contents (Los Barrios (FA-LB (2.8% w)) and Teruel (FA-TE (4.8% w))) were evaluated as sorbents for P_i recovery (Hermassi et al. 2017a). The selection of FA samples was based on the Ca^{2+} and Mg contents as well as the contents of toxic metallic and non-metallic traces. This made the leaching of heavy metals (Pb, Cd, Hg) and non-metals (As) from both samples low (Moreno et al. 2005).

Both FA samples consisted mainly of Al_2O_3 and SiO_2, which accounted for 73 and 83 wt% of the samples, respectively. FA-TE has a higher content of Fe_2O_3 (18.9%) and CaO (4.2%) than FA-LB (7.4% and 2.3%, respectively). Hematite and magnetite (Fe_2O_3), mullite ($Al_2Si_2O_{13}$), and quartz (SiO_2) were the main phases identified in both samples. SEM analysis has identified spherical particles containing encapsulated smaller particles (Fig. 32.1).

P_i-sorption isotherms for both FA samples revealed a dependence on pH (Fig. 32.2). The maximum P_i-sorption capacities were 38.8 ± 3.4 and 19.1 ± 1.7 mgP-PO_4/g for FA-TE and FA-LB, respectively, when the pH values exceeded 7. At pH 8, the maximum uptakes were 59.5 ± 4.3 and 54.1 ± 3.7 mgP-PO_4/g for FA-TE and FA-LB, respectively. The P_i-sorption capacities measured here are in the same order of magnitude as those reported by Chen et al. (2006). They have reported the P_i immobilisation capacity in FA samples with the CaO content from 2 to 5 wt% using an initial P_i concentration of 1 g P-PO_4/L.

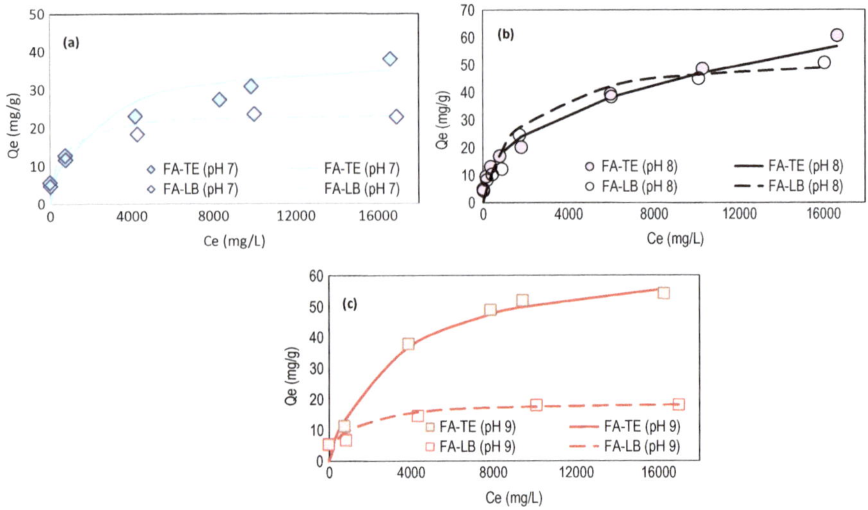

Fig. 32.2 P_i-sorption isotherms at different pH and those predicted by the Langmuir model for FA-TE and FA-LB at (**a**) pH = 7, (**b**) pH = 8, and (**c**) pH = 9 (dots, experimental data; line, the predicted values) (Hermassi et al. 2017a)

Given that $H_2PO_4^-$ and HPO_4^{2-} are the predominant species of P_i at pH values between 7 and 9, two main sorption mechanisms can be postulated for the P_i sorption on FA:

1. Surface complexation with ≅AlOH and ≅FeOH functional groups of Al and Fe oxides:

$$\text{Labile complexes with} \cong MOH_2^+ : \cong MOH_2^+ + H_2PO_4^- / HPO_4^{2-} \leftrightarrow \cong MOH_2 + H_2PO_4^- / HPO_4^{2-} \quad (32.1)$$

$$\text{Inner sphere complexes with} \cong MOH : \cong MOH + H_2PO_4^- / HPO_4^{2-} \leftrightarrow \cong MH_2PO_4^- / HPO_4^{2-} + OH \quad (32.2)$$

where M represents Al or Fe.

2. Formation of Ca-P_i minerals on the FA samples:

$$HPO_4^- / HPO_4^{2-} + CaO(s) \leftrightarrow Ca - P_i\left(Ca_5(OH)(PO_4)_3(s), CaHPO_4(s)\right) \quad (32.3)$$

Cheung and Venkitachalam (2000) have reported that the removal of P_i by FA containing high- and low-calcium contents was associated with Ca-P_i precipitation. Johansson and Gustafsson (2000) have proposed the formation of amorphous calcium P_i and octacalcium P_i as the major P_i-removal mechanism, suggesting the direct formation of hydroxyapatite as the predominant removal mechanism.

XRD analysis of FA samples revealed the presence of brushite at pH 7 and 8, while the presence of Ca-P_i minerals was not detected at pH 9. This was attributable to the formation of undetectable nanocrystals or amorphous structures because the removal rate at pH 9 was faster than those at pH 7 and 8 as indicated by the kinetic analysis. Generally, precipitation processes with fast kinetics produce less-crystalline solids. The formation of Ca-P_i is thermodynamically favoured under the studied conditions; Hap is a more stable phase than brushite which is considered as a precursor. However, as the reaction proceeds on the microporous FA structure where the oversaturation of Ca^{2+} ion can be avoided by the slow dissolution of CaO(s), brushite can be formed and stabilised. Thus, the main P_i-removal process can be postulated by Eqs. (32.4 and 32.5):

$$Ca(OH)_2(s) + HPO_4^{2-} \leftrightarrow CaHPO_4(s) + 2OH^- \quad \log K_{298K} = 32.2 \tag{32.4}$$

$$Ca(OH)_2(s) + H_2PO_4^- \leftrightarrow CaHPO_4(s) + H_2O + OH^- \quad \log K_{298K} = 22.2 \tag{32.5}$$

The time-course data exhibited that the concentrations of P_i and Ca^{2+} decreased concomitantly with the increase of pH values. At high P_i concentrations, the sorption rates of both FA samples were quite similar as they were controlled by the presence of excess P_i relative to Ca^{2+} provided by the FA. The sorption process could be divided into the first, faster step and the second, slower one. Initially, P_i rapidly reached the boundary layer to interact with dissolved Ca^{2+} ions from the $Ca(OH)_2$(s) grains, and then, it slowly diffused through the boundary-layer film onto the FA particle. At that time, P_i removal was coupled with $Ca(OH)_2$(s) dissolution, which supplied Ca^{2+}ions needed for the brushite formation (Barca et al. 2012). The evolution of Ca^{2+} for both FA samples in the kinetic experiments with 0.1 and 0.5 gP-PO_4/L revealed that the total Ca concentration followed a profile very similar to that of the total P concentration. For FA-LB, the P_i removal was achieved mostly in the first 120 min: 40% at pH 7 and 80% at pH 9. In contrast, FA-TE exhibited a lower sorption rate so that longer contact times (more than 15 h) were required to reach equilibrium. These differences were related to the composition of the FA samples, namely, the contents of SiO_2, Al_2O_3, and mullite which could improve their hydrophilic properties.

Olsen et al. (1954) have suggested that the bicarbonate extraction is a suitable method for predicting the plant availability of P_i in calcareous soils where the main role of $NaHCO_3$ in P_i extraction is decreasing the Ca^{2+} activity by forming $CaCO_3$(s). The P_i availability data in 0.5 M $NaHCO_3$ are plotted in Fig. 32.3 as a function of the extracted amount of P_i from FA (mgP-PO_4/g FA). For both FA samples, the P_i extraction ratio was in the range from 8 to 30 mgP-PO_4/g FA.

Partial extraction of 20–70% has been reported in a single-extraction trial for both FA samples. In the presence of excess of bicarbonate, the labile Pi fraction

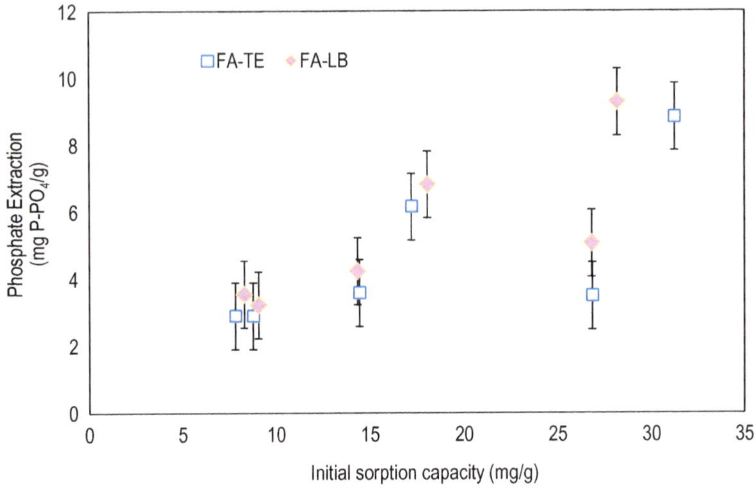

Fig. 32.3 P_i extraction using mixture of $NaHCO_3$ (0.2 M) from loaded FA-TE and FA-LB samples. (Hermassi et al. 2017a)

(P-KCl) can be displaced by bicarbonate ions, and partial brushite dissolution is achieved according to Eq. (32.6):

$$CaHPO_4(s) + HCO_{3^-} \leftrightarrow CaCO_3(s) + H_2PO_{4^-} \; \log K_{298K} = -1.3 \quad (32.6)$$

The sorbed P_i on FA samples has been demonstrated to dissolve in solutions containing moderate to high bicarbonate concentrations similar to those expected in basic soils characterised by a high content of calcareous rocks (e.g. limestone) where other Ca-P_i minerals such as Hap are insoluble and thus show limited plant availability. The use of P_i-containing mineral-based sorbents as soil amendments may be advantageous when other agronomic benefits are expected due to the provision of other plant nutrients or the enhancement of the moisture-holding capacity of soil.

32.2.2 Synthetic Zeolite Powder Produced from Coal Fly Ash

Since coal FA is rich in SiO_2 and Al_2O_3, it can be transformed by the chemical treatment into zeolite-like crystalline materials. The synthesis of zeolites is attracting attention as an effective use for coal FA, possibly because of its similar composition to the volcanic material that serves as the precursor of natural zeolites. The substitution of Si by Al atoms in the crystal framework leads to extra negative charges that must be balanced by surrounding counterions (e.g. Na^+, K^+, Ca^{2+}, and Mg^{2+}) which can be easily exchanged for other surrounding cations in contact with aqueous solutions. By applying several synthesis methods, different families of zeolites have

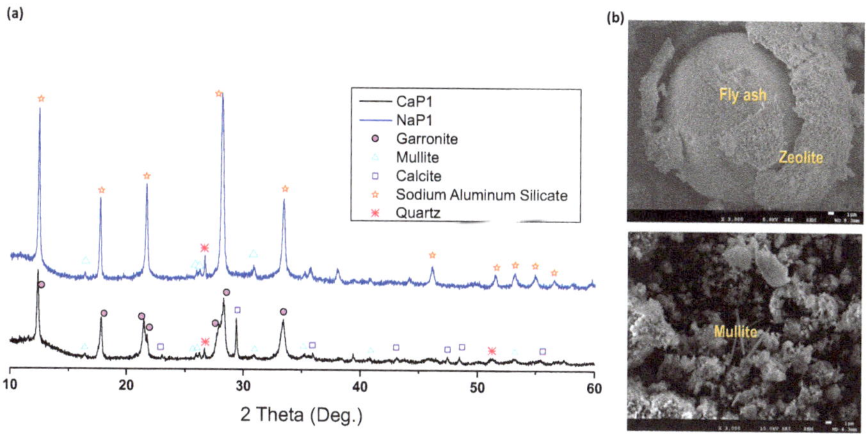

Fig. 32.4 (**a**) XRD patterns of NaP1-NA and its modified zeolite CaP1-NA and (**b**) SEM images of NaP1-NA and CaP1-NA. (Hermassi et al. 2016a)

been synthesised from coal FA. However, few have been successfully converted into pure-phase zeolites (Zhou et al. 2014; Ansari et al. 2014).

Querol et al. (2007) have demonstrated the economic and technical viability of synthesising Na^+-zeolite (NaP1-NA) under mild hydrothermal conditions without using templates. NaP1-NA was evaluated for the removal of toxic metals from acid mine drainage and brines because of its unique ion-exchange and water-sorption properties, which are attributable to its high porosity, surface area, cation exchange capacity (CEC), and unusual framework flexibility (Cama et al. 2005). NaP1-NA also has high capacity to sorb NH_4^+ and K^+, and it has been evaluated as a slow-release fertiliser. However, the sorption of oxyanions as P_i is not favoured by the zeolite structure. Hermassi et al. (2016a) have evaluated both NaP1-NA, synthesised from coal FA, and its Ca-modified form (CaP1-NA) as sorbents for P_i recovery from aqueous solution. The conversion of FA involves three stages: (1) dissolution of Al and Si from FA, (2) deposition of aluminosilicate gel on the FA surface, and (3) crystallisation of zeolite from aluminosilicate gel (Murayama et al. 2002). The three Al- and Si-containing phases of the FA are (1) amorphous aluminosilicate glass, (2) quartz, and (3) mullite. Aluminosilicate glass is the largest component and is the most unstable in the hydrothermal environment, and therefore, it exhibits the highest rate of dissolution (Querol et al. 2001) and makes the largest contribution to the produced zeolites. The X-ray diffraction (XRD) patterns of the Na zeolite (NaP1-NA) and the modified Ca-zeolite (CaP1-NA) are shown in Fig. 32.4a. Garronite ($Na_6(Al_6Si_{10}O_{32})\cdot 12H_2O$), mullite ($Al_2Si_2O_{13}$), and a trace of quartz (SiO_2) were the main phases identified for NaP1-NA sample. The XRD patterns of CaP1-NA indicated the presence of mullite, quartz, calcite, and garronite ($NaCa_{2.5}(Si_{10}Al_6)O_{32}\cdot 14H_2O$) as the predominant phases. The SEM observation (Fig. 32.4b) showed that NaP1-NA and CaP1-NA coat the FA. It is also observed that the coating of NaP1-FA is not complete. The chemical compositions of both

Table 32.1 Average chemical composition of zeolitic sorbents NaP1-NA and CaP1-NA by using FSEM-EDX

	C	O	Na	Mg	Al	Si	K	Ca	P	Fe
NaP1-NA (%)	11.61	50.59	7.56	0.52	8.08	15.11	1.84	1.44	0.38	2.86
CaP1-NA (%)	12.53	51.27	0.97	0.56	8.04	14.58	1.66	7.38	0.43	2.57

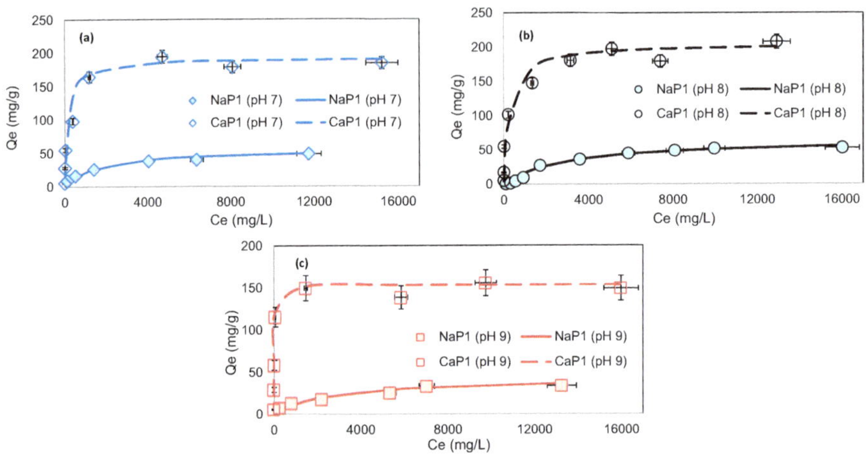

Fig. 32.5 P_i-sorption isotherms and predicted uptake by the Langmuir isotherms for CaP1-NA- and NaP1-NA-modified zeolitic material at (**a**) pH = 7, (**b**) pH = 8, and (**c**) pH = 9 (dots, experimental data; line, the predicted values) (Hermassi et al. 2016a)

zeolitic materials are listed in Table 32.1. It is indicated that NaP1-FA and CaP1-NA contained mainly Al_2O_3 and SiO_2 which accounted for a total of 74%. The modification of Na-zeolite to CaP1-NA considerably increased the Ca content (from 1.4% to 7.4% (83 mg Ca/g)). Accordingly, the Na^+ content decreased significantly (from 7.6 to less than 1%) because of treatment with $CaCl_2$ and the partial exchange of sodium and calcium ions in the zeolite structure.

The P_i-sorption isotherms of CaP1-NA and NaP1-NA at different pH values (7, 8, and 9) are shown in Fig. 32.5. The P_i-loading capacity of CaP1-NA is higher than that of NaP1-NA, e.g. 203 ± 11mgP-PO_4/g, compared with 57 ± 5 mg P-PO_4/g at pH 8. The larger sorption capacity of CaP1-NA is associated with its Ca content and high availability for reaction, mainly at the ion-exchange sites, while the lower sorption capacity of NaP1-NA is related to unreacted Ca in the form of CaO. It should be noted that the initial BET surface area increased during zeolite modification from 6.3 to 13.6 m^2/g. Thus, the higher Ca content and larger surface area enhanced the P_i sorption.

According to the removal patterns exhibited by NaP1-NA and CaP1-NA zeolitic materials, the sorption of P_i ions, mainly $H_2PO_4^-$ and HPO_4^{2-} that occur in the expected pH range (7–9), may follow two postulated mechanisms shown below:

(a) Surface complexation with ≅AlOH and ≅FeOH functional groups originally present as Al and Fe oxides or in the zeolitic structure:

1. Outer-sphere complexes with $\cong MOH_2^+$ surface groups

$$\cong MOH_2^+ + H_2PO_4^- / HPO_4^{2-} \leftrightarrow \cong MOH_2^+H_2PO_4^- / HPO_4^{2-} \tag{32.7}$$

2. Inner-sphere complexes with ≅MOH surface groups

$$\cong MOH + H_2PO_4^- / HPO_4^{2-} \leftrightarrow \cong MH_2PO_4^- / HPO_4^{2-} + OH^- \tag{32.8}$$

Formation of Ca-P_i minerals with Ca(II) ions present on the zeolite through two main reactions:

1. Ca-P_i minerals with Ca(II) present on the zeolitic material as CaO(s) and $CaCO_3$(s):

$$\begin{aligned} & H_2PO_4^- / HPO_4^{2-} + CaO(s) / CaCO_3(s) \\ & \quad = Ca - P_i \left(CaHPO_4(s) \text{ or } Ca_5(OH)(PO_4)_3(s)\right) \end{aligned} \tag{32.9}$$

2. Ca-P_i minerals with Ca(II) ions occupying the ion-exchange groups of the zeolitic structure:

$$(\cong ZO^-)_2 Ca^{2+} + H_2PO_4^- + 2Na^+ \leftrightarrow 2(\cong ZO^- Na^+) + CaHPO_4(s) + H^+ \tag{32.10}$$

$$5(\cong ZO^-)_2 Ca^{2+} + 3H_2PO_4^- + 10Na^+ \leftrightarrow 10(\cong ZO^- Na^+) + Ca_5(OH)(PO_4)_3(s)) + 7H^+ \tag{32.11}$$

where $\cong ZO^-$ represents the anionic groups of the zeolite structure.

XRD patterns revealed the formation of Ca-P_i in the form of brushite as the major phase at pH 7 and 8 and as the minor phase at pH 9. Similar P_i removal and brushite formation have been reported by Pengthamkeerati et al. (2008) with Ca-zeolites and by other researchers (e.g. Lu et al. (2009) with FA). The formation of brushite instead of Hap is associated with the prevalence of kinetic control. In fact, the reactive crystallisation of brushite has been described to occur through five successive stages: (1) spontaneous Hap precipitation, (2) complete dissolution of Ca and Hap growth, (3) initial appearance of brushite nuclei, (4) coexistence of Hap and brushite, and (5) transformation of Hap into brushite and subsequent brushite growth.

The speciation results of CaP1-NA and NaP1-NA zeolitic materials are shown in Fig. 32.6. The easily exchangeable speciation (KCl-P) associated with exchange reactions (Eq. (32.7)) accounts for up to 20% for NaP1-NA and less than 10% for

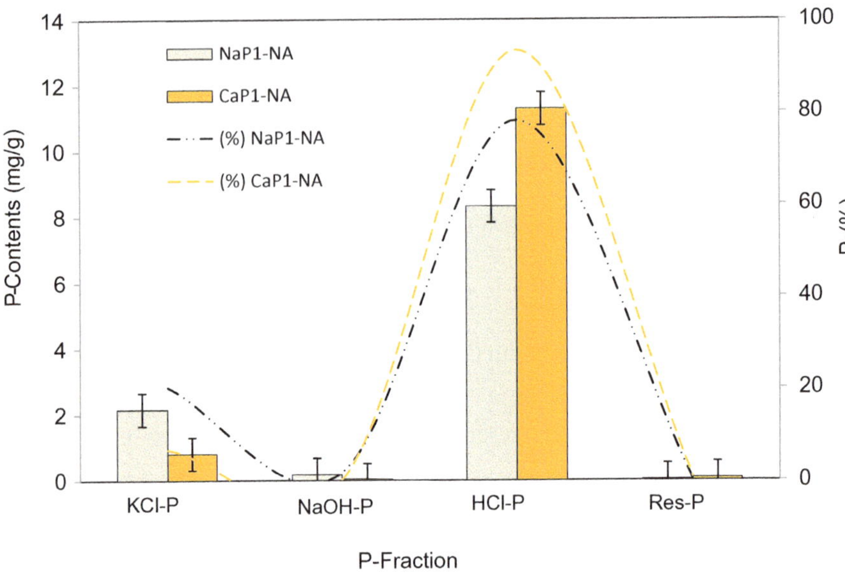

Fig. 32.6 P_i speciation of the NaP1-NA- and CaP1-NA-loaded samples (Hermassi et al. 2016a)

CaP1-NA. The dissolved species obtained using NaOH solutions (NaOH-P) reveal that the P bound to the hydrated metal oxides (inner-sphere and outer-sphere complexes described by Eqs. (32.7) and (32.8)) makes only a small contribution for NaP1-FA and no contribution for CaP1-NA. In NaOH solutions, the$\cong MOH_2^+$ and $\cong$MOH groups are deprotonated, and the excess OH ions disrupt the$\cong MH_2PO_4^-$/ HPO_4^{2-} complexes. The HCl-P speciation (in line with laboratory experiments reported by other studies (Wang et al. 2012)) associated with Ca-P_i mineral phases accounts for up to 80% for NaP1-NA and more than 90% for CaP1-NA. The species produced by the residual speciation contributed less than 1% for both zeolites.

The P_i desorption achieved using a 0.1 mol/L $NaHCO_3$/Na_2CO_3 solution increased as the amount of P_i on the zeolitic material decreased. Partial desorption (30–70%) was reported for CaP1-NA, whereas values of 10–70% were observed for NaP1-NA in a single extraction trial. These results are in agreement with the speciation results obtained in using bicarbonate. Indeed, during labile speciation (P-KCl), P_i anions will be displaced by bicarbonate, and partial brushite dissolution will be achieved as increasing the pH increases the solubility of brushite up to one order of magnitude. It can be concluded that P_i sorption on CaP1-NA is relatively irreversible and that the bonding between the active sites and the sorbed P_i is quite strong.

32.3 Simultaneous Removal of P and N by Powdered Modified Zeolites

Simultaneous removal of NH_4^+ and P_i from diluted streams (e.g. secondary effluent from a conventional activated sludge process) can be achieved using inorganic sorbents like natural zeolites (Yin and Kong 2014). The NH_4^+ removal is favoured by the high CEC of natural zeolites (Xie et al. 2013). However, they exhibited poor performance for removing anions such as P_i (Ji et al. 2015). Hence, it is necessary to modify the zeolite by incorporating hydrated metal oxides (e.g. Fe, Al, Mn) with complexing properties (inner- and outer-sphere complexes with P_i ions). It is also needed to avoid incorporating cations such as Ba (II), Ca (II), and Mg (II) which form low-solubility P_i minerals (Figueiredo and Quintelas 2014). The resulting exhausted sorbents could be applied to farmland as a fertilising material and also as a soil improver. The methodology to modify a granular natural zeolite in a sodium form into Al (III), Fe (II)), and Mn (II) forms, and its use in sorption and desorption cycles has been described previously (Guaya et al. 2015, 2016, 2017).

Guaya et al. (2016) have evaluated the simultaneous removal of NH_4^+ and P_i from the secondary-treated wastewater using impregnated Al, Fe, and Mn forms of a powder natural zeolite in its potassium form. The natural zeolite (NC) (Zeocem Co., Slovak Republic) was mainly identified as clinoptilolite. Traces of cristobalite and mordenite were also identified by XDR analysis (Fig. 32.7). The natural zeolite did

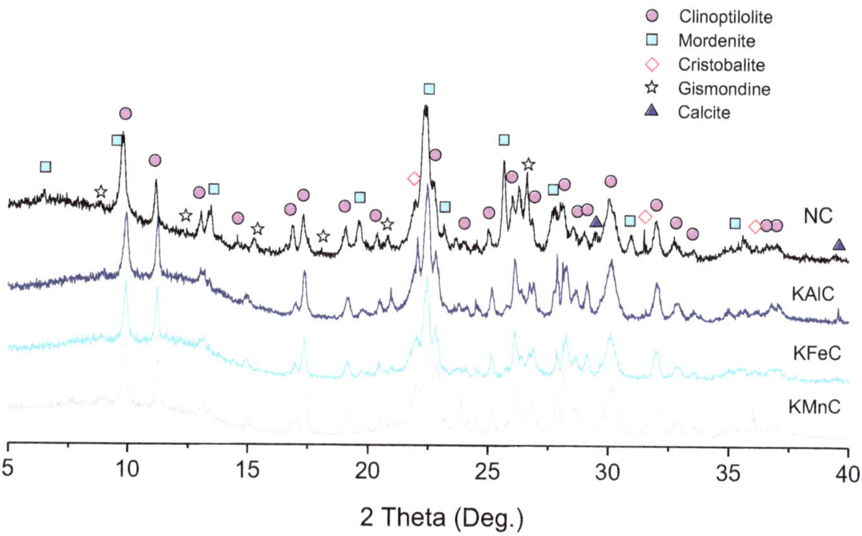

Fig. 32.7 X-ray diffractograms of a natural zeolite (NC) and the hydrated metal oxide forms (KAlC, KFeC, and KMnC) (Guaya et al. 2016)

Table 32.2 Relative atomic percentages measured by EDX of natural and potassium-modified zeolites by impregnation with HMO (Al, Fe, and Mn) (Guaya et al. 2016)

Sample	O	Na	Mg	Al	Si	K	Ca	Fe	Mn
NC	57.9 ± 3	0.3 ± 0.1	0.4 ± 0.1	5.3 ± 0.4	29.7 ± 2	2.9 ± 1	1.9 ± 0.3	1.6 ± 0.2	–
KALC	46.6 ± 1	<loq[a]	<loq[a]	5.6±	14.7 ± 2	3.4 ± 1	<loq[a]	<loq[a]	–
KFeC	42.4 ± 3	<loq[a]	<loq[a]	2.7±	15.6 ± 3	3.4 ± 1	<loq[a]	8.9 ± 1	–
KMnC	47.8 ± 3	<loq[a]	<loq[a]	3.8±	21.9 ± 5	3.4 ± 1	<loq[a]	<loq[a]	1.6 ± 0.2

[a]loq: limit of quantification

not exhibit a highly pure and crystalline nature. There was no significant attenuation of the peak intensity of the modified forms of NC zeolite, revealing the absence of changes on the structure of NC sample after the modification with Al, Fe, and Mn hydrated oxides. The absence of new mineralogical phases and no significant shift in signal peaks suggested that K^+ and remaining non-precipitated Al^{3+}, Fe^{3+}, and Mn^{2+} ions caused no modification on the raw zeolite structure (Benaliouche et al. 2012).

The chemical compositions of the natural and modified zeolites are shown in Table 32.2. The three modified zeolites revealed a reduction in the contents of Na, Mg, and Ca and a slight increase in the K content compared to those of the natural zeolite (NC). The clinoptilolite platelike morphology was characterised by networks of crystal clusters with cavities and entries to the channels inside the framework.

The removal of NH_4^+ can be described by an ion-exchange reaction with K^+ (Eq. (32.12)):

$$ZO-K^{+}+NH_{4}^{+}\leftrightarrow K^{+}+ZO-NH_{4}^{+} \tag{32.12}$$

The selectivity of the exchange process is mainly affected by the ionic charge and ionic radius, although the Stokes hydration ionic radius for both ions is similar (130 nm). The difference in selectivity for the exchange of K^+/NH_4^+ is enough to assure a high removal efficiency for NH_4^+ (Murayama et al. 2002). The influence of pH on NH_4^+ sorption capacity of the K-modified forms KAlC, KFeC, and KMnC is plotted in Fig. 32.8.

The NH_4^+-sorption capacity reached a maximum between pH 4 and 7, which is near below the pHpzc of the sorbents. However, it was observed a progressive reduction of NH_4^+-sorption capacity above pHpzc, since the concentration of NH_4^+ decreases above pH 7 due to its conversion into the non-protonated form (NH_3). The Al- and Fe-hydrated oxide forms revealed similar pH dependence behaviour on P_i sorption. The highest value of sorption capacity was below the pHpzc of these sorbents so that the presence of positive charges favoured the anion sorption. However, the reduction of P_i-sorption capacity at pH 2 seems to be connected to the conversion of charged P_i species (e.g. $H_2PO_4^-$) to non-charge H_3PO_4. In the pH range from 4 to 11, near and above the pHpzc of these zeolitic modified sorbents, the decrease of P_i sorption was attributed to the existence of negative surface-charged species. Moreover, for the manganese zeolite, the capacity of P_i sorption was low in the pH

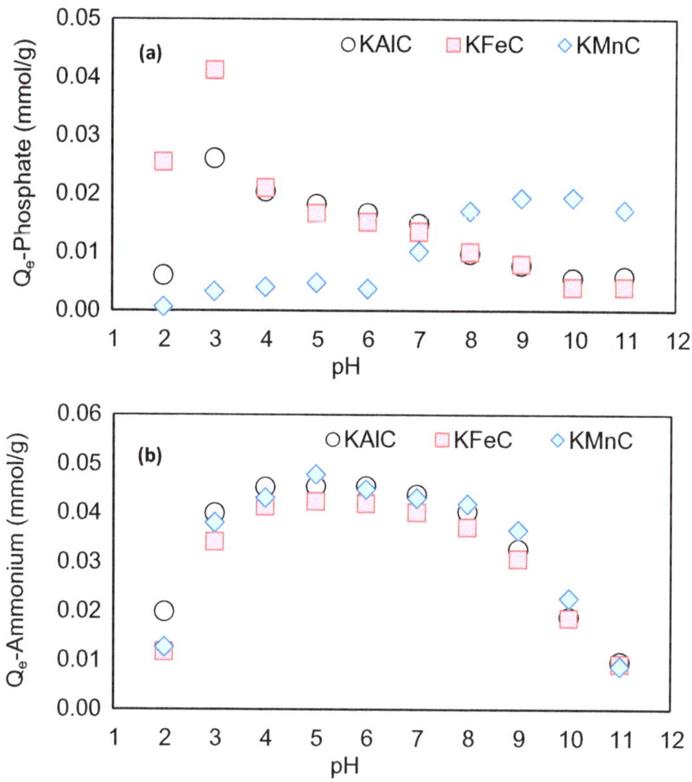

Fig. 32.8 Effect of pH on the removal of **a** P_i and **b** NH_4^+ on modified zeolites KAlC, KFeC, and KMnC (Guaya et al. 2016)

range from 2 to 6 but increased with increasing pH 7–10. The P_i oxyanion ($H_2PO_4^-$ and HPO_4^{2-}) sorption occurred through the formation of complexes with HMO functional groups (>MOH).

The NH_4^+-sorption rates were comparable for the three modified zeolites which reached equilibrium in only 15 min, whereas the P_i-sorption rates were lower, and more than 60 min was required to reach equilibrium. This suggests that the ion exchange reaction between NH_4^+/K^+ occurred faster than complexation reactions of P_i ions. It is also likely that NH_4^+ cations can access more easily to the negative sites than to the surface hydroxide groups on the zeolite particles. Sequential chemical fractionation analysis revealed that the major fraction of P absorbed by the zeolites was associated with inorganic form. This was in accordance with the fact that the secondary-treated wastewater sample contained P mostly in the inorganic forms. The most abundant P fraction was the biological active HCO_3^- fraction, accounting for 40% of total P. The amount of P in the NaOH fraction, which is associated with the formation of Fe-Al-Mn hydroxide minerals, accounted for 32% of total P. The HCl extractable fraction, associated with Ca-P_i and Mg-P_i, contained 27% of total P.

The H_2SO_4/H_2O_2 fraction accounted for only 2–3% of total P. Thus, the sequential fractionation of the loaded modified zeolite revealed the existence of an important fraction of biological active P. Furthermore, the recovered P product can be applied as fertiliser in P_i-deficient soil. Cation exchange sites occupied by K^+ are selectively exchangeable with NH_4^+ and to a lesser extent with Na^+, Ca^{2+}, and Mg^{2+}. Thus, they can be used as a slow-release K^+ and NH_4^+ fertiliser.

32.4 P_i Removal by Sorption-Filtration Processes

The sorption technology has rarely been applied to the simultaneous recovery of N and P from wastewater. The removal of P_i from industrial and domestic effluents (pH values of 6–9) by using powder inorganic sorbents is mostly performed using agitated reactors. Few studies have reported a hybrid process integrating sorption on powder sorbents and membrane filtration. Yildiz et al. (Yildiz 2004) have demonstrated that membrane filtration was more efficient than classical batch separation for recovery of P_i by FA. More recently, a hybrid sorption-ultrafiltration (UF) process for P_i recovery using a hydrated ferric oxide (HFO)-based agglomerated sorbent has been shown to be effective in reducing the P_i concentration of secondary-treated wastewater below 0.1 mg P/L (Zelmanov and Semiat 2014). The hybrid membrane (UF)-sorption system has been operated in dead-end flow or cross-flow modes (Hui et al. 2013). Currently, most relevant studies have focused their attention on the integration of powdered activated carbon with UF (Hermassi et al. 2001).

Hermassi et al. (2016b) have integrated P_i adsorption using powdered Ca-activated zeolite (CaP1) with membrane separation using a hollow-fibre UF. The membrane UF-powder activated zeolite (UF-PAZ) hybrid system (Fig. 32.9) employs a cross-

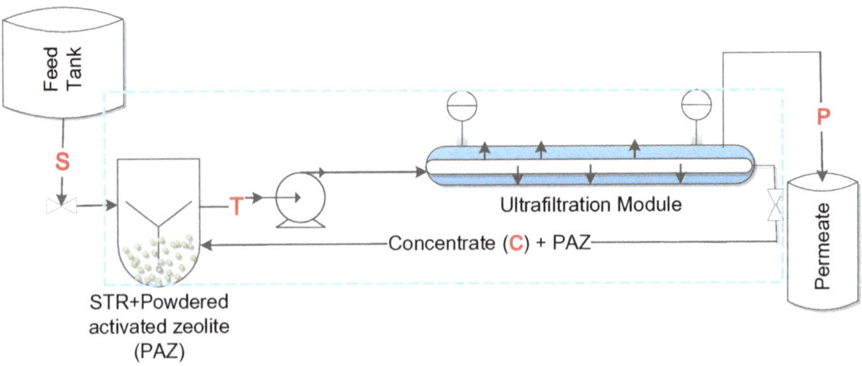

Fig. 32.9 Schematic of the hybrid membrane UF-sorption system including an ultrafiltration hollow-fibber module, the feed P(V) stream (S), a stirred-tank reactor (STR), the stream leaving the tank (T), and the concentrate (C) and the permeate (P) streams from the membrane module with a flow rate of Q (m^3/s)

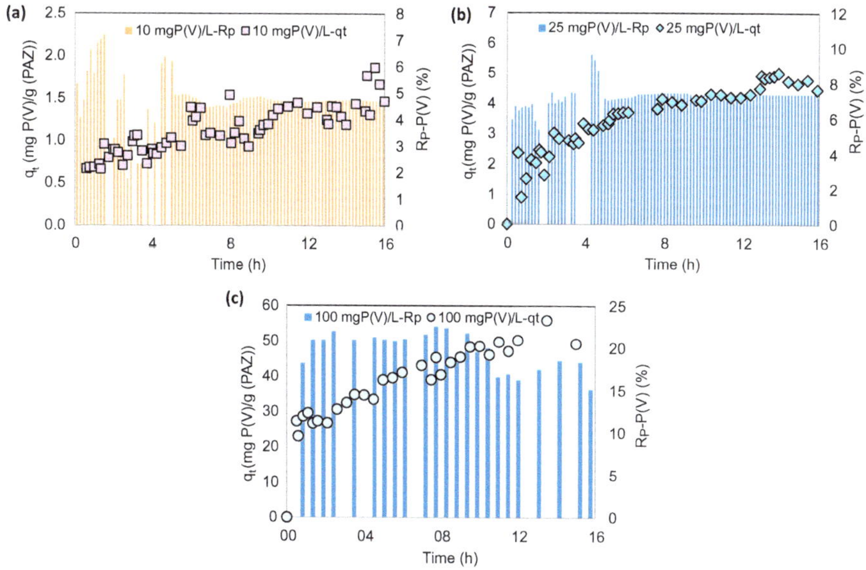

Fig. 32.10 Effect of the initial P_i concentration on the P_i-sorption capacity q(t) profile and removal efficiency percentage (Rp) during the filtration cycle with 2.5 gPAZ/L for (**a**) 10 ± 1, (**b**) 25 ± 2, and (**c**) 100 ± 6 mg/L). (Hermassi et al. 2016b)

flow UF module consisting of 100 hollow fibres with a molecular weight cut-off of >100,000 Da. The membrane fibres consisted of a blend of hydrophilic polyvinylpyrrolidone and polyethersulfone with an inner diameter of 0.8 mm and a length of 1.0 m, corresponding to a surface area of 0.25 m^2.

The P_i-removal efficiency (Rp-P(V)) remained fairly constant during the filtration cycle for the evaluated concentrations with a P_i-sorption capacity of 1.8 ± 0.2, 5.7 ± 0.3, and 47.2 ± 2 mg P /g(PAZ) for 10, 25, and 100 mg P/L, respectively. The P_i-loading values [q(t)] during filtration cycles (Fig. 32.10) increased over time, and by the end of the experiment, the PAZ was not saturated at any of the tested initial P_i concentrations. The increase of P_i concentrations increased the sorption capacity.

The removal efficiency was also higher for the higher P_i concentration. The removal efficiency could be limited by the hydrodynamic conditions and the solid/liquid ratio in the reactor. However, under conditions used in sorption-filtration experiments, the saturation was not observed, taking as a reference the values reported in the equilibrium test. Thus, the higher the P_i concentration, the higher the removal capacity achieved (Fig. 32.10). The influence of pH on P_i removal by PAZ [q(t)] and removal efficiency RP-P_i (%) at an initial Pi concentration of 10 mg P/L with a PAZ dose of 2.5 gPAZ/L at pH of 8 or 9 in buffered bicarbonate solutions are shown in Fig. 32.11.

The P_i-sorption capacity reached 1.6 ± 0.3 mg P/gPAZ (corresponding to 6 ± 0.5% removal efficiency) and 14 ± 0.5 (corresponding to 70 ± 3% removal efficiency) at

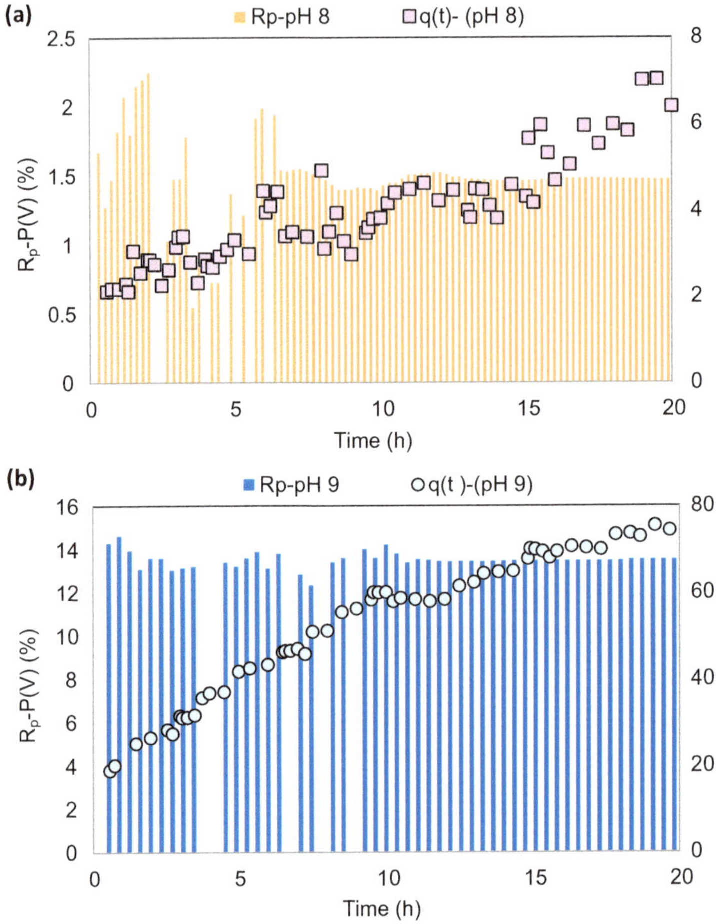

Fig. 32.11 Sorption capacity [q(t)] profile and P(V) removal efficiency [RP-P(V)] percentage as a function of filtration time by PAZ (2.5 gPAZ/L) at an initial P(V) concentration of 10 ± 1 mg/L at (**a**) pH = 8 and (**b**) pH = 9. (Hermassi et al. 2016b)

pH 8 and 9, respectively, after 20 h of filtration. During this time period, the removal efficiency remained constant for both pH values. XRD analysis of PAZ samples collected at the end of the sorption-filtration experiment revealed the presence of brushite at pH 8. However, no calcium- P_i mineral phase was detected at pH 9. The recovery of P_i using the PAZ-UF hybrid sorption-filtration process was not subjected to irreversible fouling. Fast membrane fouling and shorter filtration time were observed for the higher dose of 2.5 gPAZ/L compared with 2 gPAZ/L, mainly determined by the cake layer on the surface of the membrane, resulting in specific cake resistance values for 2.5 gPAZ/L being slightly lower than those for 2.0 gPAZ/L. The P_i-removal capacity remained constant during filtration cycles, and the

Table 32.3 Chemical composition (mg/L) and pH of model water feed used in both batch and continuous sorption-filtration experiments (UFA, UFB, and UFC). (Hermassi et al. 2017b)

	Na^+	NH_4^+	K^+	Mg(II)	Ca(II)	Cl^-	SO_4^{2-}	NO_3^-	HCO_3^-	P(V)	TOC[a]
Synthetic water feed (UF_A, UF_B) pH = 7.8 ± 0.2	505	678	51	99	500	2384	40	1	1322	56	5.2
Anaerobic digestion side stream (UF_C) pH = 7.8 ± 0.2	475	665	63	92	408	1224	4	<loq[b]	1301	48	299.5

[a]TOC (total organic carbon)
[b]loq: limit of quantification

removal efficiency increased as the initial P_i concentration increased and substantially increased when the pH was increased to 9.

The mechanism of P_i removal involves the formation of calcium-P_i precipitate. XRD analysis confirmed the P_i precipitation as brushite at pH 8 and as Hap at pH 9. The P speciation also confirmed that for all cases, the fraction HCl-P was associated with the presence of calcium-P_i minerals. The hybrid sorption-UF configuration can be used for evaluating the potential of a sorption-filtration system as a pretreatment step of water reuse schemes in conventional wastewater treatment plants.

32.5 Nutrient Recovery from Sludge Anaerobic Digestion Side Streams

The P_i-sorption capacity reached 1.6 ± 0.3 mg P/gPAZ (corresponding to 6 ± 0.5% removal efficiency) and 14 ± 0.5 (corresponding to 70 ± 3% removal efficiency) at pH 8 and 9, respectively, after 20 h of filtration. During this time period, the removal efficiency remained constant for both pH values. XRD analysis of PAZ samples collected at the end of the sorption-filtration experiment revealed the presence of brushite at pH 8. However, no calcium-P_i mineral phase was detected at pH 9. The recovery of P_i using the PAZ-UF hybrid sorption-filtration process was not subjected to irreversible fouling. Fast membrane fouling and shorter filtration time were observed for the higher dose of 2.5 gPAZ/L compared with 2 gPAZ/L, mainly determined by the cake layer on the surface of the membrane, resulting in specific cake resistance values for 2.5 gPAZ/L being slightly lower than those for 2.0 gPAZ/L. The P_i-removal capacity remained constant during filtration cycles, and the removal efficiency increased as the initial P_i concentration increased. The P_i-removal capacity was substantially improved by increasing the pH to 9, as was confirmed by the increase of the sorption rate.

The mechanism of P_i removal involves the formation of calcium-P_i precipitate. XRD analysis confirmed the P_i precipitation as brushite at pH 8 and as Hap at pH 9. The P speciation also confirmed that for all cases the fraction HCl-P was associated with the presence of calcium-P_i minerals. The hybrid sorption-UF configuration can be used for evaluating the potential of a sorption-filtration system as a pretreatment step of water reuse schemes in conventional wastewater treatment plants (Table 32.3).

N-P-K removal processes were very fast, as demonstrated by the fact that the system reached equilibrium within 10 min. This phenomenon is the result of the fast exchange of Na^+ with NH_4^+ and K^+, as reported for both natural and synthetic zeolites (Wu et al. 2006; Guaya et al. 2015). P_i and K^+ sorption kinetics are comparable with NH_4^+ exchange kinetics, although more complex mechanisms are involved (e.g. complexation and precipitation of P_i with calcium and/or potassium ions and with the ≅AlOH and ≅FeOH surface groups and Ca^{2+} ions on CaP1). In fact, Ca^{2+}/NH_4^+ exchange on CaP1 is the driving force for the formation of Ca-P_i, followed by the reaction between free Ca^{2+} and P_i when saturation is reached (Watanabe et al. 2014). The N-P-K exchange/sorption processes for the reactive mixtures are described by:

(a) NH_4^+ and K^+ cations are exchanged with Na^+ and Ca^{+2} ions from the NaP1 and CaP1:

$$ZO^-M^{n+} + nNH_4^+ / K^+ \leftrightarrow nZO^-NH_4^+ / K^+ + M^{n+} \quad (32.13)$$

where $M^{n+}$is Na^+ or Ca^{2+}.

(b) P_i removal by the reaction of P_i with Ca^{2+} from the zeolite ion-exchange groups:

$$Mg^{2+} + NH_4^+ + HPO_4^{-2} + 6H_2O \rightarrow MgNH_4PO_4{\cdot}6H_2O(s) + H^+ \quad (32.14)$$

(c) N-P-K removal with caustic magnesia (MgO) is expected through the formation of struvite ($MgNH_4PO_4{\cdot}6H_2O(s)$) or potassium struvite ($KMgPO_4(s)$):

$$Mg^{2+} + NH_4^+ + HPO_4^{-2} + 6H_2O \rightarrow MgNH_4PO_4{\cdot}6H_2O(s) + H^+ \quad (32.15)$$

$$K^+ + Mg^{2+} + HPO_4^{-2} \rightarrow KMgPO_4(s) + H^+ \quad (32.16)$$

The N-P-K sorption capacity and removal efficiency as a function of the filtration time are shown in Fig. 32.12.

The pH was maintained constant at 8.2 ± 0.3, and shorter filtration cycles were performed as the hydraulic operation limit of the UF module (2.5 g PRS/L) was approached. The removal efficiency for N-P-K remained constant during the experiment, while the sorption capacity (qt) increased. The removal efficiency reached values of 20 ± 2, 85 ± 5, and 55 ± 4%, while the maximum sorption capacities (qe) were 185 ± 8, 32 ± 3, and 23 ± 2 mg/g PRS for NH_4^+, P_i, and K^+, respectively.

The use of a tertiary mixture of sorbents did not afford any significant improvement of the system performance since the NH_4^+ removal was slightly lower, while the Pi-removal efficiency was similar to that obtained with the binary mixture. The NH_4^+ and Pi speciation results for the loaded PRS samples are shown in Fig. 32.13. The P speciation confirmed that the HCl-P fraction associated with the presence of

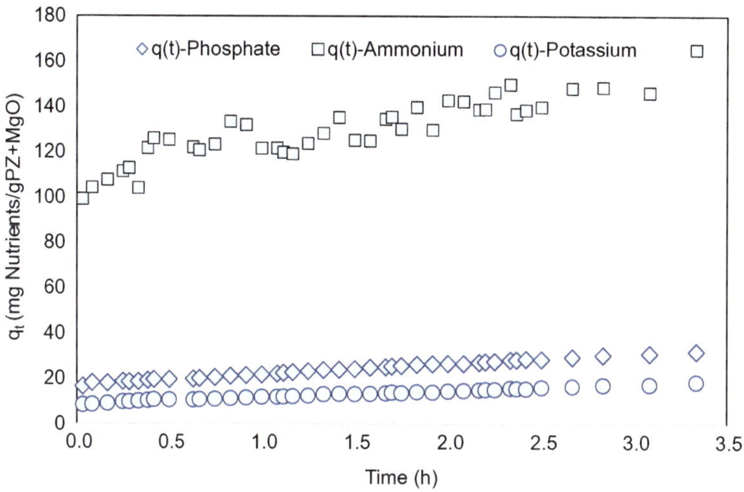

Fig. 32.12 Nutrient (phosphate, ammonium, and potassium) recovery sorption capacity (qt) as a function of time for mixtures of NaP1/CaP1/MgO (UFB) using synthetic sewage sludge anaerobic side streams as feed solutions. (Hermassi et al. 2017b)

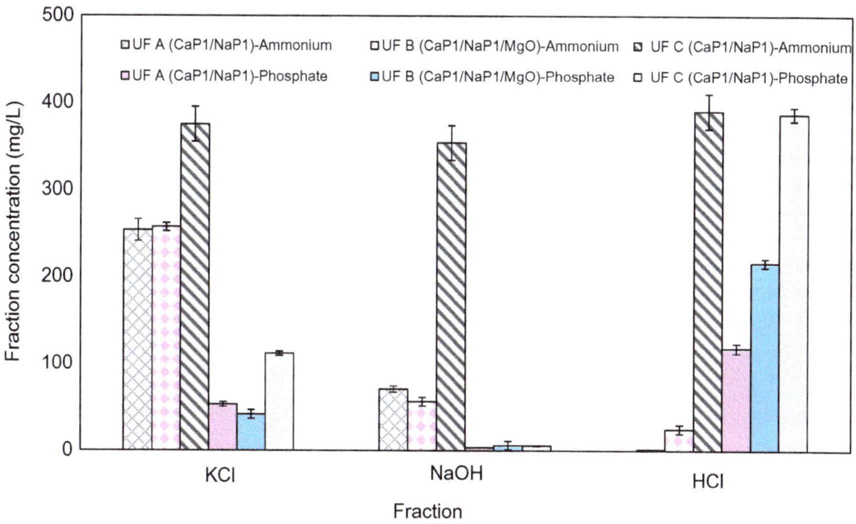

Fig. 32.13 Speciation of the PRSs for the UF experiments for phosphate and ammonium fractions. (Hermassi et al. 2017b)

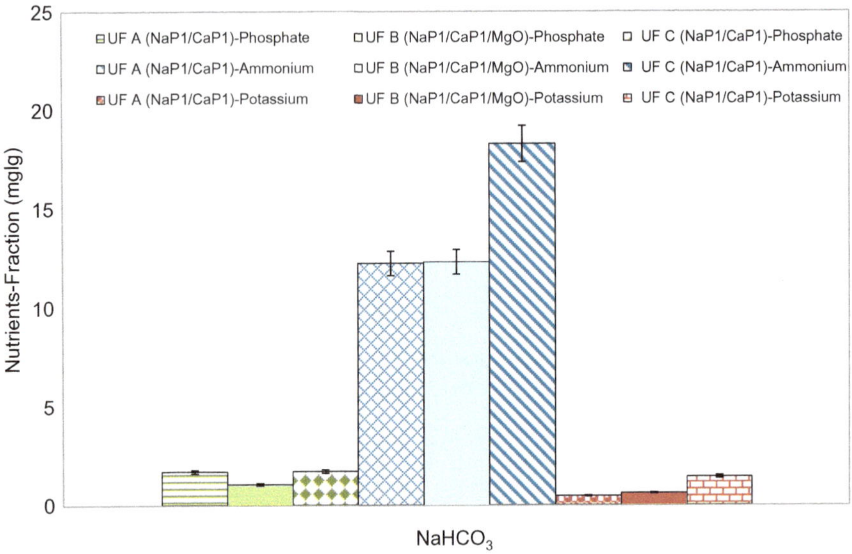

Fig. 32.14 Nutrient (phosphate, ammonium, and potassium) extraction from loaded PRSs using $NaHCO_3$ solutions (0.5 M) (Hermassi et al. 2017b)

Ca-Pi mineral species (Wang et al. 2012; Meis et al. 2012) or magnesium ammonium phosphate (struvite) was 120 mg $P\text{-}PO_4/L$ for the UFA and UFB samples and up to 370 mg $P\text{-}PO_4/L$ for UFC. Additionally, the easily exchangeable speciation (KCl-P) associated with the formation of labile complexes with Al and Fe metal oxides contributed with 41–53 mg $P\text{-}PO_4/L$ for UFB and UFA and 118 mg $P\text{-}PO_4/L$ for the UFC sample.

In the case of NH_4^+, the highest fraction of the three samples was associated with a labile exchange fraction, confirming that the main mechanism in both zeolites was ion exchange. Only in the UFC sample, where struvite was identified by XRD analysis, the fraction with HCl represented an important contribution (350 ± 50 mg $N\text{-}NH_4/L$). The Olsen method using bicarbonate extraction has been applied to predicting the plant availability of P_i in calcareous soil. The P_i-bioavailability data for 0.5 M $NaHCO_3$ solution are plotted in Fig. 32.14 as the extracted amount of P_i per mass of loaded sorbent powder (mg $P\text{-}PO_4/g$).

In the presence of excess bicarbonate, brushite and struvite dissolution occur according to Eqs. (32.17) and (32.18):

$$CaHPO_4(s) + HCO_3^- = CaCO_3(s) + H_2PO_4^- \quad \log K_{298K} = -1.3 \tag{32.17}$$

$$NH_4MgPO_4(s) + HCO_3^- = MgCO_3(s) + HPO_4^{-2} + NH_4^+ \quad \log K_{298K} = -3.1 \tag{32.18}$$

The K-availability data revealed similar ratios from 0.5 to 1.5 mg K/g for UFC. Moreover, the maximum ratio for NH_4^+ was found in the range of 12–18.5 mg NH_4^+/g for UFA-B and UFC. These results are in good agreement with the presence of PRS in NH_4^+ and K^+ form (NaP1/CaP1/N-K) as well as minerals as struvite.

The NH_4^+, P_i, and K^+ sorbed on the PRSs samples were thus demonstrated to dissolve in solutions containing moderate-to-high bicarbonate concentrations, similar to those expected in basic soils characterised by a high calcareous rock content (e.g. limestone) and where other Ca/Mg-P_i minerals such as hydroxyapatite and struvite are insoluble and thus offer limited plant availability.

32.6 Conclusions

The sorption technology using fly ash and its derivative zeolites has been applied to recover single or multiple nutrients from wastewater. The mixture of chemically modified zeolites enables to simultaneously remove NH_4^+, P_i, and K^+ from wastewater. The mechanism underlying the sorption was associated with ion exchange, complexation, and mineral formation on the structure of the sorbents. The presence of hydrated metal oxides as well as Ca^{2+} and Mg^{2+} ions has been shown to play a key role in the recovery of nutrients by sorption. The use of coal combustion fly ash and powder Ca-activated zeolite (CaP1) as sorbents has demonstrated that P_i could be removed mainly as brushite rather than Hap, although Hap is most thermodynamically stable among Ca-P_i precipitates. The brushite formation has an advantage over Hap because of the P_i bioavailability for plant when applied to soil.

The concentration of nutrients varies widely from a few mg/L (e.g. the secondary effluent from conventional activated sludge process) up to the order of g/L (e.g. the side stream of anaerobic sludge digestion). The presence of dissolved organic matter in wastewater can also influence the performance of inorganic sorbents on nutrient removal. Depending on the composition of waste stream to be treated, the blend ratio of sorbents needs to be tuned to optimise the nutrient removal and recovery. The recovered products could be used not only as a fertilising material but also as a soil improver, since the zeolite-based particles have a structure capable of improving the water retention capacity of soil. Future research is needed to scale up the sorption process integrating the membrane filtration technology. In addition, the study on the possibility to transfer toxic substances from nutrient-enriched sorbents to soil is essential to recycle the nutrients for farmland.mland.

Acknowledgement This research was supported by the Waste2Product project (CTM2014-57302-R) financed by the Ministerio de Economía y Competitividad (MINECO) and the Catalan government (Project Ref. 2014SGR50), Spain.

References

Ansari M, Aroujalian A, Raisi A, Dabir B, Fathizadeh M (2014) Preparation and characterization of nano-NaX zeolite by microwave assisted hydrothermal method. Adv Powder Technol 25:722–727

Awual R, Jyo A, El-safty SA, Tamada M, Seko N (2011) A weak-base fibrous anion exchanger effective for rapid phosphate removal from water. J Hazard Mater 188:164–171

Barca C, Gérente C, Meyer D, Chazaren F, Andrès Y (2012) Phosphate removal from synthetic and real wastewater using steel slags produced in Europe. Water Res 46:2376–2348

Benaliouche F, Boucheffa Y, Thibault-starzyk F (2012) Microporous and mesoporous materials in situ FTIR studies of propene adsorption over Ag- and Cu-exchanged Y zeolites. Microporous Mesoporous Mater 147:10–16

Bhattacharjee U, Kandpal TC (2002) Potential of fly ash utilisation in India. Energy 27:151–166

Boyer TH, Persaud A, Banerjee P, Palomino P (2011) Comparison of low-cost and engineered materials for phosphorus removal from organic-rich surface water. Water Res 45:4803–4814

Cama J, Ayora C, Querol X, Ganor J (2005) Dissolution kinetics of synthetic zeolite NaP1 and its implication to zeolite treatment of contaminated waters. Environ Sci Technol 39:4871–4877

Chen J, Kong H, Wu D, Hu Z, Wang Z, Wang Y (2006) Removal of phosphate from aqueous solution by zeolite synthesized from fly ash. J Colloid Interface Sci 300:491–497

Cheung KC, Venkitachalam TH (2000) Improving phosphate removal of sand infiltration system using alkaline fly ash. Chemosphere 41:243–249

De Gisi S, Lofrano G, Grassi M, Notarnicola M (2016) Characteristics and adsorption capacities of low-cost sorbents for wastewater treatment: a review. Sustain Technol J 9:10–40

Demarco MJ, Sengupta AK, Greenleaf JE (2003) Arsenic removal using a polymeric/inorganic hybrid sorbent. Water Res 37:164–176

Evelyn Desmidt BM, Ghyselbrecht K, Zhang Y, Pinoy L, Van der Bruggen B, Verstraete W, Rabaey K (2014) Critical reviews in environmental science and technology global phosphorus scarcity and full-scale P-recovery techniques – a review. Crit Rev Environ Sci Technol 45:336–384

Figueiredo H, Quintelas C (2014) Tailored zeolites for the removal of metal oxyanions : overcoming intrinsic limitations of zeolites. J Hazard Mater 274:287–299

Grubb DG, Guimaraes S, Valencia R (2000) Phosphate immobilization using an acidic type F fly ash. J Hazard Mater 76:217–236

Guaya D, Valderrama C, Farran A, Armijos C, Cortina JL (2015) Simultaneous phosphate and ammonium removal from aqueous solution by a hydrated aluminum oxide modified natural zeolite. Chem Eng J 271:204–213

Guaya D, Hermassi M, Valderrama C, Farran A, Cortina JL (2016) Journal of environmental chemical engineering recovery of ammonium and phosphate from treated urban wastewater by using potassium clinoptilolite impregnated hydrated metal oxides as N-P-K fertilizer. J Environ Chem Eng 4:3519–3526

Guaya D, Valderrama C, Farran A, Cortina JL (2017) Simultaneous nutrients (N,P) removal by using a hybrid inorganic sorbent impregnated with hydrated manganese oxide. J Environ Chem Eng 5:1516–1525

He X, Xu W (2012) Phosphate removal using compounds prepared from paper sludge and fly ash. Environ Earth Sci 70:615–623

Hermassi M, Moreno N, Font O, Querol X, Batis NH, Cortina JL (2016a) Powdered Ca-activated zeolite for phosphate removal from treated waste-water. J Chem Technol Biotechnol 91:1962–1971

Hermassi M, Valderrama C, Gibert O, Moreno N, Font O, Querol X, Batis NH, Cortina JL (2016b) Integration of powdered ca-activated zeolites in a hybrid sorption – membrane ultrafiltration process for phosphate recovery. Ind Eng Chem Res 55:6204–6212

Hermassi M, Valderrama C, Moreno N, Font O, Querol X, Cortina JL, Batis NH (2017a) Fly ash as reactive sorbent for phosphate removal from treated waste water as a potential slow release fertilizer. J Environ Chem Eng 5:160–169

Hermassi M, Valderrama C, Gibert O, Moreno N, Querol X, Batis NH, Cortina JL (2017b) Recovery of nutrients (N-P-K) from potassium-rich sludge anaerobic digestion side-streams by integration of a hybrid sorption-membrane ultrafiltration process: use of powder reactive sorbents as nutrient carriers. Sci Total Environ 599–600:422–430

Hui KS, Chao CYH (2006) Effects of step-change of synthesis temperature on synthesis of zeolite 4A from coal fly ash. Microporous Mesoporous Mater 88:145–151

Hui B, Zhang Y, Ye L (2013) Preparation of PVA hydrogel beads and adsorption mechanism for advanced phosphate removal. Chem Eng J 235:207–214

Ji X, Zhang M, Wang Y, Song Y, Ke Y, Wang Y (2015) Journal of industrial and engineering chemistry immobilization of ammonium and phosphate in aqueous solution by zeolites synthesized from fl y ashes with different compositions. J Ind Eng Chem 22:1–7

Johansson L, Gustafsson JP (2000) Phosphate removal using blast furnace slags and opoka-mechanisms. Water Res 34:259–265

Kumar R, Pal P (2013) Turning hazardous waste into value-added products: production and characterization of struvite from ammoniacal waste with new approaches. J Clean Prod 43:59–70

Ledda C, Schievano A, Salati S, Adani F (2013) Nitrogen and water recovery from animal slurries by a new integrated ultrafiltration, reverse osmosis and cold stripping process: a case study. Water Res 47:6157–6166

Liu X, Ding C (2002) Morphology of apatite formed on surface of wollastonite coating soaked in simulate body fluid. Mater Lett 57:652–655

Liu R, Lal R (2014) Synthetic apatite nanoparticles as a phosphorus fertilizer for soybean (Glycine max). Sci Rep 4:1–6

Lu SG, Bai SQ, Zhu L, Shan HD (2009) Removal mechanism of phosphate from aqueous solution by fly ash. J Hazard Mater J 161:95–101

M YM, Yuasa A, Ariga K (2001) PAC-UF systems I: theory and modeling. Water Res 35:455–463

Ma J, Zhu L (2006) Simultaneous sorption of phosphate and phenanthrene to inorgano – organo-bentonite from water. J Hazard Mater 136:982–988

Meis S, Spears BM, Maberly SC, O'Malley MB, Perkins RG (2012) Sediment amendment with Phoslock® in Clatto reservoir (Dundee, UK): investigating changes in sediment elemental composition and phosphorus fractionation. J Environ Manag 93:185–193

Moreno RJN, Querol X, Andrés JM, Stanton K, Towler M, Nugteren H, Janssen-Jurkovicová M (2005) Physico-chemical characteristics of European pulverized coal combustion fly ashes. Fuel 84:1351–1363

Murayama N, Yamamoto H, Shibata J (2002) Zeolite synthesis from coal fly ash by hydrothermal reaction using various alkali sources. J Chem Technol Biotechnol 77:280–286

Olsen L, Cole SR, Watanabe CV, Dean FS (1954) Estimation of available phosphorus in soils by extraction with sodium bicarbonate. United States, Department of Agriculture, Washington, DC

Pengthamkeerati P, Satapanajaru T, Chularuengoaksorn P (2008) Chemical modification of coal fly ash for the removal of phosphate from aqueous solution. Fuel 87:2469–2476

Penill RP, Bustos AG, Elizalde SG (2006) Immobilization of Cs, Cd, Pb and Cr by synthetic zeolites from Spanish low-calcium coal fly ash. Fuel 85:823–832

Petersen SO, Sommer SG, Béline F, Burton C, Dach J, Dourmad JY, Leip A, Misselbrook T, Nicholson F, Poulsen HD, Provolo G, Sørensen P, Vinneras B, Weiske A, Bernal MP, Böhm R, Juhász C, Mihelic R (2007) Recycling of livestock manure in a whole-farm perspective. Livest Sci 112:180–191

Querol J, Umaña X, Plana F, Alastuey A, Lopez-soler A, Medinaceli A, Valero A, Domingo MJ, Garcia-rojo E (2001) Synthesis of zeolites from ¯ y ash at pilot plant scale. Examples of potential applications. Fuel 80:857–865

Querol X, Moreno N, Alastuey A, Juan R, Ayora C, Medinaceli A, Valero A, Productos C (2007) Synthesis of high ion exchange zeolites from coal fly ash. Geol Acta 5:49–57

Safa Jaouadi SJ, Wahab MA, Anane M, Bousselmi L (2014) Powdered marble wastes reuse as a low-cost material for phosphorus removal from aqueous solutions under dynamic conditions. Desalin Water Treat 52:1705–1715

Scholz RW, Ulrich AE, Eilittä M, Roy A (2013) Science of the total environment sustainable use of phosphorus: a finite resource. Sci Total Environ 462:799–803

Taddeo R, Kolppo K, Lepist R (2016) Sustainable nutrients recovery and recycling by optimizing the chemical addition sequence for struvite precipitation from raw swine slurries. J Environ Manag 180:52–58

Tsitouridou R, Georgiou J (1988) A contribution to the study of phosphate sorption by three Greek fly ashes. Toxicol Environ Chem 17:129–138

Wang C, Qi Y, Pei Y (2012) Laboratory investigation of phosphorus immobilization in lake sediments using water treatment residuals. Chem Eng J 209:379–385

Wang H, Qu F, Ding A, Liang H, Jia R, Li K, Bai L, Chang H (2016) Combined effects of PAC adsorption and in situ chlorination on membrane fouling in a pilot-scale coagulation and ultrafiltration process. Chem Eng J 283:1374–1383

Watanabe Y, Ikoma T, Yamada H, Stevens GW, Moriyoshi Y, Tanaka J, Komatsu Y (2010) Formation of hydroxyapatite nanocrystals on the surface of Ca–Al-layered double hydroxide. J Am Ceram Soc 93:1195–1200

Watanabe Y, Yamada H, Ikoma T, Tanaka J, Stevens GW, Komatsu Y (2014) Preparation of a zeolite NaP1/hydroxyapatite nanocomposite and study of its behavior as inorganic fertilizer. J Chem Technol Biotechnol 89:963–968

Wu D, Zhang B, Li C, Zhang Z, Kong H (2006) Simultaneous removal of ammonium and phosphate by zeolite synthesized from fly ash as influenced by salt treatment. J Colloid Interface Sci 304:300–306

Xie J, Li C, Chi L, Wu D (2013) Chitosan modified zeolite as a versatile adsorbent for the removal of different pollutants from water. Fuel 103:480–485

Yildiz E (2004) Phosphate removal from water by fly ash using crossflow microfiltration. Sep Purif Technol 35:241–252

Yin H, Kong M (2014) Simultaneous removal of ammonium and phosphate from eutrophic waters using natural calcium-rich attapulgite-based versatile adsorbent. Desalination 351:128–137

Zelmanov G, Semiat R (2014) Phosphate removal from aqueous solution by an adsorption ultrafiltration system. Sep Purif Technol 132:487–495

Zhou L, Chen Y, Zhang X, Tian F, Zu Z (2014) Zeolites developed from mixed alkali modi fi ed coal fl y ash for adsorption of volatile organic compounds. Mater Lett 119:140–142

Zhu X, Jyo A (2005) Column-mode phosphate removal by a novel highly selective adsorbent. Water Res 39:2301–2308

Chapter 33
Phosphate Separation from Aqueous Solution Using a Chitosan-Based Biodegradable Ion Exchanger

Jae-Woo Choi and Sang-Hyup Lee

Abstract To remove phosphate (P_i) from wastewater, a novel, low-cost adsorbent was developed by immobilizing copper ions (Cu^{2+}) on a naturally available biopolymer chitosan. The copper ions bound to chitosan through its complex formation with the amino and hydroxyl groups in the polymer chain. This complex formation could reduce the size of the hydrogels and thus increase the density of the natural biopolymer. The chelating interaction between nitrogen and copper enabled the intra- and inter-molecular cross-linking, thereby improving the physical and chemical stability of the polymeric ligand exchanger. Most importantly, Cu^{2+} ions could serve as the active functional group for P_i removal from aqueous solution. The polymeric ligand exchanger displayed a greater affinity with P_i rather than sulfate due to the electrostatic and Lewis acid/base interactions between immobilized Cu^{2+} and P_i, regardless of the solution pH. Batch adsorption experiments showed that the polymeric ligand exchanger had the maximum capacity of approximately 85 mg/g in P_i solutions. The polymeric ligand exchanger could also be applied to a fixed-bed column reactor, demonstrating the high performance on the P_i removal from aqueous solutions.

Keywords Phosphate removal · Polymeric ligand exchanger · Chitosan · Cross-linking agent · Adsorption

---

J.-W. Choi
Center for Water Resource Cycle Research, Korea Institute of Science and Technology, Seoul, Republic of Korea

Division of Energy and Environmental Engineering, KIST School, Korea University of Science and Technology, Seoul, Republic of Korea

S.-H. Lee (✉)
Green School, Korea University, Seoul, Republic of Korea

Center for Water Resource Cycle Research, Korea Institute of Science and Technology, Seoul, Republic of Korea
e-mail: yisanghyup@kist.re.kr

H. Ohtake, S. Tsuneda (eds.), *Phosphorus Recovery and Recycling*,
https://doi.org/10.1007/978-981-10-8031-9_33

33.1 Introduction

Phosphorus (P) is an essential nutrient for plant growth and cell functionality and widely used as fertilizer and feed additives in agriculture (Kataki et al. 2016; Kolodziej et al. 2015; Heckenmuller et al. 2014). Although P is an irreplaceable element for food production, the inefficient use of P often causes water pollution problems (known as eutrophication problems) in natural bodies of water (Wahab et al. 2011). To prevent the eutrophication in natural bodies of water, the US Environmental Protection Agency has established the strict maximum contaminant level for phosphate (P_i) in surface waters below 0.02 mg P/L (USEPA 2000). South Korea also regulates the effluent P_i concentration in wastewater between 0.2 and 2 mg/L (Nam et al. 2013) and is expected to further enforce the regulation level in the future. Therefore, increasing attention has been paid to the removal of P_i from wastewater for the control of eutrophication in Korea.

P_i can be present in four different species, H_3PO_4, $H_2PO_4^-$, HPO_4^{2-}, and PO_4^{3-}, depending on the solution pH. The dominant phosphoric ion species are $H_2PO_4^-$ and HPO_4^{2-} at the pH range of 6–8 (An et al. 2014b). Up to now, various technologies have been developed for removing P_i from wastewater (Acelas et al. 2015). Advanced physical technologies such as membrane separation, UV radiation, and electrocoagulation often require expensive apparatus and high technical skills for removing P_i from wastewater. Chemical technologies are mostly based on precipitation reactions between P_i and chemicals such as ferric chloride ($FeCl_2$), alum ($M{\cdot}Al(SO_4)_2{\cdot}12H_2O$), and lime ($CaO$ or $Ca(OH)_2$). They consume a lot of chemicals and inevitably increase the volume of sludge, thereby increasing the cost of chemicals as well as the disposal of sludge (Acelas et al. 2015; Guan et al. 2014; Loganathan et al. 2014). P_i can also be removed from wastewater using microbial activities known as enhanced biological P removal (Li et al. 2015). However, the application of biological P removal has been still limited to large-scale wastewater treatment plants (WWTP) in major metropolitan areas, particularly in Asia (Li et al. 2015).

Meanwhile, simple adsorption technologies have been developed using various adsorbents such as activated carbon, metal hydroxides, and calcium/magnesium carbonates (Loganathan et al. 2014; Han et al. 2012; Lee et al. 2012). The adsorption technologies have the potential to contribute to the efficient P_i removal from wastewater in the light of their cost-effectiveness, operational simplicity, and reusability (Choi et al. 2014). However, for the expanded use of these technologies in a full-scale wastewater treatment process, it is necessary to solve technical issues relevant to the manipulation, replacement, and management at a full scale.

Natural biopolymers such as alginate and chitosan have been investigated for pharmaceutical purposes due to their harmlessness to human health, natural abundance, high hydrophilicity, and antibacterial properties. Recently, applications of biopolymers have been expanded to the water purification and wastewater treatment sectors due to their efficiency, economic feasibility, and non-toxicity to humans (Chen and Chung 2006; Guibal et al. 1999). Chitosan has a variety of functional groups including amino, hydroxyl, and acetamine groups. It is an effective chelating polymer for capturing transition metals such as copper, nickel, and zinc. The free amino groups of chitosan have nitrogen atoms which can exhibit extremely high selectivity for transition metals through the coordination bonding (the Lewis acid/base reaction) (Kumar et al. 2009). The metal sorption capacity of chitosan is known to be determined by the p*K*a value, the degree of deacetylation, and its molecular weight, all of which are changed with deacetylation (An et al. 2014a).

Chitosan can be modified to generate different polymer forms such as powder, nanoparticle, gel bead, and fiber, depending on the specific application (An et al. 2014b). For instance, the bead-form chitosan is applicable to a fixed-bed reactor. The use of chitosan in the form of gel beads can reduce the resistance to mass transfer in the near-biopolymer region and ease the hydrodynamic limitations of using raw and particulate chitosan. The remarkable property of chitosan as a natural polymer is relevant to its solubility which is determined by the solution pH. Chitosan maintains its important physico-chemical characteristics at a neutral or basic pH and dissolves in water upon protonation of the amino group (Kildeeva et al. 2009). However, chitosan remains chemically unchanged after being prepared under acidic conditions.

A cross-linking process has been applied for chitosan-based adsorbents during linking of the chitosan polymer using the dialdehyde species such as epichlorohydrin and glyoxal (Miretzky and Cirelli 2009). This process hinders solubilization of chitosan and enables the use of chitosan as an adsorbent during water treatment. Furthermore, the cross-linked chitosan exhibits the enhanced adsorption capacity for transition metal ions (Koyama and Taniguchi 1986). Therefore, chitosan beads hold the potential to replace commercial chelating resins as a novel polymer template if the protonation of amine groups can be properly controlled.

In this chapter, we describe the development of a new chitosan-based ion exchanger for removing P_i from aqueous solutions. The study presents the development and characterization of Cu^{2+}-immobilizing chitosan and evaluates its capacity for P_i removal.

33.2 Mechanism Underlying P_i Removal

33.2.1 Adsorption

Adsorption is the adhesion of atoms, ions, or molecules from a gas, liquid, or dissolved solid to a surface. Adsorption of dissolved substances is possible due to either physical or chemical attraction. This technique is widely used for environmental applications because contaminants can be selectively removed, even if the contaminant concentration is extremely low in wastewater. The key factor controlling the degree of adsorption for a contaminant is the surface area of adsorbents. The solidity of adsorption tends to vary depending on the properties of the adsorbent and relevant contaminants. Adsorption can occur on the liquid-solid interface where the concentration gradient of a contaminant exists. In this chapter, the chitosan-based polymer is an adsorbent, while P_i is the target contaminant to be removed.

33.2.2 Ion Exchange

Strictly speaking, ion exchange and adsorption are different concepts. However, this technique is similar to the adsorption technique in the on-site application of contaminant treatment. Ion exchange is a phenomenon that exhibits reversible substitution reactions between ionic solid materials and ions in a solution. A solid material with a reversible substitution function is referred to as an ion exchanger. Ion exchangers are classified into inorganic and organic forms, most of which are based on a chemically synthesized resin. They are mostly applied in a fixed-bed reactor for the treatment of wastewater. In this case, the regeneration process is required when the ion-exchange capacity is filled with the target contaminant. In this chapter, copper-chelated chitosan beads can serve as an ion-exchange resin in a fixed-bed reactor.

33.3 Preparation of Copper-Chelated Chitosan Beads

33.3.1 Materials

Chitosan powder was obtained from Sigma-Aldrich and used without further purification. It had a medium molecular weight (250,000 g/mol) and a 75–80% degree of deacetylation from chitin. Copper chloride ($CuCl_2 \cdot H_2O$) and glutaraldehyde solution were purchased from SHOWA (Tokyo, Japan) and Sigma-Aldrich (Shanghai, China). All chemicals including KH_2PO_4, Na_2SO_4, $NaNO_3$, HCl, NaOH, and methanol were ACS grade and used without purification. All solutions were prepared with ultrapure, deionized water.

33.3.2 Preparation of Chitosan Beads

Chitosan bead preparation was conducted in two stages (Fig. 33.1): chitosan powder was dissolved in acidic solution, which was then dropped into an alkaline phase (Dai et al. 2011). 2.5% (w/w) chitosan solution was prepared by dissolving 5 g of chitosan powder in 200 ml of 1% (v/v) HCl for 12 h at ambient temperature while stirring at 100 rpm. The chitosan solution was added into 200 ml of 1 M NaOH while stirring at 120 rpm. The NaOH solution containing the chitosan beads was mixed for an additional 4 h to complete the gelation of the chitosan beads. Then, the beads were washed with deionized water until a pH of approximately 7.0–8.0 was reached (Fig. 33.1).

Copper-loaded chitosan beads were prepared using the following procedure. Washed chitosan beads were loaded with copper ion by equilibrating the beads with

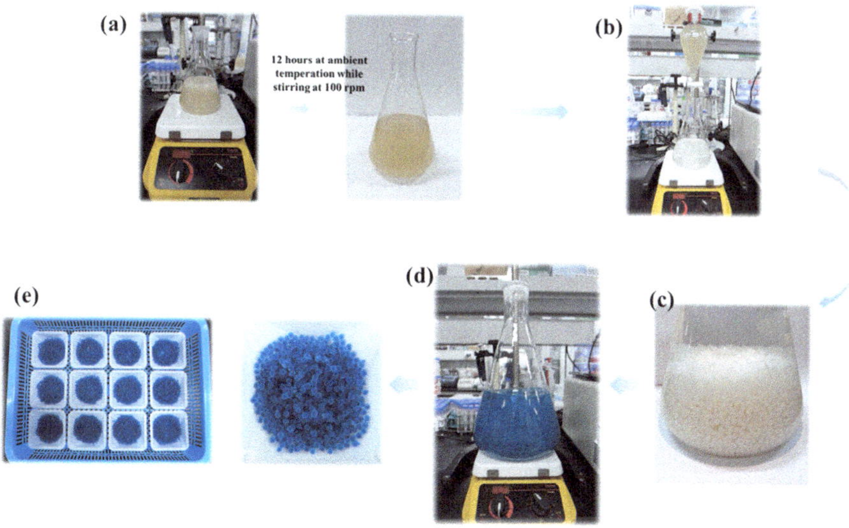

Fig. 33.1 Synthesis process of copper-loaded chitosan bead: (**a**) the preparation of 2.5% chitosan solution, (**b**) NaOH addition for generating chitosan bead, (**c**) washing with deionized water, (**d**) copper loading, and (**e**) the air-drying of chitosan beads

400 mL of a copper ion solution at 100–5000 mg/L for 24 h at pH 4–4.5. After the loading step, they were rinsed with deionized water several times. If the copper-loaded chitosan beads were not used immediately, they were stored in deionized water to prevent decomposition. Similarly, the chitosan beads cross-linked with glutaraldehyde were prepared using the same process, except that 0.1 M glutaraldehyde solution was used for cross-linking after washing with deionized water.

33.3.3 Characterization of Chitosan Beads

Compressive tests were conducted to check the physical strength of beads via a universal testing machine (WL2111, WITHLAB, Korea) operated at a 1.3 mm/min velocity pressing on the 200 N loading cell. X-ray diffraction (XRD) of chitosan powder and hydrogel before and after copper loading was investigated using a D8 advance with an LYNXEYE detector at 40 kV and 40 mA (1.6 kW) (Bruker, USA). In addition, Fourier-transform infrared spectroscopy (FTIR; Spotlight 200, PerkinElmer, USA), for a spectra range from 4000 to 380/cm, was used to assess the chemical forms of chitosan, copper ion, and glutaraldehyde before and after P_i removal.

33.3.4 P_i Adsorption Tests

Adsorption isotherm experiments were conducted by varying the initial concentration of P_i from 0 to 100 mg/L and using 0.05 mg of adsorbents in 50 mL conical tubes. These tests were conducted in P_i solution with or without adding ions such as sulfate and nitrate. The mixture was rotated at 80 rpm for 24 h, during which it reached equilibrium state, and then the concentrations of the target anions in the aqueous solution were measured. The amount of P_i removed was calculated using the following mass balance equation:

$$q_e = \frac{V\left(C_0 - C_e\right)}{M} \tag{33.1}$$

where q_e is the equilibrium mass removal of P_i (mg/g), V is the solution volume (L), C_0 and C_e are the initial and final concentrations of P_i in the mixture solution (mg/L), respectively, and M is the mass of the added adsorbent (g).

Additional batch tests were conducted to investigate the effect of coexisting ions on the removal of P_i. A 200-mL solution with varied concentrations of nitrate, P_i, and sulfate was prepared with 0.25 g of each adsorbent. The mixtures were rotated at 150 rpm on a platform shaker (JSSI-100C, Republic of Korea) under the same conditions as employed for equilibrium sorption tests.

In addition, fixed-bed column experiments were performed to assess the breakthrough behavior of P_i as well as nitrate and sulfate for copper-loaded chitosan beads. An acrylic column (30 mm in diameter and 150 mm in length) filled with copper-loaded chitosan beads was used for this testing. The flow rate was constantly maintained at 1 mL/min. The influent water contained 5.0 mg/L of P_i, 50 mg/L of sulfate, 50 mg/L of nitrate, and 100 mg/L of biocarbonate at an initial pH of 7.5 and was passed through in upward flow mode.

33.3.5 Effects of pH

As described above, the chitosan powder can be dissolved in acidic solutions. The amino group in chitosan undergoes protonation under acidic conditions, and P_i species also depend on solution pH (Snoeyink and Jenkins 1980). Therefore, the effect of solution pH on the P_i adsorption capacity was investigated by varying the solution pH from 3 to 9. Initial pH was adjusted using dilute HCl and NaOH solutions. Each sample was rotated at 80 rpm for 24 h. Then, the concentrations of copper, P_i, and TOC were determined.

33.3.6 Chemical Analyses

The concentrations of P_i, nitrate, sulfate, and chloride were analyzed with an ion chromatographer (ICS-1000, USA). The dissolved copper concentration was measured using inductively coupled plasma (ICP) spectroscopy (NexION 300D, USA). Solution pH was measured with an ORION Star A211 (Thermo Scientific, USA). Total organic carbon was analyzed to assess the chitosan concentration in solution by a Shimadzu TOC-V analyzer equipped with an ASI-5000A auto-sampler (Shimadzu, Japan).

33.4 P_i Removal Capacity

33.4.1 Characteristics of Copper-Loaded Chitosan Beads

To investigate the effects of copper ion concentrations on the physical and chemical properties of copper-loaded chitosan beads, four different initial copper ion concentrations in the range of 100–5000 mg/L were applied for copper loading to chitosan beads. Size and shape changes due to the copper loading are shown in Fig. 33.2. The results revealed no change in the shape of spherical beads with varying copper concentrations. However, the bead size decreased from 4.8 to 3.0 mm with increasing the copper concentration, although the bead swelled more or less in the 100-mg/L copper loading solution. In addition, Fig. 33.2 also shows the change of bead color after being soaked with copper solution. The content of copper in chitosan beads varied from approximately 5 to 53 mg/g when being equilibrated with the copper solutions of 100–5000 mg/L.

Chitosan has attracted considerable attention in various fields of research due to its unique characteristics. However, it has some limitations regarding full-scale applications. Specifically, issues relevant to the low mechanical strength and water solubility of chitosan under acidic conditions need to be addressed. The possibility

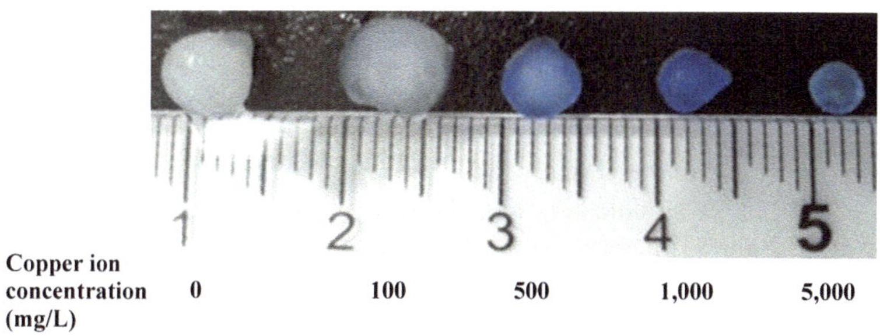

Fig. 33.2 Images showing the change of size and shape of copper-loaded chitosan bead with varying the copper concentrations. (An et al. 2014a)

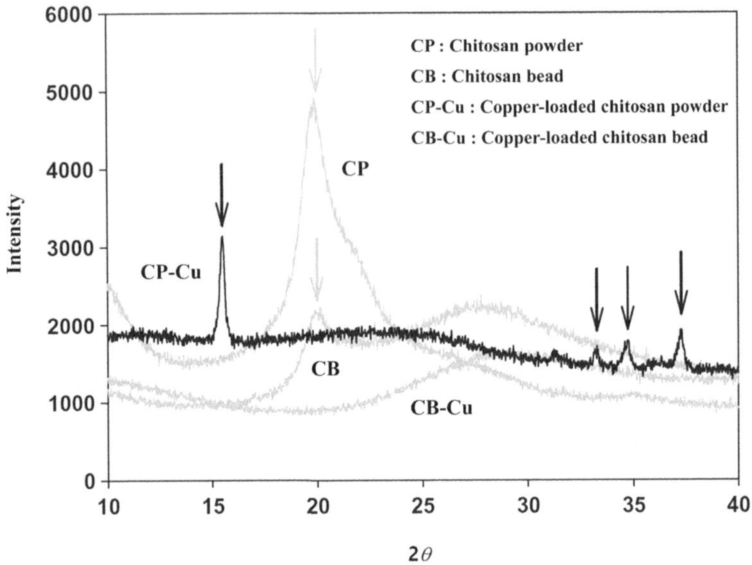

Fig. 33.3 XRD patterns of chitosan powder, chitosan beads, and copper-loaded chitosan beads (5000 mg/L of copper). (An et al. 2014a)

of improving the mechanical strength of chitosan beads by loading copper ions on the bead surface was investigated based on the ultimate tensile strength. Copper-free and copper-loaded chitosan beads showed a threshold of 1 and 11 kgf with a SD of 0.4 and 0.2, respectively. This result shows that the copper-loaded chitosan beads have the mechanical strength 12 times larger than that of copper-free chitosan beads. Cross-linking agents such as glutaraldehyde were also tested for their ability to solve the chemical stability problem. Although they could improve the chemical stability of chitosan beads, they decreased the number of free amino groups which are needed for P_i removal.

The equivalent interaction between copper and nitrogen improved the density of beads as well as the intensification and contraction of the internal structure of the chitosan bead. Figure 33.3 shows the XRD patterns characterized by scattering angles for chitosan and copper-loaded chitosan beads (An et al. 2014a). The peaks for chitosan powder and beads appeared near $2\theta = 20°$, and the intensity of the powder was higher than that of the beads. However, the peak of $2\theta = 20°$ disappeared for copper-loaded chitosan powder and beads after allowing the chelating interaction between copper and nitrogen. The peaks for copper-loaded chitosan beads appeared at 15, 33, 34, and 37, indicating that the copper crystal form is CuOH or CuO (Xiang et al. 2008). This result implies that copper is present in a molecular form and that the enhancement of mechanical strength is due to the increase in density, not crystallization.

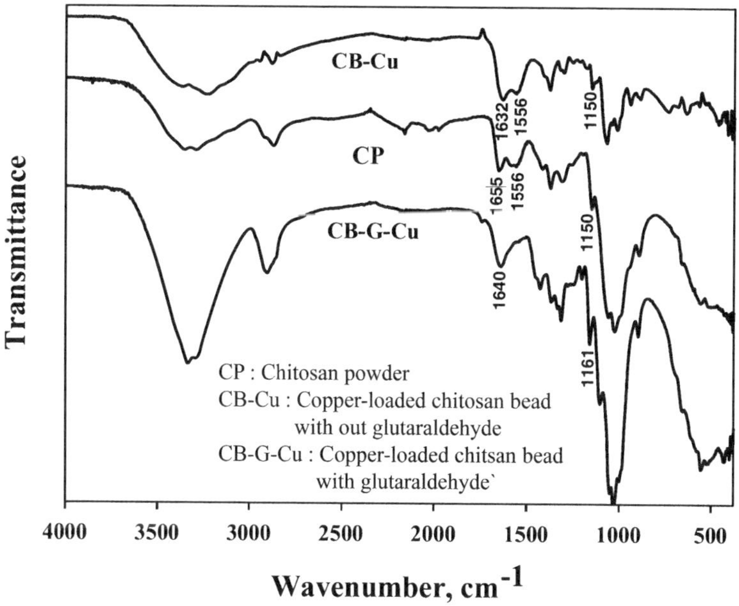

Fig. 33.4 FTIR spectra of the chitosan powder and copper-loaded chitosan beads with and without glutaraldehyde. (An et al. 2014b)

FTIR was used to analyze the chemical bonding between copper, chitosan, and glutaraldehyde. Figure 33.4 shows the FTIR spectra for raw chitosan powder and two dried copper-loaded chitosan beads with and without glutaraldehyde. The peak at 1556 cm^{-1} observed for the raw chitosan powder and copper-loaded chitosan beads indicates the stretching vibration of the amine group. It was not observed with glutaraldehyde because glutaraldehyde hampered the availability of the amine function. The peak of the O-H bond shifted from 1651 to 1629 cm^{-1} because of increasing copper complexation with water molecules (An and Zhao 2012). In addition, the saccharide structure corresponded to the asymmetric C-O-C stretching and the skeletal CO stretching vibration at 1030 and 1155 cm^{-1}, respectively.

33.4.2 Adsorption Capacity

The P_i adsorption experiments for copper-free and copper-loaded chitosan beads were conducted at pH 7 ± 0.5. Adsorption isotherms provide important information for designing the sorption process in a full-scale plant (Lee et al. 2015). The Langmuir

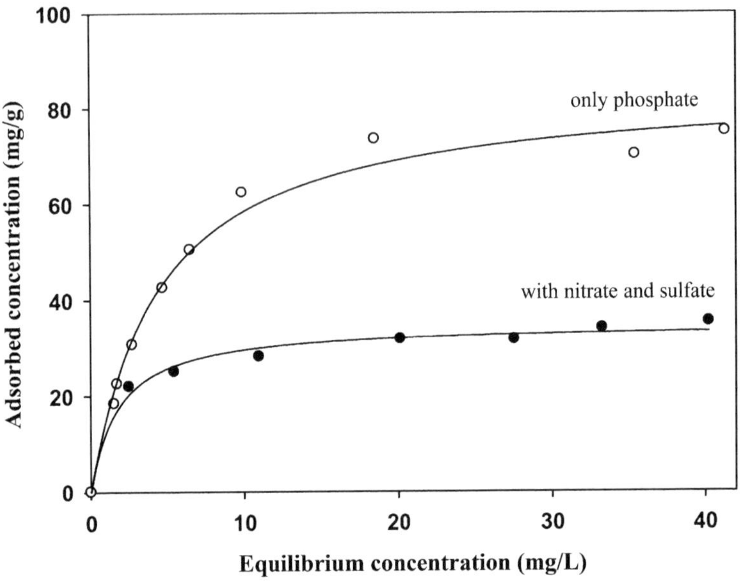

Fig. 33.5 Adsorption equilibrium isotherm of phosphate for copper-loaded chitosan beads in the absence or presence of 100 mg/L of nitrate and sulfate. (An et al. 2014a)

adsorption model was employed for the sorption data analysis. The Langmuir isotherm model was employed to interpret experimental data using Eq. 33.2:

$$Q_e = \frac{Q_{\max} b C_e}{1 + bC_e} \tag{33.2}$$

where Q_e is the equilibrium P_i uptake (mg/g), C_e is the equilibrium concentration of phosphate in solution (mg/L), $Q_{\max}$ is the maximum adsorption capacity (mg/g), and b is the Langmuir affinity coefficient of the binding sites (L/mg). The maximum adsorption capacity of P_i in the aqueous solutions with and without nitrate and sulfate ions was 85 mg/g and 34 mg/g, respectively (Fig. 33.5). The Langmuir affinity coefficient predicted that this adsorption system is favorable.

Then, fixed-bed column experiments were conducted to investigate the applicability of copper-loaded chitosan beads for adsorption of anions such as P_i, nitrate, and sulfate. Figure 33.6 shows that the concentration of chloride increased soon after the start of experiments and then decreased to 0 until a bed volume of 250. This change of chloride concentration occurred because the copper-loaded chitosan beads were prepared in the form of chloride as a counter ion, which is easily exchangeable by P_i, nitrate, and sulfate (An et al. 2014a). The breakthrough of nitrate occurred first, followed by sulfate. The effluent concentrations of nitrate and sulfate reached the same level as their influent concentrations at bed volumes of 45

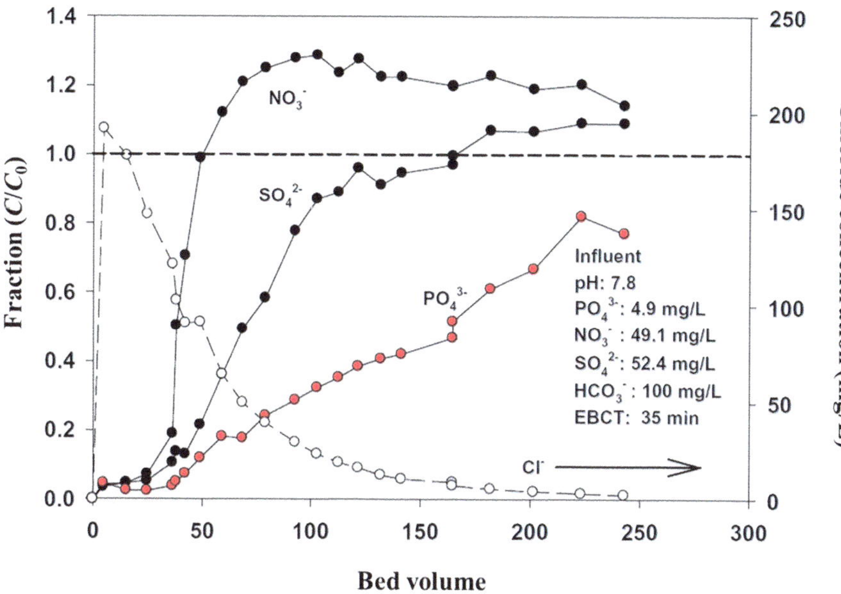

Fig. 33.6 Breakthrough profiles of P_i, nitrate, sulfate, and chloride using copper-loaded chitosan beads through a fixed-bed column. (An et al. 2014a)

and 150, respectively. The breakthrough curve of P_i occurred at a bed volume of 40. The effluent concentration of Pi increased linearly up to the level of the influent concentration after a bed volume of 250. The breakthrough sequence occurred as follows: P_i > sulfate > nitrate > chloride. Commonly, when a commercial anion exchanger is used to remove P_i with competing ions, the following affinity sequence is obtained: sulfate > phosphate > nitrate > chloride (Modrzejewska 2013).

33.4.3 pH Effect

A batch experiment was conducted to confirm the changing pattern of P_i removal efficiency with varying the solution pH, which is one of the most influential factors in P_i removal. In Fig. 33.7, the P_i removal efficiency is shown for copper-loaded chitosan beads with and without glutaraldehyde in the pH range from 4 to 9 where no significant elution of copper occurred. The P_i removal efficiency for copper-loaded chitosan beads was 20 mg/g higher in the absence of glutaraldehyde than that detected in the presence of glutaraldehyde at pH 4.0. As the pH value increased, this gap has widened to 30 mg/g. This result indicates that copper-loaded chitosan beads with glutaraldehyde were more pH-sensitive for P_i adsorption. pH 7 was optimal for P_i removal by the anion exchanger, because P_i has a p*K*a value of 7.2.

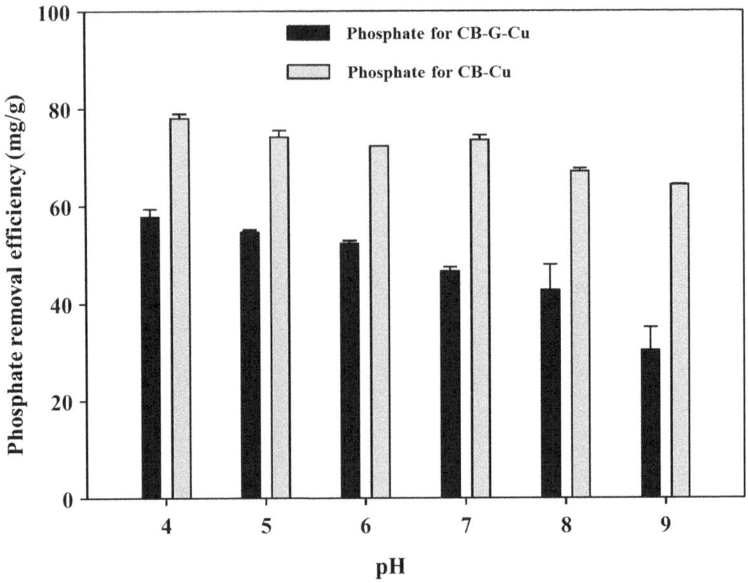

Fig. 33.7 P_i removal efficiency as a function of pH for copper-loaded chitosan beads with and without glutaraldehyde (initial phosphate concentration = 100 mg/L). (An et al. 2014b).

33.5 Conclusions

In this chapter, P_i removal by copper-loaded chitosan beads was examined. The immobilization of Cu^{2+} in the biopolymer could enhance the physical and chemical stability of chitosan. Experimental results indicated that the polymeric adsorbent based on natural chitosan was effective in the removal of P_i with a maximum adsorption capacity of 85 mg/g. Additionally, this adsorbent could exhibit the high selectivity for P_i even at pH 5. The optimal pH range was from 5 to 7. Batch and column tests confirmed the high selectivity of copper-loaded chitosan beads for P_i against sulfate, nitrate, and chloride. Fixed-bed column experiments also revealed a breakthrough sequence of P_i > sulfate > nitrate > chloride. This preference is significantly different from those observed with commercial anion-exchange resins. The copper-immobilizing chitosan beads are likely to serve as an alternative, promising, and "greener" adsorbent for the selective removal of P_i from wastewater.

References

Acelas NY, Florez E, Lopez D (2015) Phosphorus recovery through struvite precipitation from wastewater: effect of the competitive ions. Desalin Water Treat 54(9):2468–2479

An B, Jung KY, Zhao D, Lee SH, Choi JW (2014a) Preparation and characterization of polymeric ligand exchanger based on chitosan hydrogel for selective removal of phosphate. React Funct Polym 85:45–53

An B, Jung KY, Lee SH, Lee S, Choi JW (2014b) Effective phosphate removal from synthesized wastewater using copper-chitosan bead: batch and fixed-bed column studies. Water Air Soil Pollut 225:2050

An B, Zhao DY (2012) Immobilization of As(III) in soil and groundwater using a new class of polysaccharide stabilized Fe-Mn oxide nanoparticles. J Hazard Mater 211:332–341

Chen CC, Chung YC (2006) Arsenic removal using a biopolymer chitosan sorbent. J Environ Sci Health A 41:645–658

Choi JW, Kwon KS, Lee S, An B, Hong SW, Lee SH (2014) Pilot-scale test for a phosphate treatment using sulfate-coated zeolite at a sewage disposal facility. Water Air Soil Pollut 225:1835

Dai J, Yang H, Yan H, Shangguan YG, Zheng QA, Cheng RS (2011) Phosphate adsorption from aqueous solutions by disused adsorbents: chitosan hydrogel beads after the removal of copper(II). Chem Eng J 166:970–977

Guan W, Ji FY, Chen QK, Yan P, Zhou WW (2014) Phosphorus recovery using porous calcium silicate hydrate as seed crystal in form of hydroxyapatite. Mater Res Innov 18(1):43–49

Guibal E, Milot C, Eterradossi O, Gauffier C, Domard A (1999) Study of molybdate ion sorption on chitosan gel beads by different spectrometric analyses. Int J Biol Macromol 24:49–59

Han YU, Lee CG, Park JA, Kang JK, Lee I, Kim SB (2012) Immobilization of layered double hydroxide into polyvinyl alcohol/alginate hydrogel beads for phosphate removal. Environ Eng Res 17:133–138

Heckenmuller M, Narita D, Kleppere G (2014) Global availability of phosphorus and its implications for global food supply: an economic overview, Kiel working paper no. 1897. Kiel Institute for the World Economy, Kiel

Kataki S, West H, Clarke M, Baruah DC (2016) Phosphorus recovery as struvite: recent concerns for use of seed, alternative Mg source, nitrogen conservation and fertilizer potential. Resour Conserv Recycl 107:142–156

Kildeeva NR, Perminov PA, Vladimirov LV, Novikov VV, Mikhailov SN (2009) About mechanism of chitosan cross-linking with glutaraldehyde. Russ J Bioorg Chem 35:360–369

Kolodziej B, Sugier D, Kwiatkowski C (2015) Phosphorus fertilization and inflorescence removal in American ginseng culture. Ind Crop Prod 74:619–627

Koyama Y, Taniguchi A (1986) Studies on chitin X homogeneous cross-linking of chitosan for enhanced cupric ion adsorption. J Appl Polym Sci 31:1951–1954

Kumar M, Tripathi BP, Shahi VK (2009) Crosslinked chitosan/polyvinyl alcohol blend beads for removal and recovery of Cd(II) from wastewater. J Hazard Mater 172(2–3):1041–1048

Lee CG, Jeon JW, Hwang MJ, Ahn KH, Park C, Choi JW, Lee SH (2015) Lead and copper removal from aqueous solutions using carbon foam derived from phenol resin. Chemosphere 130:59–65

Lee CG, Park JA, Kim SB (2012) Phosphate removal from aqueous solutions using slag microspheres. Desalin Water Treat 44:229–236

Li H, Yang Q, Li J, Gao H, Li P, Zhou H (2015) The impact of temperature on microbial diversity and AOA activity in the Tengchong geothermal field, China. Sci Rep 5:17056

Loganathan P, Vigneswaran S, Kandasamy J, Bolan NS (2014) Removal and recovery of phosphate from water using sorption. Crit Rev Environ Sci Technol 44(8):847–907

Miretzky P, Cirelli AF (2009) Hg(II) removal from water by chitosan and chitosan derivatives: a review. J Hazard Mater 167:10–23

Modrzejewska Z (2013) Sorption mechanism of copper in chitosan hydrogel. React Funct Polym 73:719–729

Nam JH, Lee SH, Choi JW, Hong SW, An B (2013) Enhanced removal of phosphate on modified ion exchanger with competing ion. J Kor Soc Water Wastewater 27:121–128

Snoeyink VL, Jenkins D (1980) Water chemistry. Wiley, New York

USEPA (2000) Nutrient criteria technical guidance manual: rivers and streams, EPA-822b-00-002. U.S. Environmental Protection Agency, Washington, DC

Wahab MA, Hassine RB, Jellai S (2011) Removal of phosphorus from aqueous solution by Posidonia oceanica fibers using continuous stirring tank reactor. J Hazard Mater 189:577–585

Xiang JY, Tu JP, Huang XH, Yang YZ (2008) A comparison of anodically grown CuO nanotube film and Cu_2O film as anodes for lithium ion batteries. J Solid State Electrochem 12(7–8):941–945

Part VII
Biotechnology

Chapter 34
Biological Phosphite Oxidation and Its Application to Phosphorus Recycling

Ryuichi Hirota, Kei Motomura, and Akio Kuroda

Abstract Several chemical industrial processes produce phosphite (Pt) as a by-product. This Pt waste should be recycled and reused as an alternative phosphorus (P) source to reduce the demand for nonrenewable phosphate rock reserves. Nearly all organisms require inorganic phosphate (Pi) or its esters as their P source. Therefore, Pt has been considered a biologically inert P compound, hampering the development of biotechnologies that utilize the Pt waste. During the last decade, the molecular mechanisms involved in the metabolism of inorganic reduced P compounds, including Pt, have been elucidated. Pt dehydrogenase (PtxD) catalyzes the oxidation of Pt to Pi, with a concomitant reduction of NAD^+ to NADH, and thus is a promising biocatalyst for developing Pt-based applications. The initial discovery of PtxD, followed by the finding and development of PtxD enzymes with high catalytic activity and thermostability, facilitated the development of several unique biotechnological applications. These applications include (i) a dominant selective cultivation system for microorganisms and plants, (ii) a biological containment strategy for the safe use of genetically modified organisms, and (iii) a cofactor regeneration system for efficient production of chiral compounds by dehydrogenases. This section describes the emerging biotechnology applications that should contribute to the utilization of Pt as a valuable chemical.

Keywords Phosphite · Phosphite dehydrogenase · Contamination · Industrial fermentation · Engineered metabolic pathway · Cofactor regeneration · Biocontainment

---

R. Hirota (✉) · K. Motomura · A. Kuroda
Department of Molecular Biotechnology, Graduate School of Advanced Sciences of Matter, Hiroshima University, Higashi-Hiroshima, Hiroshima, Japan
e-mail: hirota@hiroshima-u.ac.jp

H. Ohtake, S. Tsuneda (eds.), *Phosphorus Recovery and Recycling*,
https://doi.org/10.1007/978-981-10-8031-9_34

34.1 Introduction

The growing consumption of phosphorus (P) fertilizers has contributed to major increases in food production in the twentieth century. However, minable P resources are being depleted at an alarming rate. At current consumption levels, easily accessible P reserves will be depleted in around 80 years (Cordell et al. 2011). However, due to the rising food demand driven by an increasing world population, consumption will not stay at current levels. Ensuring long-term availability and accessibility of P resources is critical to the future of humanity (Cordell et al. 2011). We need to use less P, use it more efficiently, and recycle the currently untapped P waste. To realize a more sustainable P resource management, we should develop new technologies to effectively recycle secondary P resources such as sewage sludge, animal manure, and industrial P waste.

Phosphite (H_3PO_3, Pt) is an inorganic reduced P compound in which the P valence is +3 (Fig. 34.1a). Although Pt is considered a rare chemical in the present oxidizing environment, it is generated as a by-product of several industrial processes (Kuroda and Hirota 2015). In particular, large quantities of Pt-containing waste are produced in electroless nickel plating and acid chloride synthesis (Morton

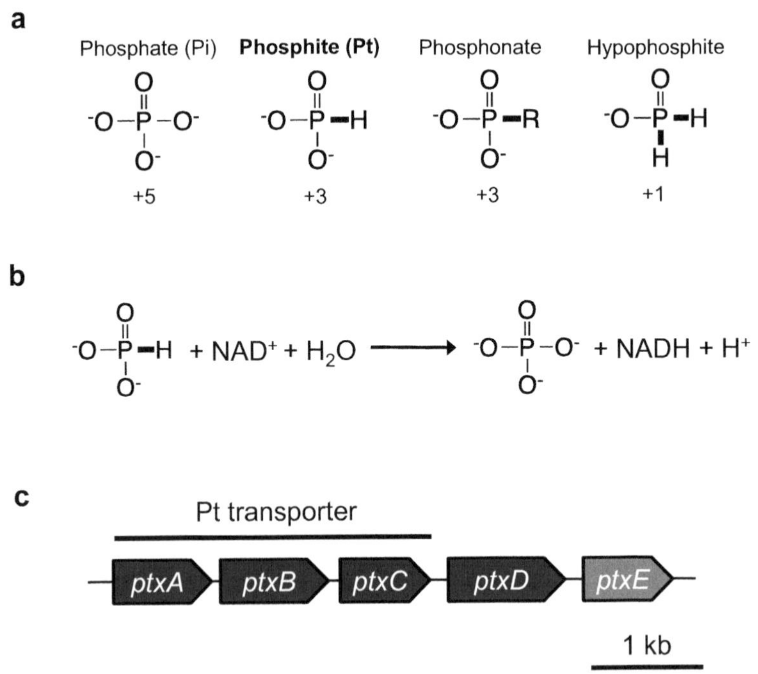

Fig. 34.1 Reduced P compounds and Pt oxidation by PtxD. (**a**) Structure of reduced phosphorus compounds. The numbers shown below are the P valence of each P compound. (**b**) Pt-oxidizing pathway by PtxD. (**c**) Operon structure of *ptxABCDE* genes

and Edwards 2005; Kuroda and Hirota 2015). In the former, Pt is produced from the oxidation of hypophosphite, which is used as a reducing agent to deposit a nickel alloy layer on a solid workpiece. In most cases, Pt-containing waste is either treated with incineration, followed by disposal, or oxidized to Pi by chemical or physicochemical reactions and then subjected to chemical or biological P removal processes (Liu et al. 2013). In Japan, approximately 30,000 tons of Pt-containing waste is disposed of annually without being reused and recycled (Hashizume et al. 2007). In acid chloride synthesis, Pt is stoichiometrically produced from phosphorus trichloride (PCl_3), which is used as a chlorinating reagent for various organic acids, including fatty acids, and the building block chemicals for pharmaceuticals. Approximately 0.7 million tons of PCl_3 is manufactured annually (Clements et al. 2010). Sustainable management of P resources requires that these industrial Pt wastes are recovered and reused.

Phosphorus (P), as a component of nucleic acids, lipids, and various cellular metabolites, is an essential nutrient for all living organisms. Nearly all biologically available P in the environment is in the form of phosphate (H_3PO_4, Pi) and its esters, in which the P valence is +5. Pt has generally been considered a form of P that is not metabolizable in living organisms. While the vast majority of organisms are indeed unable to use Pt, several bacteria have been known to assimilate it (Malacinski and Konetzka 1966; Malacinski and Konetzka 1967). In the last decade, the molecular basis for the oxidation of Pt in several Pt-oxidizing bacteria has been revealed (White and Metcalf 2007). Pt dehydrogenase (PtxD) is an enzyme that catalyzes the direct oxidation of Pt to Pi, with a concomitant reduction of NAD^+ to NADH (Costas et al. 2001). Therefore, it is of particular interest as a catalyst to convert biologically inert Pt into the essential nutrient, Pi (Fig. 34.1b), and to regenerate NADH for economically viable bioprocessing (Vrtis et al. 2002). Heterologous expression of PtxD can expand the P substrate repertoire of various organisms, enabling them to obtain competitive advantages in Pt-based culture media (Kanda et al. 2014; Shaw et al. 2016; Loera-Quezada et al. 2016). Furthermore, this provides a novel principle for creating an engineered nutrient requirement for Pt, which can be used as a robust and practical biological containment strategy for genetically modified organisms (Hirota et al. 2017).

34.2 Bacterial Enzymes Involved in Pt Metabolism

34.2.1 Pt Utilization of Bacteria

The existence of Pt-oxidizing bacteria has been known since the 1950s. Adams and Conrad (1953) observed that Pt added to soil disappeared and this was followed by a corresponding increase in Pi concentration. Several bacteria have subsequently been shown to oxidize Pt (Malacinski and Konetzka 1966; Malacinski and Konetzka 1967). Three pathways for oxidizing Pt have been discovered thus far. Firstly, a C-P lyase pathway dedicated to phosphonate degradation has also been shown to be involved in Pt oxidation (Metcalf and Wanner 1991). Several gram-negative bacteria, including *Pseudomonads* and *Enterobacter*, carry a *phn* gene operon that encodes C-P lyase (Horsman and Zechel 2017). Although the Pt oxidation mechanism involving C-P lyase has never been characterized in vitro, it is likely to be similar to phosphonate catabolism, as the deletion of *phn* abolishes Pt oxidation activity in *E. coli* (Metcalf and Wanner 1991). The second Pt-oxidizing pathway is dependent on the products of the *ptxABCDE* gene, which was initially found in *Pseudomonas stutzeri* WM88 (Metcalf and Wolfe 1998; Costas et al. 2001) (Fig. 34.1c). Of these, *ptxABC* encodes a binding protein-dependent transporter for Pt. After Pt is taken up into cells, it is oxidized to Pi by the NAD^+-dependent enzyme, PtxD. PtxE shares homology with a member of the LysR family of transcriptional regulators, but because its deletion has no effect on Pt oxidation, it is considered to be just an accessory protein. The third pathway is dependent on the alkaline phosphatase of *E. coli* (BAP). Yang and Metcalf (2004) found that BAP catalyzes the oxidation of Pt to Pi and molecular H_2; thus, BAP is a unique Pt-dependent, H_2-evolving hydrogenase. However, the specific activity of BAP for Pt oxidation is less than one hundredth of that for Pi ester hydrolysis (Yang and Metcalf 2004). Thus, Pt oxidation is not a common feature of alkaline phosphatases but a unique feature of BAP.

34.2.2 Pt Uptake Systems of Bacteria

In most *ptxD*-harboring bacteria, *ptxD* is part of an operon with the gene *ptxABC*, a binding protein-dependent transporter. Thus, PtxABC is likely to be involved in Pt transport. Direct evidence of Pt transport via PtxABC has been shown in *E. coli* MT2012 (Motomura et al. 2011), which lacks all endogenous Pi transporters, PitA, PitB, PstSCAB, and PhnCDE (Hirota et al. 2010) (Fig. 34.2). Since this strain is unable to grow on media containing Pi or Pt as a sole P source, MT2012 expressing PtxD (MT2012-*ptxD*) can be used as a host to investigate Pt (and Pi) transport abilities of heterologously expressed transporters. An MT2012-*ptxD* strain expressing *ptxABC* of *Ralstonia* sp. 4506 (Hirota et al. 2012) grows on MOPS-glucose synthetic medium containing Pt, thus confirming that PtxABC is a Pt transporter. Endogenous bacterial Pi transporters can also take up Pt, probably because the chemical structure of Pt is similar to Pi.

Expressing any of the *E. coli* Pi transporters, PitA, PitB, PstSCAB, or PhnCDE, in MT2012-*ptxD* enabled the cells to grow on MOPS media containing either Pi (MOPS-Pi) or Pt (MOPS-Pt), suggesting that these Pi transporters are capable of transporting Pt (Fig. 34.2). Low-affinity Pi transporters, PitA or PitB, supported a higher growth rate than the high-affinity Pi transporter, PstSCAB, suggesting that the Pit system takes up more Pt than the Pst system. *Pseudomonas stutzeri* WM88 also possesses an oxidation system for hypophosphite by the gene products of *htxABCDE* (Fig. 34.2). This gene operon consists of the genes for succinate-dependent hypophosphite dioxygenase, HtxA, and a binding protein-dependent hypophosphite transporter, HtxBCDE. Interestingly, P substrate analysis using the MT2012-*ptxD*

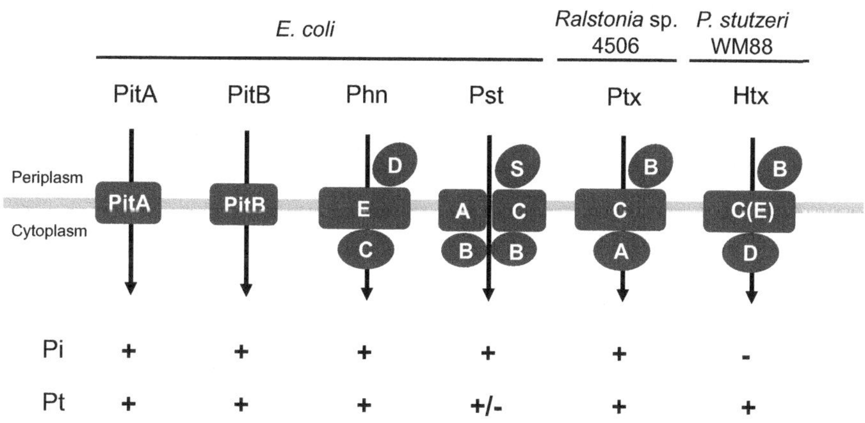

Fig. 34.2 Bacterial P uptake system. Schematic of the structures of bacterial Pi transporters. PitA and PitB are constitutively expressed transporters, whereas binding protein-dependent transporters, PhnCDE, PstSCAB, PtxABC, and HtxBCDE, are induced under Pi-limited conditions. PhnD, PstS, PtxB, and HtxB are substrate binding proteins that present in the periplasmic space. PhnE, PstAC, PtxC, and HtxCE are transmembrane proteins. Substrate preferences of each transporter are shown below. + strongly transport, +/– weakly transport, – do not transport

strain expressing *htxBCDE* showed that HtxBCDE transports hypophosphite and Pt, but not Pi, making it possible to create a nutrient requirement for Pt (see Sect. 3.3).

34.3 Biotechnology for Efficient Utilization of Pt Waste

34.3.1 Application of PtxD as a Dominant Selection Marker for Microbial Cultivation

Advances in synthetic biology and metabolic engineering have made it possible to enhance the metabolic capabilities of microorganisms to produce various useful compounds for chemical, pharmaceutical, and food industries. As the engineered strains are commercialized, microbial cultivation is increasingly performed on an industrial scale. Microbial contamination is one of the major obstacles for large-scale commercial fermentation processes. Contaminating microbes compete for nutrients in culture media, and their by-products, such as organic acids, can inhibit the growth of engineered cells, resulting in reducing yield and productivity. For instance, in the bioethanol industry, a major culprit, *Lactobacillus fermentum*, has been shown to reduce ethanol production in yeast fermentation cultures by as much as 27% (Bischoff et al. 2009). Antibiotics have been widely used in laboratory-scale experiments to suppress the growth of contaminants. However, their industrial use is costly and should be strictly regulated from an environmental health perspective, since the release of antibiotics increases the risk of the emergence of new drug-resistant microorganisms (Baquero et al. 2008; Homem and Santos 2011). Therefore, costly and energy-intensive cleaning and sterilization procedures are needed for fermentation facilities to reduce contamination risks.

An alternative method to achieve selective cultivation is to exploit the competitive growth advantages of host cells over contaminants under specific growth conditions. When a growth medium is supplemented with a nonmetabolizable form of an essential nutrient, only the cells capable of metabolizing the nutrient could grow dominantly in the culture. Since Pt is a nonmetabolizable P source for general organisms and must be oxidized to Pi prior to assimilation, the introduction of *ptxD* into cells that are incapable of utilizing Pt confers the ability to grow on a medium containing Pt as a sole P source. We introduced PtxD from *Ralstonia* sp. 4506 into the fission yeast, *Schizosaccharomyces pombe*, under the control of the strong promoter, *nmt1*. In contrast to wild-type cells, the PtxD-expressing cells could grow on Pt medium as fast as on Pi medium (Kanda et al. 2014) (Fig. 34.3a). We also demonstrated that a codon-adapted *ptxD* could confer Pt-dependent growth ability to *Saccharomyces cerevisiae*. Thus, *PtxD* expression could expand the range of metabolizable P substrates of host cells, providing them with a competitive advantage over contaminant microbes that cannot oxidize Pt. Shaw et al. (2016) applied this engineered competitive advantage to the oleaginous recombinant yeast, *Yarrowia lipolytica* NS324, and demonstrated that a PtxD-expressing *Y. lipolytica* NS324 strain outcompeted a control strain in Pt-containing medium.

E. coli has endogenous Pt-oxidizing pathways, such as Phn and BAP. However, PtxD expression in *E. coli* enhances Pt-dependent growth on MOPS-Pt, suggesting that the Pt-oxidizing activity of these pathways is lower than that of PtxD. To test whether PtxD-expressing *E. coli* is capable of outcompeting contaminant microorganisms in MOPS-Pt medium, we simultaneously inoculated wild-type *E. coli* and *Bacillus subtilis* cells at approximately 50- and 100-times the number of the PtxD-expressing *E. coli* strains, respectively. After a 48-h cultivation, we found that the cultures competing with wild-type *E. coli* and *B. subtilis* were dominated by the PtxD-expressing *E. coli* approximately, which accounted for 96% and 98% of the total bacteria, respectively (Fig. 34.3b, c). These results suggest that Pt oxidation ability could provide a competitive growth advantage to host cells, enabling selective cultivation without using sterilized growth media and equipment.

Pt is a safe chemical and its cost is far lower than the cost of antibiotics. Furthermore, since Pt is an environmentally rare chemical, the expression of PtxD would not be advantageous for the survival of recombinants outside the Pt culture. Therefore, a Pt-based cultivation system would enable cost-effective and environmentally safe cultivation of microorganisms on an industrial scale.

34.3.2 Pt Fertilization and Weed Control

Nearly 90% of P resources are used in fertilizers. However, the largest problem with Pi fertilization is that Pi is quickly immobilized in many soil types and becomes unavailable for plant uptake due to adsorption and precipitation reactions. Up to

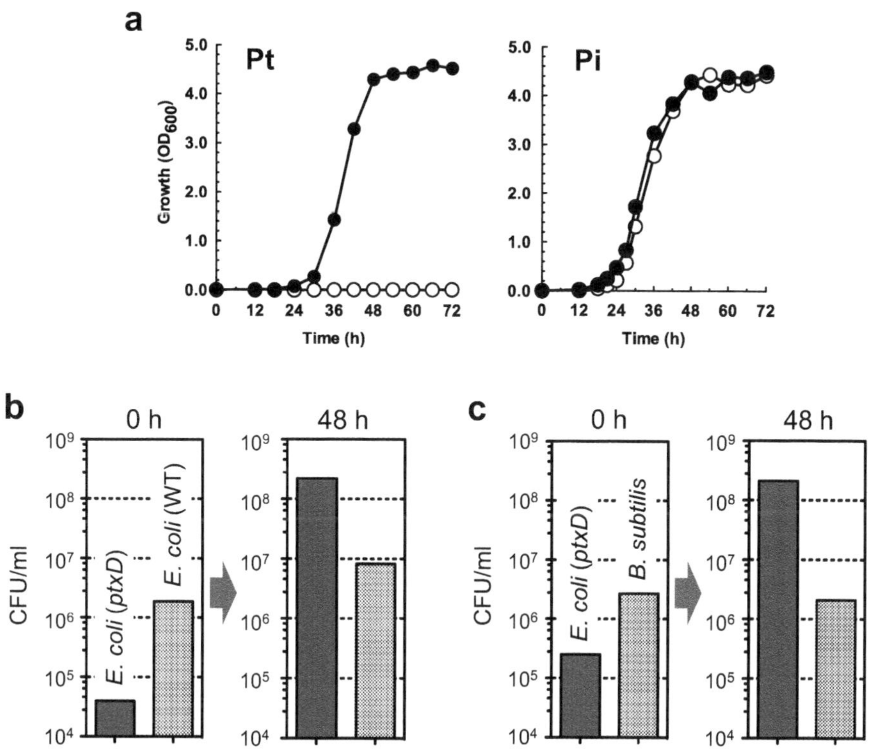

Fig. 34.3 Selective cultivation of PtxD-expressing microorganisms during cultivation. (**a**) Growth of *Sz. pombe* expressing PtxD from the episomally maintained gene on a multicopy plasmid (*filled circles*) and the control strain carrying an empty plasmid (*open circles*) cultured in synthetic EM media containing Pt (*left graph*) or Pi (*right graph*) were monitored. (**b**, **c**) Competitive culture of PtxD-expressing *E. coli* with wild-type *E. coli* (**b**) and *B. subtilis* (**c**) in MOPS-Pt medium. *Solid bars* indicate colony formation units (CFU) of PtxD-expressing *E. coli*. *Gray bars* indicate CFUs of wild-type *E. coli* (**b**) and *B. subtilis* (**c**)

80% of the supplied Pi can become immobilized in insoluble forms that are not available to plants. Compared to Pi, Pt shows higher solubility and lower reactivity with soil components (Morton and Edwards 2005; Pasek 2008). Thus, direct utilization of Pt as a fertilizer opens the way for reusing Pt waste. However, no plants are able to utilize Pt as a P source. Several attempts to express bacterial PtxD in plant hosts have been reported using *Arabidopsis* (Lopez-Arredondo and Herrera-Estrella 2012; Kuroda and Hirota 2015), tobacco (Lopez-Arredondo and Herrera-Estrella 2012), maize (Nahampun et al. 2016), and rice (Manna et al. 2016) as model organisms. In *Arabidopsis*, *ptxD* was introduced under the control of the CaMV 35S promoter. The resultant transgenic lines exhibited comparable physiological phenotypes in a growth medium containing Pt as a sole P source to those grown in Pi medium (Lopez-Arredondo and Herrera-Estrella 2012; Kuroda and Hirota 2015). Also, *ptxD*-transgenic lines could germinate directly on Pt medium, suggesting that

PtxD can be used as a selectable marker for transgenic plants (Kuroda and Hirota 2015; Nahampun et al. 2016). Interestingly, *ptxD*-transgenic *Arabidopsis* showed increased growth on Pt medium, with greater biomass production and yield compared to wild-type plants grown on Pi medium (Lopez-Arredondo and Herrera-Estrella 2012). Similar effects were also observed in rice, which showed increased shoot and root length, shoot and root biomass, and total body weight in *ptxD*-transgenic rice grown on Pt medium. Additionally, transgenic lines grown on Pt medium accumulated significantly higher chlorophyll levels and had improved photosystem II (PSII) activity than wild-type plants (Manna et al. 2016), which may account for their increased growth.

Another potential benefit of the Pt fertilization system is the ease of weed control. Lopes-Arredondo and Herrera-Estrella (Lopez-Arredondo and Herrera-Estrella 2012) carried out greenhouse growth competition experiments using a *ptxD*-recombinant tobacco plant and a grass weed (*Brachypodium distachyon*). Both the *ptxD*-recombinant tobacco plant and the *B. distachyon* weed grew poorly on unfertilized soil. Supplementation with Pi resulted in faster growth of the *B. distachyon* weed. In contrast, fertilization with Pt resulted in limited growth of *B. distachyon* but vigorous growth of the *ptxD*-recombinant tobacco plant. Application of Pt fertilizer to genetically modified (GM) crops could decrease the use of herbicides and reduce the related environmental risks. Currently, 185 million hectares of farm land is dedicated to GM crops, which is more than 10% of the world's arable land (ISAAA 2016). From the viewpoint of P security, strong expansion of agricultural production for bioenergy purposes would accelerate the depletion of P resources. The use of *ptxD*-recombinant plants in bioenergy production should increase the demand for Pt, contributing to the efficient utilization of Pt waste.

34.3.3 Biological Containment of Genetically Modified Microorganisms

Genetically modified microorganisms (GMMs) play a vital role in chemical, pharmaceutical, and food industries. Owing to recent developments in the fields of synthetic biology and genomics, there is a growing demand for functionalized GMMs that can be safely used outside of enclosed laboratory facilities. Novel applications of GMMs include the production of biofuels, bioremediation (Wright et al. 2015), and clinical treatments (Pinero-Lambea et al. 2015). One of the biggest concerns for the practical use of GMMs in the open environment is the possibility of uncontrolled proliferation that could, in a worst-case scenario, endanger public health or biodiversity. Therefore, strategies for safeguarding against the spread and proliferation of GMMs in the environment must be developed in order to realize their practical applications.

Intrinsic biocontainment is a strategy that contains host microbes to grow and survive only under specific conditions or in limited environments (Berg et al. 1975; Curtiss 1978). The development of biocontainment systems that strictly limit the proliferation of GMMs would contribute to their safe use, making biocontainment an integral part of GMM risk management. The US Department of Energy has recently required that applications for certain grants include strategies to address biocontainment, minimizing risks of potential release of engineered organisms into the environment (grant number DE-FOA-0001650). Current biocontainment strategies are mainly based on the use of toxin expression and nutrient auxotrophy. The former integrates killing switches, such as toxin/antitoxin genes (Molin et al. 1993; Wright et al. 2015), or self-destructing DNA into host cells (Caliando and Voigt 2015), which results in conditional lethality. The latter creates requirements for certain nutrients by eliminating essential gene functions, rendering the microbes nonproliferative without the addition of the nutrients (Curtiss 1978; Ronchel and Ramos 2001). Several synthetic auxotrophy strategies that render cells dependent on artificial compounds have recently been reported (Lopez and Anderson 2015; Mandell et al. 2015; Rovner et al. 2015). However, they usually require highly complicated genetic modifications and use expensive chemicals as required nutrients, limiting their practical use as containment systems on an industrial scale.

Given the expanded P substrate repertoire of PtxD-expressing cells and the scarcity of Pt in the environment, we conceived an idea to redesign P metabolism for biocontainment. If a strain expressing PtxD and a Pt transporter were engineered to be unable to take up Pi, its growth would rely solely on the availability of Pt. However, this strategy requires a P uptake system that specifically takes up Pt. As described in Sect. 2.2, unlike other P transporters, HtxBCDE of *P. stutzeri* was found to take up Pt or hypophosphite, but not Pi (Fig. 34.2). This finding motivated us to develop a strain that uses Pt as a required nutrient. After the disruption of all endogenous Pi and organic Pi transporters in an *Escherichia coli* strain expressing HtxBCDE and PtxD, Pt uptake and oxidation became the only means to obtain Pi (Fig. 34.4a). As expected, the resultant strain (RN1008) was not able to grow on any culture media other than MOPS-Pt (Hirota et al. 2017) (Fig. 34.4b). Interestingly, in the absence of Pt, the viability of RN1008 dropped steeply to below 0.0001% after 7 days. Furthermore, no RN1008 cells were detected after 2 weeks when five trillion cells were plated on a nonpermissive growth medium (2xYT). These results suggest that this strategy exerts an intrinsic containment effect by Pt dependency and an active containment effect, similar to a "killing switch," by an unknown mechanism. This strategy offers the highest security level among the reported strategies (Hirota et al. 2017). It should also be noted that such an extremely high containment level was achieved with a single strategy. Considering the simplicity of the required genetic modifications and the low cost of Pt, this strategy offers an effective biocontainment system for various microbial hosts. Employing a culture using Pt as the P source would also contribute to effective management of biological contamination.

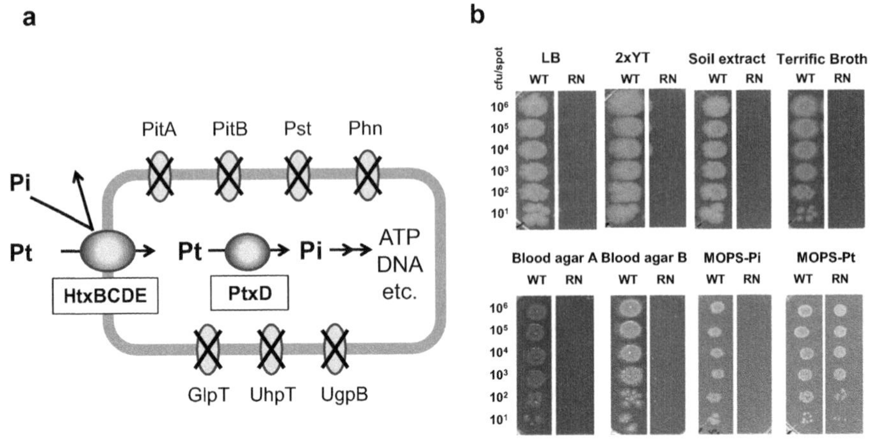

Fig. 34.4 Biocontainment strategy using the engineered Pt nutrient requirement. (**a**) A schematic of the engineered P metabolic pathway for biocontainment. Dependency on Pt is created by disruption of endogenous Pi (PitA, PitB, PstSCAB, PhnCDE) and organic Pi transporters (GlpT, UgpB, UhpT) and exogenous expression of HtxBCDE and PtxD. HtxBCDE takes up Pt and hypophosphite, but not Pi or organic Pi compounds. PtxD expression confers Pt oxidation activities. (**b**) Growth of RN1008 on diverse types of growth media

34.4 Other Biotechnological Applications

34.4.1 NAD(P)H Regeneration

Cofactors are required for many potentially useful enzymes with applications in industrial bioprocessing. Nicotine amide cofactor, NAD(P)H, is required as a cofactor of dehydrogenases that are used for chiral synthesis of building block chemicals for pharmaceuticals (Weckbecker et al. 2010). However, the stoichiometric use of cofactors is not practical due to their high cost. PtxD catalyzes the nearly irreversible oxidation of Pt to Pi, with the concomitant reduction of NAD to NADH. Owing to the large change in free energy of this reaction ($\Delta G^{\circ\prime} = -63.3$ kJ/mol), this enzyme provides a stronger NADH regeneration system than any other enzyme, including formate dehydrogenase and glucose dehydrogenase (Relyea and van der Donk 2005). In addition to the favorable equilibrium constant, PtxD uses an inexpensive Pt substrate and produces benign Pi as a by-product, conferring practical advantages over other cofactor regeneration systems. Therefore, PtxD has great potential as an industrial biocatalyst for NADH regeneration. Thus far, several PtxD proteins have been biochemically characterized, and their performance as NADH regeneration enzymes has been investigated. Since the PtxD enzyme originally discovered in *P. stutzeri* ($PtxD_{PS}$) is heat labile, McLachlan et al. (2008) created a PtxD mutant (12× mutant) with a 23,000-fold higher half-life of thermal inactivation at 45 °C compared with the parent enzyme (McLachlan et al. 2008). Another thermostable PtxD was isolated from the mesophilic soil bacterium, *Ralstonia* sp. 4506

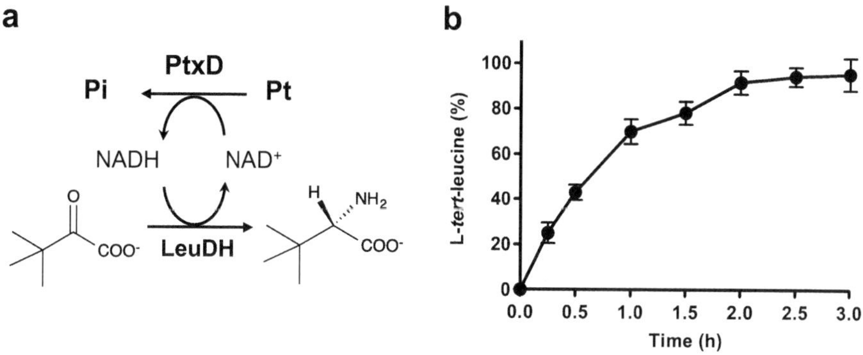

Fig. 34.5 NADH regeneration system using PtxD. *A*. l-*tert*-leucine production from trimethylpyruvate by LeuDH, coupled with NADH regeneration by $PtxD_{R4506}$. A reaction solution containing 50 mM trimethylpyruvate neutralized with ammonia, 75 mM Pt, 0.5 mM NAD^+, leucine dehydrogenase, and $PtxD_{R4506}$ was incubated at 45 °C

(Hirota et al. 2012). The purified PtxD from this bacterium ($PtxD_{R4506}$) shows maximum activity at 50 °C, and its half-life of thermal inactivation at 45 °C is 73 h, which is approximately 3100-fold greater than that of $PtxD_{PS}$. Moreover, $PtxD_{R4506}$ showed approximately 2.4-fold and 5.5-fold greater catalytic efficiency ($K_{cat}/K_{m,\,Pt}$) than the 12× mutant and $PtxD_{PS}$, respectively. In addition, $PtxD_{R4506}$ is more tolerant than $PtxD_{PS}$ to several inhibitory compounds, such as sulfite, nitrate, and sodium chloride. Thus, $PtxD_{R4506}$ is likely to retain high catalytic activity under various reaction conditions in practical use.

To test the applicability of $PtxD_{R4506}$ as an NADH regeneration system, the PtxD reaction was coupled with the production of l-*tert*-leucine, an important chiral building block used in the pharmaceutical industry. Leucine dehydrogenase (LeuDH) from *Bacillus* sp. catalyzes the reductive amination of trimethylpyruvate to l-*tert*-leucine, with the concomitant oxidation of NADH to NAD^+ (Fig. 34.5a). Fifty millimolar trimethylpyruvate is almost fully converted to l-*tert*-leucine in the presence of 0.5 mM NAD^+ and 75 mM Pt, indicating that $PtxD_{R4506}$ regenerates NADH approximately 100 times during the reaction (Hirota et al. 2012) (Fig. 34.5b). In order to expand the applicability of this cofactor regeneration system to be coupled with an $NADP^+$-dependent enzyme, rational design to alter cofactor specificity was successfully performed (Woodyer et al. 2005). The replacement of two amino acid residues located in the Rossmann-fold domain of PtxD resulted in more than 1000-fold higher catalytic efficiency for $NADP^+$ reduction compared to $PtxD_{PS}$. The applicability of the NADPH regeneration system using this engineered PtxD was demonstrated in the xylitose reductase-catalyzed synthesis of xylitol and the alcohol dehydrogenase-catalyzed synthesis of (R)-phenylethanol. These results demonstrate that this system is more effective than the NADPH regeneration system relying on formate dehydrogenase from *Pseudomonas* sp. 101 (Johannes et al. 2007). As another cofactor regeneration system, Wang et al. (2017) engineered $PtxD_{R4506}$ to use the nonnatural cofactor, nicotinamide cytosine dinucleotide (NCD), to generate reduced NCD (NCDH). This synthetic cofactor regeneration system enabled the transfer of reducing energy to an engineered NCDH-dependent malic enzyme, demonstrating a pathway-specific control system for metabolic engineering. In these cofactor regeneration systems, the use of Pt waste would reduce the cost of cofactor regeneration and contribute to sustainable use of P resources.

34.5 Conclusions

P is an essential nutrient for all organisms and is indispensable to various industries essential for human society. P has no substitute for crop production. To prepare for its inevitable depletion, P-containing waste should be recovered and recycled. Industrial P waste is a suitable target for P recycling, as it contains a relatively high concentration of P. However, due to the lack of high value-added biotechnological applications, Pt waste has not been reused efficiently. In this chapter, we summarized recent development of several novel methodologies that could contribute to efficient utilization of Pt waste in the fields of industrial microbial fermentation, agriculture, and bioprocessing. These technologies would also contribute to reduced energy consumption and cost of bioprocessing, improve fertilization efficiency for crops, and expand the potential applications with increasing the safety of GMMs. Thus, Pt-based biotechnologies should make a major contribution to the sustainable development of human society.

Acknowledgments This work was supported by Grant-in-Aid for Scientific Research (B) (grant number 22360346, 16H04899), Grant-in-Aid for Scientific Research (C) (grant number 25450102), and Grant-in-Aid for Exploratory Research (16K14889) from JSPS KAKENHI and Advanced Low Carbon Technology Research and Development Program (ALCA) of Japan Science and Technology Agency (JST).

References

Adams F, Conrad JP (1953) Transition of phosphite to phosphate in soils. Soil Sci 75:361–371
Baquero F, Martinez JL, Canton R (2008) Antibiotics and antibiotic resistance in water environments. Curr Opin Biotechnol 19(3):260–265. https://doi.org/10.1016/j.copbio.2008.05.006
Berg P, Baltimore D, Brenner S, Roblin RO, Singer MF (1975) Summary statement of the Asilomar conference on recombinant DNA molecules. Proc Natl Acad Sci U S A 72(6):1981–1984
Bischoff KM, Liu S, Leathers TD, Worthington RE, Rich JO (2009) Modeling bacterial contamination of fuel ethanol fermentation. Biotechnol Bioeng 103(1):117–122. https://doi.org/10.1002/bit.22244
Caliando BJ, Voigt CA (2015) Targeted DNA degradation using a CRISPR device stably carried in the host genome. Nat Commun 6:6989. https://doi.org/10.1038/ncomms7989
Clements A, Dunn M, Firth V, Hubbard L, Lazonby J, Waddington D (2010) The essential chemical industry. Chemical Industry Education Centre at The University of York, York

Cordell D, Rosemarin A, Schroder JJ, Smit AL (2011) Towards global phosphorus security: a systems framework for phosphorus recovery and reuse options. Chemosphere 84(6):747–758. https://doi.org/10.1016/j.chemosphere.2011.02.032

Costas AM, White AK, Metcalf WW (2001) Purification and characterization of a novel phosphorus-oxidizing enzyme from *Pseudomonas stutzeri* WM88. J Biol Chem 276(20):17429–17436. https://doi.org/10.1074/jbc.M011764200

Curtiss R 3rd (1978) Biological containment and cloning vector transmissibility. J Infect Dis 137(5):668–675

Hashizume K, Naito K, Oka H, Okumura H (2007) The situation and future of electroless nickel plating. Hyomengijutsu (Jpn) 58(2):87–91

Hirota R, Kuroda A, Kato J, Ohtake H (2010) Bacterial phosphate metabolism and its application to phosphorus recovery and industrial bioprocesses. J Biosci Bioeng 109(5):423–432. https://doi.org/10.1016/j.jbiosc.2009.10.018

Hirota R, Yamane ST, Fujibuchi T, Motomura K, Ishida T, Ikeda T, Kuroda A (2012) Isolation and characterization of a soluble and thermostable phosphite dehydrogenase from *Ralstonia* sp. strain 4506. J Biosci Bioeng 113(4):445–450. https://doi.org/10.1016/j.jbiosc.2011.11.027

Hirota R, Abe K, Katsuura ZI, Noguchi R, Moribe S, Motomura K, Ishida T, Alexandrov M, Funabashi H, Ikeda T, Kuroda A (2017) A novel biocontainment strategy makes bacterial growth and survival dependent on phosphite. Sci Rep 7:44748. https://doi.org/10.1038/srep44748

Homem V, Santos L (2011) Degradation and removal methods of antibiotics from aqueous matrices – a review. J Environ Manag 92(10):2304–2347. https://doi.org/10.1016/j.jenvman.2011.05.023

Horsman GP, Zechel DL (2017) Phosphonate biochemistry. Chem Rev 117(8):5704–5783. https://doi.org/10.1021/acs.chemrev.6b00536

International Service for the Acquisition of Agri-biotech Application (ISAAA) (2016) Biotech crop highlights in 2016. http://www.isaaa.org/resources/publications/pocketk/16/. Accessed 9 July 2017

Johannes TW, Woodyer RD, Zhao H (2007) Efficient regeneration of NADPH using an engineered phosphite dehydrogenase. Biotechnol Bioeng 96(1):18–26. https://doi.org/10.1002/bit.21168

Kanda K, Ishida T, Hirota R, Ono S, Motomura K, Ikeda T, Kitamura K, Kuroda A (2014) Application of a phosphite dehydrogenase gene as a novel dominant selection marker for yeasts. J Biotechnol 182–183:68–73. https://doi.org/10.1016/j.jbiotec.2014.04.012

Kuroda A, Hirota R (2015) Environmental biotechnology for efficient utilization of industrial phosphite waste. Global Environ Res 19(1):77–82

Liu P, Li C, Liang X, Xu J, Lu G, Ji F (2013) Advanced oxidation of hypophosphite and phosphite using a UV/H_2O_2 process. Environ Technol 34(13–16):2231–2239. https://doi.org/10.1080/09593330.2013.765917

Loera-Quezada MM, Leyva-Gonzalez MA, Velazquez-Juarez G, Sanchez-Calderon L, Do Nascimento M, Lopez-Arredondo D, Herrera-Estrella L (2016) A novel genetic engineering platform for the effective management of biological contaminants for the production of microalgae. Plant Biotechnol J. https://doi.org/10.1111/pbi.12564

Lopez G, Anderson JC (2015) Synthetic auxotrophs with ligand-dependent essential genes for a BL21(DE3) biosafety strain. ACS Synth Biol 4(12):1279–1286. https://doi.org/10.1021/acssynbio.5b00085

Lopez-Arredondo DL, Herrera-Estrella L (2012) Engineering phosphorus metabolism in plants to produce a dual fertilization and weed control system. Nat Biotechnol 30(9):889–893. https://doi.org/10.1038/nbt.2346

Malacinski G, Konetzka A (1966) Bacterial oxidation of orthophosphite. J Bacteriol 91(2):578–582

Malacinski G, Konetzka W (1967) Orthophosphite-nicotinamide adenine dinucleotide oxidoreductase from *Pseudomonas fluorescens*. J Bacteriol 93(6):1906–1910

Mandell DJ, Lajoie MJ, Mee MT, Takeuchi R, Kuznetsov G, Norville JE, Gregg CJ, Stoddard BL, Church GM (2015) Biocontainment of genetically modified organisms by synthetic protein design. Nature 518(7537):55–60. https://doi.org/10.1038/nature14121

Manna M, Achary VM, Islam T, Agrawal PK, Reddy MK (2016) The development of a phosphite-mediated fertilization and weed control system for rice. Sci Rep 6:24941. https://doi.org/10.1038/srep24941

McLachlan MJ, Johannes TW, Zhao H (2008) Further improvement of phosphite dehydrogenase thermostability by saturation mutagenesis. Biotechnol Bioeng 99(2):268–274. https://doi.org/10.1002/bit.21546

Metcalf WW, Wanner BL (1991) Involvement of the *Escherichia coli phn* (*psiD*) gene cluster in assimilation of phosphorus in the form of phosphonates, phosphite, Pi esters, and Pi. J Bacteriol 173(2):587–600

Metcalf WW, Wolfe RS (1998) Molecular genetic analysis of phosphite and hypophosphite oxidation by *Pseudomonas stutzeri* WM88. J Bacteriol 180:5547–5558

Molin S, Boe L, Jensen LB, Kristensen CS, Givskov M, Ramos JL, Bej AK (1993) Suicidal genetic elements and their use in biological containment of bacteria. Annu Rev Microbiol 47:139–166. https://doi.org/10.1146/annurev.mi.47.100193.001035

Morton SC, Edwards M (2005) Reduced phosphorus compounds in the environment. Crit Rev Environ Sci Technol 35(4):333–364. https://doi.org/10.1080/10643380590944978

Motomura K, Hirota R, Ohnaka N, Okada M, Ikeda T, Morohoshi T, Ohtake H, Kuroda A (2011) Overproduction of YjbB reduces the level of polyphosphate in *Escherichia coli*: a hypothetical role of YjbB in phosphate export and polyphosphate accumulation. FEMS Microbiol Lett 320(1):25–32. https://doi.org/10.1111/j.1574-6968.2011.02285.x

Nahampun HN, Lopez-Arredondo D, Xu X, Herrera-Estrella L, Wang K (2016) Assessment of *PtxD* gene as an alternative selectable marker for agrobacterium-mediated maize transformation. Plant Cell Rep. https://doi.org/10.1007/s00299-016-1942-x

Pasek MA (2008) Rethinking early earth phosphorus geochemistry. Proc Natl Acad Sci U S A 105(3):853–858. https://doi.org/10.1073/pnas.0708205105

Pinero-Lambea C, Ruano-Gallego D, Fernandez LA (2015) Engineered bacteria as therapeutic agents. Curr Opin Biotechnol 35:94–102. https://doi.org/10.1016/j.copbio.2015.05.004

Relyea HA, van der Donk WA (2005) Mechanism and applications of phosphite dehydrogenase. Bioorg Chem 33(3):171–189. https://doi.org/10.1016/j.bioorg.2005.01.003

Ronchel MC, Ramos JL (2001) Dual system to reinforce biological containment of recombinant bacteria designed for rhizoremediation. Appl Environ Microbiol 67(6):2649–2656. https://doi.org/10.1128/AEM.67.6.2649-2656.2001

Rovner AJ, Haimovich AD, Katz SR, Li Z, Grome MW, Gassaway BM, Amiram M, Patel JR, Gallagher RR, Rinehart J, Isaacs FJ (2015) Recoded organisms engineered to depend on synthetic amino acids. Nature 518(7537):89–93. https://doi.org/10.1038/nature14095

Shaw AJ, Lam FH, Hamilton M, Consiglio A, MacEwen K, Brevnova EE, Greenhagen E, LaTouf WG, South CR, van Dijken H, Stephanopoulos G (2016) Metabolic engineering of microbial competitive advantage for industrial fermentation processes. Science 353(6299):583–586. https://doi.org/10.1126/science.aaf6159

Vrtis JM, White AK, Metcalf WW, van der Donk WA (2002) Phosphite dehydrogenase: a versatile cofactor-regeneration enzyme. Angew Chem Int Ed Eng 41(17):3257–3259. https://doi.org/10.1002/1521-3773(20020902)41:17<3071::AID-ANIE3071>3.0.CO;2-U

Wang L, Ji D, Liu Y, Wang Q, Wang X, Zhou YJ, Zhang Y, Liu W, Zhao ZK (2017) Synthetic cofactor-linked metabolic circuits for selective energy transfer. ACS Catal 7(3):1977–1983. https://doi.org/10.1021/acscatal.6b03579

Weckbecker A, Groger H, Hummel W (2010) Regeneration of nicotinamide coenzymes: principles and applications for the synthesis of chiral compounds. Adv Biochem Eng Biotechnol 120:195–242. https://doi.org/10.1007/10_2009_55

White AK, Metcalf WW (2007) Microbial metabolism of reduced phosphorus compounds. Annu Rev Microbiol 61:379–400. https://doi.org/10.1146/annurev.micro.61.080706.093357

Woodyer R, Wheatley JL, Relyea HA, Rimkus S, van der Donk WA (2005) Site-directed mutagenesis of active site residues of phosphite dehydrogenase. Biochemistry 44(12):4765–4774. https://doi.org/10.1021/bi047868c

Wright O, Delmans M, Stan GB, Ellis T (2015) GeneGuard: a modular plasmid system designed for biosafety. ACS Synth Biol 4(3):307–316. https://doi.org/10.1021/sb500234s

Yang K, Metcalf WW (2004) A new activity for an old enzyme: *Escherichia coli* bacterial alkaline phosphatase is a phosphite-dependent hydrogenase. Proc Natl Acad Sci U S A 101(21):7919–7924. https://doi.org/10.1073/pnas.0400664101

Chapter 35
Heatphos Process for Recovering Phosphorus from Bio-P Sludge Before Anaerobic Digestion

Noboru Takiguchi, Akio Kuroda, Hisao Ohtake, and Satoshi Tsuneda

Abstract Sewage sludge generated in Bio-P processes (also known as enhanced biological phosphorus removal processes) typically contains 3–5% of its dry weight as phosphorus (P). Approximately 60–80% of the P is stored as polyphosphate (polyP) which is a heat-labile polymer of inorganic phosphate (P_i). This chapter describes a simple technology, named "Heatphos," for leaching polyP from Bio-P sludge by heating at 70 °C for about 1 h and recovering P_i from the leachate by the addition of $CaCl_2$ without needing to adjust pH to a high value. Heating sludge for polyP leaching can also improve the digestive efficiency and thus biogas productivity in the subsequent anaerobic sludge digestion at both mesophilic (typically 35–40 °C) and thermophilic (50–60 °C) temperatures. The heat energy required for polyP leaching can be supplied from biogas generated by anaerobic sludge digestion. The demonstration plant with a capacity of treating Bio-P sludge of 0.36 m^3/day has showed that P could be recovered from Bio-P sludge at a rate of approximately 10 kgP/day (c. 3.5 tP/a).

Keywords Polyphosphate · Heatphos · Bio-P sludge · Anaerobic sludge digestion

N. Takiguchi
College of Science and Technology, Kanazawa University, Kanazawa, Japan

A. Kuroda
Graduate School of Advanced Sciences of Matter, Hiroshima University, Higashi-Hiroshima, Hiroshima, Japan

H. Ohtake (✉) · S. Tsuneda
Phosphorus Atlas Research Institute, Waseda University, Shinjuku-ku, Tokyo, Japan
e-mail: hohtake@bio.eng.osaka-u.ac.jp

H. Ohtake, S. Tsuneda (eds.), *Phosphorus Recovery and Recycling*,
https://doi.org/10.1007/978-981-10-8031-9_35

35.1 Introduction

Microorganisms can effectively capture inorganic phosphate (P_i) from dilute streams and accumulate excess P_i in the form of polyphosphate (polyP) (Kulaev and Vagabov 1983). Biologically synthesized polyP is a linear polymer of P_i linked by high-energy phosphoanhydride bonds having a chain length of up to 1000 residues or more. PolyP can serve as a phosphorus (P) source for the biosynthesis of nucleic acids, phospholipids, lipopolysaccharides, and various cytoplasmic solutes in microbial cells under P_i starvation conditions (Hirota et al. 2010). The ability of microorganisms to take up P_i from dilute streams and accumulate polyP provides the basis for biological P_i removal in wastewater treatment processes (Sedlak 1991). For instance, Bio-P processes, which are also known as enhanced biological P removal processes, are well established for P_i removal from sewage and have been widely practiced in full-scale wastewater treatment plants (WWTP). In Bio-P processes, sludge microorganisms release P_i in the anaerobic tank but remove more P_i in the subsequent aerobic tank than that they released in the anaerobic tank (Sedlak 1991). Although the detailed mechanism still remains unclear, the alternate anaerobic and aerobic cycles allow sludge microorganisms to accumulate polyP intracellularly, thereby removing typically 80–90% of P_i from sewage at WWTP. Hence the process performance relies primarily on the ability of sludge microorganisms to accumulate polyP.

Sewage sludge generated in Bio-P processes typically contains 3–5% of its dry weight as P. Approximately 60–80% of the P is stored as polyP which has a heat-lability characteristic of phosphoanhydride bonds (Kornberg 1995). This chapter describes a simple technology, named "Heatphos," for leaching polyP from Bio-P sludge by heating at 70 °C for about 1 h and recovering P_i from the leachate by the addition of $CaCl_2$ without needing to adjust pH to a high value. Heating sludge for polyP leaching can also improve the sludge digestibility and thus the biogas productivity in the subsequent anaerobic sludge digestion at both mesophilic (typically 35–40 °C) and thermophilic (50–60 °C) temperatures. The heat energy required for polyP leaching can be supplied by biogas generated by anaerobic sludge digestion.

35.2 PolyP Leaching and Recovery

35.2.1 PolyP Leaching from Microbial Cells

Microbial cells form polyP granules (PPGs) of various sizes in the cytoplasm, when they accumulate excess P_i intracellularly (Fig. 35.1a). PPGs are one of the first subcellular structures found in various bacteria (Kornberg 1995). PPGs, which are randomly distributed in the cytoplasm, can be detected as highly refractive, black electron-dense particles with transmission electron microscopy (TEM). High levels of P can be detected in cell areas containing PPGs by energy-dispersive X-ray

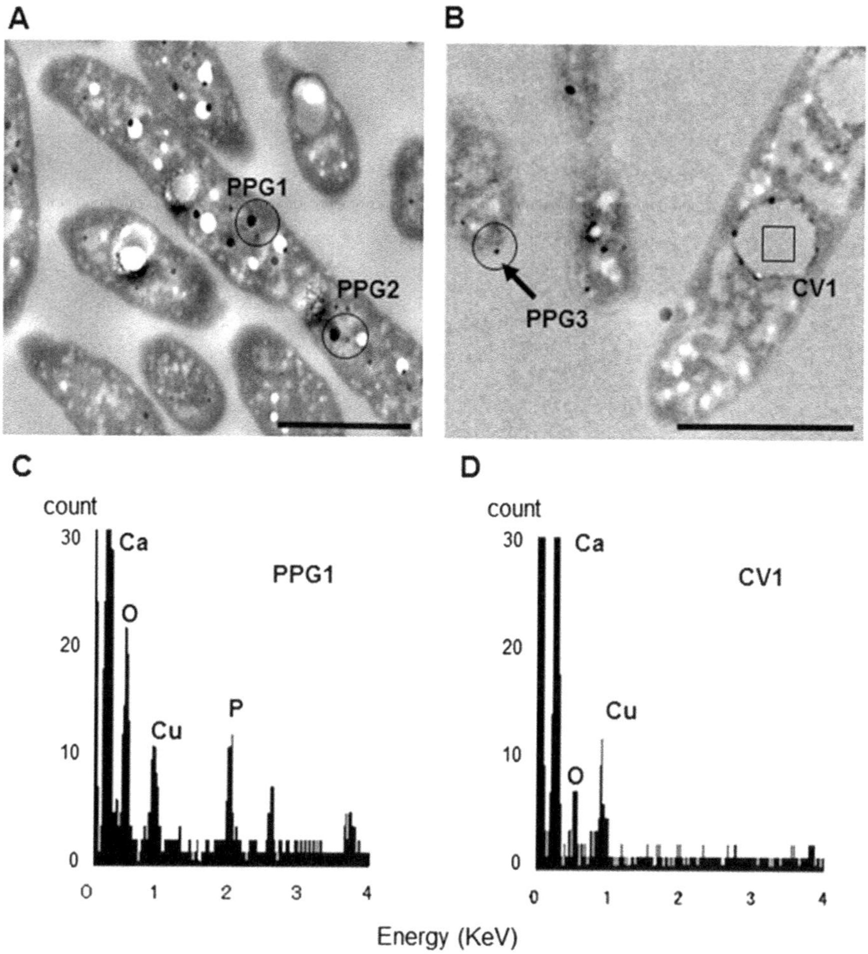

Fig. 35.1 Transmission electron microscopy (TEM) images of *Escherichia coli phoU* mutant cells before (**a**) and after (**b**) heating at 70 °C for 5 min (scale bar = 0.25 μm) and EDX spectra of areas containing a polyP granule (PPG) (C) and a cytoplasmic void (CV) (D). The *E. coli phoU* mutant was used as a polyP-accumulating model organism (Kashihara et al. 2010)

(EDX) analysis (Fig. 35.1c). PolyP can be readily released from polyP-accumulating cells by heating at 70 °C for a short period of time (Kashihara et al. 2010). Heating of the cells at 70 °C induces cell voids resulting from plasmolysis and cytoplasmic aggregations (Fig. 35.1b). Although large cell voids are formed in the cytoplasm, no significant P is detectable in cytoplasmic voids (Fig. 35.1d). PPGs are rather perceivable around the surface of the heated cells (Fig.35.1b). Obviously, the sizes of PPGs become significantly small compared to those detected with unheated cells, indicating the occurrence of PPG degradation in the cytoplasm.

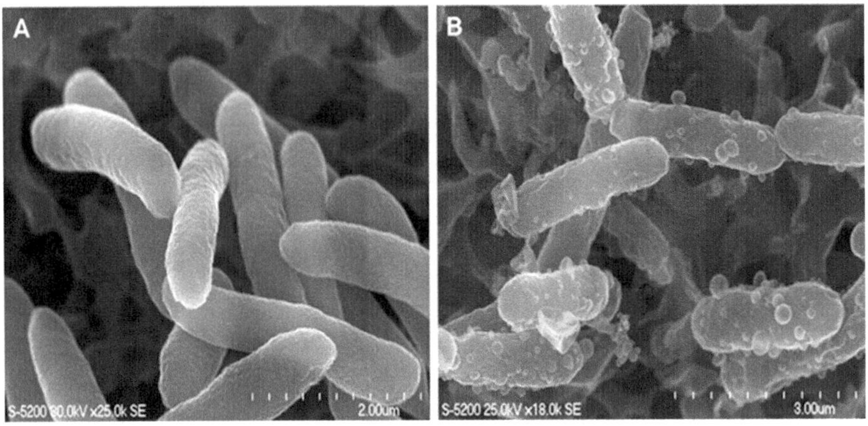

Fig. 35.2 Scanning electron microscopic (SEM) images of *E. coli phoU* mutants before (**a**) and after (**b**) being heated at 70 °C for 15 min

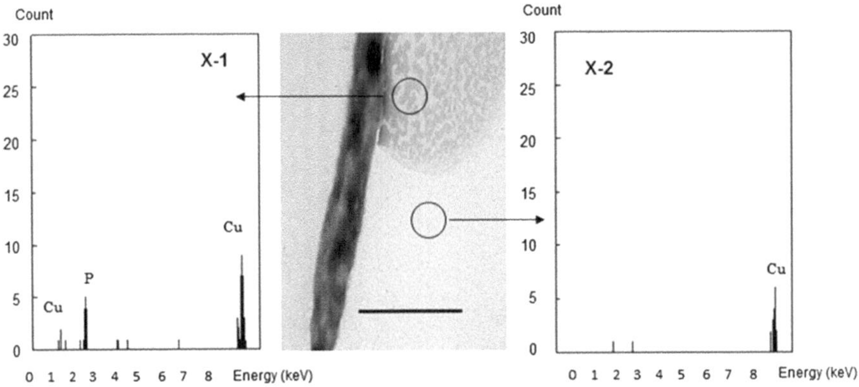

Fig. 35.3 TEM and EDAX analyses of *E. coli phoU* mutants after heating at 70 °C for 15 min. Different areas (indicated by *circles*) of unstained and unfixed samples were analyzed for chemical composition. X-1, EDAX spectrum of a cytoplasmic leakage (visible as a smear near the bacterial cell); X-2, EDAX spectrum of a reference area

Heating of microbial cells at 70 °C causes the formation of blebs of various sizes on the cell surface (Fig. 35.2). The bleb formation is observed soon after the start of heating. Although considerable irregularity in the shape of the microbial cells is observed, no detectable cell lysis occurs at 70 °C. Meanwhile, cytoplasmic leakage becomes visible near the heated cells as the heat treatment proceeds (Fig. 35.3). A considerable amount of P can be detected in the cytoplasmic leakage using TEM and EDX analyses. PolyP is likely to be leaked from heated cells through the openings of the ruptured cell envelope. Heating of microbial cells at 70 °C leads to significant structural alterations such as plasmolysis, cytoplasmic aggregation, cell

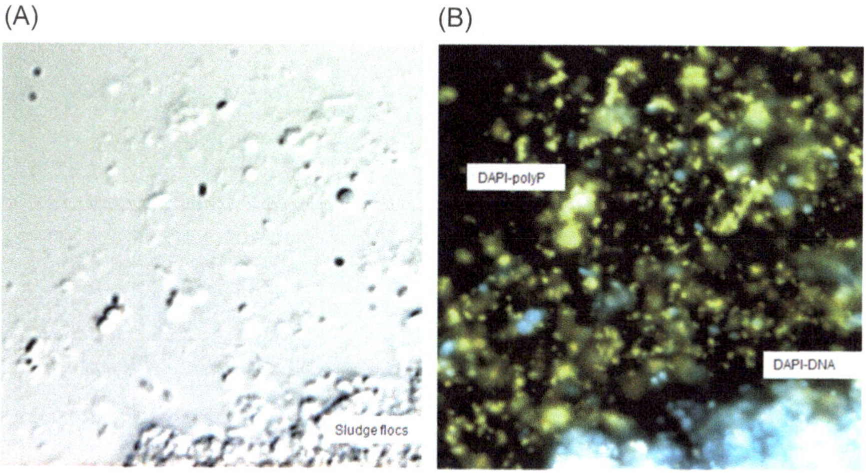

Fig. 35.4 PolyP release from Bio-P sludge. The Nomarski image of Bio-P sludge was taken after heating at 70 °C for 1 h (**a**). The DAPI-polyP fluorescence was detected with the heat-treated sludge at 525 nm (*yellow*) under ultraviolet irradiation (**b**). The DAPI-DNA fluorescence (*bluish white*) was detected at 470 nm

void formation, wall disruption, and cytoplasmic leakage. These events are likely to cause polyP leaching from microbial cells during heating at 70 °C.

35.2.2 PolyP Leaching from Bio-P Sludge

PolyP accumulated by Bio-P sludge can be rendered visible using 4′,6-di-amidino-2-phenylindole (DAPI). The DAPI-polyP fluorescence is detected at 525 nm by UV light (Fig. 35.4). After heating Bio-P sludge at 70 °C for 1 h, DAPI-polyP fluorescence can be detected outside sludge flocs. By contrast, DAPI-DNA molecules, which are visible at 470 nm by UV light, remain nearly unchanged inside sludge flocs, suggesting that polyP is specifically released from Bio-P sludge by heat treatment. Although the detailed mechanism remains to be clarified, it seems likely that the cell membrane of sludge microorganisms is partially disrupted by heating so that polyP could diffuse out of the cytoplasm as described in Sect. 2.1. The rate and extent of polyP release are clearly dependent on temperature. For example, approximately 90% of P can be released from Bio-P sludge by heating at 70–90 °C for 1 h. However, both the rate and extent of polyP release markedly decrease at temperatures lower than 50 °C. The rate of polyP degradation to P_i is also dependent on heating temperature. Approximately 20% of polyP is degraded to P_i at 70 °C in the leachate by 1 h, while more than 60% of polyP disappears at 90 °C. Importantly, heating at temperatures 70–90 °C does not decompose sludge flocs and therefore does not significantly change their settleability compared to unheated sludge

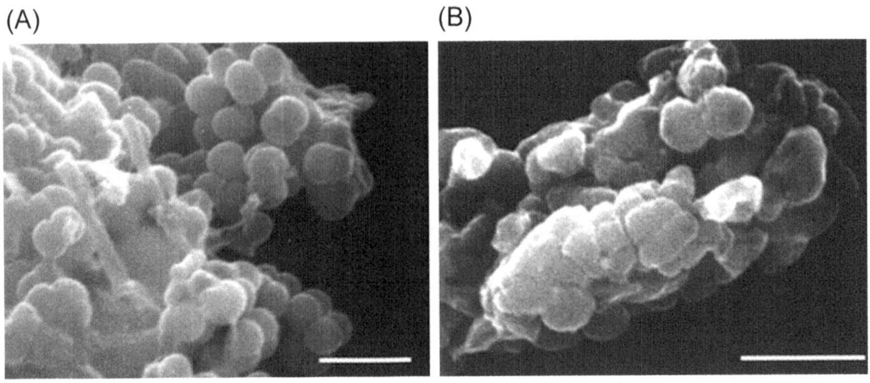

Fig. 35.5 SEM images of Bio-P sludge before (**a**) and after (**b**) heating at 90 °C for 2 h (scale bar = 3 μm)

(Fig. 35.5). Actually, heated sludge showed good settling property, and thus it could be easily separated by free sedimentation. It has also been observed that heat-induced polyP release from Bio-P sludge is strongly inhibited by the presence of metal cations such as Mg^{2+}, Fe^{3+}, and Al^{3+}. In this respect, SEM and EDAX analyses have showed that P_i can form poorly soluble chemical complexes with Al, Mg, and Fe at the surface of sludge biomass (Takiguchi et al. 2004).

The chain length of released polyP can be determined by polyacrylamide gel electrophoresis (PAGE) (Kuroda et al. 2002). $^{32}P_i$-labeled polyP can be detected in the leachate soon after the start of heating Bio-P sludge at 70 °C. Initially, the released polyP has a chain length of approximately 100–200 P_i residues. However, the chain length of polyP gradually decreases due to polyP hydrolysis. After 2 h of heat treatment, the fraction of polyP longer than 100 P_i residues disappears. Meanwhile, cyclic trimetaphosphate becomes detectable concomitantly with the degradation of polyP. Since trimetaphosphate is stable at 70 °C, its amount gradually increases as the heat treatment proceeds. It has been reported that trimetaphosphate can be generated from polyP by heating it in the presence of Mg^{2+} (Osterberg and Orgel 1972). The released polyP is almost completely degraded to P_i when it is treated with exopolyPase (PPX) (Kuroda et al. 2002).

35.2.3 Recovery of Released PolyP

PolyP released from Bio-P sludge is easily precipitated by the addition of $CaCl_2$ at room temperature without adjusting pH (Kuroda et al. 2002). Floc formation is visible soon after the addition of 50 mM $CaCl_2$ to the leachate. Addition of $CaCl_2$ precipitates about 65% of total P (T-P) during the first 30 min and approximately 75% of T-P by 2 h. The Ca-P precipitation is significantly influenced by the concentration of polyP in the leachate. When polyP is completely degraded to P_i, Ca-P precipitation is negligible unless pH is adjusted to 10 or higher. Interestingly, P_i can be precipitated with $CaCl_2$ in the presence of polyP at pH 7.7 (Kuroda et al. 2002). Trimetaphosphate is not precipitated with $CaCl_2$ even at pH 10. Releasing polyP, but not P_i, can offer a distinct advantage in recovering P from the leachate because it requires less chemicals compared to P_i precipitation and there is no need to adjust pH to a high value. Elemental analyses show that the recovered products typically contain about 16% and 18% of their dry weight as P and Ca, respectively. These values are comparable to those of mined P_i rock that typically contains 8–13% P and 26–33% Ca. The recovered products also contain about 4% of their dry weight as carbon. The present technology has been referred to as "Heatphos" (Takiguchi et al. 2003).

35.3 Heatphos Process

The feasibility of Heatphos process has been examined using a pilot-scale plant (Fig. 35.6). The pilot test plant was built at the Higashi-nada WWTP located in Kobe City, Japan. The pilot plant consisted of a Bio-P process for removing P_i from sewage and the downstream process for P_i recovery from Bio-P sludge. The pilot plant influent was obtained from the primary clarifier of the full-scale WWTP and introduced into the anaerobic tank of the pilot plant at a rate of approximately 100 m^3/day. The volumes of the anaerobic and aerobic tanks were 9 and 23 m^3,

Fig. 35.6 The pilot-scale test plant which was built at the Higashi-nada WWTP in Kobe

respectively. The level of mixed liquor suspended solids (MLSS) in the aerobic tank was maintained at about 2.0 g/L, while that of the return sludge was approximately 6.0 g/L. About 50% of thickened sludge was returned from the settling tank to the anaerobic tank. The sludge retention time (SRT) was about 4.5 days. Bio-P sludge was generated at a rate of approximately 10–20 kg MLSS/d. The Bio-P sludge was heated at 70 °C for about 1 h in a 1-m^3 capacity heating tank. PolyP-rich liquor was separated from sludge solids in a flotation tank. Finally, P was recovered by the chemical precipitation with $CaCl_2$.

Approximately 90–95% of influent T-P could be removed from sewage throughout the 1-year operating period. The P content of Bio-P sludge in the aeration tank reached approximately 3.5–4.5% on a dry weight basis. The polyP content reached 2.0–3.0% (approximately 50–70% of T-P in Bio-P sludge) about 2 months after the start of operation. This showed that about 50–70% of influent T-P was stored in the form of polyP by Bio-P sludge. Approximately 50–60% of cellular P could be released from Bio-P sludge by treatment at 70 °C for 1 h. Since about 30–50% of intracellular P is incorporated in large molecules such as nucleic acids and phospholipids which cannot be released from Bio-P sludge simply by heating at 70 °C, this observed level was likely the upper limit of P leaching from Bio-P sludge.

In laboratory experiments, about 65% of the released P could be precipitated with $CaCl_2$ at a Ca:P ratio of 2:1 without pH adjustment (Kuroda et al. 2002). However, in the pilot tests, only about 3% of T-P was precipitated under these conditions. Unexpectedly, approximately 70% of the released P was detected as P_i, but not polyP, in the precipitation tank of the pilot-scale plant. It remains unsolved why polyP was degraded to P_i in the leachate in the pilot-scale plant. However, the degradation of polyP to P_i was likely to inhibit the Ca-P precipitation without pH adjustment in the pilot-scale plant. PolyP precipitates with Ca at neutral pH more efficiently than P_i, because the p*K*a value of polyP is lower than that of P_i. When pH was adjusted to 11, 80% of T-P was precipitated with $CaCl_2$ even at the stoichiometric Ca-P ratio of 1:1. Overall, approximately 40% of T-P could be recovered from sewage in the pilot tests by adjusting pH to 11 in the precipitation tank. To realize P recovery without needing to adjust pH, it seems necessary to prevent polyP hydrolysis as is the case in laboratory experiments.

The applicability of Heatphos process at a full scale has been examined using a large demonstration plant which was installed at the Ashida-gawa WWTP in Fukuyama City, Japan (Fig. 35.7). Bio-P sludge was processed in the demonstration plant at a rate of approximately 360 kg/day, which was 20 times greater than that of the pilot-scale test plant at the Higashi-nada WWTP. Bio-P sludge was condensed approximately threefold in a sludge flotation tank. After being passed through a heat exchanger, the condensed sludge was heated at 70 °C for 1 h in a heating tank in continuous mode. To recover heat energy, the heated sludge was once again passed through the heat exchanger and cooled to atmospheric temperature. After centrifugation, the released P compounds were precipitated by adding $CaCl_2$ to the centrate at pH 11.

The influent T-P to this demonstration plant ranged from about 4.0 to 6.0 mg P/L throughout the 1-year period of operation. After the start-up period of about 30 days,

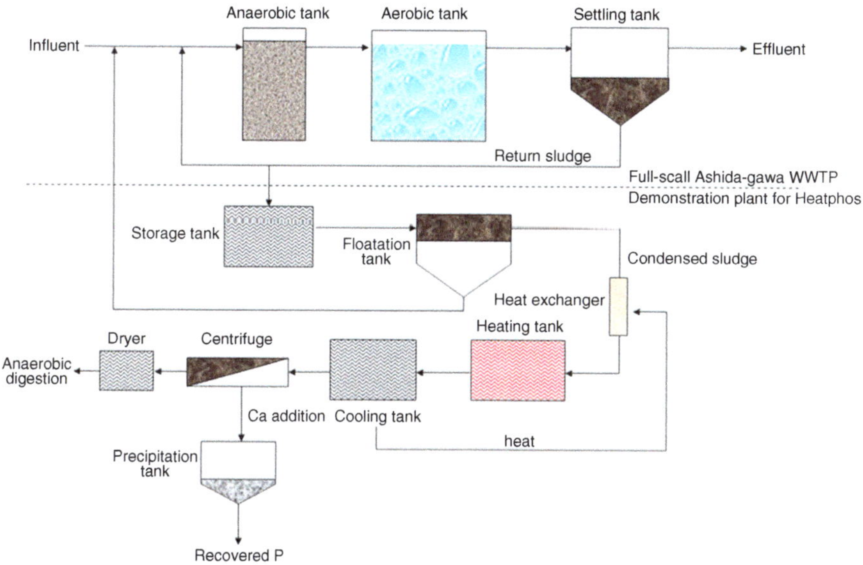

Fig. 35.7 Schematic diagram of the full-scale test plant for Heatphos process. The full-scale test plant was installed at the Ashida-gawa WWTP in Fukuyama City of Japan

the rate of P removal dramatically increased, and the effluent T-P decreased down to less than 1.0 mg P/L. After then, the P removal of 90% or higher was achieved throughout the operating period. The demonstration test plant could recover P from Bio-P sludge at a rate of approximately 10 kgP/day (c. 3.6 t/a). Overall, the full-scale demonstration plant could recover about 50% of P from sewage and save the energy requirement for P recovery by approximately 80% compared to the pilot-scale plant at the Higashi-nada WWTP.

The usability of recovered P has been examined for the production of by-product P_i fertilizer (Ohtake et al. 2008). By-product P_i fertilizer needs to contain at least 15% citric acid-soluble P_i (C-P_2O_5) according to the Fertilizer Control Act of Japan. In addition, the total nitrogen (T-N) content must be lower than 1% on a dry weight basis. Recovered P products typically contained 31% C-P_2O_5, 15% organics (estimated as ignition loss), and 1–2% of T-N on a dry weight basis. To make by-product P_i fertilizer, recovered P products were calcined at 550 °C for 2 h. The calcined product typically contained 28% C-P_2O_5 and no significant amounts of T-N and organics.

The possibility of using the calcined product as a raw material for the manufacture of phosphoric acid has also been examined. It has been shown that mineral elements such as Fe, Mg, Al, Na, and K restrict the solvent extraction of P_i after dissolving calcined product in H_2SO_4. If the total content of these minerals exceeds 2 wt%, the solvent extraction of P_i from the slurry becomes difficult. To overcome this, the calcined product was blended with powdered P_i rock before being subjected to the wet acid treatment. To generate phosphoric acid (75% P_2O_5), the blend ratio

Table 35.1 Percent reduction of sludge mass (MLSS) by anaerobic digestion

	% reduction[a]	
Sludge	Mesophilic	Thermophilic
Untreated	49.9 ± 1.1	53.8[b]
Heat-treated	40.5 ± 3.2	42.5[b]

[a]Mean values ± SD in three separate experiments
[b]Mean values of two separated experiments

of calcined product with P_i rock needs to be 10% or lower. However, low-grade phosphoric acid (less than 35% P_2O_5) can be made from the 1:4 mixture of calcined P product and P_i rock. This low-grade phosphoric acid is usable for the manufacture of P_i fertilizer.

35.4 Anaerobic Sludge Digestion

Anaerobic digestion has been widely used to reduce the volume of sewage sludge in WWTP. It is one of the most technically mature and cost-effective processes to reduce volume, odors, and pathogens in sewage sludge. The methane gas generated by anaerobic sludge digestion can be used for supplying the thermal energy requirement in WWTP. Needless to say, the thermal energy requirement is an important economic factor for operating Heatphos process at a full scale. Anaerobic digestion may be a promising means of supplying the thermal energy requirement for P recovery from Bio-P sludge. It should also be noted that Bio-P sludge often causes struvite deposit problems in anaerobic sludge digestion (Borgerding 1972). The possibility of the occurrence of struvite-clogging problem generally increases with increasing the P content of sewage sludge. Accordingly, P recovery from Bio-P sludge prior to anaerobic digestion may have a potential advantage for preventing struvite deposit problems in sludge manipulation processes.

In laboratory experiments, heating Bio-P sludge at 70 °C for 1 h decreases the concentration of MLSS by approximately 25–30%. Mesophilic anaerobic digestion (35 °C) of heat-treated sludge further reduces the concentration of MLSS by approximately 40% (Table 35.1). Together, the overall reduction in MLSS is approximately 65–70% which is about 15–20% high compared to that observed with unheated Bio-P sludge. Similar results are obtained by thermophilic digestion of heat-treated sludge (53 °C). However, there was no significant difference in the productivity of methane (mmol CH_4/g MLSS) between heat-treated and untreated Bio-P sludges during anaerobic digestion. Although a significant amount of organic materials is released from Bio-P sludge during heat treatment, the concentration of MLSS was also reduced by the heat treatment. This may explain that similar levels of methane productivity (mmol CH_4/g MLSS) were detected with untreated and heat-treated Bio-P sludges.

The potential of struvite precipitation is related to the concentrations of P_i, NH_4^+, and Mg^{2+} in the sludge liquor (Lowenthal et al. 1994). Since the concentration of NH_4^+ in anaerobic sludge digester liquor is relatively high, either P_i or Mg^{2+} concentrations are likely to limit the rate of struvite formation. Heat-treated sludge releases far less P_i into the anaerobic sludge digestion liquor than did untreated sludge (Takiguchi et al. 2004). For example, after anaerobic digestion, the P_i concentration reached up to 34 mg PO_4-P/L for unheated sludge, which was fourfold higher than that observed with heat-treated sludge. The concentration of NH_4^+ in the digester liquor from untreated sludge was also approximately 1.5-fold higher than that from heat-treated sludge after 20 days of digestion. Hence, it is likely that P recovery prior to anaerobic digestion has a potential advantage for controlling struvite deposit problems. Heat treatment has often been used as a conditioning process for sewage sludge before anaerobic digestion (Muller 2001). There have been many reports demonstrating that heat treatment prior to anaerobic digestion improves the digestibility of sludge and thus the productivity of methane (Stuckey and McCarty 1984; Kepp et al. 2000).

35.5 Conclusion

This chapter described a simple technology, referred to as Heatphos, for leaching polyP from Bio-P sludge by heating at 70 °C for about 1 h and recovering P_i from the leachate by the addition of $CaCl_2$. Heating sludge for polyP leaching has the potential to improve the sludge digestibility and thus to increase the biogas productivity in the subsequent anaerobic digestion. The heat energy required for polyP leaching can be supplied from recycling biogas from anaerobic digesters. In addition, Heatphos process is likely effective in controlling struvite deposit problems in anaerobic sludge digestion.

It is well known that polyP-rich activated sludge releases P_i, but not polyP, back into the solution when it is subjected to anaerobiosis (Bond et al. 1999). When compared with the anaerobic release of P_i, the heating method shows much better performance for releasing P from Bio-P sludge. Unlike anaerobic P_i release from Bio-P sludge, Heatphos process does not need the addition of a readily decomposable organic acid, for example, acetate, to release P. Bio-P processes can avoid consuming a lot of chemicals and correspondingly reduce the cost of sludge disposal. In addition, it is achievable in existing activated sludge processes with a minimum modification on the operating regimes. The alternate anaerobic-aerobic cycle can be installed by introducing an anaerobic zone ahead of an aerobic stage. In a conventional WWTP, this can be readily realized by shutting off aeration at the upstream part of an aeration tank and returning the sludge to this region from a secondary settler. Hence, the present technology can offer a simple option to recover P in a reusable form from wastewater.

References

Bond PL, Keller J, Blackall LL (1999) Anaerobic phosphate release from activated sludge with enhanced biological phosphorus removal. A possible mechanism of intracellular pH control. Biotechnol Bioeng 63:507–515

Borgerding J (1972) Phosphate deposits in digestion system. J Water Pollut Control Fed 44:813–819

Hirota R, Kuroda A, Kato J, Ohtake H (2010) Bacterial phosphate metabolism and its application to phosphorus recovery and industrial bioprocesses. J Biosci Bioeng 109:423–432

Kashihara H, Kang B, Omasa T, Honda K, Sameshima Y, Kuroda A, Ohtake H (2010) Electron microscopic analysis of heat-induced leakage of polyphosphate from a *phoU* mutant of *Escherichia coli*. Biosci Biotechnol Biochem 74:865–868

Kepp U, Machenbach I, Weisz N, Solheim OE (2000) Enhanced stabilization of sewage sludge through thermal hydrolysis – three years of experience with a full scale plant. Water Sci Technol 42:89–96

Kornberg A (1995) Inorganic polyphosphate: toward making a forgotten polymer unforgettable. J Bacteriol 177:491–496

Kulaev IS, Vagabov VM (1983) Polyphosphate metabolism in micro-organisms. Adv Microb Physiol 24:83–171

Kuroda A, Takiguchi N, Gotanda T, Nomura K, Kato J, Ikeda T, Ohtake H (2002) A simple method to release polyphosphate from activated sludge for phosphorus reuse and recycling. Biotechnol Bioeng 78:333–338

Lowenthal RE, Kornmuller URC, van Heerden EP (1994) Modelling struvite precipitation in anaerobic treatment systems. Water Sci Technol 30:107–116

Muller JA (2001) Prospects and problems of sludge pretreatment process. Water Sci Technol 44:121–128

Ohtake H, Omasa K, Honda K (2008) Bio-based production of rock phosphate for resource recycling and environmental protection. J Biotechnol 136S:5602

Osterberg R, Orgel LE (1972) Polyphosphate and trimetaphosphate formation under potentially prebiotic conditions. J Mol Evol 1:241–248

Sedlak RI (1991) Phosphorus and nitrogen removal from municipal wastewater, 2nd edn. Lewis Publishers, New York

Stuckey DC, McCarty PL (1984) The effect of thermal pretreatment on the anaerobic biodegradability and toxicity of waste activated sludge. Water Res 18:1343–1353

Takiguchi N, Kuroda A, Kato J, Nukanobu K, Ohtake H (2003) Pilot plant tests on the novel process for phosphorus recovery from municipal wastewater. J Chem Eng Jpn 36:1143–1146

Takiguchi N, Kishino M, Kuroda A, Kato J, Ohtake H (2004) A laboratory-scale test of anaerobic digestion and methane production after phosphorus recovery from waste activated sludge. J Biosci Bioeng 97:365–368

Printed by Printforce, the Netherlands